Blockchain, Smart Contracts and Distributed Ledger Technologies in the Built Environment

Other related titles:

You may also like

- PBBE008 | Shakir | AI in the Built Environment and Construction | contracted 2023
- PBBE002 | Clarke | Applied Building Performance Simulation | 2024
- PBSE024 | Nguyen | Proof-of-Stake for Blockchain Networks: Fundamentals, challenges and approaches | 2024
- PBBE003 | Al-Kodmany | Sustainable High-Rise Buildings: Design, technology, and innovation | 2022

We also publish a wide range of books on the following topics:
Computing and Networks
Control, Robotics and Sensors
Electrical Regulations
Electromagnetics and Radar
Energy Engineering
Healthcare Technologies
History and Management of Technology
IET Codes and Guidance
Materials, Circuits and Devices
Model Forms
Nanomaterials and Nanotechnologies
Optics, Photonics and Lasers
Production, Design and Manufacturing
Security
Telecommunications
Transportation

All books are available in print via https://shop.theiet.org or as eBooks via our Digital Library https://digital-library.theiet.org.

IET BUILT ENVIRONMENT SERIES 07

Blockchain, Smart Contracts and Distributed Ledger Technologies in the Built Environment

Key concepts, technologies, and applications

Edited by
Mohamad Kassem, Abel Maciel and Daniel M. Hall

The Institution of Engineering and Technology

About the IET

This book is published by the Institution of Engineering and Technology (The IET).

We inspire, inform and influence the global engineering community to engineer a better world. As a diverse home across engineering and technology, we share knowledge that helps make better sense of the world, to accelerate innovation and solve the global challenges that matter.

The IET is a not-for-profit organisation. The surplus we make from our books is used to support activities and products for the engineering community and promote the positive role of science, engineering and technology in the world. This includes education resources and outreach, scholarships and awards, events and courses, publications, professional development and mentoring, and advocacy to governments.

To discover more about the IET please visit https://www.theiet.org/.

About IET books

The IET publishes books across many engineering and technology disciplines. Our authors and editors offer fresh perspectives from universities and industry. Within our subject areas, we have several book series steered by editorial boards made up of leading subject experts.

We peer review each book at the proposal stage to ensure the quality and relevance of our publications.

Get involved

If you are interested in becoming an author, editor, series advisor, or peer reviewer please visit https://www.theiet.org/publishing/publishing-with-iet-books/ or contact author_support@theiet.org.

Discovering our electronic content

All of our books are available online via the IET's Digital Library. Our Digital Library is the home of technical documents, eBooks, conference publications, real-life case studies and journal articles. To find out more, please visit https://digital-library.theiet.org.

In collaboration with the United Nations and the International Publishers Association, the IET is a Signatory member of the SDG Publishers Compact. The Compact aims to accelerate progress to achieve the Sustainable Development Goals (SDGs) by 2030. Signatories aspire to develop sustainable practices and act as champions of the SDGs during the Decade of Action (2020-2030), publishing books and journals that will help inform, develop, and inspire action in that direction.

In line with our sustainable goals, our UK printing partner has FSC accreditation, which is reducing our environmental impact to the planet. We use a print-on-demand model to further reduce our carbon footprint.

Published by The Institution of Engineering and Technology, London, United Kingdom

The Institution of Engineering and Technology (the "**Publisher**") is registered as a Charity in England & Wales (no. 211014) and Scotland (no. SC038698).

First published 2025

The Institution of Engineering and Technology
Futures Place,
Kings Way,
Stevenage, Herts, SG1 2UA,
United Kingdom

www.theiet.org

British Library Cataloguing in Publication Data
A catalogue record for this product is available from the British Library

ISBN 978-1-83953-834-6 (hardback)
ISBN 978-1-83953-835-3 (PDF)

Typeset in India by MPS Limited
Printed in the UK by CPI Group (UK) Ltd, Eastbourne

Cover image: Andriy Onufriyenko/Moment via Getty Images

The IET Book Series on Advances in Distributed Computing and Blockchain Technologies

Call for authors' page.
The objective of this book series is to enlighten researchers with the current trends in the field of blockchain technology and its integration with distributed computing and systems. Blockchains show riveting features such as decentralization, verifiability, fault tolerance, transparency, and accountability. The technology enables more secure, transparent, and traceable scenarios to assist in distributed applications. We plan to bring forward the technological aspects of blockchain technology for Internet of Things (IoTs), cloud computing, edge computing, fog computing, wireless sensor networks (WSNs), peer-to-peer (P2P) networks, mobile edge computing, and other distributed network paradigms for real-world enhanced and evolving future applications of distributed computing and systems.

Topics to be covered include: distributed computing and its applications; distributed vs. decentralized systems; tools and platforms for distributed systems; performance evaluation of distributed applications; blockchain technologies; blockchains and application domains; blockchains vs. distributed systems; cryptocurrency and distributed ledger technology; blockchain and IoT; ubiquitous computing; pervasive computing; cloud, edge, and fog computing; Internet of Things (IoT) technology; security issues in IoT; software tools and platforms for implementing blockchains in IoT networks; performance optimization of blockchain technology for distributed systems; social and ethical impacts of blockchains in distributed applications; blockchain technology and sustainability.

Proposals for international coherently integrated multi-authored or edited books, handbooks, and research monographs will be considered for this book series. We do not endorse, however, call for chapter books. The chapter authors will be invited personally and individually by the book editors. Each proposal will be reviewed by the book series editors with additional peer reviews from independent reviewers. Please contact:

- Dr Brij B. Gupta, National Institute of Technology (NIT), India; e-mail: gupta.brij@ieee.org
- Prof. Gregorio Martinez Perez, University of Murcia (UMU), Spain; e-mail: gregorio@um.es
- Dr Tu N. Nguyen, Kennesaw State University of Georgia, USA; e-mail: tu.nguyen@kennesaw.edu

Book Series Blurb for back cover
The objective of this book series is to enlighten researchers with the current trends in the field of blockchain technology and its integration with distributed computing and systems. Blockchains show riveting features such as decentralization, verifiability, fault tolerance, transparency, and accountability. The technology enables more secure, transparent, and traceable scenarios to assist in distributed applications. We plan to bring forward the technological aspects of blockchain technology for Internet of Things (IoTs), cloud computing, edge computing, fog computing, wireless sensor networks (WSNs), peer-to-peer (P2P) networks, mobile edge computing, and other distributed network paradigms for real-world enhanced and evolving future applications of distributed computing and systems.

Contents

Foreword **xvii**
Preface **xxi**
About the editors **xxiii**

1 Introduction **1**
Mohamad Kassem, Abel Maciel and Daniel Hall

2 Distributed ledger technologies, blockchain and smart contracts: technical foundations and key concepts **9**
Mahir Msawil, Mohamad Kassem and David Greenwood
2.1 DLT: a definition 9
2.2 Blockchain technology 10
2.3 Blockchain types 11
2.4 Prevailing blockchain platforms 12
2.5 Smart contracts 14
2.6 Blockchain oracles 16
2.7 Benefits and capabilities of blockchain-based applications 18
2.8 How can a blockchain-based application be justified? 20
2.9 Evidence of blockchain applicability in the built environment 22
2.10 Conclusions 24
References 25

3 Socio-technical approach to blockchain adoption in construction: conceptual models and roadmaps for macro- and meso-scale implementation **31**
Jennifer Li, David Greenwood and Mohamad Kassem
3.1 Socio-technical systems theory 32
3.1.1 Conceptual foundations 32
3.1.2 STS and the adoption of technological innovations 33
3.1.3 STS within the organisation 35
3.1.4 STS approaches in the built environment literature 35
3.1.5 STS and the adoption of building information modelling 36
3.2 Blockchain as a socio-technical system 38
3.2.1 Socio-technical considerations for DLT and SCs in the built environment 40
3.2.2 The trust perspective of blockchain-based systems 41

3.3 A socio-technical model of DLT and SCs in the built environment 43
3.3.1 DLT four-dimensional model 44
3.3.2 DLT actors model 47
3.3.3 Application of the DLT four-dimensional model and the DLT actors model 47
3.4 Blockchain and DLT adoption 52
3.4.1 Adoption studies at different levels 53
3.4.2 Barriers and facilitators to blockchain adoption 56
3.4.3 Critical success factors for adoption of blockchain technologies 62
3.5 Roadmaps to adoption 63
3.5.1 Macro roadmap 63
3.5.2 Meso roadmap 67
3.5.3 Linking the meso and macro roadmaps 70
3.6 Conclusions 71
References 73

4 Enhancing information security and digital trust in construction: decentralisation and encryption approaches 83
Moumita Das, Jack C.P. Cheng and Xingyu Tao
4.1 Context and challenges 84
4.2 Data security requirements for AECO projects and the need for distributed security approaches 87
4.2.1 Data confidentiality 88
4.2.2 Digital trust 88
4.2.3 Data utility 90
4.3 A universal distributed framework for secure collaboration in AECO projects 91
4.3.1 The distributed network 92
4.3.2 Encryption modules 96
4.3.3 Selectively transparent data models 98
4.4 Case scenario 100
4.5 Conclusions and future research 103
References 105

5 Blockchain and tokenization of built assets 109
Lavinia Chiara Tagliabue, Paolo Mistrangelo and Algan Tezel
5.1 Property tokenization as a paradigm shift 111
5.2 Fractional ownership and digital innovation 113
5.3 Framework for property tokenization 114
5.3.1 Organization of the information in a property tokenization management system 118
5.3.2 Barriers to adoption: insights from the Notary Council on property tokenization 121

5.3.3 Possible business models with property tokenization platform 122
5.3.4 Technological chain for the implementation of the property tokenization 124
5.3.5 Product and process innovation 127
5.3.6 Property tokenization as a financing mechanism to upkeep built assets 128
5.3.7 Economic impact 129
5.4 Digital architecture implementation for property tokenization 133
5.4.1 Use of cloud computing 133
5.4.2 The role of cloud computing 135
5.4.3 Ethereum as a potential blockchain platform for property tokenization 135
5.4.4 A proposed technical architecture 136
5.5 Conclusions, challenges and future directions 138
References 140

6 Digital building logbooks on the blockchain: first conceptualisation and future research directions 147
Theodoros Dounas, Daniel M. Hall, Dimosthenis Kifokeris, David Christie, Jens Hunhevicz, Firehiwot Kedir, Joseph Mante, Goran Sibenik, Marijana Sreckovic, Lavinia Chiara Tagliabue and Catherine De Wolf
6.1 Background 149
6.1.1 Implementation challenges for digital building logbooks 149
6.1.2 Possible technical approaches to digital building logbooks 149
6.1.3 Distributed ledger technologies, blockchains and smart contracts 152
6.1.4 Other key enabling technologies 153
6.2 Conceptualisation 155
6.2.1 Motivation and objective 155
6.2.2 Conceptual architecture 156
6.2.3 Summary 164
6.3 Potential impacts and future research directions 164
6.3.1 Enabling a new information ‘self-infrastructuring’ backbone 166
6.3.2 Accelerating the business case for renovation and reuse 166
6.3.3 Enabling new peer-to-peer business models 167
6.3.4 Improving existing legal structures 167
6.3.5 Fostering new governmental policies, incentives and regulations 168
6.3.6 Moving towards a more circular Europe 169
6.3.7 Enabling better security and privacy of data 170
6.4 Conclusion 170
References 171

7 Blockchain for energy management in seaport infrastructure: a microgrid case study **177**

Alexander James Howe, Ioan Petri and Yacine Rezgui

7.1 Introduction 177
7.2 Related works 180
7.2.1 Renewable energy in industry 180
7.2.2 Blockchain for microgrids 183
7.3 The blockchain for energy transactions 185
7.3.1 Scenarios of analysis 188
7.3.2 Framework specification 191
7.4 Evaluation and results 192
7.5 Discussion 200
7.5.1 Comparison to related studies 200
7.5.2 Challenges and barriers 201
7.6 Conclusion 202
References 203

8 Decentralising architectural design through data governance **207**

Theodoros Dounas, Davide Lombardi and Hico McDonald

8.1 Introduction and background – crises through development 207
8.1.1 The nature of design in the AEC industry and the current political economy 208
8.2 Primitives 208
8.2.1 Blockchain, decentralised ledger technology and data storage 208
8.2.2 Decentralised autonomous organisations 210
8.3 Blockchain is a governance technology (why decentralisation matters: decentralisation primitives) 210
8.4 Interlude: a design competition as a driver for research in blockchain for architectural design 211
8.5 Decentralising architectural design: optimisation: creating architectural design decentralised autonomous organisations 218
8.5.1 Prototype 1 218
8.5.2 Prototype 2 220
8.5.3 Shape grammars and blockchain 221
8.5.4 The blockchain infrastructure: DAOstack 221
8.5.5 Prototype 3 224
8.5.6 Decentralised architectural design 225
8.5.7 DAOs and shape grammars further development 226
8.6 Building a decentralised autonomous organisation as a living lab 227
8.6.1 The architecture decentralised autonomous organisation 'archiDAO' 227
8.6.2 Stigmergy 229
8.6.3 Smart contract and token designs for stigmergy – the ArchiDAO non-fungible token 230

8.6.4 The ArchiDAO proposed structures 231
8.6.5 The ArchiDAO soulbound token 231
8.7 Discussion – contribution and conclusion 232
Acknowledgements 234
References 234

9 Blockchain in construction supply chain management 239
Liupengfei Wu, Jinying Xu and Weisheng Lu
9.1 Introduction 239
9.2 Benefits of using blockchain in construction supply chain management 241
9.2.1 Enhanced information sharing and stakeholder collaboration 241
9.2.2 Improved transparency and traceability 242
9.2.3 Enhanced security and risk management 243
9.2.4 Improved sustainability and operational performance 244
9.3 Implementing blockchain in construction supply chain management 245
9.3.1 Key considerations for implementation 245
9.3.2 Steps for successful implementation with a real-life case study 247
9.4 Prospects and challenges of blockchain in construction supply chain management 261
9.4.1 Prospects 261
9.4.2 Challenges 263
9.5 Conclusion 266
References 267

10 Blockchain for data observability: integrating BIM, digital twins and enterprise CDEs 273
Abel Maciel and Klaudia Jaskula
10.1 Introduction 273
10.2 The trajectory of CDEs 275
10.2.1 The particulars of BIM CDEs 276
10.2.2 Providing for digital twins in the CDE 277
10.3 The enterprise total lifecycle data governance 278
10.3.1 Enterprise digital twins: what is required? 279
10.4 The evolution of data architectures 280
10.5 Methodology 282
10.5.1 Levels and annotation 285
10.6 Observing provenance in enterprise CDEs: accounting from BIM and DTs 285
10.6.1 Enterprise data lineage and observability 286
10.6.2 BIM data lineage and observability 290
10.6.3 Digital twins data lineage and observability 295
10.6.4 Summary of results 299

10.7 Prospecting for enterprise digital twin data architectures 302
10.7.1 Functions 302
10.7.2 Sequences 302
10.7.3 Attributions 303
10.8 Conclusion 305
References 306

11 Blockchain and contract administration in construction 313
Mahir Msawil, David Greenwood and Mohamad Kassem
11.1 Introduction 313
11.2 Construction contract administration as a multi-functional system 314
11.3 Challenges facing CCA (causes and effects of poor CCA) 317
11.4 Blockchain technology and smart contracts in the context of CCA 320
11.5 Mapping retrieved studies to the identified CCA functions 322
11.5.1 Blockchain applications for control-orientated CCA functions 323
11.5.2 Blockchain applications for coordination-orientated CCA functions 330
11.5.3 Blockchain applications for adaptation-orientated CCA functions 334
11.6 Research agenda: challenges and opportunities to blockchain-enabled CCA 338
11.6.1 Technology-related challenges 338
11.6.2 Process-related challenges 339
11.6.3 Policy-related challenges 341
11.6.4 Society-related challenges 341
11.7 Insights and conclusions 343
References 346

12 Legal risks for blockchain applications in the built environment: a legal perspective 353
Gavin P. Johnson
12.1 Introduction 353
12.1.1 Background on blockchain technology 354
12.2 Legal risks associated with blockchain applications 360
12.2.1 Overview of legal risks 361
12.2.2 Contracts, smart legal contracts and smart contracts 361
12.2.3 Intellectual property rights (IPR) 365
12.2.4 Data protection, privacy and cybersecurity 367
12.2.5 Financial regulation 369
12.2.6 Commercial 373
12.2.7 Jurisdictional and cross-border legal issues 375
12.2.8 Summary of legal risks associated with blockchain applications 379

12.3 Cases, case studies and examples of legal risks 379
12.3.1 Real-world examples of legal risks and their consequences 379
12.3.2 Case studies of blockchain applications that have faced legal challenges 383
12.3.3 Summary of legal risk cases and examples 391
12.4 Best practices and solutions to mitigate legal risks 391
12.4.1 Overview of best practices 392
12.4.2 Strategies for risk mitigation 393
12.4.3 Regulatory compliance, legal considerations and practical tips 395
12.4.4 Smart contract design and dispute resolution mechanisms 398
12.5 Conclusion 402
12.5.1 Summary of legal risks 402
12.5.2 Outlook on legal risks 403
12.5.3 Final thoughts and recommendations 405
References 406

13 Conclusions: cross-cutting themes, challenges and recommendations 411
Mohamad Kassem, Abel Maciel and Daniel Hall
13.1 Synthesis of chapters 412
13.2 Cross-cutting themes 416
13.2.1 Blockchain as a systematic challenge 416
13.2.2 Blockchain and socio-technical environments 417
13.2.3 Blockchain integration with digital technologies 417
13.2.4 Data security and digital trust 418
13.2.5 Tokenisation and democratisation of access 419
13.2.6 Collaborative and decentralised governance 419
13.2.7 Sustainability and circular economy 420
13.3 Key challenges 420
13.3.1 Legal and regulatory challenges 420
13.3.2 Scalability and performance limitations 421
13.3.3 Integration with legacy systems 421
13.3.4 Data quality and reliability 421
13.3.5 Fragmentation of standards and protocols 422
13.3.6 Ethical and social implications 422
13.3.7 Stakeholder resistance and adoption barriers 422
13.4 Ecosystem barrier to blockchain diffusion in the built environment 423
13.5 Recommendations 424
13.6 Final reflection 424
References 428

Index 429

Foreword

The objective of *The IET Book Series on Advances in Distributed Computing and Blockchain Technologies* is to enlighten researchers with the current trends in the field of blockchain technology and its integration with distributed computing and systems. Blockchains show riveting features such as decentralization, verifiability, fault tolerance, transparency and accountability. The technology enables more secure, transparent and traceable scenarios to assist in distributed applications. We bring forward the technological aspects of blockchain technology for Internet of Things (IoTs), cloud computing, edge computing, fog computing, wireless sensor networks, peer-to-peer networks, mobile edge computing and other distributed network paradigms for real-world enhanced and evolving future applications of distributed computing and systems.

The built environment stands at the crossroads of transformation, driven by the need to enhance efficiency, foster sustainability and embrace innovation. As the complexities of urbanization, climate change and technological evolution converge, new paradigms are required to meet these challenges head-on. Among these emerging paradigms, blockchain and distributed ledger technologies (DLTs) have garnered significant attention for their potential to revolutionize the architecture, engineering, construction and operations (AECO) sectors. It is with great enthusiasm and anticipation that we introduce this seminal volume, *Blockchain, Smart Contracts and Distributed Ledger Technologies in the Built Environment: Key Concepts, Technologies, and Applications.*

This book is a timely and necessary contribution to the discourse on how blockchain and related technologies can reshape the built environment. The chapters within offer a rich tapestry of knowledge, insights and applications that collectively illuminate the path forward for integrating blockchain into the AECO sectors. Each chapter underscores the multifaceted opportunities and challenges inherent in deploying these technologies, while providing actionable strategies for overcoming barriers to adoption.

At the heart of this volume is the recognition that blockchain – a technology originally associated with cryptocurrencies – has evolved into a versatile tool with far-reaching implications. Its attributes of decentralization, immutability, transparency and security make it uniquely suited to address long-standing inefficiencies and fragmentation within the construction and built environment sectors. Moreover, its integration with other transformative technologies such as IoT, building information modelling (BIM) and Digital Twins creates possibilities for enhanced collaboration, smarter decision-making and more resilient infrastructures.

Chapter 1, penned by the book editors, Mohamad Kassem, Abel Maciel and Daniel Hall, sets the stage by offering an incisive introduction to the themes explored throughout the book. It provides a foundational understanding of blockchain and its relevance to the built environment, framing the technology as a catalyst for systemic change.

Chapter 2 delves into the technical foundations and key concepts of blockchain, smart contracts and DLTs. Authored by Mahir Msawil, Mohamad Kassem and David Greenwood, this chapter is an essential read for anyone seeking to grasp the fundamental building blocks of these technologies.

The socio-technical dimensions of blockchain adoption are explored in Chapter 3, where Jennifer Li, David Greenwood and Mohamad Kassem present conceptual models and roadmaps for implementation at macro and meso scales. This chapter underscores the importance of a holistic approach that considers not only technological aspects but also organizational and societal factors.

Information security and digital trust – critical concerns in the built environment – are addressed in Chapter 4 by Moumita Das, Jack CP Cheng and Xingyu Tao. Through an examination of decentralization and encryption approaches, the authors provide valuable insights into how blockchain can enhance security and trust in construction workflows.

The tokenization of built assets, as explored in Chapter 5 by Lavinia Chiara Tagliabue, Paolo Mistrangelo and Algan Tezel, represents a transformative application of blockchain. By enabling the fractional ownership and trading of assets, tokenization has the potential to democratize access to investment opportunities and streamline asset management.

Chapter 6 introduces the concept of digital building logbooks on the blockchain, a novel idea that promises to revolutionize the lifecycle management of buildings. Theodoros Dounas and his team provide a forward-looking perspective on the potential for blockchain-enabled logbooks to enhance data accessibility, traceability and sustainability.

The application of blockchain in energy management, particularly in seaport infrastructure, is the focus of Chapter 7. Alexander James Howe, Ioan Petri and Yacine Rezgui present a compelling case study on the use of blockchain for microgrid management, highlighting its potential to drive efficiency and resilience in energy systems.

Chapter 8 takes a bold step into the realm of decentralizing architectural design through data governance. Theodoros Dounas, Davide Lombardi and Hico McDonald explore how blockchain can empower stakeholders to take greater control of design processes, fostering innovation and inclusivity.

In Chapter 9, Liupengfei Wu, Jinying Xu and Weisheng Lu examine blockchain's role in construction supply chain management. The authors provide a thorough analysis of how blockchain can improve transparency, traceability and efficiency in complex supply chains.

The integration of blockchain with BIM, digital twins and enterprise CDEs is explored in Chapter 10 by Abel Maciel and Klaudia Jaskula. This chapter highlights the potential for blockchain to enhance data observability and interoperability, enabling seamless collaboration across project stakeholders.

Legal implications are an unavoidable consideration in the adoption of blockchain technologies. In Chapter 12, Gavin P. Johnson provides a critical analysis of the legal risks and challenges associated with blockchain applications in the built environment, offering guidance for navigating this complex landscape.

The book concludes with a synthesis of cross-cutting themes, challenges and recommendations in Chapter 13. The book editors, Mohamad Kassem, Abel Maciel and Daniel Hall bring together the diverse insights presented throughout the book, charting a course for future research and practice.

The significance of this book lies not only in its comprehensive exploration of blockchain's potential but also in its balanced and critical approach. By addressing both the opportunities and limitations of blockchain, the authors provide a nuanced perspective that will resonate with researchers, practitioners and policymakers alike.

We commend the book editors and contributors for their exceptional work in bringing this volume to fruition. Their collective expertise and dedication have resulted in a landmark publication that will undoubtedly serve as a cornerstone for future advancements in the field. It is an honour to write this foreword and to witness the transformative potential of blockchain and DLTs being realized in the built environment.

Dr Brij B. Gupta
Director, International Center for AI and
Cyber Security Research and Innovations (CCRI)
Distinguished Professor, Department of Computer Science and
Information Engineering (CSIE)
Asia University, Taichung, Taiwan

Dr Tu N. Nguyen
Assistant Professor, Department of Computer Science
Director of the Next-generation Computing and
Network Systems Laboratory
Kennesaw State University, USA

Preface

Blockchain and distributed ledger technologies (DLTs) are increasingly recognised for their potential to address complex challenges across various sectors. Within construction and the built environment, these technologies are beginning to show promise not only in tackling persistent issues such as inefficiencies, fragmented processes and trust deficits but also in reshaping business models, enabling new forms of organising and driving value creation. Construction and the built environment, known for their significant economic and environmental impact, face growing pressures to adopt innovative, sustainable and collaborative approaches.

In this context, this book brings together a range of perspectives to explore how blockchain can contribute to addressing the critical challenges of the construction and built environment sectors. This book was conceived to bridge a critical gap in the discourse surrounding blockchain's role in the built environment. While the potential of this technology is often discussed in abstract terms, this volume seeks to ground the discussion in both theoretical insights and practical applications. By bringing together leading academics and industry practitioners, we aim to provide a comprehensive exploration of blockchain's foundational concepts, technical attributes and real-world implications for the architecture, engineering, construction and operations (AECO) sectors.

Our motivation in compiling this volume stems from a shared recognition of the urgent need to modernise the built environment and the promising, yet often misunderstood, potential of blockchain to support this transformation. The construction sector's global significance, its substantial environmental footprint and intricate socio-technical ecosystems present unique challenges and opportunities for blockchain adoption. This book was written not just to highlight blockchain's potential but also to critically examine its limitations and implementation challenges.

This book is intended for a diverse audience including researchers, industry professionals, policymakers and educators. It offers value to those seeking to understand blockchain's technical underpinnings, its application in construction workflows and the broader socio-technical and regulatory contexts influencing its adoption. Whether you are a seasoned blockchain developer, a construction industry stakeholder exploring digital innovations or a student delving into the intersection of technology and the built environment, this book provides insights tailored to your needs.

The chapters are structured to guide readers through a progressive understanding of blockchain in the built environment, starting with foundational knowledge that introduces blockchain's technical principles and explores its integration with other technologies such as IoT, building information modelling and digital twins. Subsequent chapters focus on applications including information management, supply chain management, energy trading, property tokenisation, collaborative decentralised design, digital building logbooks, contract administration and the legal perspective. The concluding chapter synthesises the book's cross-cutting themes, addresses adoption barriers and proposes actionable strategies for adoption. Readers are encouraged to engage with the chapters selectively or sequentially, depending on their focus or expertise. For those new to blockchain, beginning with the foundational chapters is recommended, while practitioners may find the case studies and practical insights particularly valuable.

This book would not have been possible without the contribution of our esteemed lead authors and their co-authors. We are deeply grateful for their contribution, insightful chapters and unwavering commitment to advancing knowledge in this field.

Mohamad Kassem
School of Engineering, Newcastle University, UK

Abel Maciel
Bartlett School of Architecture, University College London, UK

Daniel Hall
Department of Management in the Built Environment,
Faculty of Architecture and the Built Environment,
Delft University of Technology (TU Delft), Netherlands

15 January 2025

About the editors

Mohamad Kassem is a professor of digital construction in the School of Engineering, Newcastle University (UK), and leads the Digital Innovation in Construction and Engineering (NU-DICE) lab. He is a renowned scholar recognised for his research on the digitalisation of construction and the built environment, focusing on technical and process advancements as well as policy development.

He has secured and managed an extensive portfolio of R&D projects totalling over £12 million, funded by UKRI councils and the European Commission. With over 160 publications in journals and conference proceedings, his research has significantly influenced the field of digital construction.

Mohamad has served as a digital transformation advisor to the European Commission, the UK Foreign and Commonwealth Office, the Construction Industry Training Board, the University of Cambridge's Centre of Digital Built Britain and the governments of Brazil and Mexico, among others. His contributions have shaped digital transformation efforts in over 11 countries globally.

He has received multiple accolades, including the European Council for Computing in Construction's Scherer Award, the RISE Award for Digital Innovation in Sustainability in Construction and an Enterprise Fellowship. Additionally, he provides expert evaluations for funding bodies in the UK, the EU (Horizon 2020) and North America.

Abel Maciel is an architect and associate professor at the Bartlett Faculty of the Built Environment, University College London (UCL). He earned an engineering doctorate in virtual environments from UCL, where his research investigated the social, economic and cognitive dimensions of collaborative design, with a focus on how these processes inform both collaborative design methodologies and building information modelling (BIM) technology. Drawing on methodologies from ethnography, economics and psychology, he developed models of this multifaceted phenomenon that opened new avenues for research, earning him the Bartlett Doctorate Prize in 2014.

Abel has worked on a wide range of design typologies and scales, from product and interaction design to large-scale urban complexes, contributing to projects delivery in the architecture, engineering, construction and operations (AECO), transportation and energy sectors. He has taught at the Architectural Association, the Bartlett and Imperial College London, as well as at the École Spéciale d' Architecture in Paris.

Throughout his career, Abel has collaborated with a number of international design and engineering practices – including Foster + Partners, Zaha Hadid Architects, Heatherwick Studio, Herzog & de Meuron, Arup and Bureau Happold – to prepare competitions, develop proposals and deliver mega-projects across Asia, the Americas, Europe and the Middle East. He is the founding director of Design Computation, a London-based consultancy dedicated to digital transformation and adoption, including the delivery of BIM and digital twin solutions that leverage blockchain and AI technologies.

In addition, Abel is the founding director of the Construction Blockchain Consortium, an initiative bridging academia and industry to advance the understanding and application of distributed ledger technologies in the AECO sector. He initiated the Design Computation Wiki, an open knowledge platform for computational and emergent design research and serves as an advisor on blockchain and AI policy to the British Parliament.

Daniel M. Hall is an assistant professor in the Department of Management in the Built Environment at the Faculty of Architecture and the Built Environment, Delft University of Technology (TU Delft). The overarching theme of his research is to enhance governance, productivity and innovation in construction projects through a transformation from fragmented project delivery methods to new organizational models that integrate the supply chain of the vertically and horizontally fragmented construction industry. His work finds that in order to foster more innovation in products and processes, the AEC industry must embrace new technologies, governance strategies and organizational models.

His current research interests include industrialized construction, governance of integrated project delivery, blockchain in construction, adoption of systemic innovations, industry 4.0 and entrepreneurship in AEC. Previously he was an assistant professor of Innovative and Industrial Construction in the Department of Civil, Environmental and Geomatic Engineering of ETH Zürich.

He is founder and organizer of the Stanford Industrialized Construction Forum since 2014. His teaching philosophy is based on project-based learning and flipped classroom approaches that are implemented in his courses such as Industrialized Construction and Lean, Integrated and Digital Project Delivery. In the past, he has team-taught several courses at the Stanford d.school, including *Design Thinking Studio: Hacking Your Innovation Mindset and Adventures in Design Thinking*.

He holds several international leadership positions for the European Conference for Computing in Construction, the International Association of Automation and Robotics in Construction and the Construction Blockchain Consortium (hosted at University College London). He holds a doctor of philosophy (2017) in civil and environmental engineering from Stanford University, USA.

Chapter 1
Introduction

Mohamad Kassem[*], *Abel Maciel*[†] *and Daniel Hall*[‡]

The construction sector serves as a cornerstone of national and global economies, driving societal and economic development. It is intricately linked to the three pillars of sustainability: society, economy and environment, positioning it as a critical sector in shaping a sustainable future. The construction sector is a major consumer of natural resources and a significant contributor to global carbon emissions and waste.

Construction and built environment projects are typically characterised by hierarchical, interconnected supply chains and complex payment systems, often causing payment delays and financial strain for those positioned lower in the project hierarchy, frequently small and medium-sized enterprises. Many procurement delivery models used within the industry further reinforce this hierarchy, limiting collaboration and exacerbating these challenges. A persistent issue within the sector is information asymmetry and imperfect information, where stakeholders either lack equal access to critical data or rely on incomplete or inaccurate information. These challenges are intensified by the inherent complexity of construction projects, characterised by diverse stakeholders, complicated workflows and significant uncertainties. As a result, misunderstandings, disputes and delays are often the norm, hindering growth and development within the construction industry.

The sector's low productivity, especially when compared to other industries, remains a significant concern. Digital innovation holds the potential not only to address process inefficiencies but also to transform the industry by enabling new forms of organising innovative business models and new methods for value creation. Due to their anticipated value, digital transformation initiatives and digitalisation technologies are being increasingly promoted to address key challenges within construction and the built environment. Progress is being made, albeit at a measured pace, due to the industry's cautious approach to adoption. This measured pace is constrained by factors such as the evolving maturity of these technologies, disparities in technical expertise across the workforce and concerns regarding the financial investments required.

[*]School of Engineering, Newcastle University, UK
[†]Bartlett School of Architecture, University College London, UK
[‡]Department of Management in the Built Environment, Faculty of Architecture and the Built Environment, Delft University of Technology (TU Delft), Netherlands

Among these digital technologies, blockchain and distributed ledger technologies (DLT) present promising opportunities for transforming construction and the built environment. Early applications of blockchain and DLT are beginning to emerge and reveal their potential not only in addressing technical inefficiencies but also in enabling novel approaches to organisation, innovative business models and value creation. While still in their infancy, blockchain and DLT are evolving rapidly, presenting substantial opportunities for the sector's future development. At this critical juncture, it is essential to evaluate the implications of these technologies to fully understand their impact.

This book comes at the right time to investigate how these technologies might shape the future of construction and the built environment. To achieve this, the book brings together leading academics and practitioners from around the world to examine the intersection of blockchain and the built environment. It explores a wide range of blockchain applications, showcasing how these innovations can transform traditional processes, enhance transparency and drive innovation across multiple domains.

The book's chapters provide a comprehensive exploration of blockchain and DLT technology, progressing from foundational technical concepts to practical implementations and offering a cohesive understanding of its role as a driver for change. Covering a wide range of themes and applications – including information security in built environment projects, tokenisation of built assets, energy management, decentralised design governance, digital building logbooks (DBLs), supply chain management, contract management, legal perspective and the macro and meso implementation roadmaps – this book presents a comprehensive and multifaceted exploration of blockchain's relevance to construction and the built environment.

To provide a clear understanding of the book's structure and scope, the following sections provide an overview of its chapters. This summary outlines the key themes, applications and insights explored in each chapter, demonstrating how they collectively contribute to a comprehensive understanding of blockchain's potential in construction and built environment sector.

Chapter 2 provides a technical yet accessible overview of DLTs, focusing on blockchain as the most prominent example. The chapter thoroughly explains the blockchain's structure, including its distributed nature, cryptographic security and consensus mechanisms. Differences between permissioned and permissionless blockchains are detailed, offering examples such as Bitcoin (permissionless) and Hyperledger Fabric (permissioned).

The chapter clarifies the concepts of smart contracts and oracles, highlighting how they extend blockchain's functionality beyond cryptocurrencies to enable secure, automated transactions and data exchanges, with applications tailored to construction such as automated payments and real-time tracking of materials. By outlining the key benefits and capabilities of blockchain – including immutability, transparency and enhanced trust – the chapter presents a conceptual model for justifying blockchain-based applications. It further examines examples from the literature to illustrate blockchain's potential in addressing inherent challenges within the construction industry, setting the stage for a deeper exploration of its adoption at macro and meso levels and its framing as a socio-technical system (STS) in the following chapter.

Chapter 3 examines the adoption of blockchain and DLT in the construction industry, utilising STS theory to analyse the interplay between social and technical factors influencing adoption. The authors provide a historical overview of STS theory and establish its relevance to the built environment, particularly in addressing persistent challenges such as information inadequacy, unreliability and asymmetry among stakeholders.

The chapter introduces key conceptual models, including a four-dimensional framework that integrates the dimensions of technology, process, policy and society. This framework is employed as an analytical tool to evaluate blockchain applications and identify challenges and opportunities for implementation. Additionally, the authors present the "DLT Actors Model," which maps the roles and responsibilities of stakeholders within the blockchain ecosystem, providing a structured approach to adoption strategies. Roadmaps for macro- and meso-level implementation are also proposed, offering strategic insights into the integration of blockchain and DLT within construction workflows. These roadmaps emphasise the importance of aligning technological innovation with organisational and societal needs, addressing barriers to adoption and enhancing trust and collaboration.

Chapter 4 focuses on the growing importance of data security in the architecture, engineering, construction and operations (AECO) sector, driven by the widespread adoption of digital technologies such as building information modelling (BIM), unmanned aerial vehicles (e.g. drones) and other interconnected systems. The authors identify challenges stemming from the sector's unique contract-based and collaborative structure, often resulting in risks such as unauthorised access, data manipulation and limited information-sharing trust among stakeholders. To address these issues, the chapter introduces a decentralised framework integrating blockchain technology, inter planetary file system (IPFS) and encryption protocols to enhance data confidentiality, digital trust and utility while mitigating vulnerabilities inherent in centralised data management systems. The proposed framework combines blockchain's immutability and transparency with IPFS's decentralised storage capabilities and selective transparency techniques, enabling the secure sharing of sensitive information without compromising functionality. The chapter also explores the use of smart contracts to automate and secure workflows, such as request for information processes, demonstrating how distributed technologies ensure traceability, accountability and robustness against tampering. Recognising the barriers to adoption, particularly for smaller subcontractors, the authors suggest incentive-based approaches to foster inclusivity and explore future research avenues, including real-time data classification using domain-specific models. This chapter underscores the potential of decentralised technologies to transform information management practices in construction and the built environment, ensuring greater security, collaboration and resilience.

Chapter 5 investigates the transformative potential of blockchain technology and tokenisation in the real estate sector, focusing on how these innovations can enhance liquidity, transparency and inclusivity in property markets. The authors explore the challenges of a traditionally rigid and illiquid real estate market, particularly in countries like Italy, where regulatory barriers and a lack of digital maturity hinder investment and market fluidity. The chapter emphasises the role of property tokenisation as a paradigm shift, enabling fractional ownership, reducing entry barriers for small-scale investors and fostering broader participation.

The proposed tokenisation framework combines blockchain with GIS, BIM and smart contracts to create a decentralised and transparent system for real estate transactions. Key concepts include the creation of property tokens that represent fractional ownership of assets and their management via secure smart contracts. The integration of GIS and BIM supports precise asset localisation and detailed property information, enhancing decision-making and compliance. The authors highlight the socio-economic benefits of property tokenisation, including increased financial inclusion for underprivileged groups, opportunities for energy-efficient retrofitting and mechanisms to address housing affordability. Business models such as crowdfunding platforms, fractional ownership exchanges and real estate investment services are discussed, demonstrating the potential for innovative revenue streams and market expansion. Barriers to adoption, including regulatory uncertainties, technological infrastructure limitations and resistance to change, are addressed alongside potential solutions such as standardised protocols and collaboration with legal bodies like the Notary Council. The chapter concludes with a call for further research to refine the integration of tokenisation into real estate markets, underscoring its potential to drive sustainable development, reduce transaction inefficiencies and democratise property ownership.

Chapter 6 explores the concept of DBLs as a solution for managing building lifecycle data, aligning with the European Union's twin transitions of green and digital transformation. DBLs are presented as dynamic repositories capable of capturing detailed data on material composition, energy usage and operational performance, supporting transparency, informed decision-making and regulatory compliance within the built environment. The chapter identifies several barriers to implementing DBLs, including insufficient user incentives, high administrative costs and challenges related to data interoperability, privacy and security. Existing technical approaches, such as open BIM, centralised platforms and linked data, are critically assessed, revealing limitations in addressing long-term data storage, ensuring interoperability and fostering stakeholder participation.To overcome these challenges, the authors propose a blockchain-based approach, leveraging blockchain's inherent characteristics of trust, immutability and decentralised peer-to-peer transactions. They present a conceptual architecture for a blockchain-enabled DBL system comprising three integrated layers: a decentralised data layer, a logic and services layer driven by smart contracts to automate updates and manage ownership and a user layer featuring crypto-economic incentives to encourage active participation and data contributions. By combining legacy databases with decentralised systems such as IPFS, the proposed framework is proposed as a potential solution to ensure long-term data resilience while integrating existing datasets.

The chapter emphasises the ability of blockchain-enabled DBLs to foster circular economies, streamline building retrofits and enhance data security and privacy through advanced cryptographic mechanisms. The authors propose avenues for future research, including the development of legal frameworks tailored to blockchain adoption, refinement of business models to support sustainable practices and exploration of advanced blockchain technologies, such as zero-knowledge proofs, to balance privacy and transparency.

Chapter 7 investigates the application of blockchain technology in enhancing energy management within seaport infrastructure, focusing on a microgrid case study at the Port of Milford Haven in Wales. The chapter contextualises the growing energy demands of seaports, driven by refrigeration, air conditioning and other energy-intensive activities, which collectively contribute significantly to greenhouse gas emissions. The authors examine blockchain's role in decentralising energy management, particularly using Ethereum-based smart contracts.

The chapter introduces a blockchain-enabled microgrid framework that supports renewable energy integration, leveraging smart contracts for energy transfers and pricing while ensuring tamper-proof record-keeping. Using real-world data, two scenarios are analysed to evaluate the scalability and economic feasibility of the proposed framework. Scenario 1 focuses on five buildings within the port utilising locally produced renewable energy, while Scenario 2 scales up to include 200 additional community households. The analysis highlights significant cost implications tied to Ethereum's gas prices and the computational requirements of smart contracts. Although the system demonstrates potential for reducing energy costs and enhancing local energy resilience, the volatility of gas and cryptocurrency prices presents challenges to widespread adoption.

The chapter discusses key barriers, including the need for user training, hardware upgrades and ensuring the integrity of energy sources within the grid. It emphasises the importance of further research into private blockchain and centralised alternatives to overcome cost and scalability challenges.

Chapter 8 examines the potential of blockchain technology to decentralise architectural design processes by introducing innovative governance models and enabling collaborative, non-hierarchical approaches. The authors critique the traditional architectural design process as both autographic and extractive, arguing for a shift towards an allographic discipline where architecture becomes a decentralised collective endeavour. This transformation aligns with emerging technologies such as blockchain, IoT and AI within the context of the Fourth Industrial Revolution.

The chapter highlights blockchain's capacity to facilitate decentralised governance through mechanisms such as smart contracts, decentralised autonomous organisations (DAOs) and distributed data systems like IPFS. A series of prototypes are presented, demonstrating the integration of blockchain with architectural design tools like computer aided design and BIM to create systems for design collaboration, optimisation and decision-making. These include the development of shape grammars and DAOs as platforms for collaborative problem-solving and design governance. The authors discuss the application of stigmergy, a biological concept of self-organisation, to architectural collaboration, proposing decentralised feedback loops and collective decision-making frameworks as alternatives to traditional hierarchies. Then, they introduce ArchiDAO, a blockchain-enabled organisation designed to support architectural design, research and learning, emphasising principles of equity, transparency and sustainability.

The chapter provides insights for transforming architectural practice into a more inclusive and sustainable discipline through decentralised design workflows. It highlights the potential for blockchain to address issues of trust, transparency and collaboration in architectural design while fostering innovation and democratising the creation of the built environment.

Chapter 9 investigates the role of blockchain technology in transforming construction supply chain management (CSCM) by tackling longstanding challenges, including inefficiencies, limited transparency and trust deficits among stakeholders. The authors introduce a blockchain-based CSCM system, outlining its design, architecture and workflows. A real-world case study of cross-border modular construction between Hong Kong and Mainland China illustrates the practical implementation of the system. The case study highlights how blockchain can improve data traceability, ensure the immutability of records and facilitate trust among stakeholders by creating a shared, tamper-proof ledger. Key considerations for implementing blockchain in CSCM, such as stakeholder engagement, cost-benefit analysis, scalability, interoperability and legal compliance, are also discussed in depth.

Blockchain's integration with technologies such as IoT and BIM is examined as a means to improve logistics tracking and material provenance. The chapter also identifies emerging opportunities, including digital identity management and DAOs, which have the potential to automate processes and enhance collaboration within CSCM. However, the chapter acknowledges significant barriers to blockchain adoption in CSCM, including regulatory uncertainty, resistance to change, scalability limitations and data privacy concerns. These challenges underscore the need for industry-wide collaboration, policy development and further research to unlock blockchain's full potential. The chapter concludes by emphasising the importance of stakeholder readiness, robust legal frameworks and scalable solutions for successful blockchain adoption in CSCM.

Chapter 10 explores the integration of blockchain technology within BIM, digital twins (DT) and enterprise common data environments (CDEs) to enhance data observability, traceability and governance in large-scale infrastructure projects. The chapter draws on a case study of a significant UK infrastructure initiative, using participatory observation and literature analysis to contextualise the challenges and opportunities of enterprise-level data integration. The authors highlight the limitations of traditional data management systems, particularly in handling fragmented and heterogeneous data across diverse stakeholders and systems. Blockchain technology is proposed as a transformative solution, offering immutable data lineage, enhanced transparency and decentralised control. Key applications include creating tamper-proof audit trails, managing data provenance and ensuring seamless interoperability between BIM and DT data within enterprise CDEs. A conceptual framework for enterprise data integration is presented, emphasising three core elements: decentralised data layers, logic and service layers and user layers enabling stakeholder-specific access and collaboration. The framework integrates blockchain with legacy systems, leveraging hybrid architectures such as IPFS for distributed storage.

The chapter also introduces the concept of "golden records" to maintain consistent, authoritative data versions and discusses the importance of aligning enterprise data strategies with ISO 19650 standards for BIM and CDE workflows. Practical insights into addressing scalability, cost and user adoption challenges are provided, alongside recommendations for future research, including the integration of advanced technologies like AI and quantum computing.

Chapter 11 examines the integration of blockchain technology into construction contract administration (CCA) to address critical issues associated with traditional contract management, including inefficiencies, disputes and the mismanagement of contractual obligations. The chapter conceptualises CCA as a multi-functional system critical for project success, focusing on its core mechanisms: control, coordination and adaptation. Highlighting the challenges of poor CCA – including delays, cost overruns and disputes – the authors highlight the potential of blockchain to mitigate these issues through traceability, immutability and automation.

The authors perform a structured analysis of the blockchain's applicability to key CCA functions, such as financial management, performance monitoring and document handling, using examples from the *FIDIC Red Book* (2017 edition). They propose blockchain-enabled smart contracts to automate payment processes, improve transparency in decision-making and ensure accurate real-time data for contract provisions. Use cases such as automated interim payments and performance monitoring with integrated IoT and blockchain systems demonstrate the technology's potential to streamline workflows and minimise disputes.

The chapter identifies barriers to adoption, including technical challenges, the complexity of encoding contractual logic and the need for alignment with existing procurement frameworks. It recommends prioritising future research that fosters collaboration between scholars and contract drafting committees to develop scalable and context-specific solutions.

Chapter 12 presents a detailed legal perspective for the application of blockchain in construction and built environment applications. The chapter focuses on the intersection of physical assets, such as real estate and construction, with digital innovations like blockchain and tokenisation. The author acknowledges blockchain's transformative potential but argues that its adoption is limited by jurisdictional inconsistencies, evolving regulatory landscapes and operational risks. The author distinguishes between smart contracts (automated execution) and smart legal contracts (binding agreements with coded elements), addressing challenges like ambiguity in contract formation and confidentiality risks.

The chapter examines tokenisation's implications for fractional ownership, asset entanglement and intellectual property rights, highlighting conflicts between blockchain's immutability and data protection laws such as the General Data Protection Regulation. To address compliance challenges, the author proposes solutions including cryptographic protocols and segregated identity systems. The chapter examines financial regulations applicable to cryptocurrencies, tokenised assets and payment systems, with a focus on frameworks such as the UK's Financial Services and Markets Act and the EU's Markets in Crypto-Assets Regulation. It also addresses legal risks associated with token classification and compliance with Anti-Money Laundering and Know Your Customer regulations. Finally, the chapter explores cross-border legal challenges by comparing regulatory practices in jurisdictions such as Switzerland and Singapore. It examines practical applications of blockchain in areas like payment systems, DT and supply chain management, highlighting its potential to enhance transparency, automate compliance and ensure data authenticity.

The chapter provides practitioners, developers and legal professionals with guidance on integrating legal compliance into blockchain applications to achieve operational resilience and long-term success.

This concluding chapter summarises the key insights presented throughout this book. It identifies and analyses recurring themes, challenges and opportunities that emerge from the preceding chapters, aiming to connect individual contributions to a broader discourse on blockchain's role in addressing inefficiencies in the AECO industry.

The chapter begins by summarising the key contributions of each chapter, highlighting blockchain's foundational principles, its socio-technical implications and its application in domains such as supply chain management, CCA, property tokenisation and integration with DT and IoT systems. By doing so, the chapter underscores how this book bridges theoretical frameworks with practical applications.

The chapter identifies several cross-cutting themes, including blockchain's capacity to enhance transparency, decentralise governance, support lifecycle management and enable tokenisation for inclusive and sustainable practices. Challenges to blockchain adoption are analysed in detail, including legal and regulatory hurdles, scalability and performance limitations, integration with legacy systems and stakeholder resistance. These barriers are contextualised within the fragmented and complex nature of the built environment, underscoring the need for coordinated efforts to address them. The chapter also explores the dynamics of innovation ecosystems, identifying how open, closed and hybrid models influence the scalability and diffusion of blockchain in the AECO sector.

The chapter concludes with actionable recommendations adapted from the World Economic Forum's blockchain framework, tailored to the AECO sector. These include strategies for stakeholder engagement, regulatory alignment, pilot projects and the development of interoperable and scalable systems.

Ultimately, the conclusion calls for a collective effort among academia, industry and policy to realise blockchain's potential in the AECO sector. It positions blockchain not as a standalone solution but as an integral component of a broader effort to modernise the built environment. The chapter concludes with a reflection inviting stakeholders to regard blockchain as an enabler for fostering innovation, inclusivity and resilience in the AECO sector.

Chapter 2

Distributed ledger technologies, blockchain and smart contracts: technical foundations and key concepts

Mahir Msawil[*], *Mohamad Kassem*[†] *and David Greenwood*[*]

This chapter focuses on introducing blockchain coupled with its associated smart contracts as a particular and prominent instance of distributed ledger technologies (DLT). The chapter begins with a recent definition of DLT. Next, it discusses blockchain technology while referring to its two prominent types (i.e. permissionless and permissioned). Subsequently, the notions of smart contracts and oracles are explained and the key benefits and capabilities of blockchain are presented. A section is devoted to a conceptual model of how a blockchain-based application can be justified, and examples from literature are analysed to pave a logical path towards the subsequent chapters.

2.1 DLT: a definition

A DLT system has recently been defined as:

> 'a system of electronic records that (i) enables a network of independent participants to establish a consensus around (ii) the authoritative ordering of cryptographically-validated ('signed') transactions. These records are made (iii) persistent by replicating the data across multiple nodes, and (iv) tamper-evident by linking them by cryptographic hashes. (v) The shared result of the reconciliation/consensus process – the 'ledger' – serves as the authoritative version for these records'. [1: p. 24]

As this integrative definition implies, the design of DLT systems is orientated towards operating in an adversarial environment [1,2] with 'the presence of malicious actors ... who undermine the system by using it in ways it was not intended

[*]Department of Architecture and Built Environment, Northumbria University, UK
[†]Department of Engineering, Newcastle University, UK

for' [1: p. 22]. Both time and experience have proven that only a limited number of systems and industries can be immune to such adversarial environments.

2.2 Blockchain technology

The blockchain is the prominent instance of DLT [3]. The technology offers innovative approaches to digitally executing processes, sharing and storing corresponding transaction records among individuals or organisations [4]. Blockchain technology integrates multiple advanced technologies (e.g. peer-to-peer (P2P) networks, cryptography, distributed ledger and consensus mechanism) and technological infrastructure (e.g. cyber security protocols). It is a shared transactional database that is used to record transactions, and it runs on a distributed P2P network [5]. In essence, it is a chain of sequentially linked digital blocks containing a set of verified transactions with their encrypted data.

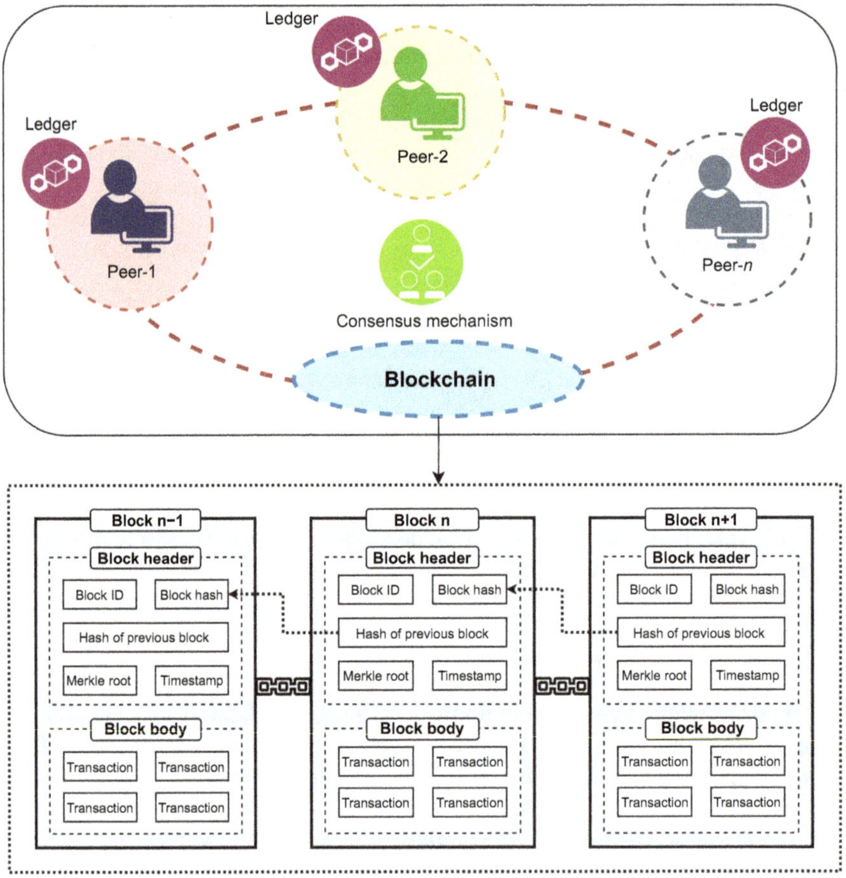

Figure 2.1 A generic view of a blockchain

Put another way, blockchain is a form of cryptography that links together blocks of information in a secure chain, where each piece of information is given specific timestamp and transaction data that is practically impossible to hack or alter [6]. This means that any data being input to the blockchain must be inserted only once and can never be altered, preventing any tampering or human error as the information gets shared throughout the network of parties who need to access it [7]. This sequence of transactional data revolves around a specific defined asset and its updated states throughout its lifecycle [8]. Within the context of blockchain, assets can be either tangible and/or intangible. For example, a car can be represented on the ledger as an asset, but so can a document, or a piece of art, or a vote. Other examples that can be represented are certificates or even a construction project.

Figure 2.1 presents a generic view of a blockchain highlighting key elements of a block and the connection between different blocks. A block consists of a block header and block body, as illustrated in Figure 2.1. The block is generated by underlying cryptographic algorithms to ensure integrity, security and immutability of recorded transactions. The blocks are appended to the chain in a linear, incremental manner with cryptographic connections. Each block references the hash of the previous block. A block body contains a single or many transactions depending on the data size and block capacity. Each transaction is digitally signed and validated using underlying cryptographic algorithms to achieve authentic security of the transaction creator and to timestamp the validated transactions. The creation of digital signatures and validation strictly follow specific consensus mechanisms that are coded and act as the core of any blockchain network. While it is recognised that the terminology used here may sound unfamiliar to some readers, by the end of this chapter this terminology will become clearer.

2.3 Blockchain types

Blockchain networks are generally categorised into two types based on identity management and permission to access the platform [9]:

Permissionless platforms allow any actor to access the platform and create transactions without any checks. This feature enables a high number of actors to join the platform. Thus, this type is less suitable for inter-organisational applications, and it is more aligned with business undertaking. Examples of this type are the Bitcoin and Ethereum platforms.

Permissioned platforms are formed by identified actors and organisations. Hence, they restrict access and the creation of transactions to known actors with high level of digital control mechanisms. Owing to this restricted accessibility feature and lower number of authorised actors, they are more suitable for inter-organisational applications. They are sometimes called consortium blockchain platforms. The best known permissioned blockchain is Hyperledger Fabric.

Decision trees have been proposed for selecting an appropriate type for a given process in construction [10,11]. Similarly, Nanayakkara *et al.* [12] offer a platform selection methodology that may be applied in construction projects at an inter-organisational level. Both types will be further explained with examples.

2.4 Prevailing blockchain platforms

Blockchain was first publicly introduced in 2009, when Satoshi Nakamoto developed and operated the Bitcoin blockchain which became the first blockchain network [13]. The Bitcoin blockchain operates an electronic P2P cash transfer system without the need for financial intermediaries [13]. There is evidence that the underlying concepts and techniques that Bitcoin blockchain relies on can be traced back to 1980s and 1990s in the form of cryptography and secure digital signatures used for payment and document versioning [14].

The key drivers for reviving these ideas in the form of blockchain were the need for transparency, trust and softening coercive power coupled with an increase cash transfer speed and elimination of traditional banking systems. Bitcoin is a permissionless type of blockchain, meaning it allows anyone with an online account (known as a wallet) to execute, trace and verify transactions on the network without the need for authorisation [15]. A limitation of Bitcoin blockchain is that execution is enabled by a scripting language that is not Turing complete,[1] making it inflexible for supporting applications beyond electronic P2P cash transfer system. In other words, it is inflexible to scale to other applications beyond the electronic cash which was subsequently termed 'cryptocurrency'.

To address this inflexibility, Ethereum was proposed as an open-source blockchain platform in a white paper [16]. Ethereum uses the cryptocurrency Ether to pay transaction fees on the platform, but it also has capabilities beyond the operation of cryptocurrency. Like Bitcoin, Ethereum is a permissionless blockchain platform. However, it is developed with a built-in Turing-complete programming language (i.e. Solidity) that operates on Ethereum virtual machine (referred to as 'EVM').

Ethereum thus enables a novel feature termed *smart contract* to execute persistent computational rules [17]. This novel feature allows anyone to encode business rules in the form of smart contracts and create adaptable decentralised applications (DApps) [18]. Such an innovative platform allows any group of organisations or individuals to create their own rules of transactional data including ownership and formats. With this development in mind, a range of new theoretical (and to some extent practical) blockchain-based DApps have emerged for various use cases in many industries including decentralised file storage, decentralised finance (DeFi) applications, supply chain management, energy and carbon emissions tracking. However, this permissionless form has been argued to nullify blockchain's applicability to some industries and systems due to raised concerns such as identity management, definition of roles and policies, technical performance, data confidentiality of sensitive transactions and its digital operation reliance on Ether cryptocurrency to use the network [15,19].

To address these concerns, Linux Foundation initiated the Hyperledger projects[2] with the aim of developing a variety of permissioned blockchain designs that

[1] https://www.bitstamp.net/en-gb/learn/blockchain/what-is-turing-complete/

[2] https://www.linuxfoundation.org/resources/case-studies/hyperledger

suit the requirements of various industries and systems beyond finance and cryptocurrencies [20]. With this permissioned network design, digital accessibility is only granted to pre-authorised organisations or individuals. The most prominent type of permissioned blockchain is the open-source Hyperledger Fabric project [19,21]. Hyperledger Fabric is a digital implementation protocol that was specifically devised to execute permissioned blockchain-based business processes [22]. In essence, it operates on the digital design philosophy of a permissionless blockchain (e.g. Bitcoin and Ethereum) while accommodating rising demands of enterprise businesses planning to leverage blockchain capabilities [12,15].

According to its official documentation [23], Hyperledger Fabric blockchain-governed applications are executed through a 'chaincode' that contains a set smart contract functions (SCFs) for each specific business process. Table 2.1 demonstrates a comparison between Ethereum and Hyperledger Fabric.

Table 2.1 Comparison of Ethereum and Hyperledger Fabric

Criterion	Ethereum	Hyperledger Fabric
Governance	Decentralised model	Consortium model
Consensus mechanism	Proof-of-stake (as of April 2024)	Endorsement policies: reliable, replicated, redundant and fault-tolerant based on crash fault tolerant as of April 2024 (with plans to move to Byzantine fault tolerant).
Cryptocurrency	Ether (gas transaction fee)	Not applicable
Permission	Permissionless	Permissioned based on membership service provider and certificate authority
Privacy	Limited	Configurable
Smart contracts	Pre-specified rules	Chaincode that contains a set of smart contract functions and coded parameters
Programming language	Solidity	Mainly JavaScript, Typescript and Go
Scalability	Moderate	High
Technical performance	Low due to high decentralisation	High due to the consensus mechanism with relatively limited number of authorised peer nodes
Certainty of operation cost	Moderate certainty due to fluctuation of gas transaction fee (Ether)	High certainty due to the absence of gas transaction fee
Development complexity	Moderate due to the availability of libraries and defined endpoints	High due to underlying digital components and infrastructure (e.g. Docker)
Ideal use cases	Decentralised finance (DeFi), public transaction settlement, tokenisation	Business-to-business (B2B) data transactions, inter-organisational processes, managerial and contractual processes

Source: Adapted from [25].

In contrast to Ethereum, Hyperledger Fabric provides authorised identity management, definition of roles, elimination of cryptocurrency use, reduced requirement for computational energy, scalable technical performance, flexibility of using own contractual governance and data confidentiality of transactions [12,24]. The technical development and deployment environments of both Ethereum and Hyperledger Fabric-based applications are recognised to be complex [25]. In general, blockchain-based platforms can be implemented by experienced developers or in the form of Blockchain as a Service offered by tech companies such as IBM[3] or Software as a Service offered by emerging startups, for example, AstraKode.[4] Regardless of the blockchain type and its development, the core elements to operate any blockchain-based application is the definition of 'digital assets' and rules in the form of smart contracts that operate a given blockchain-based application.

2.5 Smart contracts

The term 'smart contract' can be misleading, particularly in the context of built environment literature. Rather than implying an entire set of contractual conditions that are 'smart', in the context of blockchain, the term derives the work of Nick Szabo, who, in 1990s described a smart contract as a computerised transaction protocol that executes the terms of a contract [26]. Executing such a smart contract can be achieved on both blockchained or non-blockchained applications [4]; for example, the contractual transaction executed in the process of bike renting in smart cities [27] could be a non-blockchained application.

The association of smart contracts with blockchain was popularised by the development of Ethereum blockchain in 2014 [16]. Solidity as a programming language for Ethereum blockchain used the term [27], though in hindsight, the co-founder, Vitalik Buterin, of Ethereum publicly stated on X (formerly known as Twitter) in 2018: 'I quite regret adopting the term smart contracts. I should have called them something more boring and technical, perhaps something like persistent scripts'.[5] In contrast to the first category envisioned by Nick Szabo in 1990s, executing business contractual processes on blockchain through smart contract codes is transparent and trustworthy. The reason for this is that blockchain enables business partners to inspect and agree to the codes and the input data method before their deployment [27].

There is now a consensus among scholars that a blockchain-based smart contract can be described as a logical structure of computer code that self-executes upon the fulfilment of certain conditions (inputs) to enable changing the state of a digital variable by triggering a transaction event (outputs) while benefiting from blockchain characteristics [28,29]. In line with the stages of a given process, the

[3]https://www.ibm.com/products/blockchain-platform-hyperledger-fabric
[4]https://www.astrakode.tech
[5]https://x.com/VitalikButerin/status/1051160932699770882

conditions may be programmed based on contractual conditional statements or customised rules that prescribe specific actions in response to data/information input [30]. This is to ensure that smart contracts and their execution supporting methods reflect paper-based contractual processes and conditions.

In the context of traditional construction contracts, it has been proposed that blockchain-based smart contracts enable automation and self-enforcement of contractual conditions [28,29] and processes [4,31]. In theory, both can be converted into smart contract codes to reflect a full construction contract that can be operated and executed on blockchain [32]. But this conversion is influenced by the complexity embedded in contractual conditions. This complexity varies according to the business contractual terms of a given use case and its context in respect of the technical, social, process and policy dimensions (see Chapter 3). In addition, the codifiability of contractual conditions in the traditional contract must be assessed [33]. Table 2.2 highlights the lifecycle of a smart contract development through four phases along with their key activities and actors involved.

Feeding the input data/information into SCFs can be achieved using on-chain and/or off-chain sources. In the former, the required information can be automatically retrieved from previous transactions immutably stored in the blockchain. In the off-chain, so-called 'oracles' are required.

Table 2.2 The development lifecycle of a smart contract

Phase	1. Planning and creation	2. Coding and testing	3. Deployment	4. Execution
Key activities	– Identifying process and asset – Analysing the codifiability – Converting into rules – Reviewing and revising	– Coding the rules – Verifying code – Testing code – Approving code	– Deploying approved smart contract code on the selected blockchain platform – Initialising the deployed smart contract code	– Auto-executing smart contract code once triggered – Generating transactions for the blockchain-based process – Updating the state of the digital asset – Terminating the lifecycle if no longer needed
Key actors	Representatives of organisations Contract experts Blockchain developers	Programmers Blockchain developers Representatives of organisations	Blockchain developers Representatives of organisations	Representatives of organisations Technical administrators

2.6 Blockchain oracles

The execution of the blockchain-based smart contracts relies on real-world information to a great extent. In the context of blockchain, oracles are agents that connect blockchain to the external world and vice versa [34] and can transfer this information to the smart contract [35] and from it to the real world [36].

Oracles play the role of identifying and collecting off-chain data and forwarding this data for on-chain processing through smart contracts. In turn, resulting transactions are recorded on the blockchain [36]. For instance, if a smart contract on a blockchain-based application requires information about the status of a given 'digital asset', an oracle can provide that information to the network. Based on the provided information to the coded parameters, the on-chain smart contract self-executes, thereby recording a blockchain transaction that contains the updated information on that 'digital asset' at that point in time.

Therefore, oracles play an essential role in the operation of blockchain-based applications and hence their types have been grouped into categories [37]. There exists a wide range of different types of oracles with varying degrees of technical design. The suitability of an oracle depends on the blockchain-based application requirements for data integrity and validity considering that for a single application multiple types of oracles may be required [34]. Of relevance to this chapter, three types of oracles are discussed: human, software and hardware oracles within the context of construction projects.

Human oracles are human agents entrusted to manually providing input data to the smart contracts for blockchain transaction processing. Such oracles may be design managers, project managers, project controls engineers, quantity surveyors, or maintenance and operation engineers.

Software oracles rely on existing digital platforms and solutions implemented in construction projects. These include building information management (BIM)-based solutions, artificial intelligence (AI)-based tools, electronic document management systems or weather data websites.

Hardware oracles directly feed SCFs with data from sensors or other similar devices. A hardware oracle may collect data from different hardware devices and then pre-process it in a suitable format for a blockchain-based application. A good example of this is the Weather Ledger project.[6]

In any blockchain-based application environment, these oracles may be designed to work in tandem or in isolation with varying degrees of reliability and deployment. A blockchain-based application that requires human oracles classifies it as a semi-automated application, while one relying only on software and hardware oracles can be regarded as fully automated [28]. In real-world applications, a blockchain-based solution involves two ends: the back end and front end. The back end mainly consists of the ledger and the code of the smart contract. The front end enables an

[6]https://www.digicatapult.org.uk/wp-content/uploads/2021/11/Weather_Ledger_WP5_report-v2.pdf

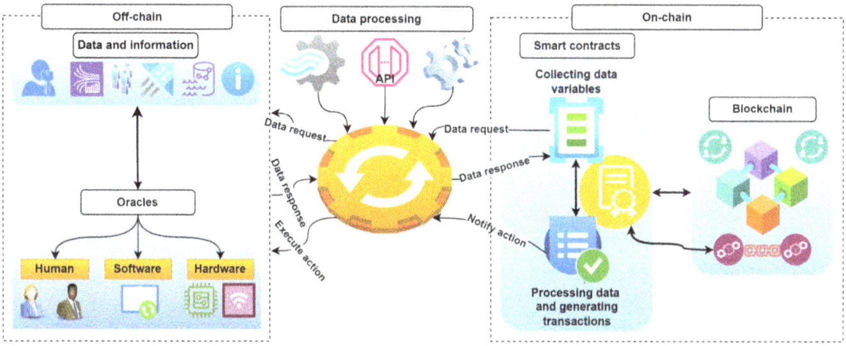

Figure 2.2 Processing of data between off-chain and on-chain (adapted from [38])

oracle to interact with the solution through a user interface and an application programming interface (API). Figure 2.2 shows an abstraction of off-chain and on-chain interaction through oracles and their digital relationships.

In what follows, the key components along with their digital flow are described.

Off-chain: this component represents the physical world outside the blockchain where data and information are generated while received by oracles for processing and forwarding to smart contracts through interfaces and APIs. This off-chain component is implicitly governed by contractual arrangements and specific mechanisms that regulate the interactions among the involved organisations and their respective three types of oracles.

Data processing: this component performs a range of digital operations including requests, responses, invocation, queries and updates. This component hinges on database servers and APIs. It serves multiple technical functions. First, it enables human oracles to interact with the blockchain. This interaction can be to invoke a SCF by entering the required data variables or to query/view the blockchain ledger. Second, it enables software and hardware oracles to trigger a SCF in response to the occurrence of an event or query the blockchain ledger at regular coded intervals (e.g. each 24 h) or invoke a specific function when pre-defined conditions are met based on programmed conditional statements. In addition, this set of middleware serves to return results retrieved from the ledger (on-chain) to human and software oracle. Furthermore, the database updates itself for subsequent digital operations (e.g. automatic invocation). This component acts as a digital bridge between the off-chain side and the on-chain side.

On-chain: the design of this component follows the design philosophy of smart contracts and blockchain as theoretically discussed above and will be further elaborated on in subsequent chapters.

Operationalising such technical reference connects the involved organisations in the simultaneous information and time dimensions while each organisation is ensured that the records are validated and immutable. The challenge is the case of incorrect information that can result in erroneous transactions immutability recorded on the blockchain. The counterargument here is that the human and software oracles providing input data are both identified in the recorded transactions. On the part of human oracles, this increases the level of due-diligence and accountability in respect of data inputs. The need for increased due-diligence assessment fundamentally stems from the prospective awareness that a SCF cannot be invoked without the input of all required data variables while taking into account that there will neither be an opportunity to change any input data nor to revoke recorded resultant transactions due to immutability. This awareness will substantially reduce erroneous data inputs and, hence, increase the likelihood of keeping the blockchain input error free. With this in mind, the data variables can also be obtained from available digital solutions as software oracles. It is envisaged that existing digital solutions implemented on construction projects and built assets can act as software oracles to feed SCFs with required data variables. Yet, an unintended limitation of current available digital solutions is that they tend to treat their digitalised processes as being independent which are often interdependent. Hence, blockchain-based applications integrate those solutions (deployed off-chain) with their corresponding on-chain ledger and SCFs through APIs and databases that can technically be developed. In turn, SCFs are interconnected on blockchain platforms with the capability of calling each other as well as calling APIs to obtain and feed data from relevant processes. Subsequently, the various cumulative transactions can be queried by human agents or software agents.

As a result of the possibility interconnecting blockchain and smart contracts with oracles in general and existing software and hardware oracles in particular, the technology has evolved into a general-purpose technology that has been investigated at research level in many different domains with the aim of enhancing processes to address inherent challenges [31,38]. One reason for its widespread diffusion at the research level is its unique benefits created through the integration of its core components. Put another way, this integration results in offering a set of unique benefits that has not been observed in a single technology earlier.

2.7 Benefits and capabilities of blockchain-based applications

Regardless of its type, blockchain technology is widely considered as a technical innovation that is expected to bring about changes in many industries that could be both incremental and radical [33,39]. Why may this be the case? This section attempts to offer an answer to this question through the demonstration of blockchain tangible benefits as achieved from its core components. Figure 2.3 illustrates the benefits and capabilities of blockchain while conceptually pointing to the core digital components that enable the realisation of these capabilities.

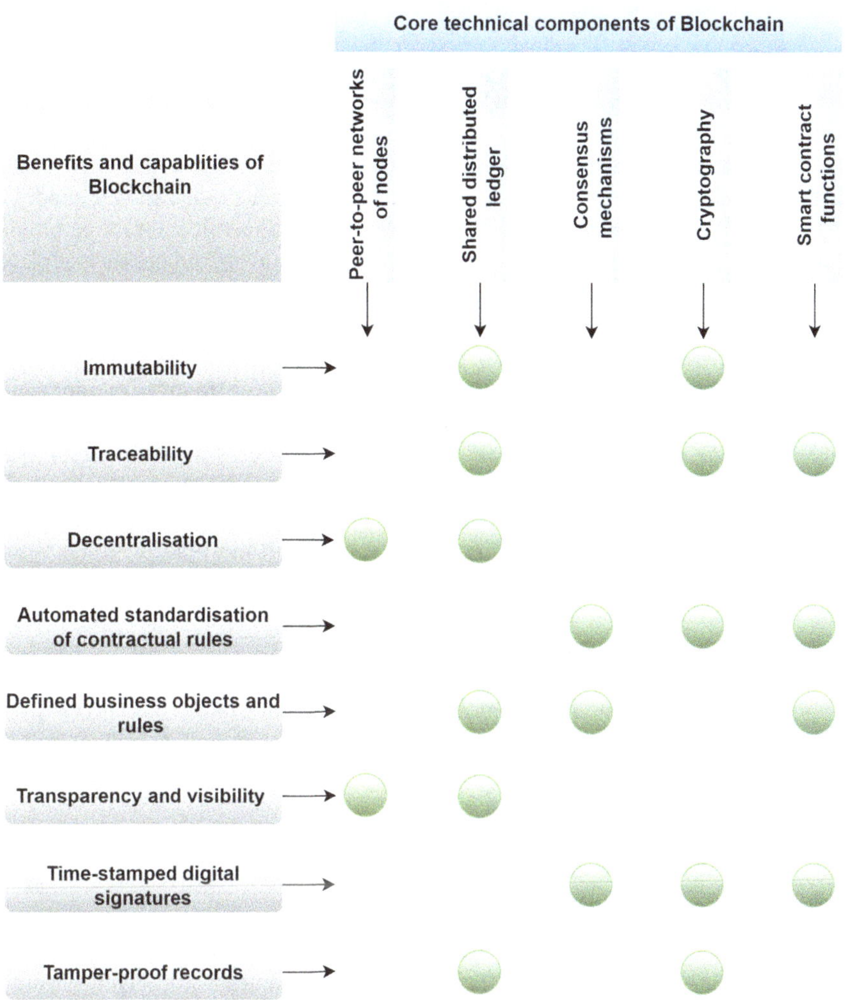

Figure 2.3 A conceptual map between blockchain components and benefits (source: authors)

It should be reiterated that prospective benefits include immutability of stored transactional data, traceability of completed and remaining tasks, instant accessibility to stored data, information symmetry of executed tasks, streamlined processes, digitally defined roles, full or partial automation of contractual statements and time-stamped signatures [4,6,8,40]. In addition to these first-order benefits, second-order benefits can be achieved including increased accountability, audibility, efficiency and effectiveness of business processes, transparency and tackling lack of trust. Together, these benefits and capabilities contribute to addressing inherent challenges including opportunism, information asymmetry and uncertainty that are rooted in business processes of many industries [41,42] including processes in the built environment and construction. In addition, it is widely reported that blockchain-based solutions instil trust in the execution of inter-organisational processes.

Trust as an abstract relational governance mechanism has been widely proposed as a solution to address a wide range of challenges that arise in adversarial environments that suffer from lack of trust, and on the basis that a higher level of trust increases the efficiency and effectiveness of inter-organisational processes [43]. Hence, lack of trust has been often postulated as a justification for employing blockchain-based application [4] in which trust in people and processes is replaced by trust in code and immutable automation and recording of transactions on the distributed ledger [7,44]. This results in eliminating or at least reducing causes and effects of adversarial environments, thus installing automatic trust [45].

As previously mentioned, blockchain technology offers innovative approaches to digitally executing processes, sharing and storing corresponding transaction records among individuals or organisations [4]. This occurs in a P2P digital business environment using a distributed ledger. As a result, there is recently an emerging theoretical notion that predicts the move to governance by blockchain in the future. This notion is equally noted by scholars in the field of organisations [33,46] and built environment [47]. However, Lumineau *et al.* [33] underline the importance of knowing the underlying mechanisms of blockchain-based smart contract codes that reflect the business rules associated with a given process. On a global scale, blockchain technology has attracted the attention of many research institutions and international organisations including the United Nations (UN)[7] and applied in various cases beyond Cryptocurrency, for example, by Volvo,[8] and the recent Tokyo 2021 and Paris 2024 Olympics.[9]

2.8 How can a blockchain-based application be justified?

This section provides a generic structure that can act as a communication mechanism between stakeholders of the built environment in its various fields and researchers as well as blockchain developers. It illuminates how a given system or process can be transitioned into a blockchain-based application. Figure 2.4 illustrates this generic structure to transition from a problem space to a solution space.

In the problem space, a given process or system is identified as a potential use case that is faced by complexity coupled with challenges in its practical context. Accordingly, the benefits and capabilities offered by blockchain technology can be mapped to the identified challenges. As a result, the adoption of blockchain technology that can address the analysed complexity and challenges can be justified. Following the justification of using blockchain as a potential solution, the logical flow moves to the solution space.

[7]https://www.unjiu.org/sites/www.unjiu.org/files/jiu_rep_2020_7_english.pdf
[8]https://spectrum.ieee.org/volvo-ex90-battery-passport
[9]https://businessday.ng/technology/article/blockchain-ai-at-the-olympics-the-new-winning-combination/

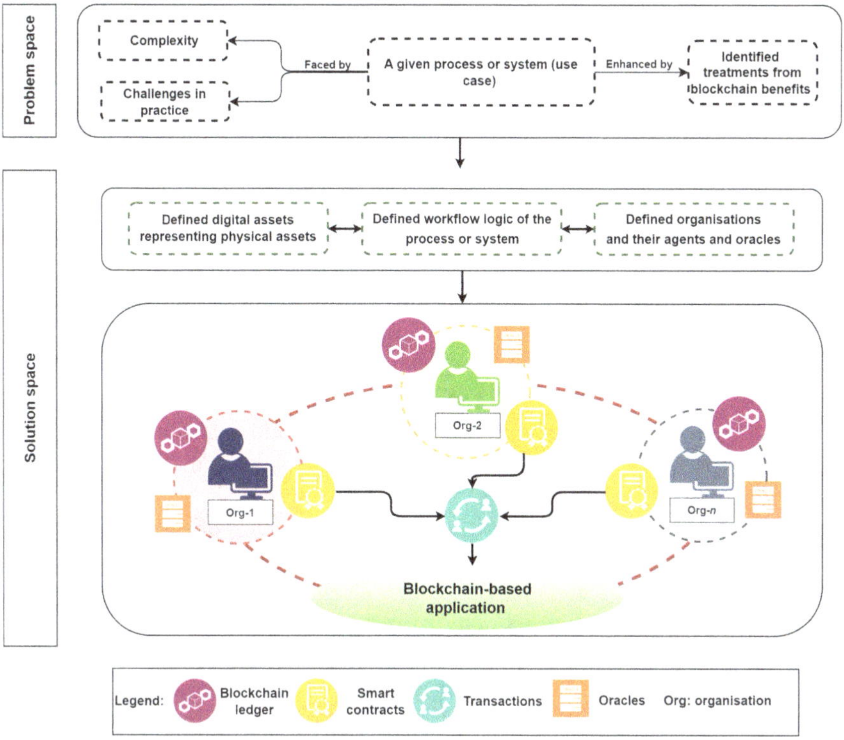

Figure 2.4 An illustration of a transition from a problem space to solution space (source: authors)

In the solution space, the first step involves the conversion of the tangible/intangible assets into 'digital assets'. For example, a particular piece of construction equipment and building materials can be represented as 'digital assets'. Likewise, a design document can be represented as a 'digital asset'. Along the same line, monetary value of progress payments and other contractual processes can be abstracted as a 'digital asset'. A 'digital asset' has a unique identification token with its own properties that digitally change on the blockchain through SCFs. This change takes place according to a defined workflow that consists of multiple stages based on current practices or new procedures. For example, a tendering process can be reflected in a blockchain-based application. Similarly, the design process of a built asset can be modelled as a workflow. In a similar vein, procurement and supply of building materials from the manufacturer to construction site involve many stages that can be converted into a digital workflow. In addition, all organisations and their respective actors along with their oracles are connected to a single digital platform enabled by blockchain-based applications.

In the transition process from the problem space to the solution space, considerable time should be spent in the problem space. This time gives opportunities to advance understanding of inter-organisational processes in developing blockchain-based solutions as discoveries may arise in the development discourse. These opportunities include the examination and analysis of task interdependence, coordination and contractual processes with the aim of deriving variables of processes and systems along with their respective actors that need to be coded in the SCFs. Such opportunities can be based on deep discussions of motivations for adopting blockchain.

As Figure 2.4 shows, a blockchain-based application is operated by connecting organisations' peers in a distributed manner consistent with their respective obligations and tasks to manage a given 'digital asset' through its workflow stages. The SCFs are deployed on the blockchain platform of the digital node of each peer. Invoking a given SCF sets the entire digitalised process in motion by acquiring off-chain input data corresponding to each required task from oracles and converting the data to on-chain transactions. As previously mentioned, a transaction carries information that reflects a corresponding completed obligation of any defined task of the 'digital asset' life cycle stages of a process or system. Each peer registered on the blockchain-based solution needs to fulfil its allocated obligations. This results in sequential changes of states in the creation of the 'digital asset' that may represent a variety of processes and systems. These flowing elements are hosted on all the digital nodes of the peers in a distributed and unified manner and digitally stored on the decentralised ledger that contains all transactions.

This blockchain-based approach chains and integrates the peers enabling them to have complete status of the 'digital asset' at any given time with a history of fulfilled obligations and visibility of remaining obligations. With the representation of physical assets in the forms of 'digital assets', the changes throughout the life cycle of any 'digital asset' from its initial state to its created state are automatically and immutability recorded on the ledger through a series of digital transactions. Compared to traditional approaches, this way results in simultaneous symmetry and instant visibility of performed and remaining obligations placed on the organisations and their respective agents. At the same time, concerns that arise about altering recorded data are eliminated due to the immutability of the ledger achieved through cryptography. Thus, trust in the executed tasks emerges as a by-product. As previously stated, the purpose of such solutions is to leverage the demonstrated blockchain benefits with the aim of improving processes and systems in the built environment in its various fields while maintaining the essence of current practices as much as possible.

2.9 Evidence of blockchain applicability in the built environment

This section offers scholarly evidence of the workability and applicability of blockchain technology to transform various processes and systems across the built environment. Table 2.3 presents studies that offer a proof-of-concept of blockchain's potential benefits in a range of built asset phases and application.

Table 2.3 Examples of blockchain-based proof-of-concept studies

Reference	Phase	Application	Platform	Oracles	Smart contract functions (SCFs)
[48]	Tendering	Decentralised secured tendering	Permissionless (Ethereum)	Human and processing oracles	10 SCFs reflecting stages in the tender process
[49]	Design	Lightweight blockchain-as-a-service prototype to enhance BIM design efficiency.	Permissioned (Hyperledger Fabric)	Human	3 SCFs: information sharing, permissions and revision management
[50]	Supply chain	Decentralised material management system for information sharing and recording	Permissioned (Hyperledger Fabric)	Software (BIM) and human	8 SCFs across three stages (purchase order, material tracking, inventory management)
[51]	Records	A secure document management system to integrate project participants and secure irreversible documents and records	Permissioned (Hyperledger Fabric)	Human	2 SCFs for design documentation in a BIM environment
[52]	Financial	Financial management (traditional procurement)	Permissioned (Hyperledger Fabric)	Software (BIM) and human	3 SCFs to interconnect parties' financial obligations
[53]	Pollution control	A blockchain-enabled system to automatically monitor and record on-site noise pollution	Permissioned (Hyperledger Fabric)	Internet of Things (IoT) sensors	Pseudo-code for detecting noise level through IoTs and recording on ledger.
[54]	Operation	Thermal performance monitoring post construction	Permissionless (Ethereum)	Sensors	SCFs interact built assets through oracles to initiate payment performance incentives
[55]	Managing carbon	Carbon management of materials and product certification	Permissionless (Ethereum)	Human	SCFs for data on material origins, energy consumption and certification.
[56]	Circular design	An integrated BIM-blockchain solution for circular construction	Permissionless (Ethereum)	BIM libraries and human	SCFs enable stakeholders to record salvageable and reusable items via BIM models

This set of proofs of concept is believed to be implementable and verifiable in real-life scenarios. Hence, such proofs of concept can be tested through pilot projects. With this evidence available, it can be inferred that the interconnected implementation of such decentralised and distributed applications is a real possibility that will address many of the dominant challenges within various fields of the built environment.

It is widely believed that owners of built environment assets should be the parties that initiate such solutions and implement them in practice with the aim of governing their built assets by blockchain technology. It should be noted here that in case such initiatives take place, prospective adopters (e.g. owners of built assets) shall make it explicit *ex-ante* that a given process will be executed using a blockchain-based application. Such explicit expression prevents potential disputes *ex-post*, because supply chain actors (e.g. suppliers and contractors) may not be familiar with blockchain technology and its implications or may not have sufficient qualified resources to operate it. For instance, a contractor may need to employ highly skilled staff for the purpose of operating such applications and, therefore, should have reflected their respective cost in its bid *ex-ante*.

That said, it should be noted that the adoption and implementation of such applications will give rise to other types of challenges. Good examples of such challenges can be found in Chapter 11. With novel research methods (e.g. Design Science Research, Action Research and Engaged Scholarship) that involve collaboration efforts between academia and practice while embracing fresh looks, dominant and emerging challenges can be addressed. To this effect, it can be inferred that the need for research will continue while giving rise to the acceptance and adoption of the technology. Theoretically, a blockchain-based process or system can support a high level of efficiency and effectiveness in relation to processes and systems. This is achieved by converting the process from its traditional operationalisation approach to a blockchain-based process, as explained above.

2.10 Conclusions

The positive influence of blockchain technology and associated smart contracts on the built environment may be evident at multiple levels. At the inter-organisational level, it is believed that managerial and contractual processes will become optimally efficient and effective resulting in faster delivery of built assets. The main reason for this is that time consumed in administrative tasks will diminish alongside the reorientation of efforts towards the delivery of built assets. This is of course believed to reduce associated administrative costs across all phases, including the design, construction and operation. It is further envisioned that across the construction phase, practices of procurement and contractual arrangements as well as materials supply will be transparent and traceable. During the operations of built assets (buildings and infrastructure), blockchain-based platforms can act as a governance connecting the wider available technologies (e.g. Internet of Things and maintenance software) into one trusted single source of data necessary for operating and maintaining these assets.

Rather than focusing on the potential for disruption, however, practitioners, policymakers and academics should look at the positive influence that these blockchain-based developments may have for the built environment. Joint efforts with blockchain developers should be established rather than taking separate undertakings. Such joint efforts will ensure that current practices are not radically disrupted, and blockchain-based solutions can be incrementally introduced taking into account 'tried and tested' practices. It is recognised that this is a fast-moving discourse across industries in general and the built environment in particular. Despite the progress achieved to date on this discourse at both research and industry levels, there remains a high level of uncertainty surrounding the operational functionality of the technology. It can be concluded with a foresight that this technology coupled with its associated smart contracts notion will likely cause various magnitude of incremental and radical changes at both the construction project and industry levels.

References

[1] Vergne JP. Decentralized vs. distributed organization: Blockchain, machine learning and the future of the digital platform. *Organization Theory* 2020; 1:1–26. https://doi.org/10.1177/2631787720977052.

[2] Nguyen LD, Bröring A, Pizzol M, and Popovski P. Analysis of distributed ledger technologies for industrial manufacturing. *Scientific Reports* 2022; 12:18055. https://doi.org/10.1038/s41598-022-22612-3.

[3] Beck R, Avital M, Rossi M, and Thatcher JB. Blockchain technology in business and information systems research. *Business and Information Systems Engineering* 2017;59:381–4. https://doi.org/10.1007/s12599-017-0505-1.

[4] Ante L. Smart contracts on the blockchain – A bibliometric analysis and review. *Telematics and Informatics* 2021;57. https://doi.org/10.1016/j.tele.2020.101519.

[5] Ostern NK, Holotiuk F, and Moormann J. Organizations' approaches to blockchain: A critical realist perspective. *Information and Management* 2022;59(7). https://doi.org/10.1016/j.im.2021.103552.

[6] Chong AYL, Lim ETK, Hua X, Zheng S, and Tan C-W. Business on chain: A comparative case study of five blockchain-inspired business models. *Journal of the Association for Information Systems* 2019;20(9):1310–39.

[7] Swan M, and de Filippi P. Toward a philosophy of blockchain: A symposium: Introduction. *Metaphilosophy* 2017;48:603–19. https://doi.org/10.1111/meta.12270.

[8] Zavolokina L, Bauer-Hänsel I, Hacker J, and Schwabe G. Organizing for value creation in blockchain information systems. *Information and Organization* 2024;34(3). https://doi.org/10.1016/j.infoandorg.2024.100522.

[9] Nawari NO, and Ravindran S. Blockchain and the built environment: Potentials and limitations. *Journal of Building Engineering* 2019;25. https://doi.org/10.1016/j.jobe.2019.100832.

[10] Hunhevicz JJ, and Hall DM. Do you need a blockchain in construction? Use case categories and decision framework for DLT design options. *Advanced Engineering Informatics* 2020;45. https://doi.org/10.1016/j.aei.2020.101094.

[11] Li J, Greenwood D, and Kassem M. Blockchain in the built environment and construction industry: A systematic review, conceptual models and practical use cases. *Automation in Construction* 2019;102:288–307. https://doi.org/10.1016/j.autcon.2019.02.005.

[12] Nanayakkara S, Rodrigo MNN, Perera S, Weerasuriya GT, and Hijazi AA. A methodology for selection of a blockchain platform to develop an enterprise system. *Journal of Industrial Information Integration* 2021;23. https://doi.org/10.1016/j.jii.2021.100215.

[13] Nakamoto S. Bitcoin: A peer-to-peer electronic cash system. *Decentralized Business Review* 2008. Available at: https://bitcoin.org/bitcoin.pdf.

[14] Tripathi G, Ahad MA, and Casalino G. A comprehensive review of blockchain technology: Underlying principles and historical background with future challenges. *Decision Analytics Journal* 2023;9. https://doi.org/10.1016/j.dajour.2023.100344.

[15] Helliar CV, Crawford L, Rocca L, Teodori C, and Veneziani M. Permissionless and permissioned blockchain diffusion. *International Journal of Information Management* 2020;54. https://doi.org/10.1016/j.ijinfomgt.2020.102136.

[16] Buterin V. *Ethereum: A next-generation smart contract and decentralized application platform*. 2014. https://www.weusecoins.com/assets/pdf/library/Ethereum_white_paper-a_next_generation_smart_contract_and_decentralized_application_platform-vitalik-buterin.pdf.

[17] Watanabe H, Fujimura S, Nakadaira A, Miyazaki Y, Akutsu A, and Kishigami J. Blockchain contract: Securing a blockchain applied to smart contracts. *Proceedings of the 2016 IEEE International Conference on Consumer Electronics (ICCE)*; 2016, p. 467–8. https://doi.org/10.1109/ICCE.2016.7430693.

[18] Hu T, Liu X, Chen T, *et al.* Transaction-based classification and detection approach for Ethereum smart contract. *Information Processing and Management* 2021;58. https://doi.org/10.1016/j.ipm.2020.102462.

[19] Hang L, and Kim DH. Optimal blockchain network construction methodology based on analysis of configurable components for enhancing Hyperledger Fabric performance. *Blockchain: Research and Applications* 2021;2. https://doi.org/10.1016/j.bcra.2021.100009.

[20] Aggarwal S, and Kumar N. Hyperledger. *Advances in Computers* 2021;121:323–43.

[21] Antwi M, Adnane A, Ahmad F, Hussain R, Habib ur Rehman M, and Kerrache CA. The case of HyperLedger Fabric as a blockchain solution for healthcare applications. *Blockchain: Research and Applications* 2021;2. https://doi.org/10.1016/j.bcra.2021.100012.

[22] Androulaki E, Barger A, Bortnikov V, *et al.* Hyperledger Fabric: A distributed operating system for permissioned blockchains. *Proceedings of the*

13th EuroSys Conference, EuroSys 2018, vol. 2018, Association for Computing Machinery, Inc; 2018. https://doi.org/10.1145/3190508.3190538.
[23] HLF. https://hyperledger-fabric.readthedocs.io/en/release-2.5/ 2023. https://hyperledger-fabric.readthedocs.io/en/release-2.5/ (accessed December 25, 2023).
[24] Guggenberger T, Sedlmeir J, Fridgen G, and Luckow A. An in-depth investigation of the performance characteristics of Hyperledger Fabric. *Computers and Industrial Engineering* 2022;173. https://doi.org/10.1016/j.cie.2022.108716.
[25] Capocasale V, Danilo G, and Perboli G. Comparative analysis of permissioned blockchain frameworks for industrial applications. *Blockchain: Research and Applications* 2022;4(1):100113. https://doi.org/10.1016/j.bcra.2022.100113.
[26] Szabo N. Formalizing and securing relationships on public networks. *First Monday* 1997;2. https://doi.org/10.5210/fm.v2i9.548.
[27] Vigliotti MG. What do we mean by smart contracts? Open challenges in smart contracts. *Frontiers in Blockchain* 2021;3. https://doi.org/10.3389/fbloc.2020.553671.
[28] Mason J. *Innovating Construction Law: Towards the Digital Age*. London: Routledge; 2021.
[29] Cummins J, and Clack CD. Transforming commercial contracts through computable contracting. *Journal of Strategic Contracting and Negotiation* 2022;6:3–25. https://doi.org/10.1177/20555636211072560.
[30] Dixit A, Deval V, Dwivedi V, Norta A, and Draheim D. Towards user-centered and legally relevant smart-contract development: A systematic literature review. *Journal of Industrial Information Integration* 2022;26. https://doi.org/10.1016/j.jii.2021.100314.
[31] Lu W, Wu L, and Zhao R. Rebuilding trust in the construction industry: A blockchain-based deployment framework. *International Journal of Construction Management* 2021;23(8):1405–16. https://doi.org/10.1080/15623599.2021.1974683.
[32] Adel K, Elhakeem A, and Marzouk M. Decentralizing construction AI applications using blockchain technology. *Expert Systems with Applications* 2022;194. https://doi.org/10.1016/j.eswa.2022.116548.
[33] Lumineau F, Wang W, and Schilke O. Blockchain governance-A new way of organizing collaborations? *Organization Science* 2021;32:500–21. https://doi.org/10.1287/orsc.2020.1379.
[34] Lu W, Li X, Xue F, Zhao R, Wu L, and Yeh AGO. Exploring smart construction objects as blockchain oracles in construction supply chain management. *Automation in Construction* 2021;129. https://doi.org/10.1016/j.autcon.2021.103816.
[35] Zheng Z, Xie S, Dai HN, *et al.* An overview on smart contracts: Challenges, advances and platforms. *Future Generation Computer Systems* 2020;105:475–91. https://doi.org/10.1016/j.future.2019.12.019.

[36] Hassan A, Makhdoom I, Iqbal W, Ahmad A, and Raza A. From trust to truth: Advancements in mitigating the blockchain oracle problem. *Journal of Network and Computer Applications* 2023;217. https://doi.org/10.1016/j.jnca.2023.103672.

[37] Pasdar A, Lee YC, and Dong Z. Connect API with blockchain: A survey on blockchain oracle implementation. *ACM Computing Surveys* 2023;55(10):208. https://doi.org/10.1145/3567582.

[38] Ølnes S, Ubacht J, and Janssen M. Blockchain in government: Benefits and implications of distributed ledger technology for information sharing. *Government Information Quarterly* 2017;34:355–64. https://doi.org/10.1016/j.giq.2017.09.007.

[39] Perera S, Nanayakkara S, Rodrigo MNN, Senaratne S, and Weinand R. Blockchain technology: Is it hype or real in the construction industry? *Journal of Industrial Information Integration* 2020;17. https://doi.org/10.1016/j.jii.2020.100125.

[40] Guo L, Liu Q, Shi K, Gao Y, Luo J, and Chen J. A blockchain-driven electronic contract management system for commodity procurement in electronic power industry. *IEEE Access* 2021;9:9473–80. https://doi.org/10.1109/ACCESS.2021.3049562.

[41] Schmidt CG, and Wagner SM. Blockchain and supply chain relations: A transaction cost theory perspective. *Journal of Purchasing and Supply Management* 2019;25. https://doi.org/10.1016/j.pursup.2019.100552.

[42] Ahluwalia S, Mahto RV, and Guerrero M. Blockchain technology and startup financing: A transaction cost economics perspective. *Technological Forecasting and Social Change* 2020;151. https://doi.org/10.1016/j.techfore.2019.119854.

[43] Anthony Jnr B. Distributed ledger and decentralised technology adoption for smart digital transition in collaborative enterprise. *Enterprise Information Systems* 2021;17(4):465–98. https://doi.org/10.1080/17517575.2021.1989494.

[44] Fu Y, and Zhu J. Trusted data infrastructure for smart cities: A blockchain perspective. *Building Research & Information* 2021;49:21–37.

[45] Qian X, and Papadonikolaki E. Shifting trust in construction supply chains through blockchain technology. *Engineering, Construction and Architectural Management* 2021;28:584–602. https://doi.org/10.1108/ECAM-12-2019-0676.

[46] Pelt R van, Jansen S, Baars D, and Overbeek S. Defining blockchain governance: A framework for analysis and comparison. *Information Systems Management* 2021;38:21–41. https://doi.org/10.1080/10580530.2020.1720046.

[47] Hunhevicz J, Dounas T, and Hall DM. The promise of blockchain for the construction industry: A governance lens. *Blockchain for Construction, Blockchain Technologies*. Singapore: Springer; 2022.

[48] Ahmadisheykhsarmast S, Senji SG, and Sonmez R. Decentralized tendering of construction projects using blockchain-based smart contracts and storage systems. *Automation in Construction* 2023;151. https://doi.org/10.1016/j.autcon.2023.104900.

[49] Tao X, Das M, Zheng C, *et al.* Enhancing BIM security in emergency construction projects using lightweight blockchain-as-a-service. *Automation in Construction* 2023;150. https://doi.org/10.1016/j.autcon.2023.104846.
[50] Basheer M, Elghaish F, Brooks T, Pour Rahimian F, and Park C. Blockchain-based decentralised material management system for construction projects. *Journal of Building Engineering* 2024;82. https://doi.org/10.1016/j.jobe.2023.108263.
[51] Das M, Tao X, Liu Y, and Cheng JCP. A blockchain-based integrated document management framework for construction applications. *Automation in Construction* 2022;133. https://doi.org/10.1016/j.autcon.2021.104001.
[52] Elghaish F, Pour Rahimian F, Hosseini MR, Edwards D, and Shelbourn M. Financial management of construction projects: Hyperledger fabric and chaincode solutions. *Automation in Construction* 2022;137. https://doi.org/10.1016/j.autcon.2022.104185.
[53] Zhong B, Guo J, Zhang L, Wu H, Li H, and Wang Y. A blockchain-based framework for on-site construction environmental monitoring: Proof of concept. *Building and Environment* 2022;217. https://doi.org/10.1016/j.buildenv.2022.109064.
[54] Hunhevicz JJ, Motie M, and Hall DM. Digital building twins and blockchain for performance-based (smart) contracts. *Automation in Construction* 2022;133. https://doi.org/10.1016/j.autcon.2021.103981.
[55] Xu Y, Tao X, Das M, *et al.* A blockchain-based framework for carbon management towards construction material and product certification. *Advanced Engineering Informatics* 2024;61. https://doi.org/10.1016/j.aei.2023.102242.
[56] Elghaish F, Hosseini MR, Kocaturk T, Arashpour M, and Bararzadeh Ledari M. Digitalised circular construction supply chain: An integrated BIM-blockchain solution. *Automation in Construction* 2023;148. https://doi.org/10.1016/j.autcon.2023.104746.

Chapter 3

Socio-technical approach to blockchain adoption in construction: conceptual models and roadmaps for macro- and meso-scale implementation

Jennifer Li[*], *David Greenwood*[†] and *Mohamad Kassem*[‡]

Researchers have advocated [1] that blockchain has the power to disrupt all industries and sectors, including healthcare [2], information and communication technology [3], real estate [4], education [5] and energy markets [6]. Such disruption is defined by Millar *et al.* [7, p. 1] as 'change that makes previous products, services and/or processes ineffective'. There is clear resonance here with the situation in the construction industry as shown by repeated criticisms of the ineffectiveness of its products [8], its services [9], its processes [10] and its fundamental business models [11]. Against this background, the disruptive potential of blockchain for the construction industry has been considered promising [12]. The likely extent of any disruption remains to be seen given the still-nascent nature of the technology and that, historically, construction has been one of the least technologically developed sectors [13]. However, many of the sector's historical challenges are rooted in problems with information: its inadequacy [14], its unreliability [15] and its asymmetry between stakeholders [16]. On the other hand, the availability, transparency, verifiability, consistency and immutability of information are among the fundamental benefits that blockchain offers [17].

The extent of adoption of any innovative technology is critical in harnessing its full potential, particularly in sectors such as construction and the built environment. Successful technology adoption, which extends beyond initial acceptance and piecemeal implementation, is essential for realising the full benefits of technological innovations. This is pertinent in the case of blockchain, the integration and widespread adoption of which depend on orchestrating an intricate interplay of technical and social factors. These include the readiness to adopt the technology and its diffusion [18], its potential to redistribute power from centrally controlled

[*]Department of Management in the Built Environment, Technical University of Delft, The Netherlands
[†]Department of Architecture and Built Environment, Northumbria University, UK
[‡]School of Engineering, Newcastle University, UK

institutions to individuals [19] and its significant implications for social responsibility [20]. Socio-technical systems (STS) theory is a well-accepted approach to understanding these interactions as well as their post-adoption implications for society and its institutions.

This chapter introduces STS theory, starting with a brief history of its development and why the theory is particularly relevant when considering the adoption of technological innovations. Next, literature on the application of STS theory in the built environment context is examined before the focus turns to how the adoption of blockchain, distributed ledger technology (DLT) and the smart contract (SC) functions within it can be viewed through an STS lens. Then, socio-technical models relating to blockchain, DLT and SCs are introduced before being applied to consider their adoption in the construction sector. Finally, attention is given to a suggested roadmap to support its adoption.

3.1 Socio-technical systems theory

The social perspective is at the root of the STS theories that evolved from the work of the Tavistock Institute for Human Relations in the UK. Founded in 1947, and initially specialising in theories of mental health in the workplace, the Institute built a reputation for action research into (successively) industrial relations, quality of working life, workplace democracy and decision-making, and, ultimately, into the proper functioning of industrial organisations within dynamic environments.

Led by psychologist Eric Trist, researchers in the Institute applied social scientific methods in field-based industrial projects starting with their investigations into the use of new technologies to increase coal production. It was further developed and applied to other sectors during the 1950s and 1960s by Trist and his Tavistock colleagues Fred Emery, Kenneth Rice, Eric Miller and Elliott Jaques. STS represented a paradigm shift in the understanding of work organisations and a break from the 'technocratic bureaucracy' of the nineteenth and early twentieth centuries.

3.1.1 Conceptual foundations

Several conceptual strands provided the fundamental influences in the origin, development and application of STS. The first was *industrial psychology* with its focus on the responses of individuals and groups to the technological demands of their working environment. Not only were such issues considered to be highly relevant, but, as Trist himself observed, 'The idea of separate approaches to the social and the technical systems of an organisation could no longer suffice... [STS represents] a search for criteria to obtain the best match between the technological and social components' [21, p. 10]. These sentiments remain true of the components of twenty-first-century advances such as blockchain.

The next important strand in STS was the concept of *open systems*. Originating in the biological studies of Ludwig von Bertalanffy, Open Systems Theory concerns the interaction between individual systems and with their surrounding environments. This is of clear relevance to an industry like construction, where the flow of information across organisational boundaries is not only crucial but also subject to failures that underlie many of the industry's problems.[1]

Fundamental to the application of STS theory is the issue of change – political, economic, social or technical – in the organisation or sector's external environment. Advances in technology create the need for change at an organisational level due to the profound effect they can have. When work systems are viewed as open systems, they can more readily recognise, reflect and potentially accommodate the impact of a changing environment, while optimising the functioning of the STS through adaptive organisational design. As Trist explained, 'Sociotechnical research needs to monitor emergent alternatives [and] establish criteria for making choices more explicit' [21, p. 53]. This is a clear message for the emergent technologies that confront the construction industry. The following sections consider the socio-technical aspects of technological innovations in general, first on industrial sectors and then on the organisations within them. This is followed by reflection on STS theory in built environment literature generally, on the industry's adoption of digital technologies (exemplified by building information modelling, BIM) and then on the prospects for its adoption of blockchain.

3.1.2 STS and the adoption of technological innovations

Following the Tavistock's lead, researchers have examined how an STS approach can facilitate the understanding of the adoption of technological innovation. Innovation and its adoption can be viewed on different levels. Trist [21] conceived of three: 'macrosocial systems' or 'domains', such as industrial sectors; 'whole organisation systems' that include individual corporate businesses or public agencies; and 'primary works systems' that are subsystems of the latter in the form of specialist groupings within an organisation. This is particularly resonant with the construction industry (the *domain*), the organisations within it (*whole organisation systems*) and their projects (*specialist groupings*). This section will examine STS approaches to the adoption of technological innovation by sectors; the next, by individual organisations.

Recent thinking suggests that stable socio-technical systems within sectors can be disrupted by emerging technological innovations, leading to periods of transition before a new form of stability is established. This 'multi-level perspective' (MLP) model has three interacting 'levels' and three time 'phases' as shown in Figure 3.1, adapted from a publication by Geels [23].

At the model's top level is the *socio-technical landscape* representing an exogenous environment (including political, economic, social, technological and regulatory elements) that is beyond the direct control of the individual actors (e.g. individual, group, organisation) in any system. Different elements of the landscape can change slowly (e.g. through cultural influences) or more dynamically (e.g. with a sudden change in political or economic circumstances). At the model's mid-level

[1]A fuller exposition of how the concepts of systems theory underpin STS can be found in Ropohl [22].

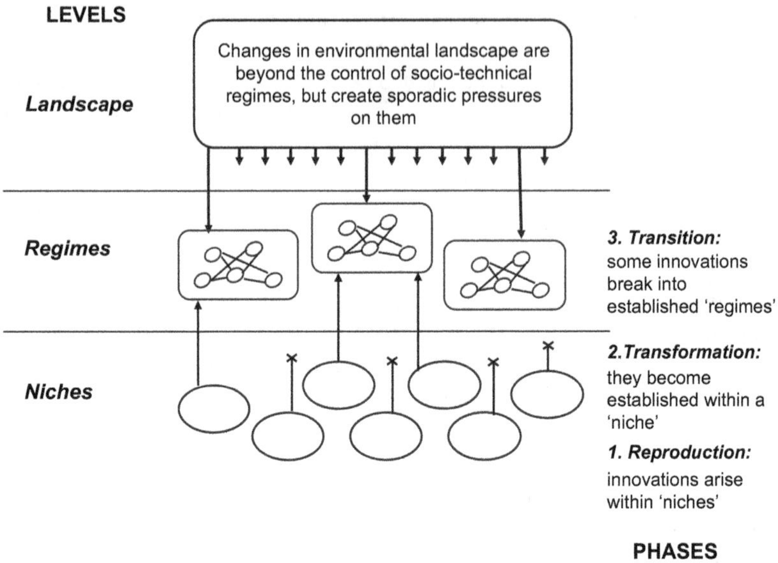

Figure 3.1 A model of change in socio-technical regimes, adapted from Geels [23]

are the *socio-technical regimes*. These are based on the accepted 'cognitive routines' of the actors within systems with their established roles, networks, work practices, rules and knowledge. Such routines enjoy the benefits of being established and understood, but they can act as barriers to what is outside their experience. The third level is that of *technological* spaces where new concepts can emerge and mature in an environment protected from normal market forces (e.g. in a funded research laboratory). Such niches are conducive to improvisation, learning, prototyping and development, but have yet to 'break through' and survive in the exposed competitive world of the *regimes*.

In the MLP model, this three-level formulation is combined with three incremental time 'phases' of change – *reproduction*, *transformation* and *transition* – that represent the potential for an interactive dynamism between the three levels. In the *reproduction* phase, innovations emerge at the niche level. Some of these might be sufficiently consistent with the prevailing socio-technical regime to be readily accepted by it: the regime benefits but remains largely unchanged. The more radical and potentially disruptive innovations may (or may not) survive and become established (the *transformation* phase) within the protected setting of their niche, remaining there as long as they continue to clash with the dominant regime. The third (*transition*) phase will only occur (if at all) when circumstances change in the *landscape* to cause a tipping point that permits the innovation to successfully challenge and disrupt the established regime. The regime then re-stabilises in its new form and the previously established knowledge, roles, networks, work practices and rules are reconfigured.

The MLP model has been convincingly employed to explain how and why major technological innovations of the past have become established [24] and how more recent transitions have faced 'regime resistance', as in the case of low-carbon transitions [25]. Indeed, similar dynamics of adoption and resistance feature in an examination of the organisational complexities of adopting blockchain technology in a variety of different sectors (i.e. 'regimes') including finance, the arts and food supply [26]. Although there has been no similar analysis relating to the construction industry, there is clear and recent evidence of its applicability; for example, in the growing acceptance of the need for digitised 'passports' for buildings and the materials within them that has been driven by the environmental landscapes of carbon reduction and 'circularity' [27,28] and regulatory control by the state such as the traceable 'golden thread' of digital information demanded by the Hackitt Review [29].

3.1.3 STS within the organisation

We now turn to the level of the individual organisations within a sector. To survive within their sector, organisations must remain viable, and to prosper they must seek competitive advantage. A key element in doing so is the adoption of new technologies [30]. According to STS theory, a successful organisation should be ready to respond to its external environment by suitably matching its technological and social components. There have been many examinations of what these technological and social components are. Among the simplest explanations is that of Leavitt who, in one of the seminal works on organisational change [31], envisaged a model (sometimes known as Leavitt's 'diamond') with four independent but interacting critical factors, namely: *task*, *structure*, *technology* and *people*. Similarly, in their book on *The Toyota Product Development System*, Morgan and Liker [32] define the three STS subsystems as *people*, *processes* and *technology*. More complicated analyses have appeared in the literature, but all share the central principle that such STS components should not be considered in isolation.

STS approaches have been particularly evident in studies of the implementation of information technology (IT). For example, researchers at the Human Sciences and Advanced Technology (HUSAT) Research Centre have developed a 'socio technical design process' to ensure the usability of IT solutions once they have been implemented [33,34]. The influential conceptual work of Wanda Orlikowski [35, p.16] shows similar socio-technical overtones in her 'structurational' model of technology that comprises *human agents*, *technology* and the *institutional properties of organisations*. She concurs with the HUSAT researchers in her insistence that consultation with users during the development of a technology is critical for ensuring that it will be fit for its intended use.

3.1.4 STS approaches in the built environment literature

From its origins in coal mining, the influence and application of STS theory have since been extended to many different industry sectors and environments. These include textile manufacture (in India), food production and processing, motor car production (in the UK and the US), chemical plants and power stations (worldwide). This section examines its presence in organisational theories that relate to the construction and operation of the built environment.

The successful exploitation and diffusion of innovations (DOIs) are not guaranteed by their initial emergence, recognition and early adoption [36]. Organisations involved in the construction, management and maintenance of the built environment have been traditionally regarded as slow to innovate [37]; indeed, the very concept of innovation in the sector is the subject of debate [38,39]. However, as the built environment fundamentally exists to meet the needs of society, it is a natural step to adopt a socio-technical approach to examining the sector's existing technological systems or the development of future ones. Consequently, in academic literature there are many examples of the application of STS analysis to built environment issues. Rohracher [40], for example, used a socio-technical lens to evaluate different strategies for the construction of sustainable buildings. Kazi [41] and his contributors applied a socio-technical perspective to knowledge management in the construction industry. In the same edited book, Wild [42] delved into the earliest STS construction studies from the 1960s to argue that the sector's economic instability undermines its ability to properly manage knowledge. Brown and Vergragt [43] highlighted the need for social 'deep learning', alongside technological advances, to achieve sustainable outcomes in buildings. Schweber and Harty presented case studies from France, the UK and the US to demonstrate the socio-technical negotiations between the 'content, meaning and use of technology' [44, p. 657]. The focus of Baxter and Sommerville [45] was on information systems and the impact of computer support on work design; that of Newton [46] was innovation in urban infrastructure systems and that of Chiu [47] and colleagues was on low-carbon domestic retrofitting. However, the interest in STS and the volume of such work has increased in the wake of BIM, as we consider in the next section.

3.1.5 STS and the adoption of building information modelling

In the decade between 2010 and 2019, publications on BIM and the related digitalisation of construction information have been prolific, as evidenced by a bibliometric analysis of the related scientific literature, which also reveals how perceptions of BIM have developed from model construction and visualisation to whole-life optimisation of built assets and information exchange with other new technologies [48]. The diversity of perspectives on BIM is exemplified by challenges to some of its early precepts: for example, the incremental catalogue of so-called 'BIM dimensions' has been criticised as a set of misleading metaphors (beyond the fourth dimension they should be properly called 'properties') [49]. There has been scepticism about the true extent and depth of BIM adoption [50] and discussion of the related question as to whether BIM has yet been a disruptive innovation or simply a sustaining one [51].

What emerges is the view that BIM is an innovation that is evolving [52] into an 'expanding building information modeling (BIM) ecosystem' [53, p. 2] and a recognition of BIM as a catalyst: 'a dominant enabler of construction innovation' [54, p. 1580]. The clear evidence of this is in the access to the group of new technologies – characterised by Construction 4.0 [55] – that has been enabled by the digitisation of information afforded by BIM. The exploitation of these technological innovations, as that of BIM itself, has attracted socio-technical consideration; some examples follow.

Sackey followed his doctoral thesis [56] with an article with colleagues Tuuli and Dainty that explored the practicality of the BIM implementation process through an STS lens [57]. Merschbrock *et al.* [58] analysed the contrasting outcomes of two hospital construction projects where BIM-based construction networks had been adopted in terms of their differing responses to dealing with socio-technical misalignment. In Alreshidi *et al.* [59], a socio-technical perspective was taken to examine BIM governance and data management through cloud-based collaboration. In their global qualitative study, Oesterreich and Teuteberg [60] used the familiar STS dimensions of people, information, processes and technology as their 'lens' to investigate the reluctance of many organisations to adopt BIM, concluding that 'the barriers to BIM adoption are primarily rooted in the social behaviour of the actors as well as the social arrangements of the construction industry rather than technical issues' [60, p. 420].

A further example of the influence of actors' behaviour is provided by an article based on two residential projects in the Netherlands [61]. In these, the use of BIM and its implications for collaboration were perceived differently by the actors in each project, and with different results. In *Project A*, BIM was treated as a set of software tools that produced discipline specific artefacts that ultimately needed integrating. Workflows were linear, sequential and collaboration followed an 'over-the-wall' pattern that was 'transactional and technical'. In *Project B*, collaboration was more communicative and model development and integration more 'organic'. It should be noted that Project B was characterised by 'long-term partnerships', whereas Project A was a transactional 'multi-party contract'; the inference, though not explicitly drawn, might be that this was a factor that impacted the way BIM was adopted.

Concepts related to STS are also evident in a comparison of BIM implementation at the *intra*-organisational (micro) and *inter*-organisational (meso) levels on three case study projects, again in the Netherlands [62]. Drawing on the work of Dubois and Gadde (see [37], above) these cases were (typically, for construction projects) regarded as 'loosely coupled systems' with little or no shared knowledge between the individual actors involved. The 'sophistication' of BIM implementation on each project was found to be related to the relative adoption level (based on Rogers' diffusion stages – see [36], above) of a 'key organisation' (typically, the main contractor). Where the key organisation's BIM adoption was simply a response to market demand, its project implementation was makeshift, and inconsistent where 'BIM visions were not well diffused across all hierarchical levels' [62, p. 9] of the key organisation.

Anderson and Ramalingam [63] address the challenges presented by the global virtual teams that are increasingly to be found on large construction projects.

Using an experimental project that involved collaborating students at universities in the US and India, the authors adopted a socio-technical approach to the drafting of an appropriate BIM Execution Plan with positive results. In theory, a BIM-based approach to delivery can improve communication and collaboration between project stakeholders. In Mani *et al.* [64] the socio-technical model of Leavitt [31] discussed above was used in two case studies (public and private sectors) to better understand how change can be managed in BIM-based projects. The final example of the use of STS in BIM-based literature is a review article that addresses the perceived shortfall in exploiting the full benefits of BIM through the construction and management of built assets throughout their lifecycle [65]. The results confirm the basic tenets of STS, that technological or social systems approaches are insufficient when used in isolation.

Despite the evidenced benefits of adopting a socio-technical approach to technological innovation, STS approaches are not widely used. Baxter and Sommerville [45] offer reasons for this including a lack of understanding of STS theory, leading to challenges in applying STS methods; the misalignment between STS and the technical issues associated with engineering; and challenges in individual interaction with technological systems.

3.2 Blockchain as a socio-technical system

In this section, the characterisation of blockchain as an STS is discussed. Then, some of the contributions STS theory has had on early research into the use of blockchain and SCs are considered.

Blockchain is characterised by several scholars as an STS [see, e.g., 1,66,67] for its ability to facilitate social relations. Shin *et al.* [68] state this is due to its interconnection 'between humans, technology and the social milieu' [68, p. 9] that drives reconsideration of 'entrenched premises as to privacy, security, transparency, distributed decision, and integrity' [68, p. 10]. It is a technological system firmly rooted in its use *by* people *in* society such that the focus of evaluation should be centred 'on the relationship between technology and its contextual environment' [69, p. 248]. Blockchain technology's popularity arose from the introduction of Bitcoin, the first globally established cryptocurrency that initiated its first transaction in January 2009 [70]. In its desire to challenge centralised power structures, Bitcoin was intended as a vehicle to demonstrate '(1) the technical puzzle of preventing opportunistic actors from double-spending digital "coins" and (2) an ideology of online privacy in the digital age' [66, p. 51]; economic gains were not the primary driver. Initially, Bitcoin proved blockchain successful as a technology, but so far it has not truly offered a viable alternative to 'fiat currency' due to its latency to process transactions in comparison to existing transaction processors such as Visa and the challenges governments find in regulating its use. In line with STS theory, a blockchain ecosystem such as Bitcoin can be deconstructed into operating platforms, data repositories, industry, governance, practice and users where it is clear that social dimensions take precedence over technical ones [68].

Such an ecosystem allows anyone to participate in the network and consensus mechanisms (e.g. the proof-of-work (PoW) algorithm) successfully deter malicious actors from taking over the system [71]. However, because anyone could participate, it was not able to prevent those with first-mover advantage from exploiting the system by developing monopolies that would dominate; thus, 'both market incentives and computing properties have driven the protocol to a more centralized topology' [72, p. 19] and this has changed Bitcoin 'from a hobbyist sport to a capital-intensive professional activity' [72, p. 22]. While the technological part of the Bitcoin protocol cannot be changed to adapt to this exploitation of power to maintain centrality, a socio-technical approach to the factors external to the system is required to address how markets and policies can hold the system to account and help maintain its integrity.

Blockchain can provide security to a system to ensure integrity of data, confidentiality can be added through cryptography, user privacy can be ensured at several levels as required, as well as offering transparency, resilience and cost efficiencies [73]. Nabben [67] reframes the concept of blockchain providing security for users to 'people security' where the needs and expectations of blockchain users are prioritised. Nabben goes on to explain that when blockchain is employed to facilitate macrosocial interactions, they in fact function as institutions themselves through the encoded rules imposed upon them. This is reinforced by Hayes [66, p. 68] who describes blockchains 'as systems of accounting; organizational forms of contracting in lieu of firms; and as institutions in their own right capable of ordering social behavior'. Challenges to facilitate people security are raised around issues of central organisations having control over blockchain-based systems, particularly where private platforms are engaged; and personally identifiable data being accessible and hackable that once recorded could be used for malicious intent. It is important, therefore, that the intention of users is considered when developing blockchain-based systems [73]. Nabben [67] defines three roles of people in the blockchain ecosystem: (1) developers of the open-source code required by the network; (2) participators in the network to validate transactions in those blockchains with consensus mechanisms; and (3) governors of the network through voting, community discussions and proposals for change within the system. Nabben misses end-users of the network from these roles who may not be involved in any of the three above, and who arguably represent society for whom blockchain exists. In addition, while these roles are apparent in public blockchains, private blockchains do not require them in the same way where centralised organisations establish governance and participation rules limiting security and privacy of users. Private blockchains are being utilised over public blockchains by centralised institutions where the latter would ensure security and privacy of individuals' data more effectively [67], but the former offers more control over those building or initiating use of the platform.

In a review by Selvanesan and Satanarachchi [74], it was found that the majority of benefits to using blockchain technologies were attributed to the socio-technical domain. The authors failed to provide a definition for what they mean by socio-technical and it seems the benefits have been miscategorised; the socio-technical benefits in the paper are more specific to the technology and the other domains include socio-technical factors. The benefits included anonymity, high security, auditability, veracity, increased quality and decentralised cooperation. The remaining benefits were attributed to the industrial domain (execution of SCs, improved trust, improved customer experience) and the organisational domain (transparency, disintermediation). It could be argued that the industrial and organisational benefits are actually socio-technical benefits for all except execution of SCs. Moreover, there is no discussion of overlap of these domains. While the benefits are valid (albeit miscategorised), this demonstrates the lack of understanding of STS that Baxter and Sommerville [45] present as a barrier to its widespread use.

Bauer *et al.* [75, p. 10:2] define blockchain as an 'inter-organisational system (IOS)', that is, the interconnections between organisational networks made possible through an information system [76]. Blockchain is often discussed as a facilitator of collaboration for its ability to support such organisations in development, management, governance and operations of information systems [75]. Bauer *et al.* [75] identify several socio-technical motives for using blockchain consortia as technology innovation motives, technology enabled and governance motives, information tokenisation motives, business goal motives and political positioning motives. The task-technology fit (TTF) model is applied by Chaudhuri *et al.* [77] to evaluate the suitability of blockchain to meet customer needs. They identify social capabilities as

> 'empathising with customers, and customer communication and stakeholder management' and technical capabilities as 'development of tracking and data login systems, system design to facilitate implementation and handle exigencies, development of own middleware and systems, development of automated quality assurance and ability to evaluate different platforms' [77, pp. 8–9].

3.2.1 Socio-technical considerations for DLT and SCs in the built environment

Many potential applications of blockchain in the construction industry have been discussed in literature beyond that of cryptocurrency such as information management, compliance checking and governance (see, e.g., Perera *et al.* [13], and Li and Kassem [17] for detailed reviews), such that challenges in construction could benefit from new economic incentive models to bring about better efficiency and reduce the power of the few (e.g. tier 1 contractors who use project funds to cashflow their businesses).

The construction industry's main challenges are rooted in social problems, most notably, lack of trust leading to lack of collaboration and information sharing [17]. Rarely are the challenges attributed to technology unless discussing the lack of digitalisation in the sector, which is then attributed to resistance to change [78] – another social issue not a technical one. Blockchain proposes to be a solution to these social challenges, not by addressing the root cause (i.e. lack of trust), but by removing the issue of trust between collaborating parties shifting it away from people towards the technology [79]. For this to be successful, blockchain systems must be stable and secure [80] to deliver what they promise. The challenge remains that technological systems are often developed with a top-down approach that considers the technology's potential, rather than being developed from a more appropriate bottom-up approach that accounts for the perspectives of the system's users [75]. This underscores the importance of investigating blockchain through a socio-technical lens, rather than viewing it solely as a technological solution, to ensure its successful adoption.

With regard to STS theory and blockchain in construction, the most notable contribution is that of Li [1] who proposed a socio-technical framework for the implementation of DLT in the construction sector. Section 3.4 presents the socio-technical models that form the basis of the framework. In addition, several authors consider the socio-technical perspective as key to answering questions about the future of blockchain use cases such as Hunhevicz *et al.* [81] who posed several socio-technical questions for the future of a self-owning house. Sigalov *et al.* [82] considered a framework for blockchain-based SCs as an STS. Van Groesen and Pauwels [83] consider the construction industry to be ready for a socio-technical shift that DLT and SCs can bring. Yu *et al.* [84] examine the state of the nation on integration of BIM and blockchain using an STS lens, specifically focusing on process, technical and contextual dimensions. They highlight the necessity of collaboration at an interdisciplinary level that brings together acceptance at an industry level, and empirical investigations to demonstrate the possible gains of BIM and blockchain integration for construction. Kifokeris and Koch [85] apply sociomateriality to their research on construction supply chains, which is a socio-technical approach that considers the inseparability of the social and material aspects of digital innovations.

While literature specifically focusing on STS for DLT and SCs is limited, there is recognition that DLT and SCs are social systems.

3.2.2 The trust perspective of blockchain-based systems

Trust is one of the main reasons blockchain exists. Centralised organisations and power structures have exploited individuals in society by employing, for example, manipulation tactics to change opinions (e.g. Cambridge Analytica's involvement in the Brexit vote and getting Donald Trump into the White House) and maintaining user engagement with social media by pushing content that will prevent users from exiting apps so that more money can be made from advertising. The subprime mortgage crisis that caused a global recession saw executives from powerful banks protect themselves before their customers that destroyed the general public's trust in centralised systems. A key facet of blockchain is that parties can transact with one another without the need for trust or (sometimes dishonest) intermediaries [73] where cryptoeconomic governance mechanisms offer 'a new way of trust-minimized social coordination' [86, p. 7]. According to Shin and

Ibahrine [69, p. 253], technological adoption is dependent on 'how well it is understood and trusted'. Blockchains require a complete rethinking of the functioning of societies in terms of data and in a world where humans are generally resistant to change, this is a challenge.

In the case of public blockchains, the ability of third parties to unilaterally change a system's rules is eliminated [67]. In the case of the subprime mortgages, had governance been facilitated via a public blockchain with carefully coded governance rules that protect the general public from such events insofar as is possible, the outcome could have been very different. An example of decentralisation (specifically, collective action) winning over centralised financial systems is the Reddit community r/WallStreetBets that caused a short squeeze of GameStop shares in January 2021. GameStop is a high street retailer of computer games. Mancini *et al.* [87] analysed the saga: on 3 April 2020, GameStop (NYST: GME) was at an all-time low share price of $2.57. At this point, financial institutions shorted the stock with a view to profiting later when the GameStop share price dropped further. The r/WallStreetBets community used collective action to trigger a short squeeze that saw a rapid share price increase resulting in an all-time high of $483.00 on 28 January 2021. Mancini *et al.* [87] opined that r/WallStreetBets saw the potential to make a profit while also retaliating against financial institutions. This activity resulted in short sellers making significant losses who had bet on the collapse of GameStop [88]. However, the collective action taken by r/WallStreetBets reversed the collapse and while the company has not maintained the share price achieved in January 2021, its position is improved from April 2020. This represents a shift in people's feelings toward centralised systems and decentralised models that are paving the way for a step change in governance structures that blockchain could facilitate.

A construction-specific example with trust implications takes the form of the Weather Ledger, a pilot project that ran in the UK to automate weather compensation events. It was a collaborative effort between EHAB, Clyde & Co, Bam Nuttall, Ferrovial Construction, Digital Catapult and Connected Places Catapult [89]. The project aimed to automate the compensation due to the injured party in the event of a 1-in-10-year weather event (under the terms of the New Engineering Contract). During this project, such an event occurred, which was more than 100 mm of rainfall. Ultra-local weather data were collected onsite from IoT devices and SCs were deployed to automate the process. When the weather event occurred, over 100 mm of rain were measured on site, but the closest Met Office weather station measured below 100 mm of rain, which resulted in the compensation not being paid [1]. The reported benefits of the project include speeding up the process allowing resources to be redirected to other (value-adding) activities; an early warning system to enable planning of activities onsite and minimise disruptions; enhanced productivity; and enhanced communication between contracting parties, streamlined by symmetrical information. Despite the benefits, trust challenges were present in the project where project participants were not comfortable with money being automatically transferred by a SC because once you send cash, it is difficult to get it back and there was insufficient trust that the SCs were making the correct decisions. It could also be argued that alongside the trust issue, there was a perceived lack of control in this instance. A key outcome from the Weather Ledger project was that such a system, while able to make some cost savings, saw more value in the creation of a risk management system that allows contractors better access to information and therefore granted them the ability to better plan onsite activities. Trust was reported as a challenge though where, typically, preferred data oracles for weather come from reputable organisations such as the Met Office and gaining trust in something like the Weather Ledger would take some time and convincing [90]. Pan and Ning [91] assert that technology by itself is insufficient to effect change and indicated a lack of human behaviours in policies.

3.3 A socio-technical model of DLT and SCs in the built environment

Earlier sections in this chapter have demonstrated that STS theory is applicable in investigating blockchain's potential to disrupt the built environment. By considering implementation at both meso and macro scales, the socio-technical approach can be used to ensure that the needs of society can be met by the application of technology at different scales of adoption, accounting for different societal needs at each scale. Bringing in such a focus at this early stage of implementation into the built environment ensures that the socio-technical approach can provide these technologies the best chance of success from the outset rather than it being a buzzword to appear as though people and society are considered.

Shin and Ibahrine [69] proposed a socio-technical framework for blockchain (Figure 3.2) that shows a delineation between the social system encompassing

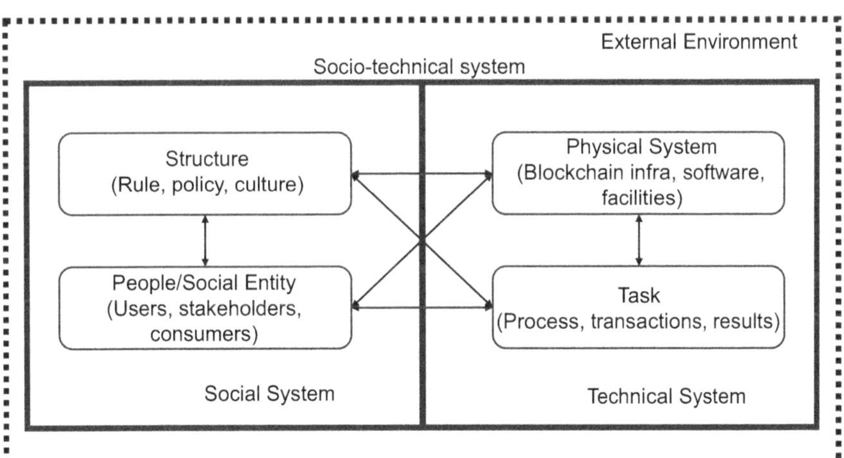

Figure 3.2 Socio-technical frame of blockchain ecosystem (recreated from [69, p. 249])

structure and people/social entity, and the technical system encompassing physical system and task. Applying the framework, the authors identified several issues in existing blockchains from the perspective of Korea, the key points of which are as follows. From the technological dimension, data entering the blockchain must be high quality reinforcing the mantra of *garbage in-garbage out*; a critical mass is required for blockchain to reach its full potential of a distributed system; blockchain immutability is at odds with the right to be forgotten laws enacted around the world; and the consensus mechanism requiring 51% or more of the network to control it is seen as a weakness by the authors given the example of Korea where opinion rigging is not difficult through online communities. From the social dimension, the security of personal data remains a concern due to risks such as hacking, breaches and fraud; however, while blockchain's transparency is sometimes perceived as limiting privacy, this depends on the specific implementation and design of the solution. Another issue relates to the lack user-friendly interfaces, which can hinder usability and limit broader acceptance. From the political dimension, governance and regulation of blockchains and associated technologies are lacking, for instance in Korea, the government is still deciding on how to classify Bitcoin – a currency or a good. The same is true of other nations.

Abdelmegid *et al.* [92, p. 579] characterise digital twins as 'complex socio-technical phenomena incorporating human-artificial intelligence interactions'. They proposed a socio-technical conceptual framework for digital twin based on a matrix that incorporates six socio-technical elements (goals and metrics; people/human factors; technology/tools; processes/practices; physical infrastructure; and culture) across three digital twin levels (digital model; digital shadow; and digital twin). A simplified illustration of the framework is given using the example of piling operations at the level of digital twin. The authors argue that a key contribution of their framework is that it provides a balance between social and technical factors with regard to the implementation of digital twin. It could be said that this is or should be a focus on all socio-technical frameworks where this chapter has emphasised the need to consider both and not just focus on one or the other.

To the best knowledge of the authors, there are no other socio-technical frameworks for blockchain-specific applications; therefore, we believe the framework presented below is the first of its kind in the field.

3.3.1 DLT four-dimensional model

Based on the findings from a literature review in a previous study [70] that explored challenges and opportunities for DLT and SC applications for the construction sector and the built environment, four dimensions were identified that combine to create an extended socio-technical framework for implementation of DLT and SCs in construction. The four-dimensional model that forms the foundation of this framework is an analytical tool that can be used to understand the ecosystem of DLT and SCs for any (proposed) application. Taking STS theory literally, the first dimensions are technology and society. The two dimensions that extend this theory are process and policy. The DLT four-dimensional model is shown in Figure 3.3 and can be used to analyse the as-is state of an application from the perspective of

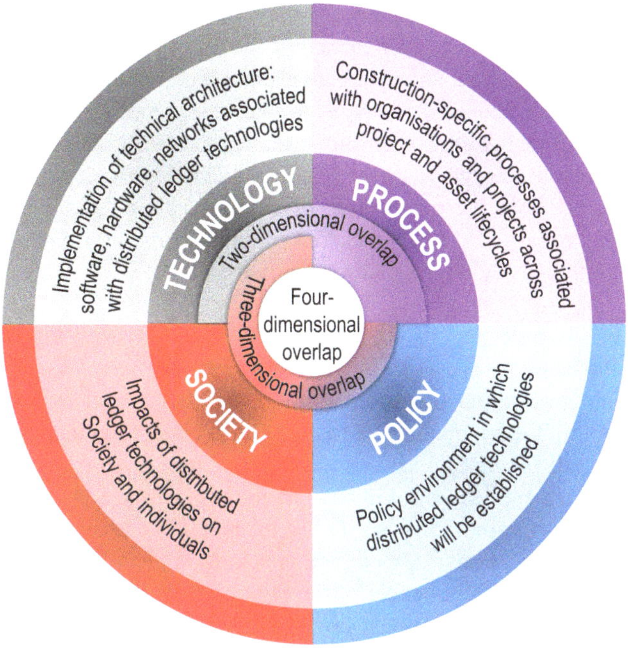

Figure 3.3 DLT four-dimensional model

challenges and opportunities where some may straddle across two, three or all of the dimensions. These overlaps represent interconnected knowledge across dimensions. Such functionality of the model allows adaptability and its longevity in a field evolving as rapidly as DLT. The four dimensions are now presented followed by an illustration of how the model can be applied to a use case.

3.3.1.1 Technology

The technical aspects of the DLT ecosystem are under consideration in the technology dimension. This includes hardware, software, networks and other technological infrastructure required for applications to run on blockchain. Depending on the blockchain platform being used, there will be different considerations for technological specifications for each application. As the current wave of blockchains evolve and new platforms become available on the market, many of the 'challenges' to the technology (e.g. interoperability, throughput, latency) will be resolved. Technological considerations for a specific platform to meet an application's requirements must include scalability, security, privacy, integration with hardware (e.g. sensors), integration with software (e.g. IoT, APIs, interoperability, information models) and data frequency requirements (e.g. real-time, near-real-time, hourly, weekly, monthly).

3.3.1.2 Process

Organisations and interorganisational collaboration are the focus of the process dimension. It considers how individuals and organisations will implement and use blockchain technologies and the impact they will have on procurement, design, construction, operations and maintenance of the built asset lifecycle, with a particular focus on changes to existing processes and the creation of new processes. It also considers the impact of changes to business models, business strategies, organisational structures and roles and responsibilities within and across organisations to take full advantage of blockchain technologies. It connects with the policy dimension from the perspective of changes to regulations that will affect organisations, projects and supply chains to ensure organisations maintain compliance with industry-wide regulatory frameworks as well as requirements of clients.

3.3.1.3 Policy

The policy dimension encompasses regulations, laws, policies, standards and compliance within the application's ecosystem. This is differentiated by application on the basis that different applications will have to meet different requirements. For example, payments applications will have to comply with regulations and standards to payment structures such as those set out in the UK's Construction Act for UK-based payment applications; compliance checking applications will have to link into building regulations applicable to the region where a built asset is being constructed or renovated. Regulations, laws, policies and standards are at different stages of progression across the globe where, for example, the United Arab Emirates who decided to fully embrace blockchain at a state level is more advanced than many countries [93]. Many of the elements in this dimension are led by government who, therefore, have a responsibility to investigate the potential of blockchain technologies and should strive to create an environment where these technologies can thrive for the benefits of society. Investment in infrastructure would facilitate its adoption and integration into existing systems focusing on solving challenges such as interoperability that could stifle innovation if not addressed in a timely manner. It also falls to policymakers to educate the general public of the benefits and challenges of blockchain technologies alongside planning to ensure there are sufficient people with the skills to design, implement, operate and manage blockchain-based systems to ensure that resourcing is not a barrier to its adoption.

3.3.1.4 Society

Concerned with the effect of DLT on individuals in society and real-world integration, the society dimension represents the social system where benefits of blockchain technologies will be realised. The growing recognition of issues around data misuse by powerful corporations places substantial importance on this dimension. While blockchain as an STS has the potential to make improvements to the day-to-day lives of people in society, there is also the risk that these technologies can be used for malicious intent. Web3 applications are providing solutions to misuse of individual data by giving ownership back to the individual. Environmental concerns are also included in this dimension, particularly given the energy consumption of PoW protocols, where sustainability should be at the forefront of technological advancement. Taking a socio-technical approach to blockchain implementation will help to avoid compromising on privacy and hindering collaboration between parties.

3.3.2 DLT actors model

To support the DLT four-dimensional model, the DLT actors model was created. It identifies all the different actor groups in construction to help in analysing the *as-is* ecosystem to the desired *to-be* ecosystem by showing the different actors concerned in and across dimensions. This helps to unravel the complexity of new technological systems and breaks down roles and responsibilities of actors within the dimensions for the applications being analysed. The term 'actors' here is preferred over 'stakeholders' based on the premise that

> 'Actors are persons, groups, or organisations acting within a system of interest. Stakeholders, on the other hand, are persons, groups, or organisations that are (1) affected by the project, (2) interested in the project, and/or (3) able to affect the project. Stakeholders are part of the actors' group, but not all actors are stakeholders for a specific project or organisation' [94, section 3.3].

Actors include developers of blockchain-based systems, users, beneficiaries of the systems and responsible entities as set out in Table 3.1. Sixteen actor groups have been identified and are classified in terms of the dimensions in Figure 3.4 as well as the extent of their involvement indicated by their level of contribution: *primary* – actors who contributed directly to the development of blockchain-based systems, standards, policies, regulations and who have authority in how these systems are used and evolve over time; *secondary* – actors classed as users of the systems on a day-to-day basis but who do not necessarily contribute to the system's functionality; and *supporting* – actors who may create data that are put into the systems or who may have an interest into how they function, but do not have any authority or contribution to how the systems operate functionally or commercially. This model can support the allocation of responsibility to different actors and therefore coordinate the complementary effort of those actors with respect to adoption and diffusion of blockchain-based systems.

3.3.3 Application of the DLT four-dimensional model and the DLT actors model

To illustrate use of these models, the use case of the Weather Ledger (discussed previously in Section 3.3.2) has been analysed and the challenges and opportunities of applying blockchain technology to a potential solution mapped across the dimensions, as shown in Figure 3.5 (challenges) and Figure 3.6 (opportunities). For the challenges shown in Figure 3.5, these relate to the application of weather compensation events before applying blockchain and SC technologies (i.e. the as-is situation), while Figure 3.6 shows the opportunities of the Weather Ledger application, a blockchain-based solution to address the current challenges in weather

Table 3.1 Actors associated with DLT

Dimension	Actor	Description	Contribution
Technology	DLT system architects	Individuals and organisations who develop DLT including programmers, coders, software developers and system engineers.	Primary
	Technology providers	Individuals and organisations who develop hardware, software, networking architecture for DLT and those associated with enabling or interrelated technologies (e.g. IoT, sensors, drone technology).	Primary
	Service providers	Companies involved in providing a technical service to organisations using DLT, particularly where private ledgers are used (e.g. consultants, insurance providers, dispute resolution firms).	Secondary
Policy	National authorities and policymakers	State and local government authorities responsible for making policy, writing standards and setting regulations along with enforcing them.	Primary
	International political authorities	International groups working together to set international regulations for transactions that cross borders to promote international partnerships and to mitigate the possibility of fraud, corruption and other criminal activities.	Primary
Process	Individual construction organisations	Individual organisations operating in the construction industry including main architectural, engineering, contractor, sub-contractor and facilities management organisations.	Secondary
	Project teams	Individuals across the supply chain who specifically form the project team who have access to the ledger and who have responsibility for producing information to the ledger or consuming information from the ledger.	Secondary
	Clients	Individuals or organisations, public and private, who commission construction projects with access to information on the ledger regarding their project.	Secondary

(Continues)

Society	Individual users	Individuals who use DLT day-to-day either through performing transactions or by providing data to be uploaded to the ledger.	Supporting
	Social groups	Groups of individuals with an interest in the impact of DLT at a societal level (e.g. regarding energy consumption, privacy, security, creation of a value-driven economy, ensuring societal needs are being met by technological solutions).	Supporting
Technology–policy overlap	DLT councils	Stakeholder groups of DLT tasked with approving changes to software, data in the ledger and ensuring technology and operations comply with regulations and who have the power over how DLTs function in general.	Primary
Technology–society overlap	Miners	Individual miners, mining pools and mining organisations operating as nodes and running the peer-to-peer network with an interest in the state of the technology and the level of energy required to run the network (in the case of proof-of-work).	Secondary
4D overlap	Industry associations	Professional associations who represent the interests of individuals and organisations operating in the construction industry.	Supporting
	Supply chain	Organisations that make up the supply chain for the construction industry that are concerned with technical elements of the system regarding tracking and updating ledgers; impacted upon regarding international politics and regulations where supply chains cross borders; have a responsibility to operate in a sustainable manner; and who must follow processes as set by industry standards and clients.	Secondary
	Educational institutions	Universities and other educational institutions conducting research in the field and developing programmes to train and upskill people in DLT.	Secondary
	Communities of practice	Groups of individual practitioners with an interest in a specific area of DLT (e.g. interoperability, privacy, speed).	Supporting

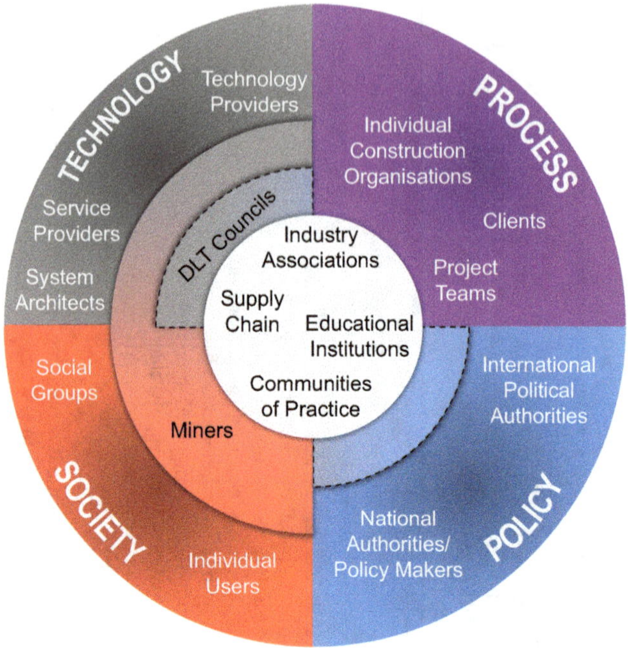

Figure 3.4 DLT actors model

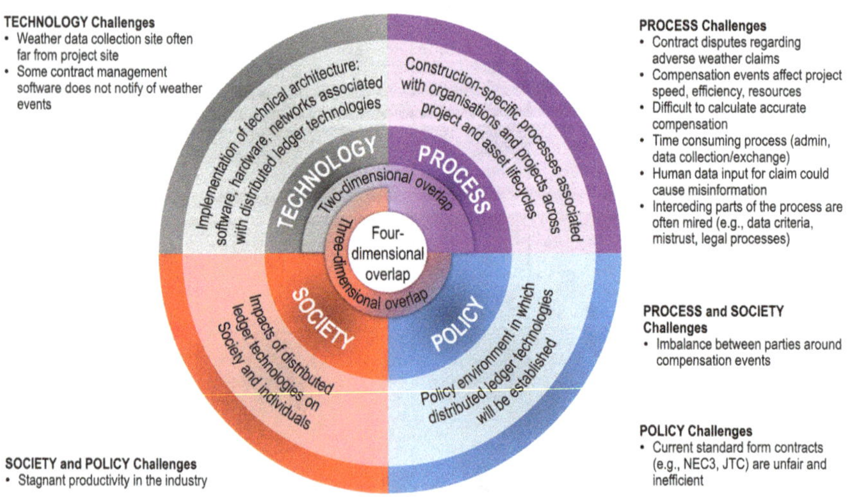

Figure 3.5 Challenges for weather compensation events under current practices

compensation events. These weather compensation events challenges and opportunities of the weather ledger application were identified previously [1], but not applied to the DLT four-dimensional model as they have been here.

Applying the DLT actors model, the actor groups acting within this system of interest (i.e. the Weather Ledger) are identified to be those shown in Table 3.2. The descriptions indicate which specific actor is modelled based on publicly available literature available on the Weather Ledger (e.g. [90]). A larger, more complex project would likely have actors from more groups as indicated in Table 3.1.

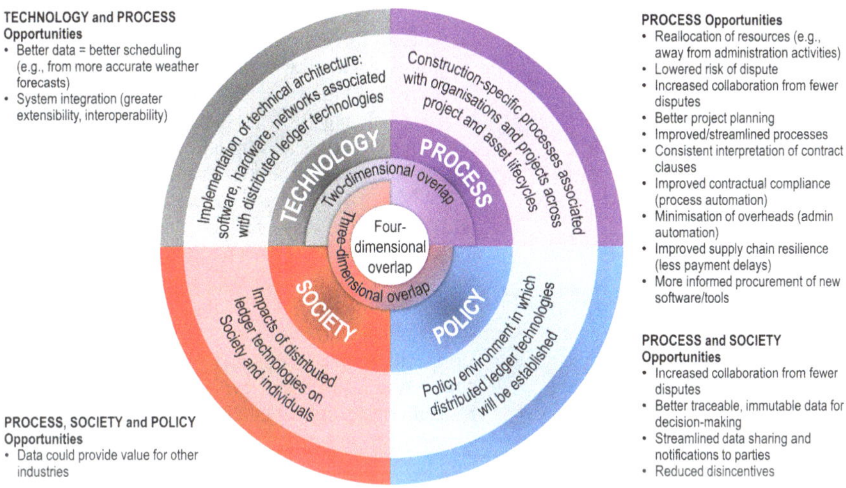

Figure 3.6 Opportunities for a blockchain-based Weather Ledger application

Table 3.2 Actor groups within the Weather Ledger ecosystem

Dimension	Actor	Description	Contribution
Technology	DLT system architect; technology provider; service provider	In the case of the Weather Ledger, Ehabitation and Digital Catapult represented these three actor groups as the designer, developer and provider of the technological solution.	Primary
Policy	National authorities and policymakers	In this actor group, Clyde & Co engaged in the Accord Project represents the smart legal contract framework adopted by the Weather Ledger that links to the New Engineering Contract under which the pilot studies were managed.	Primary

(Continues)

Table 3.2 (*Continued*)

Dimension	Actor	Description	Contribution
Process	Individual construction organisations	BAM Nuttal and Ferrovial Construction represent the construction organisations involved in the project who provided knowledge and insights to support development of the Weather Ledger system.	Secondary
4D overlap	Supply chain	Organisations that make up the supply chain for the construction industry that are concerned with technical elements of the system regarding tracking and updating ledgers, impacted upon regarding international politics and regulations where supply chains cross borders, have a responsibility to operate in a sustainable manner and who must follow processes as set by industry standards and clients.	Secondary
	Industry associations	Connected Places Catapult offered connections, skills and expertise from a construction perspective to the project, particularly around standardisation.	Supporting

Using these models allows application developers to understand the challenges of the current situation so they better understand the environment they are trying to improve. Seeing the potential benefits of a blockchain-based application as the solution can act as a way to get buy-in from investors and actors required to support its development and adoption.

3.4 Blockchain and DLT adoption

Despite blockchain being in existence since 2009, its mainstream adoption is low [95]. Literature on adoption of blockchain is therefore limited [96] and more so when the literature is delineated to construction. Available studies on this theme offer a multifaceted landscape examining blockchain at various levels including market, consortia, organisation, project and supply chain levels. Existing studies have employed various theoretical models to investigate blockchain and DLT adoption. The first subsection analyses the key blockchain and DLT adoption studies, separating them according to the level investigated and the theoretical models used, and summarising their key findings. The second subsection presents barriers and enablers to blockchain and DLT adoption across the four dimensions of technology, process, policy and society.

3.4.1 Adoption studies at different levels

3.4.1.1 Market level

This level examines blockchain and DLT adoption from an industry-wide perspective, considering factors such as market demand, competition, the overall maturity of the technology within the construction industry and the role of key market macro players such as the government. Studies at this level emphasise the need for a socio-technical perspective to recognise that technology implementation is not solely a technical endeavour but is deeply intertwined with social, organisational and environmental dynamics. Understanding these interconnected aspects is essential for successfully navigating the complexities of market-level blockchain and DLT adoption in the construction industry and beyond. The main study addressing adoption at market level by Li *et al.* [70] is extensively covered in this chapter.

Other studies at this level have focused on the interplay between market factors and institutional and technical factors. Janssen *et al.* [26] provide a framework to understand blockchain adoption from an institutional and market perspective. The authors use a combination of the institutional theory and market theories to identify three key dimensions influencing adoption: institutional pressure, market conditions and technical characteristics. Their study highlights that institutional pressures, such as regulatory mandates and market norms, significantly drive blockchain adoption. Furthermore, they argue that the technical complexity of blockchain can hinder adoption unless supported by a conducive market environment [26].

Sadeghi *et al.* [97] investigated the adoption of DLT using the ordinal priority approach. In their study, the interest and demand from clients and project owners are identified as a key enabler for the widespread adoption of blockchain and DLT. Market demand for digital innovation within the considered industry is slow and the demand for blockchain solutions in the construction industry has yet to reach a level that would drive widespread adoption [97]. The network effect, whereby a critical mass of companies successfully implements DLT, creating a combination of mimetic and normative pressures within the construction market is still absent. Educating the demand side including clients and project owners about the benefits of blockchain and DLT in construction can drive adoption [98]. Some of the key recommendations to drive blockchain and DLT adoption at this level include market-level interventions such as educational programs, incentive schemes and clarity about the impact of DLT adoption on market structure, competition and innovation in the construction industry [70].

Other market-level adoption studies have looked at the role of blockchain and DLT in enhancing data quality within the construction industry and decision-making. Hunhevicz *et al.* [99] conducted a feasibility study within the Swiss construction industry to demonstrate how blockchain could incentivise high-quality data sets throughout project lifecycles. They emphasised the importance of data completeness, correctness and structure for SC execution and overall project success [99]. Lombardi and Dounas [100] and Hunhevicz *et al.* [86] investigated the potential of decentralised autonomous organisations (DAOs) to transform collaboration and decision-making within the construction industry, envisioning DAOs as platforms for decentralised governance, funding, and execution of projects, enabled through SCs and tokenisation.

In a systematic review of 30 papers across several industries, *trust* was identified as the biggest barrier to blockchain adoption with *perceived cost, social influence, facilitating conditions, performance expectancy, effort expectancy, information security, privacy, complexity, compatibility, and top management* also being commonly identified [96].

Studies have also explored the role of key market macro players such as the government. Clear and supportive government policies and regulations are crucial for fostering blockchain adoption. Addressing legal uncertainties surrounding data security, SC enforceability and data privacy can increase confidence and encourage investment in blockchain solutions [101]. In addition, the establishment of clear guidelines and regulations for the taxation of blockchain-based transactions and reporting requirements for projects using blockchain technology is essential for wider adoption. This is particularly challenging in cross-border transactions where different countries have varying classifications and legal definitions [12].

3.4.1.2 Consortia level

Bauer *et al.* [75] studied the reasons why business consortia choose to adopt blockchain technology beyond its technical characteristics. The authors argue that while blockchain's potential in various domains is evident, the broader socio-technical reasons for its engagement are not well articulated. The researchers employed a multiple-case study approach involving 53 interviews with stakeholders from 19 different blockchain consortia. They identified 19 different motives for engaging with blockchain technology, categorised into broader themes based on the information system artefact reference framework: technology innovation motives (i.e. experimentation, opportunity seeking, addressing domain-specific problems); technology enabling motives (i.e. decentralisation, uniqueness, control); governance necessity motives (i.e. collaboration, commons, trust); information tokenisation motives (i.e. tokenised assets and services); business goal motives (i.e. innovation, efficiency, customisation, domain-specific solutions, capability development); and political positioning motives (i.e. financial means, marketing innovativeness, establishing and maintaining partnerships). This implies that to leverage the benefits of blockchain, collaboration between multiple organisations within the same industry is needed, which in turn requires establishing governance frameworks that ensure fair participation, decision-making and data ownership among consortium members. These findings provided evidence that blockchain adoption is not just a technical decision but deeply rooted in the socio-technical dynamics of the organisation.

3.4.1.3 Supply chain level

Studies on the adoption of blockchain in the supply chain focus on understanding the enablers and drivers for adoption, as well as the technology's impact on logistics (i.e. information flows and materials). Agi and Jha [102] investigated the adoption of blockchain in the supply chain using the DOI theory and the business technology adoption model. They identified 20 enablers from the literature review organised into five groups: relative advantage, compatibility, complexity, organisational readiness and external pressure. Then, they applied the Decision-Making Trial and Evaluation Laboratory (DEMATEL) method to classify these enablers and identify their interactions. The top enablers identified are lower transaction costs, customer interest in traceability and product data, management commitment and the availability of regulatory frameworks. Similar to the findings identified at the consortia level, this study concluded the need for cooperation between supply chain organisations, a regulatory framework and data integrity.

Fosso Wamba *et al.* [103] examine blockchain's impact on supply chain performance using a technology adoption model and a supply chain performance framework. Their research, conducted in India and the US, shows that trading partner pressure and knowledge sharing are significant enablers of blockchain adoption, leading to improved supply chain transparency and performance.

Kifokeris and Koch [104] proposed a blockchain-based solution (i.e. BlogCHAIN), a permissioned blockchain based on Hyperledger Besu, to address logistics inefficiencies in construction by integrating economic, informational and material flows. Using a case study in Sweden, the authors evaluated the solution's impact on construction logistics and highlighted its potential in enhancing transparency and coordination and identified resistance from practitioners and power dynamics in logistics as key challenges hindering its adoption.

3.4.1.4 Organisation and project

Blockchain and DLT adoption studies at the organisation and project level have proliferated in recent years. At this level, studies investigate the factors influencing blockchain adoption decisions within construction firms and projects, and how organisations can integrate blockchain into their existing structures, processes and cultures. Stratopoulos *et al.* [105] investigated the factors influencing the adoption of blockchain technology among firms. Their findings indicate that increased information gathering activities correlate with higher expectations about the positive payoffs from blockchain adoption. This suggests that firms that actively seek information are more likely to consider adopting blockchain technology. They also identified a substantial increase, between 2013 and 2018, in the number of firms disclosing blockchain-related information, indicating a parallel growth in demand and supply for blockchain solutions. However, despite the rising interest and disclosures, their study concluded that blockchain adoption had not yet reached the early majority stage by the end of 2017. Purusottama *et al.* [106] found that blockchain influences how organisations create value within their business models, changing the way they deliver value to their customers and how they capture value for themselves. Studying blockchain adoption in construction firms in Australia, Malik *et al.* [107] found the factors influencing adoption include (i) the perceived technology's benefits, compatibility and complexity, (ii) the organisation's innovativeness and learning capability and (iii) the external influences such as competitiveness within the sector, government support, trading partner readiness, standard/regulatory uncertainties. A significant body of literature, encompassing studies on blockchain and other digital innovations such as BIM, employs the same classification framework for drivers and barriers known as the 'technology-organisation-environment' (TOE) model. However, these studies will not be discussed further, as they tend to yield varying results depending on the specific context such as the country or region under consideration.

3.4.2 Barriers and facilitators to blockchain adoption

3.4.2.1 Technology

This section discusses the challenges of blockchain adoption related to maturity of the technology, and its interoperability and integration with other technological systems.

Teisserenc and Sepasgozar [108] identified 74 technical challenges to blockchain adoption in the context of digital twins across IT infrastructure requirements, digital twin challenges, data requirements for digital twin, traceability and transparency, identity management, decentralisation, blockchain performances challenges, asset lifecycle complexity, BIM challenges, SC challenges and digitisation challenges. In addition, they present 10 technological factors that could support adoption of blockchain (written in the context of digital twin) including IT requirements (decentralisation, IT infrastructure and computational requirements, big data requirements); blockchain application requirements (SC integration, identify management/access control, blockchain properties and performances); and blockchain integration with other technologies (blockchain integration with: BIM, IoT, digital twin; asset lifecycle complexity). These elements are discussed further in other literature.

Maturity of blockchain has advanced substantially since 2009; however, there are still limitations. Several authors discuss scalability [109,110] where current systems are slow and therefore not susceptible to scaling easily [111], especially when they are not able to compete with the likes of Visa on throughput [112]; although Layer 2 technologies are beginning to offer viable solutions [113].

While there are still challenges working across chains and integrating platforms [111] where different platforms run different protocols [113], once solutions to this are developed, Slowman *et al.* [95] believe cross-chain protocols will actually drive interoperability. Progress is being made on addressing scalability issues through, for example, the lightning protocol that allows transactions to be performed on sidechains much faster than on the mainchain. Web3 adoption is currently low, attributed to the need for higher maturity and widespread use, and Web3 applications are currently too slow in terms of functionality [114].

Blockchains are an integrative technology that can benefit many applications; comprehensive consideration should be given to this along with any trade-offs that may have to be made [111]. Current on-chain storage is not suitable to withstand the continuous growth of blockchains [111]; however, studies have considered integration of blockchains with the InterPlanetary File System where data are stored off-chain with a fingerprint of the data stored on-chain to mitigate this [115,116].

Integration of blockchain with existing systems (e.g. payment systems, healthcare systems, IoT networks, cloud/edge computing) [111] requires technical expertise and a level of standardisation that is currently lacking [109,110]. Such integration and industry standardisation can be time-consuming and requires resources [112]. Kumar Singh and Kumar [110] suggest that the way to do this is to establish industry-wide frameworks for interoperability and standardisation to foster communication and exchange of data across platforms and systems. Consideration should also be given to technological advancements that occur outside of blockchain (i.e. with wireless networks, IoT technologies). It is important that blockchain is able to keep up with other technological advancements to prevent technology gaps when adoption becomes feasible [111].

Energy usage is an environmental concern for blockchain technologies [108] and there are concerns around energy costs associated with PoW consensus mechanisms [109]. However, this relates mainly to the early blockchains such as that of the Bitcoin protocol that runs a PoW consensus mechanism. In recent years, alternative consensus mechanisms that are more energy efficient have emerged as the technology matures such as proof-of-authority, proof-of-burn, proof-of-capacity and others [117]. For example, Ethereum originally ran on PoW but switched to proof-of-stake in 2022 that consumes 99.9% less energy, equated to a megaton of carbon each week [118].

Kumar Singh *et al.* [119] suggest tailor made platforms would aid in widespread adoption working to standards and protocols specific to construction. However, this approach must be considered with interoperability in mind so as not to create issues with cross-platform functionality, for example, when logistics or fintech must integrate with construction blockchains.

A final thought suggests that any technological solution built on blockchain should be mindful of sustainability factors (e.g. carbon footprint, energy consumption) so as not the undermine any positive progress through blockchain-based applications [119].

Data, privacy and security

Digitalisation in the built environment leads to challenges around privacy, security and protection of data that could be mitigated by blockchain as a result of its decentralised and secure nature [108]. Like most technological systems, blockchain requires data. However, there are several issues with data that are rooted in 'societal failures and inequalities' [120, para. 4] including access to quality data differs between actor groups; biases are inherent in datasets; data are not always trustworthy; despite the volume of data, accessing it effectively is challenging where they often exist in silos or are subject to hierarchical access [120]. Working in data silos, which is common in construction, limits collaboration, information sharing and management, and trust in data governance [108].

In addition, there are concerns about protecting sensitive data on a blockchain [110]. Henry *et al.* [113] discuss the challenges of blockchain immutability stressing that accurately coding SCs is essential, the challenge being the inability to fix vulnerabilities once executed. They add that in Web3, the media has reported on phishing attacks and theft of assets; however, practices are in existence to help protect against these risks. Security is a key area of consideration for development of new blockchain and Web3 applications.

Blockchain can be used to incentivise data sharing within construction projects to improve collaboration [108]. But it is essential data privacy and security are addressed when developing blockchain-based systems that should encompass encryption and access control and comply with regulations to ensure compliance [110]. Krichen *et al.* [111] consider several aspects for blockchain adoption: it must be resilient to cyberattacks and mitigate malfunctioning of systems, especially where large-scale applications are deployed; sharing of data across institutions requires access control, authentication, non-repudiation of records; potential issues in data hacks of information systems; bug detection in SCs; threats to external systems integrated with blockchain (e.g. energy grids); and high levels of authentication and authorisation are required in, for example, governance of government services.

R&D and use cases

The construction sector is limited in its research and development (R&D) efforts [1]. But it is recognised that use cases can help increase understanding blockchain and those with minimum viable products can make blockchain applications become reality much sooner [112]. Demonstrations of blockchain through case studies will be effective [109] and a 'killer app' is required that does not focus on the specificities of blockchain and Web3 technologies but on offering something of value to the user who does not care about the underlying technology [95,121]. When conducting R&D and developing use cases, evaluation of whether blockchain is actually needed should be made [111].

3.4.2.2 Policy

The majority of literature attributed to this dimension is around regulatory frameworks. Uncertainty around regulation that has perpetuated for too long is a barrier to blockchain adoption [95]. Lack of clarity makes some end-users wary of blockchain and Web3 applications despite several regulatory bodies in the US providing guidance around these [113]. Regulations play a key role in acceptance of blockchain technologies [111], but at the same time legal constraints could be a limitation to blockchain adoption in some jurisdictions [110]. Several authors suggest that more guidance is required from regulatory frameworks [112] where governments should regulate use of blockchain through policies and standards [109].

Additional challenges around regulatory works include supervising and managing financial transactions, particularly across international borders; regulating decentralised information management systems; uncertainties in many countries regarding blockchain regulation; upgrading or overhauling existing systems, for example, to allow peer-to-peer trading that is not standard practice in the energy domain [111]. Current frameworks based on centralised processes are misaligned with the decentralised blockchain paradigm; practices around certifications are described as substandard; and there are misalignments between blockchain's immutability and the General Data Protection Regulation (GDPR) [108].

Kumar Singh *et al.* [119] suggest '[a]n international cooperative agreement accepted universally by all nations can [...] be signed' and while this could solve some of the challenges to adopting blockchain, it is a highly optimistic approach to have all nations collaborate on such matters. The timing is problematic due to ongoing geopolitical tensions, competing national interests and the varying levels of technological maturity and regulatory readiness across nations, making such an agreement both impractical and potentially irrelevant to the immediate priorities of blockchain implementation.

In the UK, the Financial Conduct Authority enacted regulations around the financial promotions of cryptoassets meaning there are now restrictions on organisations' mass marketing of cryptoassets products in the UK regardless of the jurisdiction from which they originate [122]. While this makes blockchain adoption more restrictive, it offers some clarity and therefore something tangible with which to comply and could provide a springboard with other countries to follow suit [95]. Henry *et al.* [113] suggest that working on small test cases and projects is the way to go such that when regulatory frameworks are enacted they are ready to scale up quickly.

3.4.2.3 Process

Within process, the impact of blockchain adoption on organisations is considered in the context of investments and costs, and business and governance models.

Investment and costs

Blockchain technologies have high implementation costs, which presents a barrier to their adoption; however, the long-term financial benefits (e.g. efficiencies) could be attractive to some stakeholders [108]. Implementing distributed systems in small and medium-sized enterprises (SMEs) and start-ups is costly and could be technologically challenging [111]. Many, however, may not have sufficient funding required for initial investment of blockchain-based applications; they may not see or give weight to the proposed benefits of these systems; and lack of performance measurements and benchmarking could make demonstration of return on investment challenging [110]. The added barrier for Web3 applications goes against the current 'free to use' advertisement-based internet as they require some form of financial implication for participation [114].

Olawumi *et al.* [109] believe investment at an organisational level and greater investment generally will drive blockchain adoption in construction [119]. Blockchain requires investment for integration into existing systems and changing/upgrading infrastructure, training and recruiting skilled developers [111]. Performing cost-benefit analyses to identify and quantify the benefits of adopting blockchain can help with informed decision-making [110]. But the question remains as to who should fund this investment [119].

Two studies show potential for blockchain-based applications from a cost-benefit analysis perspective. A simple cost-benefit analysis was performed by Kang *et al.* [123] that identified costs associated with the blockchain system and the training of actors in its use. The authors indicate that '[a]lthough implementing Blockchain demands considerate investment, stakeholders in construction projects can benefit from the implementation' [123, p. 456]. A more detailed study performed a cost-benefit analysis on 'digitalising and automating documentation flows in cross-border supply chains compare[d] to a conventional centralised relational database solution' [124, p. iii]. The analysis favoured the use of blockchain in both financial and non-financial factors as a result of lowered costs, greater financial benefits and somewhat greater utility over the conventional system.

Business and governance models

Trust is required in the implementation of new business and governance models that may be difficult, especially where such models do not align with existing practices [110]. Web3 is challenging traditional business models where organisations set objectives and direction to one where co-creation is the focus and decentralisation offers employees a way to be part of the governance structure rather than just being governed by it. This also works to integrate customers more through better relationship-building through, for example, tokenisation of goods and services [113].

An example of a successful business model is from Everest who provides access to financial services in Asia, Africa and South America to people who previously had limited or no access. Everest offers an Ethereum-based application for payments using a multi-currency wallet and identifies users via biometrics to facilitate micro-transactions, land claims and medical records [125]. Such a system enables users access to finance without requiring a bank account, just a mobile phone equipped with a crypto wallet. Seventy-three Nepalese people received money via Sikka, a similar Ethereum-based micro-finance app, during a disaster that allowed exchange of tokens via SMS through associating users' wallets with their mobile number [126]. Another example is that of TradeLens, a joint venture between Maersk and IBM who developed an enterprise blockchain for supply chain shipping. While TradeLens is a successful example of an enterprise blockchain on a public network, the project itself is considered a failure because many of the shipping organisations needed for the network to be successful were unwilling to collaborate [127].

3.4.2.4 Society

In this dimension, literature focuses on trust and acceptance of blockchain technologies as well as the need to achieve a critical mass through offering a suitable user experience.

Kumar Singh and Kumar [110] discuss several barriers that are specific to construction regarding lack of trust among stakeholders in the technology's ability to be secure and reliable; construction has a cultural barrier of resistance to change to adopt and adapt new processes; and there is limited collaboration and communication between stakeholders that will compound the adoption of blockchain where multi-party collaboration is required. In addition, limited understanding of blockchain generally and its lack of use cases presents a barrier to adoption [112] and limits informed decision-making for adoption.

A critical mass is needed to see a real impact from blockchain adoption [112]. However, user experience needs to improve for it to be more accessible to the masses [95] where current Web3 systems lack ease of use [114]. Making blockchain and Web3 more user friendly and minimising the complexities of private key management [113] will aid in adoption. Saldivar *et al.* [128] conducted a study to assess user friendliness of two blockchain-based applications with users proficient in IT. Their results showed that even with people highly competent in IT, several difficulties were faced, specifically, vocabulary and concepts used were complicating operation of the apps as they were 'highly-technical' and often rooted financial terminology, possibly resulting from blockchain's mainstream origins being in cryptocurrencies; and 'blockchain metaphors and transaction-mediated interactions challenge established mental models, imposing heavy workloads on users' [128, p. 37]. Such mental models were heavily influenced by Web 2.0 that are at odds with 'new paradigms and concepts emerging around the blockchain' [128, p. 41]. To achieve a critical mass, there is a need for sufficient skills in the sector and current supply is not meeting the undeniably increasing demand needed for blockchain and Web3 applications; universities are offering blockchain-based education and organisations are upskilling to address this [113]. Organisations should invest in education and training to enhance the technical expertise of their workforce [110]. An example where this has been successful is regarding BIM in the UK. Shojaei *et al.* [129] reported on these programmes where organisations were tasked with the responsibility of investing in the training and development of their workforce as set out in the Construction Leadership Council's Industry Skills Plan 2021–2025 [130]. A priority within the programme was 'to change from conventional methods of working to modern practices using BIM' [129, p. 400]. Training was delivered to employees based on level of competence for both generic and role-specific competencies. The success of the latter was in part due to detailed role descriptions setting out what the employee was required to know to complete their job. Of critical importance was the approach taken towards suppliers in their network who were trained as if they were in-house employees with a focus on 'growing together' [129, p. 400] to ensure successful upskilling across the supply chain.

Moving on, there needs to be some protections in place for the users; this is where regulatory frameworks come in that can support innovation so people who want to exploit the technology know what they can and cannot do with it [95]. It is suggested that integrating blockchain into existing processes will demonstrate tangible benefits to the potential userbase leading to its confidence and acceptance as a foundation towards more advanced use cases. For example, it can be integrated into existing processes as an assurance layer creating blockchain-based governance [86]. Walmart in the US leveraged blockchain to increase traceability of the food supply chain where pre-blockchain, it took over 6 days to obtain the data and only 2.2 s when the data were made available via blockchain [131]. The blockchain provided not only assurance of data but also increased traceability and accessibility. Active participation of industry practitioners [109] and increasing awareness and satisfaction of stakeholders regarding the benefits blockchain can bring as well as obtaining support from those in higher management positions will likely drive adoption [109]. Day [121] suggests presenting user appropriate terminology that is understandable by the average person, and demonstration of Web3 applications, rather than just talking about them will help gain a critical mass, along with the 'killer' Web3 DApp.

3.4.3 Critical success factors for adoption of blockchain technologies

Sun *et al.* [132] identified 22 critical success factors (CSFs) of blockchain in construction through scientometric analysis and systematic literature review of 78 peer-reviewed journal papers. The five most populous CSFs from the study were discussed as a decentralised system; transparency in data information across the construction lifecycle; ensuring data immutability; increasing data security and reliability; and full traceability. These CSFs focus more on the characteristics that blockchain can bring to construction application rather than the factors that can drive adoption of the technology.

Ameyaw *et al.* [133] applied a more holistic view and identified trialability, relative advantage, competitive advantage and compatibility as CSFs for adoption of blockchain-based SCs. Regarding trialability, they found that potential users will benefit from the opportunity to try out the technology on real-world project data before committing to full adoption. This gives the benefit of identifying bugs, risks and failures in the system before deploying SCs to real-world projects. Regarding relative advantage, this focuses on the advantages a new system will bring in comparison to its predecessor, while competitive advantage is about the benefits and gains an organisation will receive from adopting the new technology along with their ability to compete with rival organisations in the same industry. The final factor is the extent to which the new technology is compatible with the organisation's ethos and needs.

Li *et al.* [134] applied the TOE framework to analyse the factors pertinent to blockchain adoption. Like Ameyaw *et al.* [133], they identified compatibility, top management support and relative advantage as the top three factors, but found that complexity, trialability and trading partner pressures were not statistically significant in their evaluation. As this study's findings were based on a questionnaire and hypotheses derived from literature, despite not finding the latter factors statistically significant, they are important factors to consider with regard to blockchain adoption. The reasons for not being statistically significant in Li *et al.*'s [134] study related to considerations such as trialability not being relevant because cost is an inhibiting factor and therefore organisations do not consider trialling the technologies.

Hunhevicz *et al.* [86] proposed a stepped model towards the adoption of blockchain in construction. Step 1 focuses on integration with existing processes to foster a foundation from which to build up and gain user acceptance; step 2 moves towards cryptoeconomics incentives to drive better collaboration and new business models based on tokens and SCs; and step 3 aims at more coordinated activities that are built on blockchain-based governance mechanisms possibly driven by DAOs.

Taherdoost [135] assessed several technology adoption models for their suitability for blockchain. Those evaluated included theory of planned behaviour; technology acceptance model (TAM); extension of TAM; DOI; unified theory of acceptance and use of technology; TTF model; and TOE.

3.5 Roadmaps to adoption

The classification of society into micro, meso and macro scales helps understand individuals, organisations and society as a whole. The need for two distinct roadmaps was identified [1]: one for macro-scale implementation (the construction sector) and one for meso-scale implementation (DLT and SC applications within the sector). While the micro scale was excluded (to be considered at a later stage), the meso scale bridges the gap between micro and macro, offering a framework for rule establishment. Both roadmaps are intended to be live artefacts to be updated by the authors on a periodical basis as blockchain technologies advance and blockchain-based solutions are adopted across the construction sector. It is intended that reviews will take place yearly and updates made accordingly. This will be done through conducting research activities such as interviews and surveys with industry practitioners and academics working in the space inside and outside of construction accounting for the fact other sectors may advance faster.

Although developed for the construction sector, these roadmaps can be adapted for other sectors and technologies by adjusting sector-specific regulations and considerations. This adaptability recognises the need for integration with external systems and acknowledges the importance of the human element in STS.

Both roadmaps outline tasks and activities assigned to specific actors or groups, as detailed in Table 3.1.

3.5.1 Macro roadmap

The macro roadmap (Figure 3.7) aims to prepare the construction sector for the successful implementation of DLT- and SC-based applications. It operates alongside the meso roadmap, focusing on broader readiness rather than specific applications. Appointed by a national authority, a taskforce (TF) oversees eight overarching actions spanning several years. This roadmap is dynamic, with funding allocated for its implementation and a detailed delivery plan established accordingly. While some actions may have already begun by national authorities when this roadmap becomes available in the public domain, the roadmap supports ongoing efforts. Each action outlines tasks and responsible actors, with flexibility for prioritisation based on resources and timelines. Although certain actions, like defining scope of requirements (SoR) and mobilising a DLT/SC TF, should start first as indicated in Figure 3.7, all actions are equally important for successful adoption of DLT and SCs in construction. A description of the eight actions follows.

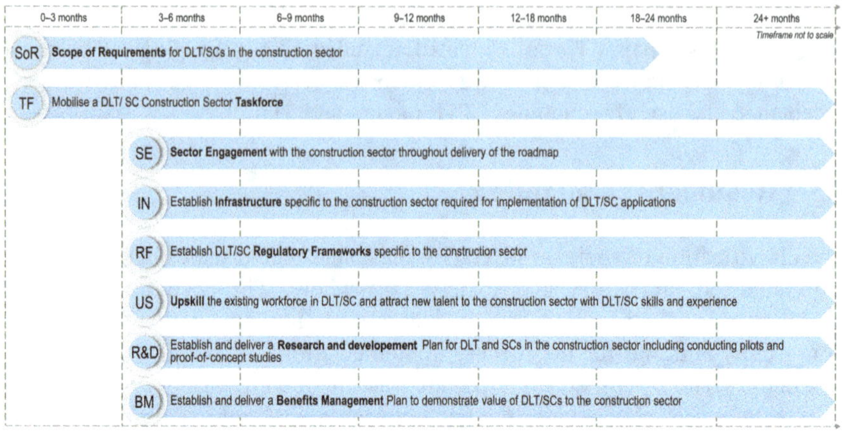

Figure 3.7 Macro roadmap to achieve readiness for DLT and SC applications in construction [1, p. 151]

3.5.1.1 Scope of requirements (SoR): scope of requirements for DLT/SCs in the construction sector

One of the most important actions in this roadmap, the SoR sets the direction for the delivery of the roadmap in terms of what a national authority wants to achieve with regard to their implementation of blockchain technologies in their region. This action involves needs identification, identifying funding streams, establishing a TF to deliver the remaining actions, aligning the roadmap with other national strategies, setting objectives and milestones, evaluating the potential benefits of blockchain technologies in their respective construction Industry and identifying resource requirements for delivery of the remaining actions. The actors involved in this action include national authority who instigates and takes ownership of readying the blockchain ecosystem; TF to drive the roadmap; system developers and software developers who will be central to creating the infrastructure required for readying the blockchain ecosystem; industry associations and technologists who can advise on requirements for the ecosystem; and users representing the end-users of the ecosystem to ensure that readiness is aligned to industry needs.

3.5.1.2 Taskforce (TF): mobilise a DLT/SC construction sector taskforce

Running in parallel with SoR, mobilising the TF takes place as an initial action in the roadmap. The appointed TF should be in place throughout the delivery of the macro roadmap driving it to a point of self-perpetuation where a dedicated TF is no longer required to maintain momentum. The TF should include representatives from across the sector who have a vested interest in its delivery to ensure the required commitment is obtained. The TF will be responsible for obtaining commitment from the actor groups required in delivery of the other actions and for specifying use cases to demonstrate benefits of blockchain technologies and to push for regulation. The actors involved in this action include national authority responsible for appointing the TF; and the TF itself.

3.5.1.3 Sector engagement (SE): conduct sector engagement with the construction sector throughout delivery of the roadmap

Sector engagement (SE) focuses on engagement with all relevant actors across the construction industry during delivery of the roadmap. This involves identifying with whom to engage, as well as when and how to do so, supported by the DLT actors model. Running for the duration of the roadmap, SE is directly linked to the benefits management (BM) action to share successes and lessons learned. Activities include marketing, outreach, awareness-raising, and results dissemination. SE will also interact with meso projects, particularly in pilot and proof-of-concept (PoC) studies that showcase the benefits of these technologies. The actor responsible for coordinating this action is TF.

3.5.1.4 Infrastructure (IN): establish infrastructure specific to the construction sector required for implementation of DLT/SC applications

Establishing the necessary infrastructure to support DLT and SCs in construction will enable their development and implementation at a meso level. While many start-ups and SMEs possess the creativity and innovation needed to develop these applications, they often encounter implementation challenges due to inadequate infrastructure. These challenges can stem from a lack of funds or resources, inability to connect to networks or interoperability issues. Similarly, larger companies might be discouraged from developing DLT/SC-based solutions in a sector characterised by low profit margins and minimal R&D investment if the infrastructure costs are too high. Therefore, it is crucial to ensure that all aspects of DLT/SC infrastructure are well-coordinated. This includes aligning standardised smart legal contracts with regulations and ensuring compatibility with underlying DLT platforms. Such alignment will provide developers with the necessary opportunities and motivation to explore the potential of these technologies. The actors required to engage in this action include TF; system architects; industry associations; regulatory bodies.

3.5.1.5 Regulatory frameworks (RF): establish DLT/SC regulatory frameworks specific to the construction sector

While many actions in the macro roadmap extend beyond 24 months, establishing regulatory frameworks will take even longer, up to 36 months or more, due to its bureaucratic nature. The first task will be to create a sub-TF to establish or reform regulatory frameworks for integrating DLT and SCs. If a national authority adopts this roadmap and some actions are already underway, they should conduct a gap analysis using the socio-technical framework (e.g. Figure 3.3) to benchmark their implementation plan. This analysis will determine if existing regulations need amendments or new ones are needed to support DLT and SC applications in construction, considering compliance with, for example, the construction act, building codes and data protection laws. Engagement with regulatory bodies like the International Organization for Standardization, the European Committee for Standardization, the British Standards Institution and contracting organisations like the Joint Contracts Tribunal, the New Engineering Contract and the International Federation of Consulting Engineers is crucial. New or amended regulations should address various applications, such as procurement and compliance, and be periodically reviewed to stay aligned with technological advancements, industry practices and health and safety requirements. The actors driving this action include TF, sub-TF, and policymakers.

3.5.1.6 Upskill workforce (US): upskill the existing workforce in DLT/SC and attract new talent to the construction sector with DLT/SC skills and experience

Understanding the requirements for developing, delivering and using applications is crucial for their success. Upskilling the workforce with the appropriate skills and experience will prepare the sector to adopt DLT- and SC-based applications. This action begins with developing educational frameworks and learning objectives to ensure that the right skills are being cultivated. Just as infrastructure is essential for success, so are the skills necessary for deploying and managing these applications. This effort may involve training new entrants or reskilling the existing workforce. The TF will secure funding for this initiative and establish a plan for upskilling. Skills requirements will be identified through gap analysis and should be regularly reviewed and updated to keep pace with new technologies in construction. The final tasks involve creating educational materials and delivering them through established training and educational institutions. Training programs could be offered as professional development courses or integrated into college and university curricula, initially within existing courses (e.g. construction management) and eventually as dedicated programs (e.g. digital construction). The actors involved in this action include TF; training/educational institutions; industry associations.

3.5.1.7 Research and development (R&D): establish and deliver a research and development plan for DLT/SCs in the construction sector including conducting pilots and proof-of-concept studies

This action aims to address the deficiency of R&D in construction by establishing a strong R&D strategy, particularly focusing on DLT and SC technologies. The initial task is to identify the key R&D areas for DLT and SCs that are crucial for the sector's digital transformation. This involves evaluating industry challenges and proposed DLT/SC applications alongside consulting with industry and research institutions to pinpoint the needs. From this analysis, a plan to boost R&D activity and secure funding sources is developed. Funding should come not only from public bodies but also from industry and academic sources to ensure that all stakeholders have a vested interest in successful applications.

This action will align with meso roadmap activities, where meso projects can be part of the R&D plan for pilot and PoC studies. It will continuously review the latest technologies, concepts, and theories, and also look to faster-advancing sectors for inspiration, for example, finance, supply chain. Additionally, this action will coordinate with the SE and BM actions, as the research and its results will be valuable for sustaining the roadmap's momentum and demonstrating the sector's benefits. Actors include TF, industry associations and research institutions.

3.5.1.8 Benefits management (BM): establish and deliver a benefits management plan to demonstrate value of DLT/SCs to the construction sector

The BM action is crucial for measuring the success of DLT and SC applications in construction. It involves identifying, planning, measuring and tracking benefits from start to finish, even though the final benefits may take years to materialise. The SoR action will set objectives and milestones, with the TF selecting which benefits to manage. The benefits management plan will be a dynamic document, continuously reviewed and updated by the TF to benchmark macro implementation progress. Identifying and measuring benefits will involve specifying the recipients, timing and methods. Ensuring that benefits are fairly distributed, especially early on, is key to preventing resistance from powerful stakeholders opposed to change. A robust plan will ensure ongoing monitoring and updates throughout the roadmap and beyond. Interaction with the SE action is necessary to disseminate information about benefits, driving sector-wide engagement with the roadmap. Monitoring development and adoption of new blockchain applications at the meso scale will feed into this macro action to enable benefits of the cutting-edge technological advancement to be disseminated. Such dissemination could help actors developing applications learn from benefits but also best practices and failures that will allow the overcoming to barriers to entry. Actors include TF, DLT system architects, technology providers, service providers and users.

3.5.2 Meso roadmap

The meso roadmap (Figure 3.8) aims to support the development and adoption of DLT and SC applications in the construction sector. It focuses on integrating

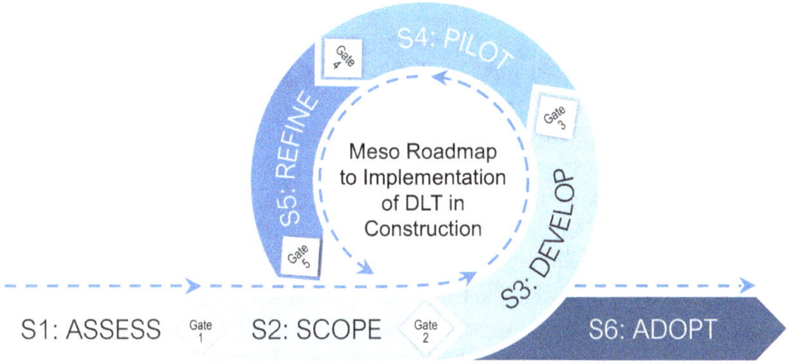

Figure 3.8 Meso roadmap for implementation of DLT and SC applications in construction [1, p. 158]

current or new technological systems with DLT and SCs. An example of such integration is the Walmart food chain that integrated blockchain to increase traceability and access to information speeding up the process significantly. While it operates independently from the macro roadmap, certain macro factors such as skills availability, regulatory environment and support for pilots and PoC studies can influence the success of new DLT/SC applications. The roadmap is depicted as six progressive stages, each ending with a decision gate for the TF to evaluate progress and potential value. This allows for flexibility and an exit strategy if necessary. Unlike the macro roadmap, the meso roadmap does not have a set timeframe, as each application will vary in time and resource requirements based on its objectives, interactions and actor groups involved. It is not a strict methodology, but rather adaptable to the needs and circumstances of each application. The meso roadmap is valuable for start-ups entering digital construction, as well as governments and institutions seeking to leverage new technologies to advance the sector and enhance business models.

3.5.2.1 Stage 1: assess

The initial phase of the meso roadmap involves evaluating the current system, processes and practices in the construction industry. This assessment examines the system's effectiveness in meeting sector-specific needs such as handling one-off projects and high fragmentation. It aims to identify existing challenges (e.g. payments, supply chain management) and opportunities for improvement. This stage also evaluates DLT and SC technologies, alongside alternatives, to help developers choose the best solution. Using models like the DLT four-dimensional model and the DLT actors model, the system is analysed from various perspectives and viewpoints. The deliverables from this stage support decision-making for the next phase. A TF is established for the application at the onset, comprising individuals committed to its success, including developers, project clients, main contractors, industry associations, policymakers and public sector representatives. The final task in Stage 1 assesses the application's value proposition, considering current and emerging technologies (e.g. 5G, DfMA) and regional differences to ensure optimal technology use and global applicability.

3.5.2.2 Stage 2: scope

After deciding to develop a new DLT-/SC-based system, the next stage involves defining the system's scope, users, impacts, and changes to existing processes. For example, automating payments may connect to procurement but remain separate. This phase helps developers identify system interactions and necessary adjustments. Extensive upfront work is crucial for success, making Stages 2 and 3 particularly demanding. Using models like the DLT four-dimensional model and DLT actors model, this stage analyses system requirements and interactions with various actors. Identified benefits can act as benchmarks for monitoring success during pilot, early adoption and mainstream phases. Links to the macro roadmap include actions such as upskilling the workforce, ensuring the identified skills for the new application align with broader training efforts or informing the macro roadmap's TF about sector-required skills.

3.5.2.3 Stage 3: develop

In the third stage of the meso roadmap, the application developer progresses through developing the technological system while addressing crucial considerations. These include ensuring regulatory compliance or designing adaptability for future regulations, and robust planning for pilot studies. BM is integral at this step, aligning with identified benefits from Stage 2 and newly identified ones. Additionally, detailed use cases are developed to encompass all intended and unintended uses of the system, ensuring user-centric design. By the end of this stage, preparations should be complete for pilot studies in Stage 4, with the system iteration possessing sufficient functionality to achieve pilot study objectives. This stage marks the beginning of an agile process within the roadmap, where iterative development, piloting and refinement lead to adoption.

3.5.2.4 Stage 4: pilot

The pilot stage in the meso roadmap is crucial for evaluating the new system's performance, demonstrating its benefits, identifying potential issues and observing user behaviour. Piloting helps reduce uncertainty associated with new technologies by testing their performance within the system, providing justification for deployment based on results. However, success during the pilot phase does not guarantee success in mainstream adoption. Nonetheless, carefully chosen evaluation criteria from Stage 3 provide developers with valuable data to increase the likelihood of mainstream adoption success. Lessons learned during piloting informs strategies to maximise success and address potential adoption obstacles.

3.5.2.5 Stage 5: refine

After the pilot stage, the application developer shifts focus to incorporating lessons learned from the pilot experiences and making necessary adjustments to refine the system and its processes before mainstream adoption, aligned to the agile mentality. The extent of effort required for this phase varies based on the application's scope and the changes needed. Detailed documentation, outlining processes, governance and guidance for the new system should be developed at this stage. Additionally, key performance metrics are devised to gauge future application success, while the benefits management plan is maintained to ensure ongoing focus on maximising application success and managing system diffusion.

3.5.2.6 Stage 6: adopt

In the final stage of the meso roadmap, the new system is integrated into the construction industry through mainstream adoption. This stage also involves early adoption of the system, emphasising the need for a diffusion roadmap to manage ongoing adoption and improvement. Efforts focus on securing funding for adoption and diffusion, as well as providing necessary training to ensure sufficient skill resources for system operation. Policymakers may play a crucial role in driving adoption, potentially through mandates or contractual clauses for publicly funded projects. For non-publicly funded projects, client engagement is vital in securing buy-in during the early adoption stages.

3.5.3 Linking the meso and macro roadmaps

These roadmaps offer pathways to advance the construction industry's digitalisation through DLT and SC implementation, aiding both researchers and industry practitioners in developing new applications and strategies. While intended to function independently, the roadmaps are interconnected to prevent misalignment between practice and policy, addressing concerns of decoupling, that is, the gap between policy and what happens in practice [136]. An example of such decoupling is the UK's mandate for BIM Level 2 on all publicly funded construction projects from 2016 onwards. Zomer *et al.* [137] conducted a longitudinal study investigating the impact of the mandate and found that it neither resulted in the planned change, nor at the expected speed. Zomer *et al.* positioned the mandate as institutional coercive pressure that resulted in a decoupling of *what* and *how* was imposed by it. They found that the reasons for such decoupling were limited willingness and ability of organisations to incorporate the change along with the perpetuating structures within the industry coupled with a resistance to change. This highlights the importance for strong alignment at the macro and meso scales for adoption of any innovation and in this paper specifically blockchain technologies. The benefit of such linkage is that macro scale oversight of the advancement of the technologies at the meso scale will highlight the needs of the industry from several perspectives (e.g. skills, regulations, infrastructure) that can be acted upon and integrated into the SoR and acted upon by the TF.

SoR: define the scope of requirements: Understanding ongoing meso-scale developments supports the sector in areas covered by the macro roadmap. It should be a requirement to identify and track progress of new blockchain applications in construction to be able to understand and respond to the needs of meso-scale developments. This will enable creation of an ecosystem with utility.

TF: mobilise a DLT/SC taskforce: A macro TF, inclusive of DLT/SC application developers, fosters an ecosystem for successful implementation, staying informed about the meso-scale roadmap. Including such representatives will ensure a direct link and keep the TF abreast of applications from concept, through to development, piloting and adoption. A challenge to this maybe that app developers may not publicise their progress (i.e. for fear of failure and/or to maintain competitive advantage).

SE: conduct sector engagement: Macro-level engagement drives adoption by informing the sector and encouraging engagement with DLT and SC projects at the meso scale. Pilots and demonstrator projects that happen at the meso scale can be widely disseminated at the macro scale. An example in practice is the BIM Task Group that was initiated in 2011 to drive BIM Level 2 adoption that was later succeeded by the UK BIM Alliance. This resulted in a greater understanding of the regulatory requirements for BIM in the UK [138].

IN: establish DLT/SC infrastructure: Infrastructure establishment supports exploration and implementation of DLT and SC applications at the meso scale. Understanding the meso scale needs at the macro scale means that appropriate infrastructure can be put in place that will support widespread development and adoption of blockchain-based applications. Without such infrastructure, advancement could be much slower where start-ups and SMEs may be presented with barriers they struggle to overcome (e.g. financial, interoperability).

RF: establish DLT/SC regulatory frameworks: Developers of applications should be part of working groups to establish regulatory frameworks. Ultimate responsibility for their enactment will lie with national authorities, necessitating awareness of macro- and meso-scale activities so as not to stifle innovation. Where appropriate, international authorities should be involved (e.g. for applications that cross borders such as cross-border payments).

US: upskill the construction sector workforce: Addressing workforce skill gaps is crucial, requiring national authorities to implement education and upskilling programs. Application owners must ensure personnel are trained to operate and train new systems. In adoption of BIM, such responsibility was put on organisations to facilitate this; however, such activity requires direction from governing bodies to ensure upskilling takes place and education and training programmes are accessible through, for example, universities and training providers. The TF should engage with industry to understand the needs and perform gap analyses to ensure that the right skills are identified and made available at the meso scale.

R&D: conduct research and development: R&D in construction is significantly lacking; therefore, efforts to increase this is required. The construction industry could adopt a programme similar to that of the Aerospace Technology Initiative (ATI), which is funded by industry and match funded by government. Within the ATI, organisations can bid for R&D funding outside of large calls such as Horizon Europe. In construction, this could encourage investment for both macro- and meso-scale projects to benefit technological advancement and the wider sector through, for example, investment in much needed infrastructure. The TF could utilise such funding streams alongside the likes of Innovate UK and Horizon Europe to focus R&D on pilots and demonstrator projects that would support advancement.

BM: establish a benefits management plan: Benefits derived from meso-level applications will inform the sector of the practical benefits of DLT and SC implementation, aiding in adoption. Provided there is macro-scale oversight of blockchain technology advancements and applications in development at the meso scale, the TF can be a conduit for disseminating information of the potential and actual benefits of blockchain in construction as they are realised from pilot studies and demonstrators.

3.6 Conclusions

This chapter focused on the adoption of DLT such as blockchain and SCs in the built environment by considering them as socio-technical innovations. In the construction industry, social factors beyond technical aspects, such as a lack of trust leading to information asymmetry and insufficient levels of collaboration, are well-documented challenges. Blockchain through streamlining processes and changing the trust relationship can alleviate some of these issues and help the sector become more efficient.

STS theory dates back to 1947 and the Tavistock Institute where, following the Second World War, research began to investigate the relationship between humans and technology. STS theory aligns technological and social components leading to better organisational agility and higher success of technological innovations. A look back at how the concept of STS has evolved at an organisational level and across the built environment offered insights into what makes a technology useful. Ultimately, this lies in resistance to technological innovations and comes from social behavioural barriers rather than poor technological systems. However, the benefits of a socio-technical approach to technological innovation are not widely seen due to lack of understanding of STS theory and, therefore, challenges in applying STS models, misalignment between STS and the technical issues associated with engineering, and challenges in individual interactions with technological systems.

Blockchain solutions are characterised as an STS for their ability to facilitate social relations such that in a blockchain-based system, the needs and expectations of users are prioritised. The socio-technical perspective could be key to answering questions about the future of blockchain use cases. The trust problem in construction operations (i.e. users can choose to trust the system instead of individuals) can be reframed through new forms of crypto economic governance mechanisms.

A comprehensive socio-technical model for the implementation of DLT applications in construction was presented. This is rooted in four dimensions of technology, process, policy and society that offer an analytical tool for understanding the environment we are seeking to change through the adoption of blockchain technologies. These dimensions are applied to understanding the challenges and facilitators of blockchain adoption along with consideration of CSFs of blockchain in construction. Two roadmaps to support implementation of blockchain applications in construction are offered. The first, a macro-scale roadmap centres on the formation of a set of requirements to be delivered by a TF that will ready the sector in a particular (geographical) region for blockchain-based applications. There are eight actions that make up the roadmap: establish the SoRs, mobilise a TF, engage with stakeholders throughout the sector, establish appropriate infrastructure to support development and adoption, establish regulatory frameworks specific to the construction industry, upskill the workforce where necessary, establish and deliver a R&D strategy, and establish a benefits management plan to demonstrate value of DLT and SCs to the sector. The second, a meso-scale roadmap is made up of six stages that focus on development of specific blockchain-based applications for construction. The agile stages are assess, scope, develop, pilot, refine and adopt. This is a progressive roadmap that is designed to interact with the macro roadmap to ensure alignment between practice and policy and to ensure a successful socio-technical approach to adoption.

These models and roadmaps are designed to support a systematic implementation of blockchain technologies in construction to achieve a steady level of adoption that will see increased productivity and alleviation of trust issues.

References

[1] Li, J.: '*A Socio-Technical Framework to Guide Implementation and Value Realisation of Distributed Ledger Technologies (DLT) in the Construction Sector*'. Doctoral Thesis, Northumbria University, Newcastle upon Tyne, 2023.

[2] Azaria, A., Ekblaw, A., Vieira, T., and Lippman, A.: 'MedRec: Using blockchain for medical data access and permission management', in *Proceedings of the 2016 2nd International Conference on Open and Big Data OBD 2016*, 2016, pp. 25–30.

[3] Kogure, J., Kamakura, K., Shima, T., and Kubo, T.: 'Blockchain technology for next generation ICT' *Fujitsu Sci. Tech. J.*, 2017, 53 (5), pp. 56–61.

[4] Turk, Ž., and Klinc, R.: 'Potentials of blockchain technology for construction management' *Procedia Eng.*, 2017, 196, pp. 638–645.

[5] Danilin, P.I., Lukin, A.A., and Reshetova, E.N.: 'Assessment organization service based on ethereum platform', in Cavalli, A.R., Petrenko, A., Pozin, B. (eds.): *CEUR Workshop Proceedings* (2017), pp. 291–294.

[6] Mengelkamp, E., Notheisen, B., Beer, C., Dauer, D., and Weinhardt, C.: 'A blockchain-based smart grid: Towards sustainable local energy markets' *Comput. Sci. Res. Dev.*, 2018, 33 (1–2), pp. 207–214.

[7] Millar, C., Lockett, M., and Ladd, T.: 'Disruption: Technology, innovation and society' *Technol. Forecast. Soc. Change*, 2018, 129, pp. 254–260.

[8] El-Sokhn, N.H., and Othman, A.A.E.: 'Project failure factors and their impacts on the construction industry: A literature review', in *Proceedings of the 10th International Conference on Civil and Architecture Engineering*, Military Technical College, Cairo, 2014, pp. 1–20.

[9] Love, P.E.D., Teo, P., and Morrison, J.: 'Revisiting quality failure costs in construction' *J. Constr. Eng. Manag.*, 2017, 144 (2), p. 05017020.

[10] Yates, J.K., and Lockley, E.E.: 'Documenting and analyzing construction failures' *J. Constr. Eng. Manag.*, 2002, 128 (1), pp. 8–17.

[11] Assaad, R., and El-Adaway, I.H.: 'Enhancing the knowledge of construction business failure: A social network analysis approach' *J. Constr. Eng. Manag.*, 2020, 146 (6), p. 04020052.

[12] Hunhevicz, J.J., and Hall, D.M.: 'Do you need a blockchain in construction? Use case categories and decision framework for DLT design options' *Adv. Eng. Inform.*, 2020, 45, p. 101094.

[13] Perera, S., Nanayakkara, S., Rodrigo, M.N.N., Senaratne, S., and Weinand, R.: 'Blockchain technology: Is it hype or real in the construction industry?' *J. Ind. Inf. Integr.*, 2020, 17, p. 100125.

[14] Krasovskii, A.N., and Krasovskii, N.N.: *Control Under Lack of Information* (Berlin: Springer Science & Business Media, 2012).

[15] Tribelsky, E., and Sacks, R.: 'Measuring information flow in the detailed design of construction projects' *Res. Eng. Des.*, 2010, 21 (3), pp. 189–206.

[16] Ivić, I., and Cerić, A.: 'Risks caused by information asymmetry in construction projects: A systematic literature review' *Sustainability*, 2023, 15 (13), p. 9979.

[17] Li, J., and Kassem, M.: 'Applications of distributed ledger technology (DLT) and blockchain-enabled smart contracts in construction' *Autom. Constr.*, 2021, 132, p. 103955.

[18] Grover, P., Kar, A.K., and Janssen, M.: 'Diffusion of blockchain technology: Insights from academic literature and social media analytics' *J. Enterp. Inf. Manag.*, 2019, 32 (5), pp. 735–757.

[19] Glaser, F.: 'Pervasive decentralisation of digital infrastructures: A framework for blockchain enabled system and use case analysis' *Hawaii International Conference on System Sciences 2017 HICSS-50*, 2017.

[20] Upadhyay, A., Mukhuty, S., Kumar, V., and Kazancoglu, Y.: 'Blockchain technology and the circular economy: Implications for sustainability and social responsibility' *J. Clean. Prod.*, 2021, 293, p. 126130.

[21] Trist, E.: *The Evolution of Socio-Technical Systems: A Conceptual Framework and an Action Research Program* (Toronto: Ontario Quality of Working Life Centre, 1981).

[22] Ropohl, G.: 'Philosophy of socio-technical systems' *Soc. Philos. Technol. Q. Electron. J.*, 1999, 4 (3), pp. 186–194.

[23] Geels, F.W.: 'Technological transitions as evolutionary reconfiguration processes: A multi-level perspective and a case-study' *Res. Policy*, 2002, 31 (8–9), pp. 1257–1274.

[24] Geels, F.W., and Kemp, R.: 'Dynamics in socio-technical systems: Typology of change processes and contrasting case studies' *Technol. Soc.*, 2007, 29 (4), pp. 441–455.

[25] Geels, F.W.: 'Regime resistance against low-carbon transitions: Introducing politics and power into the multi-level perspective' *Theory Cult. Soc.*, 2014, 31 (5), pp. 21–40.

[26] Janssen, M., Weerakkody, V., Ismagilova, E., Sivarajah, U., and Irani, Z.: 'A framework for analysing blockchain technology adoption: Integrating institutional, market and technical factors' *Int. J. Inf. Manag.*, 2020, 50, pp. 302–309.

[27] Çetin, S., Raghu, D., Honic, M., Straub, A., and Gruis, V.: 'Data requirements and availabilities for material passports: A digitally enabled framework for improving the circularity of existing buildings' *Sustain. Prod. Consum.*, 2023, 40, pp. 422–437.

[28] Treleaven, P., Barnett, J., Knight, A., and Serrano, W.: 'Real estate data marketplace' *AI Ethics*, 2021, 1 (4), pp. 445–462.

[29] Watson, R., Kassem, M., and Li, J.: 'Traceability for built assets: Proposed framework for a digital record', in Skibniewski, M.J., Hajdu, M. (eds.): *Proceedings of the Creative Construction Conference 2019, Budapest University of Technology and Economics*, 2019, pp. 496–501.

[30] Appelbaum, S.H.: 'Socio-technical systems theory: An intervention strategy for organizational development' *Manag. Decis.*, 1997, 35 (6), pp. 452–463.

[31] Leavitt, H.J.: 'Applied organizational change in industry: Structural, technological and humanistic approaches', in March, J. (ed.), *Handbook of Organizations (RLE: Organizations)* (London: Routledge, 1965).

[32] Morgan, J., and Liker, J.K.: *The Toyota Product Development System / Integrating People, Process, and Technology* (Milton Park: Taylor & Francis, 2020).

[33] Shackel, B.: 'Usability – Context, framework, definition, design and evaluation' *Interact. Comput.*, 2009, 21 (5–6), pp. 339–346.

[34] Eason, K.D.: *Information Technology and Organisational Change* (Boca Raton, FL: CRC Press, 2014).

[35] Orlikowski, W.J.: 'The duality of technology: Rethinking the concept of technology in organizations' *Organ. Sci.*, 1992, 3 (3), pp. 398–427.

[36] Rogers, E.M.: *Diffusion of Innovations* (Glencoe: Free Press, 1962).

[37] Dubois, A., and Gadde, L.-E.: 'The construction industry as a loosely coupled system: Implications for productivity and innovation' *Constr. Manag. Econ.*, 2002, 20 (7), pp. 621–631.

[38] Fernie, S., Green, S.D., Weller, S.J., and Newcombe, R.: 'Knowledge sharing: Context, confusion and controversy' *Int. J. Proj. Manag.*, 2003, 21 (3), pp. 177–187.

[39] Egbu, C.O.: 'Managing knowledge and intellectual capital for improved organizational innovations in the construction industry: An examination of critical success factors' *Eng. Constr. Archit. Manag.*, 2004, 11 (5), pp. 301–315.

[40] Rohracher, H.: 'Managing the technological transition to sustainable construction of buildings: A socio-technical perspective' *Technol. Anal. Strateg. Manag.*, 2001, 13 (1), pp. 137–150.

[41] Kazi, A.S., Hannus, M., Zarli, A., and Martens, B.: *Strategic Actions for Realising the Vision of ICT in Construction*. Strat-CON Consortium, ERA Build Project, 2016.

[42] Wild, A.: 'Uncertainty and information in construction: From the socio-technical perspective 1962–1966 to knowledge management – What have we learned?', in Kazi, A.S. (ed.), *Knowledge Management in the Construction Industry: A Socio-Technical Perspective* (Hershey, PA: IGI Global, 2005), pp. 203–224.

[43] Brown, H.S., and Vergragt, P.J.: 'Bounded socio-technical experiments as agents of systemic change: The case of a zero-energy residential building' *Technol. Forecast. Soc. Change*, 2008, 75 (1), pp. 107–130.

[44] Schweber, L., and Harty, C.: 'Actors and objects: A socio-technical networks approach to technology uptake in the construction sector' *Constr. Manag. Econ.*, 2010, 28 (6), pp. 657–674.

[45] Baxter, G., and Sommerville, I.: 'Socio-technical systems: From design methods to systems engineering' *Interact. Comput.*, 2011, 23 (1), pp. 4–17.

[46] Newton, P.W.: 'Liveable and sustainable? Socio-technical challenges for twenty-first-century cities' *J. Urban Technol.*, 2012, 19 (1), pp. 81–102.

[47] Chiu, L.F., Lowe, R., Raslan, R., Altamirano-Medina, H., and Wingfield, J.: 'A socio-technical approach to post-occupancy evaluation: Interactive adaptability in domestic retrofit' *Build. Res. Inf.*, 2014, 42, (5), pp. 574–590.

[48] Wen, Q.-J., Ren, Z.-J., Lu, H., and Wu, J.-F.: 'The progress and trend of BIM research: A bibliometrics-based visualization analysis' *Autom. Constr.*, 2021, 124, p. 103558.

[49] Koutamanis, A., Dainty, A., Kvan, T., and Turk, Ž.: 'The enigma of BIM' *Archit. Struct. Constr.*, 2023, 3 (1), pp. 1–9.

[50] Wang, J., and Lu, W.: 'A deployment framework for BIM localization' *Eng. Constr. Archit. Manag.*, 2021, 29 (1), pp. 407–430.

[51] Mathews, M.: '*A Hermeneutic Phenomenological Study into the Impact of BIM on the Social Dynamics of the AEC Professional in the Workplace*'. Doctoral Thesis, Technological University Dublin, Dublin, 2021.

[52] Olawumi, T.O., Chan, D.W.M., and Wong, J.K.W.: 'Evolution in the intellectual structure of BIM research: A bibliometric analysis' *J. Civ. Eng. Manag.*, 2017, 23 (8), pp. 1060–1081.

[53] Singh, V.: 'Digitalization, BIM ecosystem, and the future of built environment: How widely are we exploring the different possibilities?' *Eng. Constr. Archit. Manag.*, 2019, https://doi.org/10.1108/ECAM-01-2018-0004.

[54] Wang, Z., Liu, Z., and Liu, J.: 'Innovation strategy or policy pressure? The motivations of BIM adoption in China's AEC enterprises' *J. Asian Archit. Build. Eng.*, 2022, 21 (4), pp. 1578–1589.

[55] Boton, C., Rivest, L., Ghnaya, O., and Chouchen, M.: 'What is at the root of Construction 4.0: A systematic review of the recent research effort' *Arch. Comput. Methods Eng.*, 2021, 28 (4), pp. 2331–2350.

[56] Sackey, E.: '*A Sociotechnical Systems Analysis of Building Information Modelling (STSaBIM) Implementation in Construction Organisations*'. thesis, Loughborough University, 2014.

[57] Sackey, E., Tuuli, M., and Dainty, A.: 'Sociotechnical systems approach to BIM implementation in a multidisciplinary construction context' *J. Manag. Eng.*, 2015, 31 (1), p. A4014005.

[58] Merschbrock, C., Hosseini, M.R., Martek, I., Arashpour, M., and Mignone, G.: 'Collaborative role of sociotechnical components in BIM-based construction networks in two hospitals' *J. Manag. Eng.*, 2018, 34 (4), p. 05018006.

[59] Alreshidi, E., Mourshed, M., and Rezgui, Y.: 'Factors for effective BIM governance' *J. Build. Eng.*, 2017, 10, pp. 89–101.

[60] Oesterreich, T.D., and Teuteberg, F.: 'Behind the scenes: Understanding the socio-technical barriers to BIM adoption through the theoretical lens of information systems research' *Technol. Forecast. Soc. Change*, 2019, 146 (C), pp. 413–431.

[61] Papadonikolaki, E., van Oel, C., and Kagioglou, M.: 'Organising and managing boundaries: A structurational view of collaboration with building information modelling (BIM)' *Int. J. Proj. Manag.*, 2019, 37 (3), pp. 378–394.

[62] Papadonikolaki, E.: 'Loosely coupled systems of innovation: Aligning BIM adoption with implementation in Dutch construction' *J. Manag. Eng.*, 2018, 34 (6), p. 05018009.

[63] Anderson, A., and Ramalingam, S.: 'A socio-technical intervention in BIM projects – an experimental study in global virtual teams' *J. Inf. Technol. Constr.*, 2021, 26 (26), pp. 489–504.

[64] Mani, S., Ahmadi Eftekhari, N., Hosseini, M.R., and Bakhshi, J.: 'Socio-technical dimensions of BIM-induced changes in stakeholder management

of public and private building projects' *Constr. Innov.*, 2024, 24(2), pp. 425–445.

[65] Olofsson Hallén, K., Forsman, M., and Eriksson, A.: 'Interactions between human, technology and organization in building information modelling (BIM) – A scoping review of critical factors for the individual user' *Int. J. Ind. Ergon.*, 2023, 97, p. 103480.

[66] Hayes, A.: 'The socio-technological lives of Bitcoin' *Theory Cult. Soc.*, 2019, 36 (4), pp. 49–72.

[67] Nabben, K.: 'Blockchain security as "people security": Applying socio-technical security to blockchain technology' *Front. Comput. Sci.*, 2021, 2, p. 599406.

[68] Shin, D., Stylos, N., Asim, M., Park, Y., and Williams, R.B.: 'How does the blockchain find its way in the UAE? The blockchain as a sociotechnical system' *Int. J. Technol. Manag.*, 2022, 90 (1–2), pp. 122–140.

[69] Shin, D., and Ibahrine, M.: 'The socio-technical assemblages of blockchain system: How blockchains are framed and how the framing reflects societal contexts' *Digit. Policy Regul. Gov.*, 2020, 22 (3), pp. 245–263.

[70] Li, J., Greenwood, D., and Kassem, M.: 'Blockchain in the built environment and construction industry: A systematic review, conceptual models and practical use cases' *Autom. Constr.*, 2019, 102, pp. 288–307.

[71] Kinnaird, C., and Geipel, M.: 'Blockchain Technology: How the Inventions Behind Bitcoin Are Enabling a Network of Trust for the Built Environment', https://www.arup.com/publications/research/section/blockchain-technology, accessed October 15, 2024.

[72] Ekblaw, A., Barabas, C., Harvey-Buschel, J., and Lippman, A.: 'Bitcoin and the myth of decentralization: Socio-technical proposals for restoring network integrity', in *2016 IEEE 1st International Workshops on Foundations and Applications of Self* Systems (FAS*W)*, 2016, pp. 18–23.

[73] Kiš, M., and Singh, B.: 'A socio-technological analysis of the use of blockchain in financial systems', in Yang, X.-S., Sherratt, S., Dey, N., Joshi, A. (eds.): *Proceedings of 7th International Congress on Information and Communication Technology* (Cham: Springer Nature, 2023), pp. 749–759.

[74] Selvanesan, H., and Satanarachchi, N.: 'Potential for synergetic integration of building information modelling, blockchain and supply chain management in construction industry' *J. Inf. Technol. Constr.*, 2023, 28, pp. 662–691.

[75] Bauer, I., Ziolkowski, R., Hacker, J., and Schwabe, G.: 'Why blockchain: A socio-technical perspective on the motives of business consortia members to engage with blockchain technology' *Distrib. Ledger Technol. Res. Pract.*, 2023, 2 (2), pp. 1–27.

[76] Seebacher, S., and Schüritz, R.: 'Blockchain – just another IT implementation? A comparison of blockchain and interorganizational information systems', in *Proceedings of the 27th European Conference on Information Systems (ECIS2019)*, 2019, pp. 1–17.

[77] Chaudhuri, A., Bhatia, M.S., Subramanian, N., Kayikci, Y., and Dora, M.: 'Socio-technical capabilities for blockchain implementation by service

providers: Multiple case study of projects with transaction time reduction and quality improvement objectives' *Prod. Plan. Control*, 2024, 35(9), pp. 978–991.

[78] McNamara, A.: 'Automating the chaos: Intelligent construction contracts'. in *Smart Cities and Construction Technologies*. IntechOpen. 2020. https://doi.org/10.5772/intechopen.90764.

[79] Hunhevicz, J.J.: 'Exploring the potential of blockchain and cryptoeconomics for the construction industry'. Doctoral thesis, ETH Zurich, 2022. https://doi.org/10.3929/ethz-b-000575095.

[80] Ehrenberg, A.J., and King, J.L.: 'Blockchain in context' *Inf. Syst. Front.*, 2020, 22 (1), pp. 29–35.

[81] Hunhevicz, J., Wang, H., Hess, L., and Hall, D.M.: 'no1s1 – A blockchain-based DAO prototype for autonomous space', in *2021 European Conference on Computing in Construction*, 2021.

[82] Sigalov, K., Ye, X., König, M., *et al.*: 'Automated payment and contract management in the construction industry by integrating building information modeling and blockchain-based smart contracts' *Appl. Sci.*, 2021, 11 (16), p. 7653.

[83] van Groesen, W., and Pauwels, P.: 'Tracking prefabricated assets and compliance using quick response (QR) codes, blockchain and smart contract technology' *Autom. Constr.*, 2022, 141, p. 104420.

[84] Yu, J., Zhong, H., and Bolpagni, M.: 'Integrating blockchain with building information modelling (BIM): A systematic review based on a sociotechnical system perspective' *Constr. Innov.*, 2024, 24(1), pp. 280–316.

[85] Kifokeris, D., and Koch, C.: 'Blockchain in building logistics: Emerging knowledge, and related actors in Sweden', in Gorse, C., Neilson, C.J. (eds.): *Proceedings of the 35th Annual Conference of Association of Researchers in Construction Management, ARCOM*, 2019, pp. 426–435.

[86] Hunhevicz, J., Dounas, T., and Hall, D.M.: 'The promise of blockchain for the construction industry: A governance lens', in Dounas, T., Lombardi, D. (eds.): *Blockchain for Construction* (Cham: Springer Nature, 2022), pp. 5–33.

[87] Mancini, A., Desiderio, A., Di Clemente, R., and Cimini, G.: 'Self-induced consensus of Reddit users to characterise the GameStop short squeeze' *Sci. Rep.*, 2022, 12 (1), p. 13780.

[88] Jiménez, M.: 'What happened to GameStop, the company that made Wall Street tremble?', https://english.elpais.com/economy-and-business/2023-10-06/what-happened-to-gamestop-the-company-that-made-wall-street-tremble.html, accessed November 3, 2024.

[89] Digital Catapult: 'Weather Ledger – Industry insights report', https://www.digicatapult.org.uk/wp-content/uploads/2020/07/Weather_Ledger_Industry_insights_report-03.pdf, accessed September 28, 2024.

[90] Digital Catapult: 'The Weather Ledger: A practical demonstration of distributed ledger technology (DLT) and internet of things (IoT) in construction', 2021. https://www.digicatapult.org.uk/wp-content/uploads/2021/11/Weather_Ledger_WP5_report-v2.pdf, accessed January 15, 2025.

[91] Pan, W., and Ning, Y.: 'A socio-technical framework of zero-carbon building policies' *Build. Res. Inf.*, 2015, 43 (1), pp. 94–110.

[92] Abdelmegid, M., Tezel, A., Osorio-Sandoval, C., Fang, Z., and Collinge, W.: 'Conceptualizing digital twins in construction projects as socio-technical systems', in Gonzalez, V.A., Zhang, J., de Soto, B.G., and Brilakis, I. (eds.), *Proceedings of the 41st ISARC*, Lille, France, 2024, pp. 577–584.

[93] Cohen, L., and Chen, S.: 'Blockchain 2023', https://practiceguides.chambers.com/practice-guides/blockchain-2023, accessed December 5, 2024.

[94] Buser, T., Cisse, G., Crump, L., *et al.*: 'Partnering for change: Link research to societal challenges', https://tales.nmc.unibas.ch/de/partnering-for-change-link-research-to-societal-challenges-46/, accessed October 22, 2024.

[95] Slowman, A., Ajienka, T., Ajiboye, T., and Grogan, T.: *The State of Blockchain 2023* (MDRx, 2023).

[96] AlShamsi, M., Al-Emran, M., and Shaalan, K.: 'A systematic review on blockchain Adoption' *Appl. Sci.*, 2022, 12 (9), p. 4245.

[97] Sadeghi, M., Mahmoudi, A., and Deng, X.: 'Adopting distributed ledger technology for the sustainable construction industry: Evaluating the barriers using ordinal priority approach' *Environ. Sci. Pollut. Res.*, 2022, 29 (7), pp. 10495–10520.

[98] Celik, B.G., Abraham, Y.S., and Attaran, M.: 'Unlocking blockchain in construction: A systematic review of applications and barriers' *Buildings*, 2024, 14 (6), p. 1600.

[99] Hunhevicz, J., Schraner, T., and Hall, D.: 'Incentivizing high-quality data sets in construction using blockchain: A feasibility study in the Swiss industry', in *Proceedings of the 37th International Symposium on Automation and Robotics in Construction*, 2020.

[100] Lombardi, D., and Dounas, T.: 'Decentralised autonomous organisations for the AEC and design industries', in Dounas, T., Lombardi, D. (eds.): *Blockchain for Construction* (Cham: Springer Nature, 2022), pp. 35–45.

[101] Hughes, L., Dwivedi, Y.K., Misra, S.K., Rana, N.P., Raghavan, V., and Akella, V.: 'Blockchain research, practice and policy: Applications, benefits, limitations, emerging research themes and research agenda' *Int. J. Inf. Manag.*, 2019, 49, pp. 114–129.

[102] Agi, M.A.N., and Jha, A.K.: 'Blockchain technology in the supply chain: An integrated theoretical perspective of organizational adoption' *Int. J. Prod. Econ.*, 2022, 247, p. 108458.

[103] Fosso Wamba, S., Queiroz, M.M., and Trinchera, L.: 'Dynamics between blockchain adoption determinants and supply chain performance: An empirical investigation' *Int. J. Prod. Econ.*, 2020, 229, p. 107791.

[104] Kifokeris, D., and Koch, C.: 'The proof-of-concept of a blockchain solution for construction logistics integrating flows: Lessons from Sweden', in Dounas, T., and Lombardi, D. (eds.): *Blockchain for Construction* (Cham: Springer Nature, 2022), pp. 113–139.

[105] Stratopoulos, T.C., Wang, V.X., and Ye, H. (Jonathan): 'Use of corporate disclosures to identify the stage of blockchain adoption' *Account. Horiz.*, 2022, 36 (1), pp. 197–220.

[106] Purusottama, A., Simatupang, T.M., and Sunitiyoso, Y.: 'The spectrum of blockchain adoption for developing business model innovation' *Bus. Process Manag. J.*, 2022, 28 (3), pp. 834–855.

[107] Malik, S., Chadhar, M., Chetty, M., and Vatanasakdakul, S.: 'An exploratory study of the adoption of blockchain technology among Australian organizations: A theoretical model', in Themistocleous, M., Papadaki, M., Kamal, M.M. (eds.): *Information Systems* (Berlin: Springer International Publishing, 2020), pp. 205–220.

[108] Teisserenc, B., and Sepasgozar, S.: 'Adoption of blockchain technology through digital twins in the construction industry 4.0: A PESTELS approach' *Buildings*, 2021, 11 (12), p. 670.

[109] Olawumi, T.O., Ojo, S., Chan, D.W.M., and Yam, M.C.H.: 'Factors influencing the adoption of blockchain technology in the construction industry: A system dynamics approach', in Lu, X., Zhang, Z., Lu, W., and Peng, Y. (eds.): *Proceedings of the 25th International Symposium on Advancement of Construction Management and Real Estate* (Singapore: Springer, 2021), pp. 1235–1249.

[110] Singh, A. K., and Kumar, V.R.P.: 'Analyzing the barriers for blockchain-enabled BIM adoption in facility management using best-worst method approach' *Built Environ. Proj. Asset Manag.*, 2024, 14(2), pp. 164–183.

[111] Krichen, M., Ammi, M., Mihoub, A., and Almutiq, M.: 'Blockchain for modern applications: A survey' *Sensors*, 2022, 22 (14), p. 5274.

[112] Maslova, N.: 'Blockchain: Disruption and opportunity' *Strateg. Finance*, 2018, 100 (1), pp. 24–29.

[113] Henry, W., Massey, R., Hansen, B., and Davis, T.: *A Primer on Web3 Adoption for Enterprises* (New York: Deloitte, 2023).

[114] Gasser, L.: 'WEB3', in Mulder, V., Mermoud, A., Lenders, V., and Tellenbach, B. (eds.): *Trends in Data Protection and Encryption Technologies* (Cham: Springer Nature, 2023), pp. 209–213.

[115] Dounas, T., Jabi, W., and Lombardi, D.: 'Smart contracts for decentralised building information modelling'. in Köering, D. and Werner, L.C. (eds.), *Proceedings of 38th Education and research in computer aided architectural design in Europe (eCAADe)*, Berlin: 2020, pp. 565–574.

[116] Hamledari, H., and Fischer, M.: 'Construction payment automation using blockchain-enabled smart contracts and robotic reality capture technologies' *Autom. Constr.*, 2021, 132, p. 103926.

[117] Lashkari, B., and Musilek, P.: 'A comprehensive review of blockchain consensus mechanisms' *IEEE Access*, 2021, 9, pp. 43620–43652.

[118] Kessler, S.: 'The Ethereum merge is done, opening a new era for the second-biggest blockchain', https://www.coindesk.com/tech/2022/09/15/the-ethereum-merge-is-done-did-it-work/, accessed November 19, 2024.

[119] Kumar Singh, A., Kumar, V.R.P., Dehdasht, G., Mohandes, S.R., Manu, P., and Rahimian, F.P.: 'Investigating the barriers to the adoption of blockchain

technology in sustainable construction projects' *J. Clean. Prod.*, 2023, 403, p. 136840.

[120] Verhulst, S.G.: 'Information asymmetries, blockchain technologies and social change', https://sverhulst.medium.com/information-asymmetries-blockchain-technologies-and-social-change-148459b5ab1a, accessed December 2, 2024.

[121] Day, A.: '#23 Are you ready for Web3 in 2024?', https://www.linkedin.com/pulse/23-you-ready-web3-2024-click-here-find-out-anthony-day–ct9gf%3FtrackingId=tp0FMOaY5LdYhgWCXb5RUQ%253D%253D/?trackingId=tp0FMOaY5LdYhgWCXb5RUQ%3D%3D, accessed September 30, 2024.

[122] FCA: 'PS23/6: Financial promotion rules for cryptoassets', https://www.fca.org.uk/publications/policy-statements/ps23-6-financial-promotion-rules-cryptoassets, accessed October 10, 2024.

[123] Kang, K., Liu, X., Jiang, Y., *et al.*: 'Blockchain opportunities for construction industry in Hong Kong: a case study of RISC and site diary' *Constr. Innov.*, 2023, 23 (2), pp. 443–466.

[124] Wolf, D.E.: *A Cost–Benefit Analysis of Blockchain Versus Relational Databases for Digitalising and Automating Information Flows in Cross-Border Supply Chains*. Stellenbosch University, Stellenbosch, 2022.

[125] Morkunas, V.J., Paschen, J., and Boon, E.: 'How blockchain technologies impact your business model' *Bus. Horiz.*, 2019, 62 (3), pp. 295–306.

[126] Coppi, G.: 'Blockchain and microfinance: Hype or promise? | convergences', https://www.convergences.org/en/blockchain-and-microfinance-hype-or-promise/, accessed November 27, 2024.

[127] Cecere, L.: 'Tradelens discontinues operations. Why you should care', https://www.forbes.com/sites/loracecere/2022/12/05/tradelens-discontinues-operations-why-you-should-care/, accessed December 14, 2024.

[128] Saldivar, J., Martínez-Vicente, E., Rozas, D., Valiente, M.C., and Hassan, S.: 'Blockchain (Not) for everyone: Assessing the user experience of blockchain-based applications' *Extended Abstracts of the 2023 CHI Conference on Human Factors in Computing Systems*, 2023, pp. 1–8, 2022.

[129] Shojaei, R.S., Oti-Sarpong, K., and Burgess, G.: 'Leading UK construction companies' strategies to tackle BIM training and skills challenges' *Int. J. Constr. Educ. Res.*, 2023, 19 (4), pp. 383–404.

[130] Construction Leadership Council: 'Industry skills plan for the UK construction sector 2021–2025', 2021. https://www.constructionleadershipcouncil.co.uk/wp-content/uploads/2021/03/B06322_CLC_SkillsPlan_v27.pdf, accessed November 20, 2024.

[131] Sristy, A.: 'Blockchain in the food supply chain – What does the future look like?', https://tech.walmart.com/content/walmart-global-tech/en_us/blog/post/blockchain-in-the-food-supply-chain.html, accessed October 8, 2024.

[132] Sun, W., Antwi-Afari, M.F., Mehmood, I., Anwer, S., and Umer, W.: 'Critical success factors for implementing blockchain technology in construction' *Autom. Constr.*, 2023, 156, p. 105135.

[133] Ameyaw, E.E., Edwards, D.J., Kumar, B., Thurairajah, N., Owusu-Manu, D.-G., and Oppong, G.D.: 'Critical factors influencing adoption of

blockchain-enabled smart contracts in construction projects' *J. Constr. Eng. Manag.*, 2023, 149 (3), p. 04023003.
[134] Li, C., Zhang, Y., and Xu, Y.: 'Factors influencing the adoption of blockchain in the construction industry: A hybrid approach using PLS-SEM and fsQCA' *Buildings*, 2022, 12 (9), p. 1349.
[135] Taherdoost, H.: 'A critical review of blockchain acceptance models—Blockchain technology adoption frameworks and applications' *Computers*, 2022, 11 (2), p. 24.
[136] Bromley, P., and Powell, W.W.: 'From smoke and mirrors to walking the talk: Decoupling in the contemporary world' *Acad. Manag. Ann.*, 2012, 6 (1), pp. 483–530.
[137] Zomer, T., Neely, A., and Savaget, P.: 'How do projects decouple from coercive pressures? A study of decoupling in construction projects' *Int. J. Manag. Proj. Bus.*, 2024, 17 (2), pp. 247–277.
[138] Matthews, A.: 'A phased approach to national BIM adoption – EU BIM Task Group', https://eubim.eu/a-phased-approach-to-national-bim-adoption/, accessed November 30, 2024.

Chapter 4

Enhancing information security and digital trust in construction: decentralisation and encryption approaches

Moumita Das[*], *Jack C.P. Cheng*[*] *and Xingyu Tao*[*]

The safeguarding of data or data security represents a significant issue across various sectors in the contemporary digital landscape, and the Architecture, Engineering, Construction and Operations (AECO) industry is no exception [1]. Data security challenges in the AECO sector are amplified by the industry's increasing reliance on digital technologies and the growing volume of interconnected data across project lifecycles. For example, advanced technologies such as unmanned aerial vehicles (UAVs) utilised for the acquisition of high-resolution data, integrated systems like building management systems (BMS) employed for operational efficiency, and data-centric methodologies leveraging building information modelling (BIM) are facilitating the generation and exchange of substantial volumes of digital information [2]. Hence, the information disseminated in such projects contains various sensitive elements, including fiscal information, strategic project outlines, proprietary designs and intellectual property.

A recent survey has showed an alarming increase in cybersecurity threats, with over 59% of project stakeholders falling victim to cyberattacks in the years 2021 and 2022 [3]. The consequences of the lack of data security in AECO projects are numerous, including legal disputes followed by project delays, financial losses and compromises to the safety and privacy of the project stakeholders involved. Additionally, the unauthorised disclosure of intellectual property or sensitive designs can have severe consequences for a construction company's competitiveness, market position and long-term viability. Hence, the safeguarding of data confidentiality, which pertains to the protection of sensitive project information from unauthorised access, alteration and infringement, constitutes a critical dimension of data security within AECO projects [4]. In addition, studies have discovered that AECO project data is not only threatened by external attacks about breaches in data confidentiality but also other types of security issues such as 'denial of access' to project data by internal actors, as seen

[*]Department of Civil and Environmental Engineering, The Hong Kong University of Science and Technology, Hong Kong

in the Trant vs Mott McDonald case [3,5]. Therefore, adequate and well-tailored data security measures are necessary for ensuring the smooth and secure running of AECO projects.

This book chapter discusses the possible applications of decentralised technologies such as blockchain [6] and InterPlanetary File System (IPFS) [7] with standard security protocols of symmetric and asymmetric encryption for addressing the security requirements of AECO projects. Section 4.2 provides a comprehensive overview of the background and context necessary to understand the challenges and impact of data security in the AECO sector. Section 4.3 defines and discusses the three important sub-components of data security for collaborative data sharing among AECO project stakeholders. The section discusses the meaning of these sub-components in the context of AECO projects considering their unique data formats, project organisational structure and collaborative requirements. Section 4.4 presents a universal distributed framework to facilitate data security in AECO projects that include distributed technologies such as blockchain, and IPFS and security technologies such as symmetric and asymmetric encryption. The section also discusses the purpose and scope of each technology in further enhancing improving data security in AECO projects. Section 4.4 discusses an example demonstrating how an AECO project workflow may be facilitated more securely and effectively using the proposed universal distributed framework discussed in Section 4.3. Section 4.5 summarises and outlines future directions.

4.1 Context and challenges

The classical principles of cybersecurity, known as the CIA triad [8], that is, Confidentiality, Integrity and Availability are necessary but not sufficient conditions to meet the data security requirements of AECO project data using collaborative data sharing platforms, commonly termed common data environments (CDE). Due to their unique contract-bound, short-term project organisational structure, it is necessary to manifest trust among project stakeholders for sharing sensitive and time-bound project information for project collaboration, which is crucial for effective collaboration in AECO projects. Traditionally collaboration platforms (CDEs) in AECO projects have a centralised architecture where a trusted internal stakeholder or a neutral third party (such as a cloud service provider) is entrusted with hosting and managing project data. However, it is common experience in AECO projects that such a framework for data sharing often hinders the sharing of sensitive data and leads to the development of information silos, resulting in the sharing of incorrect and stale information [3].

Researchers in the field of construction management technology have explored the need for data security in AECO projects including security requirements [9] and legal consequences of security breaches [10]. Ongoing research has also identified security-related challenges such as liability and ownership in cloud-based platforms [11,12] in AECO projects. The importance and concern around information security are well recognised by the AECO sector, as reflected in current AECO information management products such as BIM 360 [13] and Aconex [14], which largely leverage cloud-based security solutions. Interestingly, the data security requirements of AECO projects are relatively unique due to their contract-based project organisation structure and collaborative requirements [15]. Typically, AECO projects involve several stakeholders, including sub-contractors, architects, engineers, facility managers and regulatory bodies, all of whom must share and access parts of sensitive project data. Consequently, the data security requirements of AECO projects are fundamentally and practically opposed to the existing security solutions, which are centralised in nature and face problems such as a single point of failure, the need to trust a local/central party and the lack of non-repudiation [16]. The existing information management platforms for AECO projects commonly use a centralised server (cloud-based or private) to host and share project information. However, this architecture may bring serious implications to project completion if the central server fails or the party in charge of the server chooses to block access considering self-interest [17]. Furthermore, the existing platforms overlook a significant concept of security, called non-repudiation, which is critical for decentralised project environments such as that of AECO projects. Non-repudiation means that a stakeholder cannot deny or refute their involvement in a transaction or the authenticity of a communication, data or action. Non-repudiation provides evidence that can be used to prove the integrity and origin of a message or transaction, thereby preventing the sender or receiver from denying their participation. This is necessary for ascertaining accountability in AECO projects [4,16,17].

Due to the project organisation structure of AECO projects, stakeholders are bound contractually only during a project, which often leads to a lack of complete 'Trust' among them [16]. Trust among project stakeholders can be improved through several approaches, including improving communication about data security, drawing up legal agreements, building a strong governance framework, educating members about mutual benefits and implementing data security-enabling technologies. In terms of data security with technology, trust or 'digital trust' [18] can be broken down into sub-components including the 'trust aspect of data confidentiality' – the confidence of users/data owners that sensitive data is protected from unauthorised access, manipulation, 'reliability' – the confidence that the system will perform as expected, 'transparency' – clear documentation and visibility into decision-making processes and 'accountability' – the confidence that the system/users can be held accountable for their actions. Guidelines for data security, particularly for the built environment, include The British Standard of Built Environment Security, ISO 19650-5:2020 [19], which provides overarching guidelines on data protection and includes data confidentiality and digital trust for AECO projects. In addition, the ISO standard outlines the concept of 'data utility', which refers to the need to maintain the usefulness of data for analysis and collaboration while implementing data security measures.

One of the crucial pieces of data in AECO projects are BIM models, which due to their visual and semantic data content pose challenges in being functional as well as secure. Das *et al.* [15] have explored the necessity of data functionality for securing BIM models for collaboration and have defined three levels of BIM security, including database-level, model-level and entity-level data security. Therefore, considering the nature of AECO projects in terms of data, project organisation structure and special requirements, this book chapter explores distributed technologies for data storage, verification, integration and collaboration with security for AECO projects. In principle, distributed technologies in collaboration with other security facilitating technologies are perfectly aligned with the needs of decentralised AECO projects as project stakeholders/data owners are geographically and organisationally distributed.

With the advent of decentralised secure technologies such as blockchain, AECO projects now have the choice to deploy a decentralised trust less system instead of having to trust a centralised party. The technology itself functions in a way that data is stored and processed collectively by multiple actors (project stakeholders), thereby obviating the need for centralisation which has its flaws and is structurally unsuited for AECO projects. Several researchers have explored the application of distributed technologies for facilitating trust for AECO projects including circular supply chains [20], integrated project delivery (IPD) [21], design management [4] and quality management [22].

Centobelli *et al.* [20] proposed a framework to integrate trust, traceability and transparency with circular supply chain processes such as recycling, redistribution and remanufacturing. A circular blockchain platform was designed involving various stakeholders like manufacturers, reverse logistics service providers, selection centres, recycling centres and landfills. The study highlights blockchain's role as a technological capability that improves control over waste movement and product return management activities in supply chains.

Elghaish *et al.* [21] proposed a framework for using blockchain to automate financial transactions in IPD projects, focusing on reimbursed costs, profit and cost savings. It facilitates trust by deploying blockchain technology, which is a decentralised ledger system. This ensures that no single participant has dominant control, and all data recorded is immutable, preventing unauthorised changes and enhancing trust among project participants. By coding financial transactions into smart contracts, the framework automates processes like reimbursed costs, profit and cost savings. This automation reduces the need for third-party involvement, thereby increasing transparency and trust. A proof of concept using IBM's Blockchain Cloud Beta 2 platform is developed to validate the framework, demonstrating its practicality and user-friendliness.

Tao *et al.* [4] developed a blockchain-based framework to enhance version management in BIM. The framework aims to provide a decentralised and immutable database model, addressing issues related to data manipulation in traditional version control systems. A key contribution of the research is the establishment of a distributed versioning environment. This environment leverages both blockchain technology and a CDE to ensure secure and efficient version control.

Elghaish *et al.* [22] discussed the integration of blockchain with IoT to enhance trust in quality management by ensuring data integrity and security. Blockchain's immutability and transparency features allow for reliable tracking and validation of quality-related data, which is crucial for maintaining trust among stakeholders in construction projects. By removing trusted intermediaries, blockchain can provide a decentralised system where all parties have access to the same verified information, reducing the potential for disputes and enhancing trust in the quality management process. The use of blockchain in quality management is further supported by its ability to automate data validation, ensuring that all quality checks and balances are consistently applied and recorded, thus fostering a trustworthy environment.

4.2 Data security requirements for AECO projects and the need for distributed security approaches

One of the objectives of this chapter is to explore the meaning of data security in the context of AECO projects and discuss how it can be enhanced in terms of data utility to facilitate collaboration and trust. Based on existing literature, Figure 4.1 illustrates the three important facets of secure collaboration in AECO projects, which are – (1) data confidentiality, (2) digital trust and (3) data utility. The following sections discuss the definition and challenges in facilitating these facets of data security measures in AECO projects.

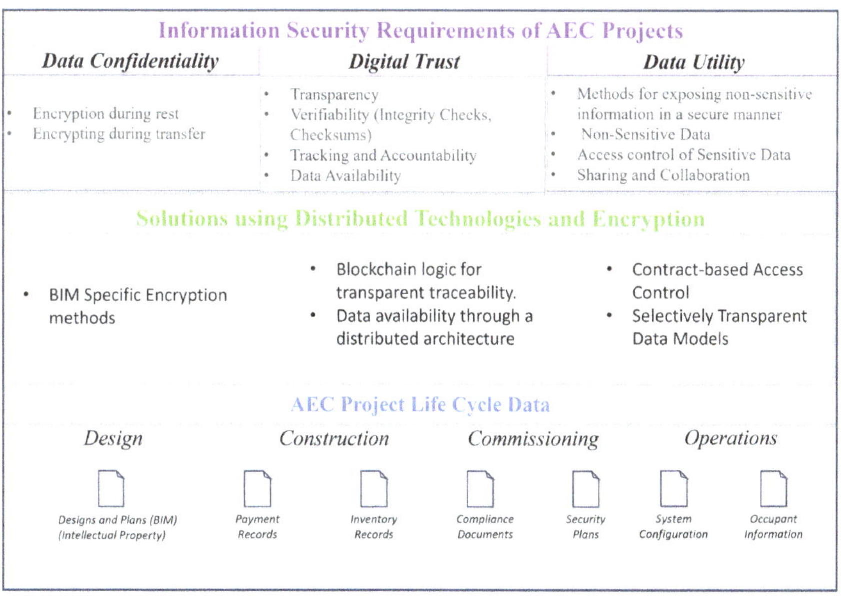

Figure 4.1 Information security requirements for AECO projects and suggested solutions

4.2.1 Data confidentiality

Data confidentiality in the context of AECO project data refers to the need to protect sensitive project information, such as BIM designs (intellectual property), plans, financial information and contracts, against unauthorised access and accidental disclosure. Data confidentiality may be categorised as data confidentiality/privacy for [15] – (1) data-at-rest and (2) data-in-transit. Confidentiality of data-at-rest refers to the need to protect project data in terms of data privacy when it is stored in storage devices. Several storage technologies, such as block-based such as Amazon EC2 [23], file-based such as Amazon S3 [24] and object-based storage systems are used to facilitate data storage in the current times. For data confidentiality for data-at-rest, encryption methods such as Advanced Encryption Standard (AES) [25] with an appropriately sized encryption key (e.g. 256 bits, as recommended by NIST [26]) may be deployed. From a data utility standpoint, an object-based storage system is most appropriate for AECO projects for a finer level of access control. Data confidentiality for data-in-transit refers to the need to protect data privacy during data transfer. Security protocols such as Transport Layer Security (TLS) [27] that facilitate privacy through a type of encryption may be deployed for secure data transfer for data transfer.

However, the challenge to data confidentiality for AECO project information is not the unavailability of robust encryption techniques. The challenge in facilitating data confidentiality arise mainly due to collaboration needs considering the type of data and contract-based (subcontracting) project organisational structure. AECO project data contains several types of data including ones for visual, contextual analytical applications. Das *et al.* [15] have conceptually defined three levels of BIM security that explore the concept of data confidentiality at different data granularities of BIM models. For example, a building model may be several entities and systems with different data ownerships. An architect may be privy to the design of a building element, for example, a façade wall, however, a contractor may own task, scheduling and material procurement information. However, if traditional file-based encryption methods are directly applied to the architectural BIM model in this case, it would render it unusable for other purposes. The challenge mainly lies in developing methods to segment segmentation-sensitive information out of BIM models while maintaining data usefulness for collaborative activities such as clashes. The second challenge is that of facilitating access control and non-repudiation. The problem of non-repudiation is directly linked to digital trust which is explained in Section 4.3.2.

4.2.2 Digital trust

Digital trust may be defined as the confidence of individuals, organisations and systems that project information is being stored and disseminated securely. The confidence that builds trust is a measure of the reliability and authenticity of digital information [28]. In other words, the confidence that builds trust depends on the methods/mechanism in place for the verification of the quality (correctness and completeness) of data or processes and the high likelihood of availability of data, when requested. In technical terms, digital trust may be broken down into data integrity, data availability, data traceability and transparency. The following sections discuss how these components of digital trust may be facilitated and improved using distributed technologies.

4.2.2.1 Data integrity

Data integrity in AECO projects refers to the correctness, completeness and freshness of project information in an environment consisting of multiple data owners. The authorised stakeholders must have access to consistent information throughout the project, considering information is shared, processed and updated in AECO projects regularly. Since AECO projects have multiple data owners who may not completely trust centralised ownership of data, a method to share data on request and a method to verify the integrity of data on its own and about other data units is important. In existing centralised CDEs in AECO projects, a method for verifying the integrity of data from data owners (such as architects, structural engineers and MEP engineers) is missing. For example, an architect may upload his/her BIM model to a CDE for coordination with structural and MEP BIM models. However, an internal malicious actor or a hacker may corrupt or modify a part of the architectural BIM model. In this case, there is no way that the MEP engineer can validate the correctness of the architectural model without strictly relying on the data storage service provider. Therefore, distributed technologies such as blockchain along with content-based data correctness proofs may be deployed to facilitate a distributed record, which is owned collectively by multiple project stakeholders and is immutable (cannot be modified / very difficult to be modified) to facilitate verification of data integrity and non-repudiation.

4.2.2.2 Data availability

Data availability refers to the accessibility and readiness of data when it is needed by authorised users or systems. In AECO projects, single points of failure can have a significant impact on project delivery causing delays, disruptions and even project failure. Hence it is very important to identify and manage these single points of failure including technical solutions and risk management strategies to ensure 'data availability'. Distributed storage technologies such as IPFS may be used to address single points of failure in collaboration with blockchain technologies to avoid data silos which are explained in Section 4.3.1.

4.2.2.3 Data traceability

Data traceability refers to the ability to track and document the origin, movement and changes made to data generated by processes and workflows, throughout its lifecycle. Traceability allows organisations to identify who had accessed, modified or transmitted data, enabling accountability for actions and decisions. AECO projects have project-specific or standard processes and workflows to facilitate project activities from design and planning to construction, commissioning and operation phases. Traditionally, traceability is facilitated by recording the lifecycle of processes and workflows in a centralised database. However, as discussed in Section 4.1, information sharing and management in an AECO project environment may not be trusted with a centralised party about the reasons for a contract-based decentralised organisation structure. In the context of AECO projects, traceability should include some additional security requirements including [29] – (i) irreversibility of business rules, that is, protection of rules/business logic driving project workflows from malicious tampering, (ii) verifiability of endorsements, that is, methods to record and verify endorsements/approvals received from authorised stakeholders and (iii) irreversibility of audit trail, that is, maintaining records of events of workflow such as endorsements, comments, rejections, in a way that cannot be tampered with by either insiders or hackers. Blockchain smart contracts, distributed ledgers and encryption protocols are suitable candidates to create automated business rules, verifiable endorsements and irreversibly record audit trails using blockchain's underlying security.

4.2.3 Data utility

Data utility in the context of data security refers to the usability of data for its intended purposes while also ensuring its protection [30]. In other terms, it is about striking a balance or trade-off between data accessibility and data confidentiality and trust (as described in Sections 4.2.1 and 4.2.2). One way to achieve this trade-off is through 'Selective Transparency' which refers to facilitating methods to disclose auditable non-sensitive information and concealing and access controlling sensitive information [16]. For example, interim payments in the AECO project are based on contractual agreements between two parties – a promiser and a promise of payment. The payments should be duly recorded, and the records should be made available to the overseeing parties (for transparency and trust building). However, keeping privacy considerations in mind, it is not necessary to disclose all the payment-related information for payment monitoring, and rather a verifiable record that substantiates the fact that the payment was made according to the agreement between the parties is sufficient. Das *et al.* [30] have explored the concept of selective transparency for secure interim payments in AECO projects using an encrypted blockchain ledger. In general, the concept of selective transparency may be applied not only to progress payment but to any domain by – (i) identifying and categorising sensitive information, (ii) applying methods to securely expose non-sensitive data and (iii) facilitating authorised access to sensitive information using appropriate access control mechanisms. For example, in interim payments, information such as bank details, paid amount and net valuation may be deemed sensitive whereas payment status, payment proof and payment disbursement date may be categorised as non-sensitive information. In this scenario, the non-sensitive information is required for payment monitoring which can be put on a commonly agreed platform such as a blockchain ledger for transparency [31], while the sensitive data may be encrypted.

Data utility also deals with access control to make authorised data accessible to authorised users. Let's consider the same example of interim payments to elaborate on this. Due to the sub-contracting nature of AECO projects, interim payments are facilitating stakeholders at various levels with the respective contractually bound parties. There are two parties in an interim payment agreement – a promiser, who has made the promise to pay, and a promise, who is entitled to payment when certain terms and conditions are met. To facilitate suitable data utility, access control for the sensitive information in this case should allow access not just to one party but a to promisor-promise duo. For this domain-specific problem, a smart contract-based access control that removes the burden of the encryption key management or key management overhead from a centralised party and is irreversible is suitable for construction projects [16]. In summary, data utility requirements for securing exposing non-sensitive information, and concealing sensitive information are domain or problem-specific and need to be investigated on a case-by-case basis. AECO project data is challenging in this regard because it not only handles documents and records but also built environment data in the form of BIM which contain visual, analytical and life cycle data.

Therefore, the three facets of data security, that is, data confidentiality, digital trust and data utility very important for collaboration in AECO project data. These three facets are interlinked with each other and can use distributed technologies such as blockchain to form a robust framework. Section 4.4 presents a conceptual framework using distributed technology such as blockchain and IPFS and security technologies such as symmetric and asymmetric encryption.

4.3 A universal distributed framework for secure collaboration in AECO projects

In Figure 4.2, a universal framework for addressing the data confidentiality, digital trust and data utility challenges (discussed in Section 4.3) for securing collaborative activities in AECO projects is presented. As shown in Figure 4.2, the system architecture of the universal framework for secure collaboration is designed to have

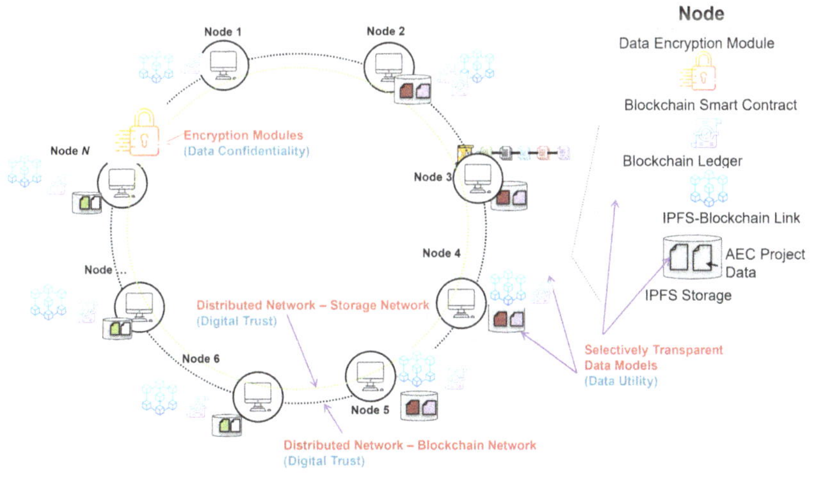

Figure 4.2 A distributed framework using distributed technologies and encryption for AECO data management and security

(as illustrated in Figure 4.2) – (1) A 'distributed network' to facilitate digital trust, (2) 'encryption modules' for data confidentiality and (3) 'selectively transparent data models' for data utility. The framework is first and foremost designed considering the decentralised organisational structure of construction projects where distributed data is required to be shared and approved by multiple parties, who may or may be data or process owners. For example, government bodies may require access to some project data for regulatory oversight in construction projects. Therefore, as shown in Figure 4.2, a distributed framework comprising interconnected nodes (computers), representing different data owners that execute project processes collectively [32] is proposed. To suit the requirements of data owners and regulatory authorities, the nodes of the proposed distributed framework perform a full or partial tasks that contributes to the final output of the framework. The framework consists of two types of nodes – full nodes and light nodes/peers [33]. The full nodes store the replica of the whole data set and participate in performing all the tasks whereas the light nodes store a part of the verified data set and may or may not participate in the processes as configured.

As shown in Figure 4.2,. two main requirements of the proposed distributed framework are – (1) data storage (according to data ownership and synchronisation) and (2) data lifecycle transparency/traceability (ensuring that the lifecycle records cannot be modified by one or more parties). As shown in Figure 4.2, a blockchain (Hyperledger Fabric [34]) and IPFS [35] are selected to facilitate the two requirements. The reason for selecting and designing the distributed network with blockchain and IPFS for construction projects is discussed in Section 4.4.1. In addition, the requirements of data utility and data confidentiality, that is, to facilitate the sharing of sensitive information in a secure yet functional manner, the design criteria for the encryption module and the selectively transparent data model, as shown in Figure 4.2, are discussed in the following sections.

4.3.1 The distributed network

4.3.1.1 The blockchain network

Blockchain is a technology that facilitates the computation of data in a distributed manner by obviating the need to trust third parties using distributed ledgers and smart contracts. Researchers have strongly advocated for the potential of blockchain in driving digital transformation, accelerating collaboration and maximising trust in the construction industry [17]. Due to the decentralised nature of blockchain, it is highly secure and resistant to tampering/manipulation because once any data (called 'blocks' in the blockchain terminology) is added to the blockchain, more appropriately called the 'distributed ledger' or 'the chain', it cannot be modified without altering all the other data blocks on the chain (as shown in Figure 4.3) which itself is duplicated across several distributed nodes. Illustrative discussions on the level of difficulty in manipulating blockchain through attacks such as selfish mining and the majority or 51% attack in the context of AECO projects can be found in existing literature [16]. The security of the blockchain network comes from the underlying distributed architecture of blockchain that necessitates permission from all the

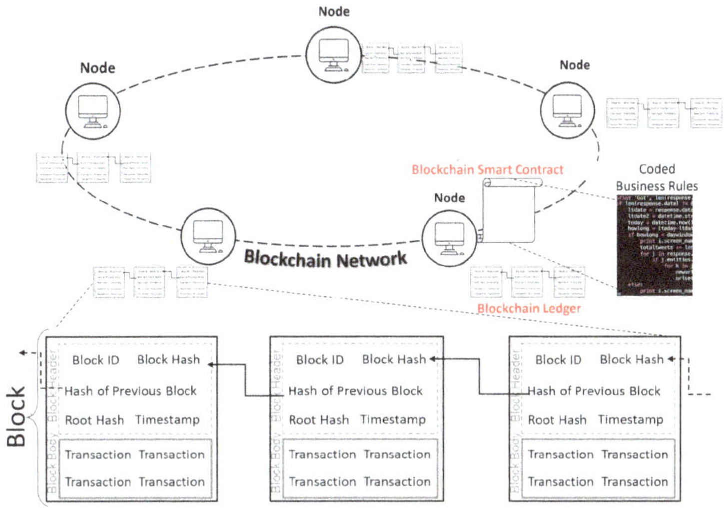

Figure 4.3 The generic distributed structure of a blockchain network

computers (called 'nodes') in the network to perform any action on the blockchain ledger, thereby making it impossible to manipulate or falsify data on the chain. Alternatives to blockchain technology includes federated systems and traditional distributed databases, that offer distinct advantages but also have notable limitations compared to blockchain. For instance, federated systems, while enhancing user privacy and control, often depend on a set of trusted nodes, which can introduce centralisation risks and reduce resilience against attacks. Traditional distributed databases, although efficient for real-time data access and management, do not inherently provide the immutability and transparency that blockchain offers, making them less suitable for applications requiring trustless environments.

Therefore, blockchain is an excellent technology for irreversible data traceability which is an important security requirement in AECO projects for building digital trust. In addition, by using the programmable feature of blockchain, better known as 'smart contracts', AECO data workflows including supply chain processes, progress payments, version control and quality control may leverage the irreversibility feature of blockchain for lifecycle recording, improve digital trust and hence reduce the risk of prolonged disputes. In research, blockchains have been explored in several areas including but not limited to supply chain traceability in off-site construction [36], digital twin technology [37] to trace records in construction projects, housing IoT systems [38], securing project interim payments [16], construction quality control [39] and tracking design changes [21]. The components of the blockchain network and their applications in AECO data management and security are discussed as follows.

Blockchain smart contracts

Smart contracts are self-executing contracts that are programmed to automatically enforce pre-defined terms and conditions before writing data to the blockchain ledger permanently. Smart contracts may be applied to many areas of AECO projects such as supply chain management, procurement management and payments. In general, smart contracts may be deployed to facilitate project workflows in compliance with relevant regulations and guidelines in a secure manner. Due to the underlying distributed blockchain infrastructure, business rules written using smart contracts can provide the security of irreversibility thereby facilitating digital trust, as discussed in Section 4.3. Smart contracts can help facilitate aspects of project workflows including the verification of data and users, enabling workflow progression (orchestration of multi-step dynamic/customised workflows), and initiating the process of recording events and users related to data (such as creation, endorsement and feedback) to the secure blockchain ledger. Later in this chapter, in Section 4.5, the application of smart contracts in facilitating process/workflow security is explained with an example.

Blockchain ledger

The blockchain ledger is a series of linked blocks of data where each block contains a record of transactions that have been verified (by smart contract rules) and added to the ledger. These blocks are linked together cryptographically and stored on multiple nodes on the blockchain network, forming a secure and tamper-proof record of all transactions on the network. The applications of the distributed ledger in managing the AECO project data range from improving data sharing and collaboration, to supply chain management and asset management. Through its decentralised system architecture (where there is no one centralised party), the blockchain ledger can be deployed to be a shared and unmodifiable ledger to record processes related to design management [40], construction progress payments [41] and facility management [42].

Distributed blockchain ledgers facilitate process security and help improve transparency and traceability, reduce fraud and ensure that products follow quality compliance standards. Due to the transparency and irreversibility of blockchain ledgers, they can help bring accountability to project stakeholders for their actions. This chapter in Section 4.5 will later discuss a scenario demonstrating the useful characteristics of transparency and irreversibility in bringing security to AECO project data.

4.3.1.2 The storage network

As discussed in Section 4.3, 'data availability' is a critical factor for facilitating digital trust in AECO projects. Therefore, several synced copies of project data must be maintained at multiple locations ensuring that all stakeholders have access and the ability to verify the correctness of the data in terms of freshness and completeness. Although blockchain is an excellent candidate for maintaining synced data across multiple locations, due to its inherent architecture, it has latency problems in handling large data sets such as BIM models and large documents in AECO projects. Similarly, the traditional protocols of data storage are HTTP/HTTPS and BitTorrent, which suffer from the limitations of high centralisation and optimisation for large files rendering them inefficient for smaller updates, as required in AECO projects. This problem can be addressed using a distributed storage system that can be linked with the blockchain ledger for maintaining irreversible information (such as version, created by and creation date) on large data sets along with a pointer to the actual data set. IPFS is a protocol that builds a content-addressable and peer-to-peer file storage system [43,44] that facilitates large-sized file storage in a distributed and verifiable manner [45]. Figure 4.4 shows the functionalities and characteristic features of IPFS. As shown in Figure 4.4(a), any AECO project data, whether it be a document or 3D model, when uploaded to the IPFS creates an 'IPFS content-based identifier' using hashing algorithms (discussed in Section 4.4.2). The 'IPFS content-based identifier' is the unique identifier of a data set that can be used to identify, download and verify the integrity (completeness) of that dataset (Figure 4.4(b)). In collaboration with blockchain, the 'IPFS content-based identifier', which is a text of a fixed size can be integrated with other metadata information such as its version, creator information and creation data information.

The distributed framework (Figure 4.2) integrated blockchain with IPFS, thereby creating a unique secure storage system suited to AECO projects, that stores all or some sensitive information locally with data owners however maintains metadata and methods of data verification using a unified blockchain framework. Sensitive data may be available to non-data owners on request who should be able to verify the authenticity of the content using information from the integrated blockchain ledger. This feature may be deployed using the concept of 'clusters' in IPFS as shown in Figure 4.5. In this configuration, peer nodes belonging to one organisation or working group are configured as private IPFS clusters of those groups/organisations. Each cluster has its unique private key that will allow access to the data in the cluster [44]. It is recommended that the data replication factor be kept at the maximum, which means that any document added to a cluster will be evenly distributed and replicated across the whole cluster. Configuring clusters, as

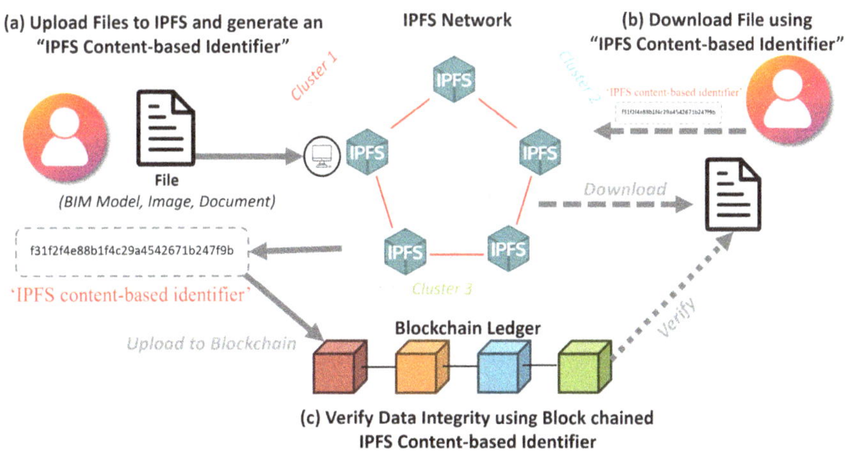

Figure 4.4 Functions and characteristic features of the interplanetary file system

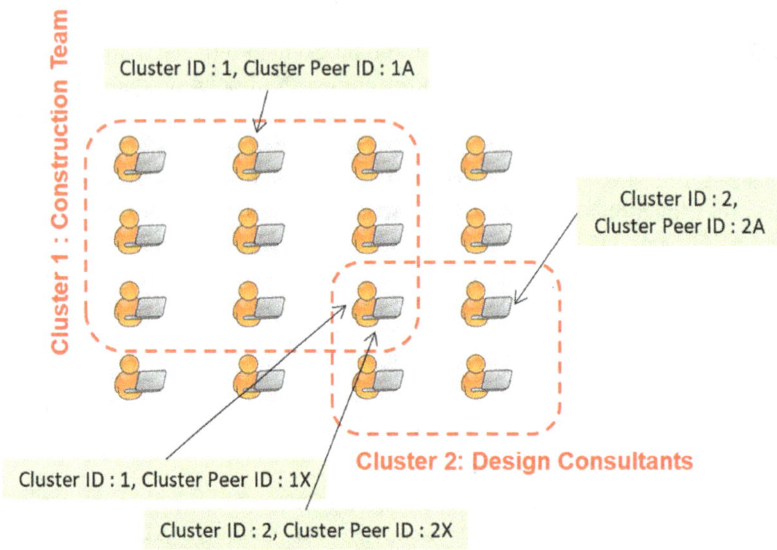

Figure 4.5 Configuration of IPFS clusters for distributed data management in AECO projects

shown in Figure 4.5 is trivial from the research perspective and can be deployed using IPFS's official documentation [44]. The IPFS network shown in Figure 4.5 is a private network and therefore will incur storage costs to its users. The alternative public IPFS network although free of costs is not suitable for construction projects because documents are public and are discarded by the network if not being downloaded by enough users to facilitate optimal use of space.

4.3.2 Encryption modules

Encryption is defined as

> the cryptographic transformation of data (plain text) into a form that conceals the original meaning of data (ciphertext) to prevent it from being known or used. If the transformation is reversible, the corresponding process is called 'decryption', which is a transformation to restore encrypted data to its original state [1].

Encryption protocols can facilitate different types of security for AECO data including data confidentiality, data integrity and user authenticity to prevent unauthorised data access and verify the authenticity of data. The three fundamental types of encryption algorithms are (1) symmetric or secret-key cryptography, (2) asymmetric or public-key cryptography and (3) one-way hash or compression functions as shown in Figure 4.6.

Symmetric cryptography is a cryptographic primitive that uses one cryptographic key (***SK***) for encrypting and decrypting data [46] (Figure 4.6). Examples of

Figure 4.6 Traditional symmetric, asymmetric and one-way hashing encryption primitives for data and user security

symmetric encryption protocols are AES [25] and Blowfish Algorithm [47]. Symmetric encryption has several applications in AECO projects ranging from encrypting construction documents to protecting intellectual property. However, a unique characteristic of AECO projects is that the project stakeholders are bound in pairwise contractual relationships and are required to share sensitive financial and personal privy to those two parties as per the terms of the contract. In such a situation, deploying symmetric encryption protocols in their original form is faced with 'key logistics security' problems, that is, the problem of securely transferring encryption keys to two or more parties who are privy to the shared data [46]. Hence, channels for private data sharing are required between certain contractually bound project stakeholders. For such special cases, the Diffie–Hellman key exchange protocol [48] may be deployed for generating a shared cryptographic key using non-sensitive ('public' key of asymmetric encryption) information transferred between two parties and sensitive information ('private' key of asymmetric encryption) generated locally. To provide more security, such shared encryption methods may be used to store sensitive information on a publicly accessible blockchain ledger to facilitate transparency yet facilitating confidentiality when dealing with sensitive information in AECO projects.

Asymmetric cryptography, however, is a cryptographic primitive that uses a pair of inversely related cryptographic keys, called a 'public key' and a 'private key' ($\boldsymbol{PK^i}$–$\boldsymbol{SK^i}$), unlike symmetric encryption. In this case, the private key, SK^i corresponds to the private identity of a project stakeholder, 'i', which should be protected by the respective participant, and can be used for creating verifiable digital endorsements of the stakeholder. However, the public key, 'PK^i' corresponds to the public identity of a project stakeholder, 'i', is sharable publicly among all project participants, and can be used for verifying the validity of the stakeholder's digital endorsement. A popular example of a robust asymmetric encryption protocol is the RSA (Rivest, Shamir and Adleman) algorithm [49] using which digital endorsements such as digital signatures can be created and verified. As shown in Figure 4.6, a message is encrypted using a project stakeholder's private key, ($\boldsymbol{SK^i}$) to create his/her digital signature, which can then be verified by any other project stakeholder by decrypting the digital signature with the signing stakeholder's publicly shared key ($\boldsymbol{PK^i}$). The credibility of digital signatures depends on the strength of the asymmetric encryption algorithm and cryptographic keys used. However, digital signatures are often susceptible to 'man-in-the-middle' attacks [50], where an attacker replaces the public keys of authentic users with his public key, poses as an authentic user and establishes secure communication, thereby infiltrating a system and learning its secrets. Therefore, asymmetric encryption may be used with blockchain technology which could act as a secure third party using its immutable distributed ledger technology.

One-way hash functions such as SHA-256 are cryptographic protocols that convert plain texts of variable lengths into unique strings of fixed lengths and can be used to facilitate data integrity in information sharing in AECO projects. For example, a one-way hash function can convert text, documents of various formats, 3D models or any kind of files into a string of a fixed length such as 'ba7816bf8f01 cfea414140de5dae2223b00361a396177a9cb410ff61f20015ad', unique to the content of the file. Due to their inherent nature, AECO projects generate multiple versions of data (such as documents and BIM models) from multiple data owners which are then shared among multiple data users. Therefore, the guarantee of completeness and freshness of information, that is, data integrity is very important. One-way hash functions in collaboration with blockchain ledgers can be very effectively used to make sure that end users always have access to correct and complete data in AECO projects.

Hence, encryption protocols can be used independently or in combination with other technologies such as blockchain and distributed storage technologies such as IPFS technologies to enable data privacy, data integrity, data authenticity and user authenticity in collaborative platforms or CDEs in AECO projects.

4.3.3 Selectively transparent data models

This section discusses how selective transparency may be incorporated into confidentiality and trust-inducing approaches to ensure that the functionality of AECO project data is not completely lost in adopting encryption and blockchain. The concept of selective transparency is best explained by the concept of partial encryption which has been well researched in the field of image encryption. Notable work includes the introduction of partial encryption for image compression [51] and format-preserving encryption of images [52]. Research close to partial encryption of AECO project data such as BIM is explored by Cai and Li [53] who propose to divide CAD models into collaborative features for co-design and confidential features (containing intellectual property) for encryption. Following the concept of partial encryption in existing literature, this chapter explores the concept of selective transparency for AECO project data. Facilitating selective transparency in AECO project data is relatively challenging because it includes BIM models that contain visual, semantic and lifecycle data. Therefore, the existing methods of file-based symmetric encryption of BIM models that are supported by existing CDEs are not suited to the collaborative needs of AECO projects. AECO projects generate data from multiple data owners who require data confidentiality in their part of the BIM model and yet have sufficient non-sensitive information to perform collaborative activities such as clash detection and revisions. Therefore, methods to facilitate selective transparency for BIM models should be further investigated.

Figure 4.7 shows the proposed conceptual method for facilitating selective transparency in BIM models. As shown in the figure, a BIM model contains both geometric and semantic information. Geometric information in BIM models refers to the graphical representation of the physical elements and spatial relationships within a building or infrastructure project. It includes the 3D geometric shapes, sizes and positions of walls, floors, ceilings, columns, doors, windows and other architectural and structural components. Geometric information enables the visualisation, coordination and clash detection of various building elements in a virtual environment. Semantic information in BIM models provides additional non-graphical data and attributes associated with the building elements. It goes beyond the geometric representation and adds meaning and context to the model. Semantic information includes data such as material properties, performance characteristics, installation details, cost data, maintenance schedules, environmental specifications and other relevant information related to the building components.

For efficient collaboration, both semantic and geometric information is necessary. Figure 4.7 introduces to concept of component-level BIM encryption. In the existing literature, this concept was introduced as 'Level 3 BIM Security' [15] which refers to the incorporation of security facilitating methods at an object level for data ownership and sharing. In other words, in a BIM model with a Security Level 3, data contributors will be able to have ownership of their respective components and sharing rights. A database-based approach for encrypting parts of information stored in BIM models was proposed [15]. For example, through a database-based approach, a cost estimator may be given access to information such as material names, unit cost and quantity while protecting the geometric information owned by the architectural firm. This kind of approach may be faced with three main issues. The first issue is that database-based hosting of BIM models is not common practice in AECO projects anymore. The second issue is that this approach does not take data 'interoperability' into account which is crucial for data

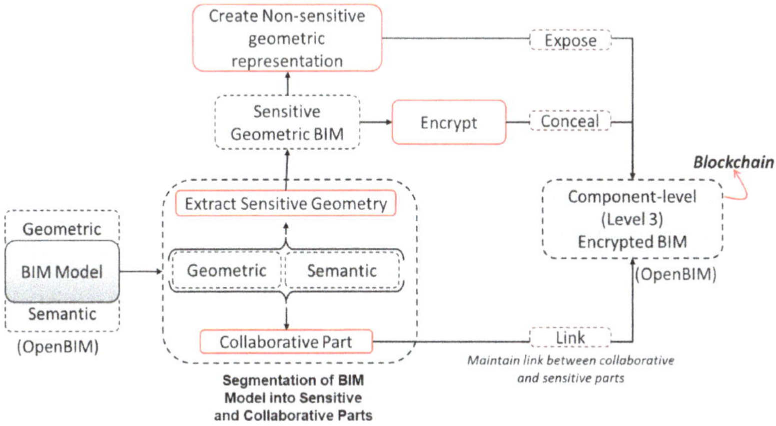

Figure 4.7 Proposed conceptual method for creating a selectively transparent BIM for data utility for collaboration

utility. The third issue is that it has limitations in facilitating selective transparency of geometric information in BIM models. AECO projects for collaborative design are required to share geometric information for better and faster design. For example, an architecture domain may have intellectual property design for some of the BIM components that they would not like to share with designers from other domains such as structural engineering. However, they might still want to share a part of their model to carry out activities such as clash detection for collaborative design. Considering the requirements of data interoperability and sharing of sensitive geometric information, Das *et al.* [54] proposed to encrypt BIM models on a component level using a method to partially expose the non-sensitive part of building geometry for functional purposes. Figure 4.7 shows a simplified method that contains two main modules, namely, 'extract sensitive geometry' and 'create non-sensitive geometric representation'. The 'extract sensitive geometry' module uses a tree-like structure or JSON-based on open-source data models of OpenBIM models such as Industry Foundation Class. IFC models follow a tree-like hierarchical data structure that is both human and machine-readable. To describe simply, IFC BIM models follow a hierarchical data structure with a 'building' (IfcBuilding) entity at the top, followed by 'storey' (IfcStorey) and 'building elements' such as wall (IfcWall), windows (IfcWindow), HVAC elements (IfcChiller) as subsequent child elements. The IFC structure also allows the grouping of elements into groups such as 'zones' and 'systems'. The 'extract sensitive geometry' uses the tree-like structure of IFC to extract sensitive eligible for encryption. Due to the collaborative nature of AECO projects, sensitive geometry may be essential for some collaborative activities and may not be completely hidden.

The 'creating non-sensitive geometric representation' module as shown in Figure 4.7 creates a simplified version of the sensitive geometry and links it back with the collaborative part of the BIM model using the tree-like data structure of IFC. For example, an IfcWall may be encrypted by first parsing the IFC tree and severing its link with its respective IfcStorey. The processes may be followed by simplification of wall geometry according to the needs of collaborative sharing and merged with the collaborative part for data sharing. For example, for collaborative activity such as clash detection, only the outermost covering may be retained as proposed by Das *et al.* [54]. However, the approach may be further refined by identifying different levels of geometric encryption for different types of collaborative activities.

4.4 Case scenario

The universal framework introduced in Section 4.4 that deploys distributed and traditional security-related technologies has several applications. For example, the framework may be deployed to facilitate AECO project workflows such as design review, tendering, change order management and progress payments. A workflow in AECO projects refers to the sequence of interconnected tasks and activities that are required to be performed to complete the different phases of an AECO project ranging from design to operation and maintenance. Project workflows are crucial for the timely and successful completion of AECO projects and hence their monitoring and management are very important. For the successful completion of workflows, it is essential to clearly define the parties responsible for each task. Alongside, a secure transparent monitoring framework to aid the management of workflows and hold individuals/parties accountable in case of a conflict is also crucial. In addition to this, the workflow logic should be clearly defined and recorded and should be ensured that no project stakeholder can alter it for self-interest.

In this section, an application scenario demonstrates how irreversible business rules and verifiable endorsements (as introduced in Section 4.2) are deployed using distributed blockchain smart contracts, ledgers and encryption protocols. As an example, let us consider the case of a design review workflow. The workflows in construction projects can be broken down into subcomponents such as sub-processes, data and roles and then organised in a logical flow based on logic as agreed in the project. Based on existing research, Figure 4.8(a) shows the example summarised components of workflows in construction projects. For example, a design review workflow may require sub-processes/events such as document creation, endorsement and feedback by various roles such as the architect, review engineer and the owner on different design documents.

These typical sub-components form the body of a workflow when arranged in a logical order with suitable inputs as shown in Figure 4.8(a). Figure 4.8(b) shows

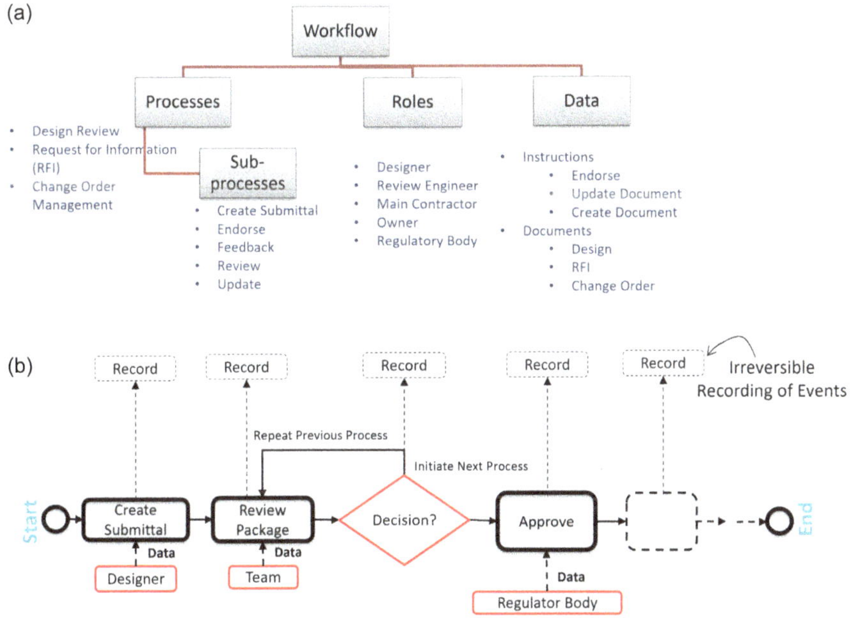

Figure 4.8 (a) Components of a workflow and (b) logical orchestration of a design review workflow

the example of a typical design review workflow which requires the orchestration of processes such as submittal creation, review and approval by different project stakeholder such as designer, main contractor and external regulatory bodies [13]. In addition to that, each sub-process must retain details such as its status of execution, time of completion and responsible parties in an irreversible record trail to enforce accountability at a later stage of the project, when and if need be. The existing centralised document management systems for AECO projects use centralised servers (hosted by private entities or cloud service providers) for facilitating workflow logic and storing audit trail records. However, leveraging the functionalities of smart contracts and distributed ledgers of blockchain technology, AECO workflows can be completely decentralised, thereby removing the need to trust a centralised entity.

To maintain an irreversible record, the workflow orchestration logic may be converted into a smart contract logic. The readers may refer to the publication titled, 'A blockchain-based integrated document management framework for construction applications' [29] by Das *et al.* published in the year 2023 for detailed examples and codes of workflow smart contract logic.

Figure 4.9 shows a detailed Submittal management workflow that consists of steps such as creation of submittal, review with official response and endorsement. The roles, that is, the responsible party associated with each sub-process are also

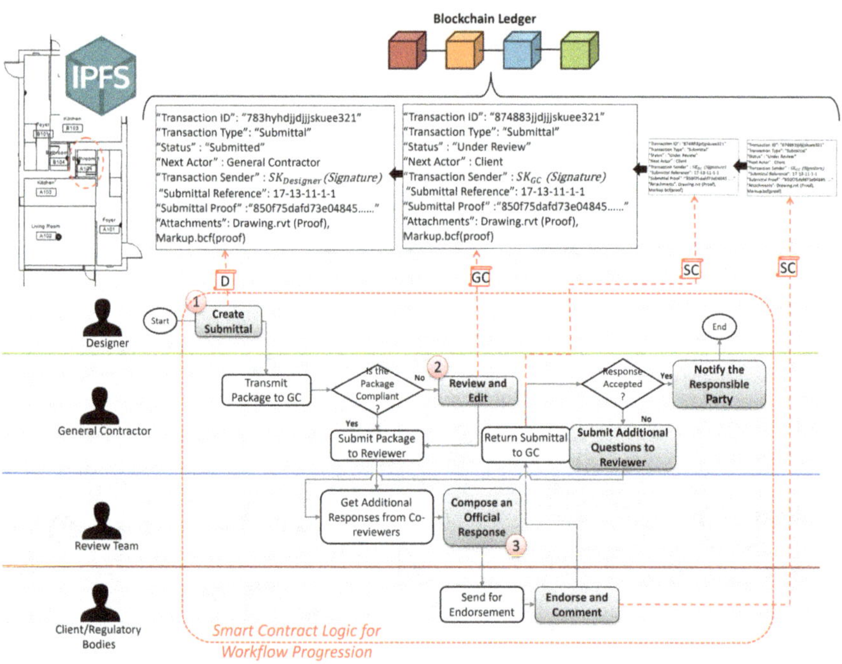

Figure 4.9 Example scenario illustrating the process of securing AECO workflows

shown in Figure 4.9. For example, Similarly, a designer or sub-contractor is responsible for creating a new submittal package, consisting of contents such as drawings, BIM models, certificates and cover letter. The other roles such as general contractor, client and regulatory bodies, and client are responsible for editing the submittal, composing an official response and providing endorsement as shown in Figure 4.9. The example also demonstrates how different digital trusts such as, the data integrity of the latest version and irreversibility of version history, user authentication, that is, certainty that only authorised users are allowed to perform actions pre-defined according to the workflow logic, irreversibility of the pre-defined workflow logic and information lifecycle transparency is facilitated by the proposed framework. As shown in Figure 4.9, a submittal manager deploys the workflow logic in which the sequence of the actions and the roles who should perform them are predefined. The workflow participants such as designer, general contractor, client and regulatory bodies who are authenticated using their credentials stored on the blockchain via their respective smart contracts. For example, the designer's smart contract and the general contractor's smart contract (depicted by 'D' and 'GC' in Figure 4.9) contain the designer's and the general contractor's cryptographic credentials (asymmetric encryption), respectively. Since the smart contracts containing the credentials are deployed on the blockchain, the credentials are irreversible, hence improving the level of security for user authentication. Figure 4.9 also shows the blockchain ledger data model which is used to record the document lifecycle through all the steps of an information approval workflow (facilitating traceability). For example, a designer may initiate the workflow by recording information such as submittal reference number, status and proofs of attachments, the next actor and due date. When the next actor that is the general contractor is notified via email, he/she may verify use the transparent blockchain ledger data to perform actions such as verifying the integrity of attachments and checking version history, thereby facilitating data integrity and process security. Therefore, using the proposed framework as demonstrated in Figure 4.9, using a smart contract, blockchain ledger and methods of correspondence such as email, an integrated and secure approach may be taken to facilitate information approval workflow and document management.

4.5 Conclusions and future research

This chapter explores data security for securely sharing project data in AECO projects. While BIM offers significant benefits in AECO projects for data representation, analysis and lifecycle recording, it also introduces security risks due to the sensitive and valuable nature of project data. This book chapter explores the unique security requirements of AECO project data including BIM, using existing research literature considering the contract-bound project-based organisational structure and collaborative needs. The traditional security requirements such as data CIA were analysed from the perspective of AECO projects and as a solution, distributed technologies such as blockchain and IPFS were recommended in collaboration with traditional security enabling technologies such as symmetric and asymmetric encryption. The universal framework deploys distributed technology to facilitate data confidentiality, digital trust and data utility. Considering the different data levels of data used in AECO projects, it implements the concept of full and light nodes. This model facilitates, integral project participants such as members of the architectural firm to act as full nodes and host the complete data set including the history of change records, BIM models and documents (as per data ownership). External bodies such as government departments may be added to the proposed universal framework as light nodes temporarily and be eligible to host a part of the data set on request to be able to support processes such as e-submissions and e-tendering. Unlike full nodes, the light nodes may only have access to update specific records (such as recording approval or rejection) and to verify the integrity of documents (requested/received from the document owners) using the document integrity proofs from the blockchain ledger made available to them. The framework also introduces the concept of selective transparency for AECO project data including records and BIM models. In general, the security requirements and the framework presented in this book chapter provide a guideline on how existing cybersecurity technologies may be best deployed to address the security requirements considering the unique nature of AECO projects.

The proposed framework assumes that the cost of its deployment will be shared collectively by all project participants. However, in AECO projects, this arrangement may not be feasible for smaller subcontractors. For these subcontractors, participating in the network as data owners – especially bearing the associated costs could be problematic. Many smaller subcontractors may lack the financial capacity or infrastructure to invest in and maintain the technological solutions required to be part of such a framework. Moreover, they may not have the technical expertise to effectively manage and protect the data, or the bargaining power to negotiate terms that would allow them to contribute as data owners within the network. As a result, expecting smaller subcontractors to bear a portion of the cost for the deployment of such a framework may create significant barriers to their participation in digital processes, limiting the inclusivity and potential benefits of the framework for all project stakeholders. Therefore, incentive-based approaches may be explored, where project participants are rewarded for providing unused storage space to the project. The means for fostering a more collaborative and sustainable participation model for all project stakeholders should be further investigated in future.

In addition, the segmentation of sensitive information to ensure both confidentiality and utility, as recommended in this chapter, presents significant challenges. This is especially true when handling real-time data from construction sites and during the facility management phase, where the dynamic nature of the information complicates the process of maintaining privacy while ensuring its usefulness for decision-making. To address this, the use of domain-specific or context-oriented deep learning models will be explored. These models would be designed to automatically identify and classify sensitive data as it is generated, enabling its segregation based on the specific confidentiality requirements of the project. This approach aims to ensure that sensitive information is handled securely and efficiently, while still supporting the operational needs of the project throughout its lifecycle.

References

[1] National Institute of Standards and Technology (NIST), *Guide to Industrial Control Systems (ICS) Security in NIST Special Publication 800-82*, USA, 2015. https://doi.org/10.6028/NIST.SP.800-82r2.

[2] Mantha, B., B. García de Soto, and R. Karri, Cyber security threat modeling in the AEC industry: An example for the commissioning of the built environment. *Sustainable Cities and Society*, 2021. 66: p. 102682. https://doi.org/10.1016/j.scs.2020.102682.

[3] Dodge Construction Network and Egnyte, *Data Resilience in Design and Construction: How Digital Discipline Builds Stronger Firms SmartMarket Brief*. 2023. https://www.construction.com/resource/data-resilience-in-design-and-construction-how-digital-discipline-builds-stronger-firms-smartmarket-brief/ (Accessed 01 January 2024).

[4] Tao, X., Y. Liu, P.K.Y. Wong, K. Chen, M. Das, and J.C. Cheng, Confidentiality-minded framework for blockchain-based BIM design collaboration. *Automation in Construction*, 2022. 136: p. 104172. https://doi.org/10.1016/j.autcon.2022.104172.

[5] England and Wales High Court, *195 Con LR 74*, 2021. Available from https://www.bailii.org/ew/cases/EWHC/TCC/2021/754.html (Accessed 15 December 2024).

[6] Narayanan, A. and J. Bonneau, *Bitcoin and Cryptocurrency Technologies: A Comprehensive Introduction*, 2016. Princeton, NJ: Princeton University Press.

[7] Benet, J. *IPFS – Content addressed, versioned, P2P file system*, 2014. Available from: https://docs.ipfs.io/ (Accessed 10 July 2023).

[8] Hasan, R., S. Myagmar., A.J. Lee, and W. Yurcik. Toward a Threat Model for Storage Systems. In *Proceedings of ACM Workshop on Storage Security and Survivability*, 2005. https://doi.org/10.1145/1103780.1103795.

[9] Singh, V., N. Gu, and X. Wang, A theoretical framework of a BIM-based multi-disciplinary collaboration platform. *Automation in Construction*, 2011. 20(2): pp. 134–144. https://doi.org/10.1016/j.autcon.2010.09.011.

[10] Olatunji, O.A., A preliminary review on the legal implications of BIM and model ownership. *Journal of Information Technology in Construction 16 (Special issue Innovation in Construction e-Business)*, 2011: pp. 687–696. http://www.itcon.org/2011/40.

[11] Parn, E.A. and D. Edwards, Cyber threats confronting the digital built environment: Common data environment vulnerabilities and blockchain deterrence. *Journal of Engineering, Construction and Architectural Management*, 2019. 26 (2): pp. 245–266. https://doi.org/10.1108/ECAM-03-2018-0101.

[12] Afsari, K., C. Eastman, and D.R. Shelden. Cloud-Based BIM Data Transmission: Current Status and Challenges. In *Proceedings of the 33rd International Symposium on Automation and Robotics in Construction*, 2016. Auburn, AL, USA. https://doi.org/10.22260/ISARC2016/0129.

[13] Autodesk. *BIM 360 Security Whitepaper (Whitepaper)*, 2018. Available from: https://bim360resources.autodesk.com/whitepapers/bim-360-security-whitepaper (Accessed 28 January 2020).

[14] ACONEX, *Cloud-Based Information Management: A Guide for IT (White-Paper)*, 2018. Oracle.

[15] Das, M., X. Tao, and J.C.P. Cheng, BIM security: A critical review and recommendations using encryption strategy and blockchain. *Automation in Construction*, 2021. 126: p. 103682. https://doi.org/10.1016/j.autcon.2021.103682.

[16] Das, M., H. Luo, and J.C.P. Cheng, Securing interim payments in construction projects through a blockchain-based framework. *Automation in Construction*, 2020. 118: p. 103284. https://doi.org/10.1016/j.autcon.2020.103284.

[17] Zheng, T., X. Tao, L. Dong, *et al.*, Distributed common data environment using blockchain and Interplanetary File System for secure BIM-based collaborative design. *Automation in Construction*, 2021. 130: p. 103851. https://doi.org/10.1016/j.autcon.2021.103851.

[18] Mubarak, M.F. and M. Petraite, Industry 4.0 technologies, digital trust and technological orientation: What matters in open innovation? *Technological Forecasting and Social Change*, 2020. 161: p. 120332. https://doi.org/10.1016/j.techfore.2020.120332.

[19] ISO, *Organization and digitization of information about buildings and civil engineering works, including building information modelling (BIM) — Information management using building information modelling — Part 5: Security-minded approach to information management*, 2020.

[20] Centobelli, P., R. Cerchione, P.D. Vecchio, E. Oropallo, and G. Secundo, Blockchain technology for bridging trust, traceability and transparency in circular supply chain. *Information & Management*, 2022. 59(7): p. 103508. https://doi.org/10.1016/j.im.2021.103508.

[21] Elghaish, F., S. Abrishami, and M.R. Hosseini, Integrated project delivery with blockchain: An automated financial system. *Automation in Construction*, 2020. 114: p. 103182. https://doi.org/10.1016/j.autcon.2020.103182.

[22] Elghaish, F., M.R. Hosseini, S. Matarneh, *et al.*, Blockchain and the 'Internet of Things' for the construction industry: research trends and opportunities. *Automation in Construction*, 2021. 132: p. 103942. https://doi.org/10.1016/j.autcon.2021.103942.

[23] Amazon Web Services, *Amazon EC2*. 2020. Available from https://aws.amazon.com/ec2/ (Accessed 10 May 2020).

[24] Amazon Web Services, *Amazon S3*. 2020. Available from https://aws.amazon.com/s3/ (Accessed 10 May 2020).

[25] Toli, I. and A. Zanoni, An algebraic interpretation of *AES 128*. In Dobbertin, H., Rijmen, V., and Sowa, A. (eds), *Advanced Encryption Standard – AES*. Berlin, Heidelberg: Springer; 2005, pp. 84–97. https://doi.org/10.1007/11506447_8.

[26] NIST, *Recommendation for Key Management Part 1, in NIST Special Publication 800-57 Part 1*, 2016. https://doi.org/10.6028/NIST.SP.800-57pt1r4.

[27] Turner, S., Transport layer security. *IEEE Internet Computing*, 2014. 18(6): pp. 60–63. http://doi.org/10.1109/MIC.2014.126.

[28] Yan, Z. and S. Holtmanns, Trust Modeling and Management: From Social Trust to Digital Trust. In *Computer Security, Privacy and Politics: Current*

Issues, Challenges and Solutions, R. Subramanian (ed.), 2008. Hershey, Pennsylvania: IGI Global. pp. 290–323.

[29] Das, M., X. Tao, Y. Liu, and J.C. Cheng, A blockchain-based integrated document management framework for construction applications. *Automation in Construction*, 2022. 133: p. 104001. https://doi.org/10.1016/j.autcon.2021.104001.

[30] Almasi, M.M., T.R. Siddiqui, N. Mohammed, and H. Hemmati, The Risk-Utility Tradeoff for Data Privacy Models. In *2016 8th IFIP International Conference on New Technologies, Mobility and Security (NTMS)*, 2016. https://doi.org/10.1109/NTMS.2016.7792481.

[31] Sigalov, K., X. Ye, M. König, *et al.*, Automated payment and contract management in the construction industry by integrating building information modeling and blockchain-based smart contracts. *Applied Sciences*, 2021. 11 (16): pp. 7653–7671. https://doi.org/10.3390/app11167653.

[32] Lamport, L. and N. Lynch, CHAPTER 18 – Distributed Computing: Models and Methods. In *Formal Models and Semantics*, J. Van Leeuwen (ed.), 1990. Amsterdam: Elsevier. pp. 1157–1199.

[33] Wormer, G.V. and M. Larkin. Document Management for Design Engineering, Construction, and Owner Operators for the New Enterprise. In *Proceedings of the Annual Pulp, Paper and Forest Industries Technical Conference (PPFIC)*, 2017. Tacoma, WA. https://doi.org/10.1109/PPIC.2017.8003856.

[34] Yuan, S. *Hyperledger Tape*, 2020. Available from: https://github.com/Hyperledger-TWGC/tape. (Accessed 25 August 2020).

[35] Benet, J. *IPFS – Content addressed, versioned, P2P file system*, 2014. Available from: https://docs.ipfs.io/ (Accessed 20 September 2021).

[36] Wang, Z., T. Wang, H. Hu, J. Gong, X. Ren, and Q. Xiao, Blockchain-based framework for improving supply chain traceability and information sharing in precast construction. *Automation in Construction*, 2020. 111: p. 103063. https://doi.org/10.1016/j.autcon.2019.103063.

[37] Lee, D., S.H. Lee, N. Masoud, M. Krishnan, and V.C. Li, Integrated digital twin and blockchain framework to support accountable information sharing in construction projects. *Automation in Construction*, 2021. 127: p. 103688. https://doi.org/10.1016/j.autcon.2021.103688.

[38] Li, C.Z., Z. Chen, F. Xue, *et al.*, A blockchain-and IoT-based smart product-service system for the sustainability of prefabricated housing construction. *Journal of Cleaner Production*, 2021. 286: p. 125391. https://doi.org/10.1016/j.jclepro.2020.125391.

[39] Sheng, D., L. Ding, B. Zhong, P.E. Love, H. Luo, and J. Chen, Construction quality information management with blockchains. *Automation in Construction*, 2020. 120: p. 103373. https://doi.org/10.1016/j.autcon.2020.103373.

[40] Cha, H. and D. Lee, Framework based on building information modelling for information management by linking construction documents to design objects. *Asian Architecture and Building Engineering*, 2018. 17(2): pp. 329–336. https://doi.org/10.3130/jaabe.17.329.

[41] Hamledari, H. and M. Fischer, Construction payment automation using blockchain-enabled smart contracts and robotic reality capture technologies.

Automation in Construction, 2021. 132: p. 103926. https://doi.org/10.1016/j.autcon.2021.103926.

[42] Singh, A.K. and V.R.P. Kumar, Analyzing the barriers for blockchain-enabled BIM adoption in facility management using best-worst method approach. *Built Environment Project and Asset Management*, 2023. 14(2): pp. 164–183. https://doi.org/10.1108/BEPAM-04-2023-0080.

[43] Muralidharan, S. and H. Ko. An InterPlanetary file system (IPFS) based IoT framework. In *2019 IEEE International Conference on Consumer Electronics*, 2019. Las Vegas, NV: IEEE. https://doi.org/10.1109/ICCE.2019.8662002.

[44] Benet, J. *InterPlanetary File System (IPFS)*. 2016. Available from: https://ipfs.io/ (Accessed 1 August 2020).

[45] Nizamuddin, N., K. Salah, M. Ajmal Azad, J. Arshad, and M. Rehman, Decentralized document version control using ethereum blockchain and IPFS. *Computers & Electrical Engineering*, 2019. 76: pp. 183–197. https://doi.org/10.1016/j.compeleceng.2019.03.014.

[46] Bellare, M., A. Desai, E. Jokipii, and P. Rogaway, A concrete security treatment of symmetric encryption. In *Proceedings 38th Annual Symposium on Foundations of Computer Science (FOCS 1997)*, 1997. Miami. pp. 394–403.

[47] De Cannière, C., Blowfish. In *Encyclopedia of Cryptography and Security*, H.C.A. van Tilborg (ed.), 2005. Boston, MA: Springer US. pp. 48–49.

[48] Nan, L. Research on Diffie–Hellman Key Exchange Protocol. In *2010 2nd International Conference on Computer Engineering and Technology*, 2010. https://doi.org/10.1109/ICCET.2010.5485276.

[49] Rivest, R.L., A. Shamir, and L. Adleman, A method for obtaining digital signatures and public-key cryptosystems. *Communications of the ACM*, 1978. 21: pp. 120–126. https://doi.org/10.1145/359340.359342.

[50] Conti, M., N. Dragoni., and V. Lesyk, A survey of man in the middle attacks. *IEEE Communications Surveys & Tutorials*, 2016. 18(3): p. 1. https://doi.org/10.1109/COMST.2016.2548426.

[51] Cheng, H. and L. Xiaobo, Partial encryption of compressed images and videos. *IEEE Transactions on Signal Processing*, 2000. 48(8): pp. 2439–2451. https://doi.org/10.1109/78.852023.

[52] Jang, W. and S.-Y. Lee, Partial image encryption using format-preserving encryption in image processing systems for internet of things environment. *International Journal of Distributed Sensor Networks*, 2020. 16(3): p. 1550147720914779. https://doi.org/10.1177/1550147720914779.

[53] Cai, X.T. and W.D. Li, Partial encryption of feature-based product models for collaborative development. *Robotics and Computer-Integrated Manufacturing*, 2020. 63: p. 101918. https://doi.org/10.1016/j.rcim.2019.101918.

[54] Das, M., X. Tao, H. Liu, Y. Xu, and J. Cheng, A Blockchain-Based Encryption Method for IFC Models for Collaborative Project Management. In *22nd International Conference on Construction Applications of Virtual Reality*, 2022. Seoul.

Chapter 5

Blockchain and tokenization of built assets

Lavinia Chiara Tagliabue[*], *Paolo Mistrangelo*[†] *and Algan Tezel*[‡]

The real estate market is experiencing an increase in investments globally [1]. Affordable housing and retrofitting in Europe are challenging by significant concerns relate to social inclusion and growth in community progress. Real estate investments are coming from European financial institutions, North American private capitals, and independent wealth funds from Middle and Far East countries. The attractiveness of investments varies across countries. For example, the northern European nations attract ten times more investments than Italy [2].

In the Italian context, this is often due to the housing market's limited potential to be included in investors' portfolios, and the build-to-sell approach which does not focus on innovative building uses. This is also largely triggered by the poor protection provided by the regulations on lease contracts, which in case of insolvency does not guarantee landlords the retrieval of their assets [3].

Germany has a different dynamic compared to Italy. Only 45% of people own a house in Germany and renting a house according to the flexible and mutable requirements of space changing over time is a commonly adopted practice [4]. Therefore, there is a broader stock of rented assets on the market and large private real estate companies possess a concentration of housing assets. Italy and Germany represent the two extremes of proprietorship distribution in the European real estate market, further stressing the resultant illiquidity related to the properties.

A different approach that can alleviate and make fluid a structurally static market [5] which at this moment exemplifies the dichotomy "property or rental," with no options in the middle is the idea of fragment the property using tokens. Additionally, operators in the sector at the national level frequently express concerns about the excessive complexity of administrative procedures for trading assets. This complexity arises, on one hand, from the need to protect both parties against fraudulent activities; however, on the other hand, it represents a significant obstacle that can, in certain instances, deter potential investment opportunities.

[*]Department of Computer Science, University of Turin, Italy
[†]Magellano Progetti, Milan, Italy
[‡]University of Nottingham, Faculty of Engineering, University Park, UK

Simplification and innovation of these methods guaranteed through asset tokenization could be an opportunity that has not been leveraged to date even in the world of cryptocurrencies. Non-fungible tokens (NFT), as an example associated with the originality certification of artworks, are emerging and in the chapter, we discuss how it could be possible to transfer this technology, with due legal restraints, to the proposed property tokenization process [6].

A "property token" is defined as a portion of the asset that describes a new property with a secondary possessor, submitted to the main one who preserves the opportunity to sell the complete asset, if required, reimbursing the property fragments in time. In the meanwhile, the secondary owner or owners can capitalize a limited sum in a lucrative investment. This has a significant social consequence as it unlocks minor and weak investors (e.g. young people, seniors, and persons with low acquiring capability) that cannot manage to pay for the investment in an entire asset.

Envisioning the property tokenization of real estate assets encompasses transforming physical properties into digital tokens notarized on a blockchain. In this process, each token represents a segment or ownership stake of the real estate asset. Tokenization implies multiple possible paybacks such as augmented liquidity, fractional ownership, and approachability to a larger variety of investors. The process includes converting the property's value, rights, and characteristics into a digital format, and generating tokens that can be bought, sold, and traded on blockchain platforms. This method intends to streamline transactions, diminish barriers to entry, and boost the global productivity of real estate markets through the application of distributed ledger blockchain technology. Tokenization is one of the opportunities made possible by the popular distributed ledger blockchain technology.

Property tokenization grants the usage of digital tokens in place of ownership and property title in the marketplace of assets, a concept deemed to have a significant potential in increasing financial fluidity, investment, and value exchange possibilities in the real estate market [7]. This new approach and business model, described in [8], incorporates environmental, economic, and social aspects of sustainability, circumventing the liquidity challenge within the current real estate assets and market. The inherently indivisible nature of real estate property, coupled with its high market value and intrinsic rigidity, makes asset investment challenging, particularly when considering both investment and disinvestment [9].

These complications result in an inert condition where assets remain locked, often inherited within families across generations. This practice excludes the economically disadvantaged from the real estate market [10] and significantly limits property ownership opportunities, particularly for newer generations. Additionally, elderly landlords face challenges when their assets require refurbishment and renovation, as these procedures are often barely affordable for retired owners who must rely solely on their retirement income.

This chapter will explore the transformative potential of blockchain technology in the real estate sector, which accounts for a substantial portion of global capital and energy consumption. The chapter emphasizes the low digital maturity of the real estate industry and discusses how blockchain applications can enhance transparency, trust, and efficiency while addressing issues such as fraud and corruption. Key topics covered in this chapter include the distinction between transaction processes and tokenization, with a focus on how digital tokens can represent real estate assets. The chapter highlights the importance of legal frameworks and governance structures that vary by jurisdiction, affecting the validity and viability of asset-backed tokens. It also addresses regulatory challenges that impede cross-border transactions and the need for clearer definitions in legislation, particularly in countries like Italy. The chapter further examines innovative business models enabled by property tokenization, including fractional ownership and crowdfunding platforms. It identifies barriers to adoption, such as legal uncertainties, security concerns, and a lack of standardization. Additionally, it discusses the integration of technologies like geographic information system (GIS) and building information modeling (BIM) with blockchain to create a transparent and efficient ecosystem for property transactions. After presenting a possible architecture for the implementation, the chapter posits that property tokenization can lead to increased liquidity in real estate markets while promoting social sustainability through reinvestment in community-focused initiatives. The discussion concludes with a call for further research and experimental observations to assess the practical benefits of blockchain in enhancing real estate digitalization.

5.1 Property tokenization as a paradigm shift

The real estate sector accounts for roughly 60% of global capital, totaling over 200 tn USD [11]. It is responsible for 36% of final energy demand and about 40% of energy-related CO_2 emissions [12], making it one of the most impactful sectors with significant potential for savings and progress. Despite its significant global economic, environmental, and social impact, the real estate sector exhibits low digital maturity compared to other industries [13,14].

Blockchain applications in the real estate sector are rapidly developing and contributing to improve transparency and trust, targeting fraud and corruption, and decreasing cost and time inefficiencies. Studies demonstrate that blockchain technology has been considered for profiting in four broad categories, such as land administration, transactions, tokenization, and real estate management [15]. Two main categories are pertinent for this chapter, namely transaction and tokenization.

The transaction category debates commercial and residential real estate negotiation where participants act as investors or consumers. The tokenization denotes the digital representation of a real estate asset as blockchain-based tokens, including platforms for real estate tokenization and crowdfunding. Blockchain applications for real estate tokenization rely on legal and governance aspects that can diverge depending on the selected country jurisdiction. Literature on this aspect demonstrates [9] that various legal schemes recognized disparities in asset-backed tokens' validity, that is, in EU countries that have diverse legal statuses, even where property rights are partial and frequently less bordered. The consequence is that the viability of global cross-border transactions is hindered, and inconsistent regulatory rules prevent the free and international exchange of security tokens, consequently weakening the possible rewards that their adoption can promote [16]. Some countries are providing regulatory definitions, for example, Italy has provided a regulation terminology definition of blockchain and smart contracts, with Article 8 of L.D no. 135/2018 [17], converted into Law no. 12/2019 [18].

The description "technologies based on distributed registers" is used for technologies and electronic protocols that adopt a shared, distributed, replicable, and simultaneously accessible register, architecturally decentralized on cryptographic bases, allowing the record, validation, updating, and storage of non-encrypted or encrypted data, which can be checked by each contributor and are non-editable. Within this context, a "smart contract" is a computer program functioning on technologies grounded on distributed registers, the execution of which automatically binds two or more parties based on the effects defined by the same.

A significant national summary [19] on real estate crowdfunding platform operators in Italy highlights that the sector is static, underdeveloped, and has strong barriers to entry. However, introducing digital tokens into real estate crowdfunding can enhance investment effectiveness and liquidity, fostering the growth of a secondary market for property exchange securities through security token offerings. The participants of the Asia Securities Industry and Financial Markets Association similarly stressed regulatory and legal uncertainty as the main challenge to tokenized security. Although some countries are beginning to clarify their regulatory agendas, it remains unclear how the token economy will achieve a comprehensive global configuration. However, it is evident that, following the pandemic, the acceptance of digital services by consumers, businesses, and governments—particularly in the Asia-Pacific region—has accelerated the demand for financial products via apps [20].

A second focus in adopting blockchain for real estate tokenization is about the intermediate structures that correspond to special purchase vehicles or funds, considered as a promising alternative to the direct tokenization of real estate ownership [9,21].

A third focus underlines the request for products, such as cryptocurrency and NFT that can increase the number of stockholders able, ready, and willing to take advantage of real estate tokens. Literature reports some practices about single-asset tokenization around the world however the main block to the dissemination of real estate security tokens is the critical absence of a centralized marketplace to empower the trade of tokens, but they are operated only on the platform where they are listed with a restricted number of members that are not able to realize a high level of liquidity [20].

The background on blockchain application and tokenization in the real estate sector underscores the need for further experimental observations [22–24] and innovative case studies to demonstrate how and whether blockchain enhances the benefits of digitalization. This includes examining how digital procedures can be adapted and transitioned to blockchain applications [25].

In the chapter, it is assumed that tokenization is an extended interpretation of the legal institution of bare ownership with a higher level of security against possible fraud using digital technologies. This assumption enables the exploration of new boundaries and parameters that support the generation of a disruptive framework for real estate activities. However, it is essential to include the perspective of the notary council that currently provides the exclusive guarantee of the reliability of property rights for real estate assets.

New methods and business models are expanding to overcome the issues affecting the asset market such as tax incentives for supporting retrofit policies for energy efficiency upgrading of buildings and push the construction sector toward virtuous practices. Additionally, new platforms eliminating third parties between demand and supply in the housing market are developing (e.g. Airbnb, Wimdu, Wunderflats, Traum-Ferienwohnung, Houzeo, Redfin, and Opendoor). The need for authentication and trust in asset market transactions makes the process lengthy and complex, requiring extensive skills and qualified practitioners to manage building compliance with frequently changing protocols. Moreover, decentralized finance (DeFi), based on the robustness of blockchain platforms is developing along with the traditional methods of notarization [26], and in line with this, platforms for property tokenization arrived in the marketplace, first in digital ecosystems such as metaverse, and now in the actual real estate market (e.g. blockchain app factory, Solidblock). In this context, the process of tokenizing an asset offers multiple benefits by converting asset value into digital tokens or securities, with ownership of the tokens recorded on blockchain [27]. This grants the asset owners capital flexibility and various options (i.e. tokens related to square meters or smaller dimensions, percentage of the asset), thus endorsing high capital flexibility. In this way the asset market is enhanced by increasing liquidity, speeding up processes through automation, lowering costs, minimizing disputes, and decentralizing data [28].

5.2 Fractional ownership and digital innovation

The real estate market is regarded as a highly lucrative segment for investment. Although it may temporarily immobilize capital, investors can be assured of an appreciation in asset value over the long term. The concept of a "fractionable" asset has the potential to eliminate various barriers and burdens for owners, thereby enhancing liquidity and attracting new participants from diverse financial and social backgrounds who are typically excluded from the market. The shift from singular ownership to a model of shared or communal property liberates financial resources for primary owners while enabling small investors to acquire economic value through fractional ownership, as defined by NFTs. These NFTs facilitate the buying and selling of ownership stakes in a fluid and dynamic marketplace. This notion underpins property tokenization that can also be directed toward enhancing asset value in an environmentally sustainable manner [29]. Furthermore, the property market can be systematized and automated through the registration of a "property token" in smart contracts, representing a title of partial ownership, thus ensuring an immutable and notarized process [30].

The proposed mechanism has the potential to serve as a disruptive financial catalyst, revitalizing asset exchange and transforming the market within a digital ecosystem. The objective is to integrate assets with tokens representing fractional ownership into a new digital cadaster for real estate, thereby establishing connections across various scales facilitated by digital tools. The urban scale can be managed through GIS [31], while the building scale can be defined by BIM, certifying the exchanged assets through a unique blockchain token.

Consequently, the real estate marketplace can be restructured using smart contracts that record a "real estate token" on the blockchain, thereby enhancing financial fluidity and creating socio-economic opportunities for marginalized groups [32].

This mechanism holds significant promise as a financial catalyst capable of revolutionizing property transactions. However, it is noteworthy that much of the existing discourse in this field remains largely theoretical or lacks integration with other pertinent technologies.

The framework for digital property tokenization, presented in this chapter, aims to promote sustainability and social inclusion through the following contributions:

- Improving the feasibility of property tokenization within the global real estate ecosystem.
- Conceptualizing an application of a property tokenization model utilizing GIS, BIM, smart contracts, blockchain, and integrated tokens.
- Demonstrating "real-world" economic evaluations of properties for tokens through linked data.
- Outlining a potential system architecture based on current IT products.

5.3 Framework for property tokenization

The implementation of property tokenization is based on the use of smart contracts to facilitate transactions on the blockchain. This process is automated through the execution of predetermined instructions in the form of executable codes, enabling the transfer of assets. To enable this process, "tokens," or digital assets representing ownership rights, are created. This concept of "digital tokenization" allows the creation of different types of virtual tokens based on specific protocols. Using the ERC-20 protocol, fungible and interchangeable tokens can be generated. In contrast, the ERC-721 protocol allows the creation of NFTs with unique metadata, providing them with distinct identification. In addition, the ERC-1238 protocol allows for the creation of nontransferable tokens that represent titles or badges owned exclusively by specific individuals [33]. Tokens can be classified as:

- Currency tokens, designed to be used as a medium of exchange and to make payments.
- Security tokens, designed to represent obligations in specific companies or projects.

- Utility tokens that allow the owner to benefit from a utility provided by the issuer.
- Asset-backed tokens, created to correspond to rights, whether proprietary or mandatory, to tangible assets [9].

Under the framework of ownership tokenization, an asset owner opts to tokenize a portion of their asset and transfer it to another party through asset-backed tokens. This process facilitates shared ownership via smart contracts that are securely transmitted on the blockchain. Asset tokenization presents several advantages, including potentially lower transaction costs, enhanced transparency regarding transactional data and issuer information, direct access for investors to both primary and secondary markets, and increased flexibility concerning assets.

In the real estate sector, this technology has the potential to foster the development of future platforms that enable cross-border transactions involving real estate assets within the European Union and globally. The global professionally managed real estate investment market was estimated to be valued at approximately US$9.6 tn in 2019 [34]. Furthermore, the proposed approach can effectively address the challenges associated with the digitization of the collaborative economy and the new economic conditions that have emerged in the aftermath of the pandemic crisis. During this period, residential markets experienced a significant slowdown, with transaction volumes plummeting by 80% due to restrictions that prevented physical visits to prospective properties [35].

To address these challenges and improve global connectivity, the use of technologies such as blockchain for digital validation of contract certificates and virtual reality for site visits can be proposed. European nations have initiated regulatory measures concerning the utilization of smart contracts and specific tokens. [36]. EU member states are required to establish national rules by March 23 to implement Regulation (EU) 2022/858 [37] in their national laws. The goal is to create a pilot scheme for market infrastructures based on distributed ledger technology (DLT).

In Italy, blockchain technology and traditional finance are moving forward similarly, thanks to the implementation of the aforementioned European regulation, which will allow the exchange of financial instruments such as stocks and bonds in the form of tokens. Recently, the Council of Ministers aligned national legislation with the requirements of EU Regulation 2022/858 through the issuance of a decree-law that lays the groundwork for compliance with European guidelines on the "tokenization" of financial assets. As a result, a new experimental regime for market infrastructures based on DLT pilot regime was established, serving as a precursor for the digitized finance sector.

Regarding the application of blockchain in cadastral registration, it is worth noting that the Swedish Land Mapping, Cadastration and Registration Authority (Lantmäteriet) has launched an experiment in collaboration with both private and public entities [38]. The experiment entails a blockchain initiative designed to validate the transfer of real estate between parties through registration in the Swedish land registry. The primary objectives of this project are to achieve efficiencies in time and cost for public administration while utilizing blockchain technology to enhance transparency and security. Initiated in 2015 and concluded in 2019, the project primarily benefited Swedish authorities by bolstering security through blockchain applications. This experiment is particularly significant due to its potential implications for developing countries that lack a centralized cadastral system. The implementation of blockchain technology could mitigate corruption risks and alleviate administrative burdens [39,40].

Graglia and Mellon [41] delineated the stages utilized by blockchain technology to document the progress of transactions within the framework of blockchain property registry adoption levels. Their analysis indicates that the Swedish experience is classified at the second level, which is below Dubai, rated at the third level. In Sweden, a permissioned blockchain replaces the traditional central database, whereas Dubai operates at the sixth level, known as the "Pangea" scheme, where rights associated with specific parcels are fragmented and managed via blockchain. The highest levels, seventh and eighth, where transactions occur without intermediaries and interoperability is achieved by integrating various blockchain registries, have yet to be implemented in any country.

Several platforms exist for asset tokenization with limited financing options, including Atlant, Smartlands, and Crowdli-token. However, the business model of these initiatives typically entails that the ownership of the tokenized asset is retained by a single company, while another entity determines its utilization and disposition. Consequently, token holders possess the right to receive profits without acquiring actual ownership of the asset [9].

While blockchain technology can serve as an automated disintermediated registry, fulfilling essential functions of traditional land registries such as reliable publication of participants, time, and events, the recognition of this digital registration as equivalent to national land registries raises public policy concerns. Policymakers must decide whether to grant legal status to digital notarization. Currently, the closest alignment between existing land registries and blockchain tokenization appears to be the most reasonable solution for harmonizing with legal systems. Parties involved in real estate tokenization could be required to include information about the blockchain notarization in the official Notarization Registry. This would enable tracking of blockchain transactions and provide data on the official token owner [22].

Italian law (art. 8-ter.3 Legge of 11 February 2019) [42] establishes that the storage of an electronic document using DLTs has the same legal effects as electronic time stamps (art. 41 Regulation 910/2014) [43]. The use of automated systems offers the advantage of reducing disputes, nevertheless the absence of notarial intervention in property tokenization means that any conflicts must be resolved through post-dispute resolution mechanisms [44]. Nonetheless, two critical challenges, maintaining coherent control over contract content and verifying the identities of the parties involved, can be effectively addressed through the implementation of the following steps:

- One approach involves incorporating metadata into the NFT or attaching a file that contains natural language contract metadata. This metadata can then be analyzed using natural language processing techniques to identify potentially abusive terms. This procedure applies not only to customer contracts related to property acquisition but also to those involving real estate more broadly.

- An alternative strategy is to embed conflict resolution mechanisms, such as mediation or arbitration, within the smart contract on the blockchain. This integration facilitates the application of cryptographic law (e.g. lex cryptographica) while ensuring alignment with existing legal frameworks.
- To verify the identities of parties on a public blockchain, digital user identities can be associated with a "sovereign identity" or authenticated by a trusted third party, such as a notary public. Additionally, the Civil Registry may serve as an oracle in this context. The decentralized system can also utilize the legally recognized electronic signature under eIDAS (electronic IDentification, Authentication and trust Services) to provide authentication that is valid across all EU member states.
- To address the necessity of certifying that parties enter into agreements free from undue influence or vices (such as coercion or intimidation), emotion recognition technologies can be integrated into the smart contract. These technologies, which are frequently employed to analyze emotional responses in product advertisements (e.g. Apple's Emotient or Emotion Research Lab), can detect customer reactions, such as happiness, surprise, anger, disgust, fear, or sadness, to identify any signs of compromised consent.

The above-mentioned property tokenization aims at introducing a financial instrument, digitally and legally operative, which allows the partial transfer of a real estate asset while maintaining at the same time the ownership and exclusive use. This mechanism enables the generation of liquidity proposing an alternative and less absolute solution than the options already available: owning or transferring an asset. The engendered liquidity promotes the opportunity to improve the real estate assets when the liquidity is reinvested in the retrofitting of buildings in a sustainable perspective. Social sustainability is also boosted through the reinvestment of part of the liquidity in social innovation and directly supporting weaker investors in trading in the real estate market. Property tokenization could be defined as an economic revolution that can produce relevant economic consequences. A dynamic property token trade market much like the stock trade market can be envisioned [28].

The recording of tokens representing fragments of assets on the blockchain, through customized smart contracts, guarantees contractual compliance and rationalizes the purchasing and trading practice [45].

The digital representation of an asset must align with the legal framework of the respective country. Currently, properties are registered at the national level within property registers, which archive acts and records associated with real estate transactions. However, these property registers do not accommodate the recording of assets through NFTs as defined by the ERC-721 protocol, which is necessary for implementing real estate tokenization on the blockchain [7].

The definition of the framework for the tokenization of property toward an inclusive and sustainable development of real estate asset markets can be organized the following steps:

- Local housing market assessment and digitization.
- Definition of the management technology infrastructure.

- Verification and validation of notarization processes.
- Processes digitalization and business model.
- Implementation and testing of the platform.
- Dissemination of the platform to extend the adoption and boost social adoption.

To facilitate the experimentation of property tokenization, it is essential to digitize real estate assets and conduct their assessment and evaluation. This process will create a comprehensive archive of resources that can be tokenized for trading in the emerging real estate property tokenization market. The digital transformation of the real estate asset cadastre is fundamental to establishing a comprehensive framework based on GIS and BIM. This framework facilitates the precise location of building units within a digital urban mapping context, ensuring adequate levels of detail. Organizing assets on a digital platform enables their tokenization and notarization on the blockchain, thereby enhancing the reliability and timely updating of asset information in urban areas, an advantage that is not trivial.

An initial test aimed at validating this assertion was conducted by linking the foundational data on city assets to the cadastral information (https://www.cert-o.com/fastmap/dbt_catasto.html). This connection was established to ensure that the data remains current and accurate. For new developments, as an example in Figure 5.1, the recent University Bocconi campus in Milan, Italy (2019–2021) was checked. The results indicate that the regional topographic database is not updated, whereas the cadastral data is current and permits the download of vector-based cadastral geometric information.

The future step will be to define from the extruded polygon the information to generate a building model in an industry foundation classes format (i.e. height, roof typology, windows distribution, and number of floors) and therefore create a direct connection from the cadastre to a BIM model of the asset. This will enable the connection from the GIS and the cadastre to the BIM definition of the asset that can be adopted to verify consistency and compliance of the construction in the tokenization platform [46].

This process is applicable and extendable beyond buildings, encompassing a wide array of built environment assets to enhance the benefits of tokenization in terms of social engagement, environmental impact, and cost reduction in urban management. These aspects are further discussed later in this section. A second critical step involves defining the technological infrastructure necessary for managing a digital ecosystem that supports the phases of identification, document management, and the activation of a marketplace for tokenization, specifically for the buying and selling of properties or their fractional interests. A management platform is essential to provide a comprehensive range of functionalities that ensure efficiency, transparency, and reliability throughout the process, thereby facilitating assistance for transaction participants and coordinating the digital cadastre.

5.3.1 Organization of the information in a property tokenization management system

The information regarding users participating in the network will undergo a systematic process that includes accreditation, collection of property documentation,

Figure 5.1 *Urban building mapping with cadastre comparison and download of the cadastral parcel*

and verification of the submitted data. This approach aims to enhance the transparency and accuracy of user profiles, and the types of properties offered through the service. Within the property tokenization platform, various modules will be organized to facilitate these processes. We envisioned the structure of the platform as follows considering the information needed for the defined purpose:

- *User master data*: Establishment of master data for participant identification and blockchain access, which includes relevant identifiers and documentation pertaining to property ownership.
- *Property master data*: Definition of property master data, encompassing location, consistency, and parceling for subsequent segmentation.
- *Digital enabler module*: The integration of properties with the GIS platform involves defining BIM parameters for consistencies and segmentations. This process includes conducting analytics on potential fragmentation scenarios tailored to user and marketplace requirements, creating associated tokens, organizing smart contracts, and preparing for notarization on the blockchain. Such comprehensive connectivity enhances the efficiency and transparency of property management, facilitating a robust framework for real estate transactions.
- *Unit/product management*: Verification and archiving of document checklists will involve the evolution of file statuses, including processing, verification, location, compartment definition and sales units designation, along with their association with tokens and smart contracts. Additionally, a statistical dashboard will be developed for unit analysis, facilitating the management of parcel deeds, acceptance or transfer of units, generation of transfer details, invoicing data generation, and storage of stock data and transactions in the blockchain.
- *Energy retrofit module*: Checklist of energy efficiency measures involves a comprehensive assessment of units and the allocation of funds for reinvestment in energy quality. This process utilizes interoperability mechanisms based on BIM to facilitate building energy modeling. By leveraging BIM-based interoperability, the evaluation of energy efficiency can be enhanced, ensuring that energy quality improvements are systematically integrated into the management of built environments. This approach not only promotes sustainability but also supports informed decision-making regarding energy investments.
- *Reporting*: The system possesses the capability to perform precise extractions from various functionalities, accommodating export in compatible formats.
- *Data warehouse*: Centralization of data involves the daily extraction of uploaded information within the platform, followed by an aggregation process that ensures data verification and standardization. This allows the monitoring of key performance indicators to facilitate statistical analysis. Additionally, properties are visualized using a GIS tool, enabling analytics on urban areas and spatial distribution of property tokenization. This comprehensive approach enhances data accessibility and supports informed decision-making regarding urban management and property development.

5.3.2 Barriers to adoption: insights from the Notary Council on property tokenization

The evaluation of the proposed framework by the Notary Council is fundamental to the successful implementation of property tokenization. Addressing existing barriers necessitates collaboration among legal, technological, and regulatory stakeholders to establish a robust framework that facilitates the secure and effective tokenization of real estate assets.

This section will explore the various obstacles highlighted by the Notary Council that impede the adoption of property tokenization. It will address regulatory uncertainties, technological challenges, market acceptance, and the need for standardized processes. The discussion will emphasize the importance of collaboration among stakeholders to navigate these barriers effectively and facilitate a smoother transition to tokenized real estate solutions.

We identified potential barriers in preliminary interviews with some components of the Notary Council to property tokenization summarized as follows:

- *Legal and regulatory challenges:* Property tokenization necessitates navigating intricate legal frameworks and ensuring adherence to existing regulations. Ambiguities or lack of clarity regarding legal aspects can hinder the seamless implementation of tokenization initiatives.
- *Security concerns:* Ensuring the security of blockchain technology and smart contracts is essential. Any vulnerabilities or breaches may jeopardize the integrity of property ownership records and transactions, potentially leading to significant repercussions for stakeholders involved.
- *Lack of standardization:* The lack of standardized processes and protocols for property tokenization may impede interoperability and present challenges for widespread adoption. This absence can lead to inconsistencies and inefficiencies that deter stakeholders from fully engaging with tokenization initiatives.
- *Resistance to change:* Traditional real estate practices, particularly those associated with notarization, have a well-established history. Resistance to change among stakeholders in the industry, including property owners, notaries, and regulatory bodies, represent a substantial barrier to the adoption of innovative approaches such as property tokenization.
- *Technology infrastructure:* A robust technological infrastructure is crucial for the effective implementation of property tokenization. Regions or entities that lack the requisite technological resources may encounter significant obstacles in adopting blockchain solutions, thereby hindering their ability to participate in this innovative framework.
- *Educational and awareness gaps:* Stakeholders, including property owners and professionals within the real estate industry, may lack awareness or a comprehensive understanding of the benefits and processes associated with property tokenization. This deficiency can impede its acceptance and adoption, limiting the potential for transformative change within the sector.
- *Market acceptance:* The market's readiness to adopt property tokenization may vary significantly. Factors such as investor sentiment, market maturity,

and perceived risks can influence the acceptance and integration of tokenized real estate assets, ultimately affecting their widespread implementation within the industry.

- *Scalability issues:* As property tokenization continues to gain traction, scalability issues may emerge. Blockchain networks must efficiently manage an increasing volume of transactions to sustain the advantages associated with tokenization. Addressing these scalability challenges is critical to ensuring that the benefits of tokenization are fully realized, and that the technology can support widespread adoption.
- *Integration with existing systems:* Integrating blockchain-based property tokenization with existing real estate and legal systems presents significant challenges. Ensuring seamless compatibility between these systems is essential for facilitating a smooth transition, as any discrepancies could hinder the effective implementation of tokenization initiatives.
- *Valuation and appraisal:* Establishing the accurate valuation of tokenized real estate assets and developing standardized appraisal methods is critical for fostering trust among investors and regulatory bodies. Without reliable valuation frameworks, the credibility of tokenization initiatives may be undermined, potentially deterring participation from key stakeholders in the market.

However, additional barriers, as well as collaborative approaches, can be identified through a fruitful dialogue with the Notary Council, nevertheless this discussion is undoubtedly complex. Furthermore, the changes in processes and the business models that may emerge require thorough investigation and definition to ensure effective implementation.

5.3.3 Possible business models with property tokenization platform

In Table 5.1 a preliminary list of possible business models is summarized. The identified business models can be merged or adapted to align with the specific objectives and offerings of a property tokenization venture. This flexibility allows for the customization of approaches to better meet the unique needs of stakeholders and enhance the overall effectiveness of the tokenization initiative.

A testing phase will be necessary to optimize the platform to ensure the effective functioning of the system across all intended functionalities. Following this, a dissemination phase will be required to expand participation in the new cooperative disruptive property tokenization trading platform. This two-phase approach will facilitate the refinement of the system while promoting broader engagement among stakeholders.

Emerging opportunities with property tokenization could be expanded considering the following processes:

- *Fractional ownership:* This model allows for ownership based on percentage shares or specific architectural characteristics, such as owning the larger room,

Table 5.1 Preliminary list of potential business models associated to property tokenization

Tokenization platforms	*Primary issuance fee:* Charge property owners or developers a fee for issuing and tokenizing their real estate assets *Transaction fees:* Earn fees for every transaction conducted on the platform, including buying, selling, and trading property tokens.
Real estate crowdfunding platforms	*Platform fees:* Charge fees for hosting property token offerings and facilitating crowdfunding campaigns. *Success fees:* Earn a percentage of funds raised as a success fee when a crowdfunding campaign reaches its goal.
Real estate investment platforms	*Management fees:* Charge investors a fee for managing and maintaining their tokenized real estate portfolios. *Performance fees:* Earn a percentage of profits generated from successful real estate investments.
Fractional ownership platforms	*Transaction fees:* Charge fees for transactions involving the fractional buying, selling, or trading of property tokens. *Subscription/membership fees:* Implement a subscription-based model for users to access and participate in fractional ownership opportunities.
Property token exchanges	*Trading fees:* Earn fees from users for trading property tokens on the exchange. *Listing fees:* Charge property owners or developers for listing their tokenized assets on the exchange.
Real estate tokenization as a service	*Licensing fees:* License the tokenization technology to real estate firms, allowing them to tokenize properties on their own. *Consulting fees:* Provide advisory and consulting services to guide businesses through the tokenization process.
Property management services	*Management fees:* Offer property management services for tokenized assets, earning fees for overseeing day-to-day operations. *Smart contract maintenance fees:* Charge fees for maintaining and updating smart contracts associated with the tokenized properties
Blockchain development and consulting	*Development fees:* Provide blockchain development services to real estate firms looking to implement tokenization. *Consulting fees:* Offer expert advice on legal, regulatory, and technical aspects of property tokenization.
Educational platforms	*Training and certification fees:* Provide educational resources, training programs, and certification courses on property tokenization for industry professionals
Data and analytics services	*Subscription fees:* Charge real estate professionals for access to data and analytics related to tokenized properties. *Customized reports:* Offer premium services for generating customized reports and insights based on tokenized real estate data.

the room with a view to the city or the quiet room. This approach enables diverse investment opportunities and caters to varying preferences among stakeholders.

- *Time-bound and time-share ownership arrangements:* These arrangements pertain to both new and existing asset stock; however, they present challenges in terms of operationalization. Effectively implementing such models requires careful planning and coordination to ensure that the interests of all stakeholders are adequately addressed.

5.3.4 Technological chain for the implementation of the property tokenization

The technological chain for property tokenization uses GIS, BIM, tokenization, blockchain, and smart contract to generate a transparent, digital, and accessible digital ecosystem, aligned with the European Digital Programme and with the experiences in Northern Europe [47]. In fact, in the Netherlands, Estonia, and Finland, GIS and BIM systems are adopted for dealing with the regeneration of urban areas and building projects as well as to implement the digital authorization processes [48]. The property tokenization is based on the possibilities enabled by blockchain platforms in terms of programming smart contracts, that is, contracts that integrate the if-then kind of verification functions and digital promises among actors [49]. Their persistence, transparency, and immutability are guaranteed by the nature of these agreements, verified and recorded on a blockchain. Blockchain platforms can be permissioned, if the owner is a single entity that acts as a guarantor of the operation, or permissionless if the operation is definite by a system of nodes that evaluate each procedure [50].

Smart contracts can be represented as digital tokens transferable in a market if they are divisible and fungible in nature. Nevertheless, if they epitomize a unique asset, such as a work of art or a specific portion of real estate, they acquire the form of NFTs and can be auctioned or traded based on their uniqueness [51].

The development of a comprehensive platform will facilitate the storage and exchange of information among the adopted digital tools and enable their seamless interaction. As a first step, the GIS platform will define the asset geometrically and geographically in the digital environment. This digital representation can be combined with a BIM model to ensure the required legal compliance and validation for potential transactions. Next, a digital fragment of the property can be linked to a token with a specific value, as defined in a smart contract. This token is then stored in the blockchain to validate and establish the legal contract between the involved parties (Figure 5.2).

The GIS model allows the association of alphanumeric information based on location [52]. For example, it enables the comparison of property sales rates in various urban areas by incorporating spatial information into the topological relationship.

In Figure 5.3, geographic information of the central area of Porta Nuova in Milan, Italy (shown in Figure 5.2), is integrated with data on the market cost of assets, which is used to determine the specific value of the property. Linked

Figure 5.2 Digital workflow for property tokenization process

information includes building type, maintenance status [53], minimum and maximum sales values [€/m^2] for different uses, and minimum and maximum rental values [€/m^2 per month] [54]. Additionally, the process of property tokenization can encompass areas with varying market values, thereby enhancing the appeal of regions with lower values that may be regarded as profitable investment opportunities. This diversification in investment potential allows for a broader range of stakeholders to participate in the market, potentially stimulating economic growth in underrepresented areas.

Consequently, the initial step entailed connecting the digital land map with databases that contain information on building status and property value. This integration creates a system that supports the necessary information framework for evaluating the market value of assets across various regions and determining the value of real estate tokens.

The application of BIM for buildings facilitates a seamless transition from the urban context, where relationships are governed by spatial constraints, to the scale

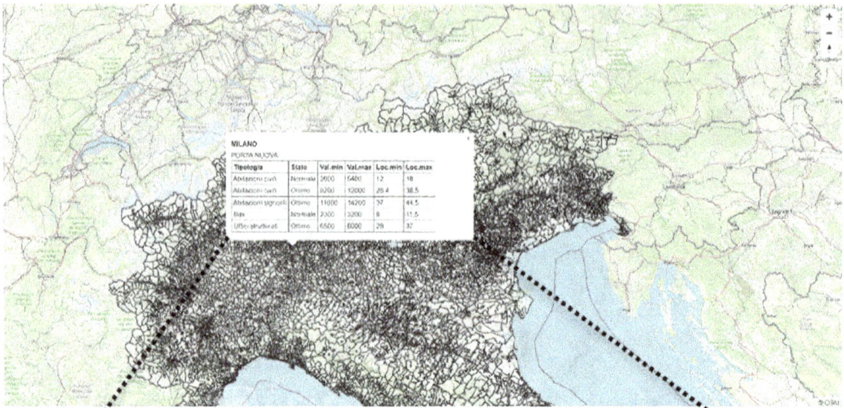

MILANO

PORTA NUOVA

Typology	Status	Min. Value	Max. Value	Rent.min	Rent.max
civil homes	Normal	3,600	5,400	12	18
civil homes	Excellent	8,200	12,000	26.4	38.5
Stately homes	Excellent	11,600	14,200	37	44.5
Garage	Normal	2,300	3,200	8	11.5
Offices	Excellent	6,500	8,000	28	37

Figure 5.3 Milan development in Porta Nuova, Italy: rates of sales [€/m^2] and lease value [€/m^2 per month] depending on conservation condition for houses, offices, and garages

of individual assets [52]. This allows a detailed definition of building units, whose dimensions can be exactly controlled, and the database can be enriched with comprehensive information regarding the property including energy certification, building registry, technical reports, ownership records, and status of the property. This enrichment of data is essential for providing a holistic view of each asset, thereby facilitating informed decision-making and effective management.

By establishing a database of real estate units within major properties and recording the current conditions and contractual terms on a blockchain, transparency, and reliability of data can be significantly enhanced. This approach eliminates the disputes and information gaps that frequently characterize traditional procedures. Such a disruptive approach holds the potential to transform the management of asset and property exchanges, addressing long-standing issues related to discrepancies and inaccuracies in property attribution and boundary definitions that are often identified years later.

Interoperability between GIS and BIM facilitates control over properties and units at both geographic and building scales, thereby creating a cadastral system registered within an immutable and accredited information chain [55]. Digital models also enable the definition of new spatial configurations, as well as the optimization and verification of compliance with regulations concerning dimensions, accessibility, air-to-light ratios, and other factors., through implementable code control procedures. In the real estate market, it is common to encounter properties for sale with significant nonconformities which are unknowingly sold to buyers. Furthermore, during the notarization phase of the sales contract, there is typically no oversight or assumption of responsibility by the selling party for any nonconformities that may arise subsequently.

The property tokenization approach is highly innovative and inclusive, proposing a decentralized revolution in real estate market management processes.

5.3.5 Product and process innovation

The concept of property tokenization seeks to encourage investment in existing buildings, thereby supporting an extended building life cycle. By enhancing liquidity for users, this approach facilitates the allocation of resources for energy efficiency upgrades, leading to cost savings and a reduced environmental impact throughout the operational life of the building. The platform is designed to facilitate investments in energy efficiency while enabling the tracking of these initiatives within a digital cadastre. BIM is employed to transfer data to energy simulation models, which allows for the evaluation and validation of proposed retrofit interventions. This process aims to minimize performance gaps and bolster economic analysis, ultimately strengthening investor confidence in the viability of such upgrades [5].

The digital transition has the potential to enhance the supply chain efficiency in green real estate investments. By establishing a digital platform for organizing geographic information, such as building location and maintenance status, enables financial and energy analyses can be conducted more effectively. This approach promotes economic, social, and energy sustainability within the community of users [56,57]. Digitizing real estate and enhancing energy efficiency not only liberates investment capacity but also fosters environmental awareness and promotes social inclusion. The implementation of blockchain technology creates a community capable of transforming the housing market, thereby encouraging the development of sustainable neighborhoods even in disadvantaged urban areas. Furthermore, the establishment of a digital cadastre paves the way for innovation within public administration by facilitating public-private partnerships.

Property tokenization presents an inclusive, digital framework for managing the processes of buying and selling real estate, thereby enhancing financial conditions, and upgrading the existing built environment. The integration of technologies for data acquisition, localization, modelling, and enrichment significantly improves transparency, dissemination, utilization, and litigation management. Enabled by blockchain technology, tokenization fosters innovation by facilitating peer-to-peer commerce and allowing for flexible representation of ownership in built assets. This approach also diminishes the risks of corruption and illicit activities [58] contributing to a more secure and efficient real estate market.

Tokenization transforms the real estate market by enhancing security, transparency, and traceability, while simultaneously lowering costs associated with intermediaries. This approach is in alignment with the broader digitization agenda, which expands the capabilities of building and urban processes through robust information infrastructure. By connecting real estate market data to urban development plans and energy retrofits, tokenization facilitates efficiency processes and serves as a comprehensive and inclusive financial instrument.

The property tokenization relies on several critical resources, including expertise in innovative technologies and digital services, as well as legal and business knowledge. Extensive research and development in GIS, BIM, smart contracts and blockchain are also essential. The future digital infrastructure will primarily consist of cloud services, software libraries, and application programming interfaces (APIs). During the prototype development phase, it is anticipated that public and permissionless blockchain agreements will be utilized. The adoption of open protocols is justified by the enhanced reliability achieved through widespread and continuous peer review of software, an assurance of quality that cannot always be guaranteed by private producers, regardless of their authority. Moreover, these resources help mitigate the risk of vendor lock-in. Such characteristics are intrinsic to the ontological structure of blockchain technology, which includes disintermediation, the absence of a singular supreme owner or manager with absolute control over the system, code transparency, widespread adoption, and the inability to be dominated by a majority group of nodes [26,59].

5.3.6 Property tokenization as a financing mechanism to upkeep built assets

More than 2 m buildings, out of the 12 mn buildings in Italy are, according to the last census, in a state of poor or very poor conservation [53]. In the period 1998–2019, tax incentives for building renovation and energy retrofit activated about 21 mn interventions, triggering investments of over 346 bn euros, with a burden on the State due to the mechanism of tax exemption of about 166 bn euros. The tax exemption, prior to the approval of the public incentive for energy retrofitting known as the Superbonus [60], was restricted to a percentage that was less than 100% of the associated costs. However, with the introduction of the Superbonus 110%, not only were all costs covered, but there was also an additional percentage allocated for credit monetization, designed to compensate for the advances provided by financial institutions. This was a strong incentive for promoting renovation supporting the historical conjuncture marked by the pandemic; nevertheless, this mechanism realistically cannot be refinanced in the medium to long term. The injection of liquidity that the Superbonus entailed in the real estate sector encouraged an overall economic recovery, as can be perceived by the national GDP predictions.

Exploring innovative financing mechanisms based on the tokenization of real estate assets represents a critical and potentially high-impact area of research. This approach can be scaled according to the age of the assets and aims to minimize government intervention in the market. By leveraging blockchain technology, tokenization can enhance accessibility and liquidity, enabling a broader range of investors to participate in real estate markets. Furthermore, it offers opportunities for more efficient capital allocation and supports sustainable development initiatives, making it a significant focus for future exploration in real estate finance. The advantages for different parties can be listed as follows:

- *Access to funds and credits*: Property owners can secure access to funds and credits without the necessity of selling significant assets, allowing them to sell only a minor fragment of an asset while retaining full ownership and usage right.
- *Opportunities for weak investors:* Less affluent investors that can benefit from new opportunities that enhance their purchasing capacity on profitable investments, contributing to future social and economic stability.
- *Benefits for private financial institutions*: Private financial institutions could leverage more effective forms of loan assurance, leading to a reassessment of mortgage institution that addresses critical issues arising from the accumulation of non-performing loans [61].
- *Environmental and societal advantages*: Renovating and revaluing the existing building stock yields benefits for the environment and common good by reducing anthropic pressure and enhancing public health.

5.3.7 Economic impact

The business model proposed by the property tokenization approach allows for a paradigm shift in real estate investments. In a country where property ownership is at 73% and rental properties are at 27%, the acquisition of a property is clearly considered an investment of unambiguous connotation for the national population [62]. The most required properties are three-room apartments (35%), two-room apartments (24%), four-room apartments (24%), 5 or more rooms (23%), and studio apartments (5%) [63]. The main percentage of loan applications come from married couples (46–45%) and singles (43–35%) that invest in housing; nevertheless, they structurally change their needs over time with a subsequently growing request for cash liquidity and additional income when retired. This phenomenon produces a societal fragility of the investors and critical loan solvency incapacity.

A financial value analysis is presented in Figures 5.4 and 5.5 where the cost of the properties in the northern Italy is calculated and visualized on the geographical map, providing detailed and reliable information about the highest values in the dense urban areas. The granularity of the information may support the analysis to realize the attributable value to tokens related to properties located in specific urban areas. Moreover, it could similarly allow governments, councils or municipalities to mandate and regulate the housing ownership prices through tokens.

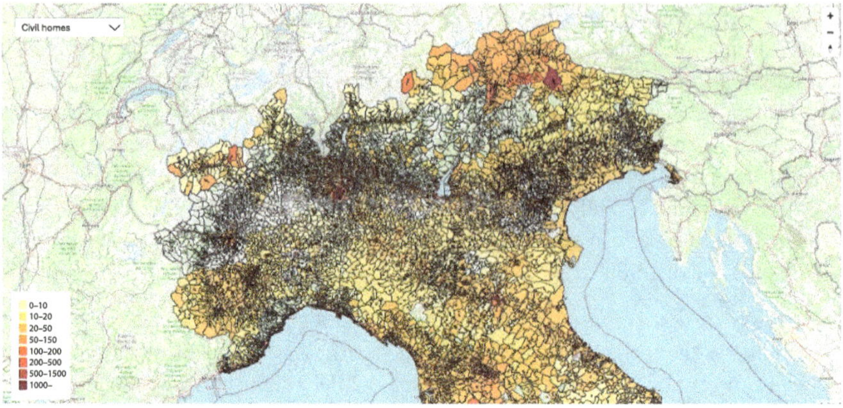

Figure 5.4 Distribution of economic value [€/m^2] in northern Italy

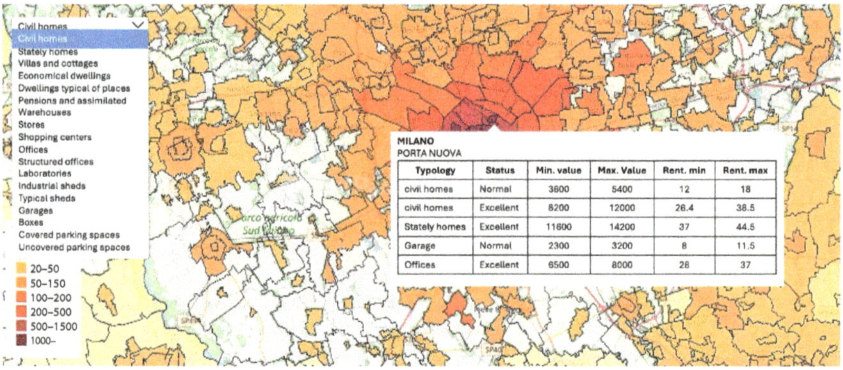

Figure 5.5 Distribution of economic value [€/m^2] in northern Italy (focus on Milan's Porta Nuova area)

The proposed approach has an ambitious however attainable objective, which is to permit the sale of a building portion while maintaining the complete availability for the owner, with the positive effect of increasing the liquidity and allowing safe investments for buyers. The competitiveness of the concept similarly relies upon the gross return of the investment, which for the residential sector is 7.6% (offices 8.6%, commercial 11.5%, garages 6.3%). This can allow the seller to upsurge his/her purchasing power and the buyer to invest with a high rate of return.

In Table 5.2 an example of value of the token for partial disposal (20%) of a property by a 60 year-old owner is calculated based on art. 14 and 17 D.Lgs n. 346 del 31/10/1990 updated by DM 18/12/2020 [64].

The calculations indicate that the annual income rate is high as an investment, with the potential for partial trading of the bare property; its value in percentage terms according to the owner's age (Figure 5.6).

Table 5.2 Example of property token: partial disposal

Input parameters	**Unit**	**Value**
Asset value	€	300,000
Disposable percentage	%	20
Owner age	years	60
Annual revaluation	%	2
Years of ownership	years	25
Output parameters	**Unit**	**Value**
Bare property value	€	24,000
Future property value	€	381,468
Revalued bare property	€	61,034
Estimated revenue	€	37,034
Annual income rate	%	9.12

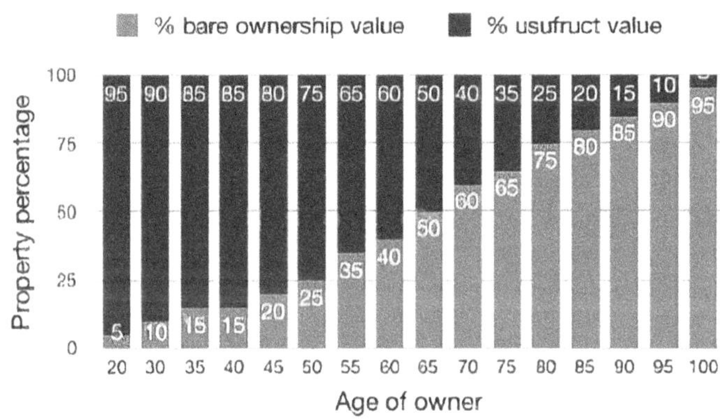

Figure 5.6 Distribution of bare property value and usufruct value related to owner age

An additional simulation has been conducted to examine a buy-back process of the property owner seeking to regain sole ownership of the asset, thereby testing the flexibility of the economic mechanism and the feasibility of the procedure (Table 5.3). These testing assumptions, integrated within an exchange platform that utilizes GIS, BIM, and blockchain technology, enhance the transparency of activities and provide the capability to monitor trends in the real estate market, ensuring long-term investment security. The calculations indicate that the annual income rate is a significant investment, with the potential for partial trading of the bare property; its value fluctuates in percentage terms according to the owner's age (Figure 5.6).

The platform, integrating the above-mentioned technologies creates a system that allows for real-time control and observation of market fluctuations, providing stakeholders with valuable insights into the dynamics of the real estate sector.

The topic of bare ownership presents a unique perspective on property rights, particularly in the context of real estate investments and financial strategies.

Table 5.3 Example of property token: buy-back

Input parameters	Unit	Value
Asset value	€	45,000
Disposable percentage	%	100
Owner age	Years	45
Annual revaluation	%	5
Years of ownership	Years	20
Loan interest	%	2
Output parameters	**Unit**	**Value**
Bare property value	€	90,000
Asset cost	€	360,000
Future property value	€	712,752
Revalued bare property	€	356,376
Revalued usufruct	€	356,376
Estimated revenue	€	266,376
Split share	€	133,188
Annual BP income	%	11.02
Annual U	%	3.24
Loan payment	€	1,821.18

Bare ownership refers to the legal title of a property held by an individual, known as the bare owner, who lacks the right to occupy or derive benefits from it. Instead, these rights are conferred to another party, called the usufructuary, who can use the property, rent it out, or otherwise benefit from it. This arrangement allows the bare owner to retain ownership while providing liquidity through the sale of bare property, often at a price significantly lower than market value.

This concept parallels current trends in property tokenization, where digital assets representing real estate can be bought and sold without necessitating full ownership transfer. Just as bare ownership allows for liquidity without losing the right to occupy the property, tokenization enables fractional ownership and investment opportunities without requiring complete control over an asset. Both mechanisms facilitate financial flexibility for owners and investors, allowing them to manage their assets more effectively while still retaining certain rights or benefits.

Moreover, bare ownership can serve as a strategic option for elderly individuals or those without heirs who seek additional income while maintaining their residence. Similarly, tokenization can empower investors to engage in real estate markets without substantial upfront capital, democratizing access to property investments. As both concepts evolve within the framework of modern financial practices, they highlight the importance of innovative approaches in managing property rights and investments in an increasingly complex economic landscape.

Initiatives related to blockchain-based property virtual gaming in the US have shown the possibility of great involvement by users through digital access to homes with a collection of $18 mn and a valuation of $300 mn [65], demonstrating a great expansion potential for this type of real estate transaction management even in a real context with actual social benefits. The credibility of the process along with the option of the relocated control, is crucial for fostering participation and inclusiveness of the initiative. Furthermore, the digitization of real estate not only enhances transparency and accountability estate but also facilitates the emergence of organizational and commercial spin-offs. By leveraging advanced technologies such as GIS, BIM, and blockchain, stakeholders can engage more effectively in decision-making processes, thereby empowering communities and promoting equitable access to real estate opportunities. This digital transformation can stimulate innovation and adaptability within the sector, ultimately contributing to sustainable economic growth and improved community well-being [66].

5.4 Digital architecture implementation for property tokenization

Creating a digital property tokenization architecture can be a very challenging task on the technical side: the infrastructure to be created includes multiple components that must be integrated and managed together. For instance, the token management component necessitates the use of blockchain technology, which inherently involves a highly complex infrastructure to manage. This involves overseeing individual nodes as well as the provisioning and maintenance of hardware and software, encompassing aspects such as backup, monitoring, and disaster recovery. Additionally, there is a critical requirement for an infrastructure that can rapidly scale to adapt to evolving business trends while effectively managing associated costs. Even possible geographic expansion of the service would mean "bringing the infrastructure" closer to customers in different geographies (e.g. to decrease latency or meet compliance requirements), and therefore maintain true replicas of the architecture. This section identifies and discusses the role of cloud computing and the deployment of a blockchain solution in the cloud.

5.4.1 Use of cloud computing

Cloud computing plays a pivotal role in the implementation of asset tokenization. By leveraging cloud services, organizations can maintain comprehensive control over their infrastructure, such as nodes and smart contracts without the complexities associated with low-level management tasks, including hardware and software setup, upgrade. Cloud computing facilitates on-demand delivery of IT resources over the Internet, employing a pay-as-you-go pricing model. This approach eliminates the need for organizations to purchase, own, and maintain physical data centers and servers, allowing them to access essential technology services, such as computing power, storage, and databases, on an as-needed basis from providers (e.g. big providers such as Amazon Web Services (AWS), Google Cloud Platform, Microsoft Azure, DigitalOcean, Alibaba Cloud, IBM Cloud, Oracle Cloud, or other options such as Kamatera, Linode, Kinsta, Vultr).

Cloud computing facilitates access to a diverse array of technologies that expedite infrastructure deployment, and enable the utilization of various resources, ranging from computing and storage services to advanced applications in the Internet of Things, machine learning, data lakes, and analytics solutions. The subsequent section will demonstrate how a cloud computing platform as example structured on AWS can significantly enhance the implementation of this infrastructure by substantially reducing the time, effort, and costs associated with deployment. Additionally, it ensures a high level of control and visibility over both the infrastructure and the underlying blockchain, capabilities that higher-level solutions do not always provide.

In the context of this chapter, it is not feasible to predict the impact of the framework on the market. Traditional on-premises resources necessitate over-provisioning the infrastructure to accommodate peaks demand, which risks implementing a system that is not future-proof. In contrast, cloud computing, enables the tokenization infrastructure to automatically adjust the resources in accordance to actual needs, thereby eliminating reliance on predictions and reducing waste associated with anticipating future business activity peaks. Additionally, cloud computing facilitates the creation of infrastructure that is more cost-effective by transitioning from fixed costs, such as those associated with data centers and physical servers, to variable costs, allowing organizations to pay only for the IT resources utilized. The application infrastructure can scale automatically based on workload demands; for instance, the platform can rapidly expand without human intervention to support hundreds of thousands or even millions of users, all while maintaining virtually no downtime. There are three main models for cloud computing. Each model represents a different part of the cloud computing stack:

- Infrastructure as a service (IaaS): provides the foundational components for cloud computing, typically granting access to networking capabilities, computing resources (whether virtual or on dedicated hardware), and data storage. IaaS offers the highest level of flexibility and management control over IT resources, closely resembling existing IT infrastructures familiar to many IT departments and developers.
- Platforms as a service: alleviates the need for organizations to manage the underlying infrastructure, such as hardware and operating systems, allowing them to concentrate on the deployment and management of applications. This shift enhances operational efficiency by eliminating concerns related to resource procurement, capacity planning, software maintenance, patching, and other routine tasks associated with application management.
- Software as a service (SaaS): delivers a complete product that is operated and managed by the service provider. Typically, references to SaaS pertain to end-user applications. With a SaaS offering, users do not need to consider how the service is maintained or how the underlying infrastructure is managed; instead, they focus solely on utilizing the software. A common example of a SaaS application is web-based email, which allows users to send and receive messages without the need to manage feature updates or maintain the servers and operating systems supporting the email service.

5.4.2 The role of cloud computing

In the context of property tokenization, blockchain technology plays a crucial role by providing a decentralized and secure framework for managing ownership and transaction records. Using AWS as an example of possible provider, we can see that the specialized tools facilitate the management of a centralized ledger database, which maintains an immutable and cryptographically verifiable record of transactions. Additionally, it provides a fully managed multi-party blockchain network that effectively eliminates the need for intermediaries, thereby streamlining processes associated with property ownership. For example, Amazon Managed Blockchain [67] significantly reduces the complexity involved in setting up blockchain networks, cutting down the time required to host Hyperledger Fabric frameworks by approximately 60%. This service simplifies network operations by supporting AWS Command Line Interface (CLI) [68], AWS CloudFormation [69], and Amazon CloudWatch [70] logs, which enhance usability and monitoring capabilities.

Utilizing AWS CLI and AWS CloudFormation enables the definition of infrastructure as code, facilitating version control of both the infrastructure and the applications in operation. This capability simplifies the deployment of similar infrastructures across different regions, such as establishing identical services in various countries or creating dedicated instances for specific clients. Furthermore, AWS supports over 70 validated blockchain solutions from various partners, encompassing all major blockchain protocols, including Hyperledger Sawtooth, Corda, DAML, Ethereum, Quorum, Blockstack Strato, RSK, and Kadena ScalableBFT. This extensive support allows organizations to select their preferred framework for deploying tokenization applications without being constrained to a specific technology. The significance of blockchain in property tokenization lies in its ability to provide a secure and transparent framework for managing ownership and transaction records. By leveraging AWS's capabilities, stakeholders can efficiently implement blockchain solutions that enhance trust and reduce the need for intermediaries in real estate transactions. The flexibility offered by AWS's cloud services not only streamlines the deployment process but also ensures that organizations can adapt their tokenization strategies to meet evolving market demands while maintaining high levels of operational control and visibility.

5.4.3 Ethereum as a potential blockchain platform for property tokenization

Ethereum [71] serves as a highly flexible platform for the development of decentralized applications (dApps), utilizing the native Solidity [72] scripting language and the Ethereum Virtual Machine [73]. Developers deploying smart contracts on Ethereum benefit from a mature ecosystem that includes extensive developer tooling and established best practices, which enhance both the development process and user experience. The availability of user-friendly wallets, such as MetaMask [74], Argent [75], and Rainbow [76], further facilitates interaction with the Ethereum blockchain and its smart contracts.

The substantial user base of Ethereum encourages developers to launch their applications on this network, reinforcing its position as a primary hub for dApps, including those related to DeFi and NFTs. Looking ahead, the upcoming Ethereum 2.0 protocol [77] aims to provide a more scalable infrastructure capable of supporting applications that require higher transaction throughput. The completion of "The Merge" on September 15, 2022, transitioned Ethereum from a Proof of Work to a Proof of Stake (PoS) consensus mechanism [78], significantly enhancing the sustainability [79] of the network by reducing energy consumption by an estimated 99.95%. This shift marks a pivotal moment in Ethereum's evolution toward becoming an eco-friendlier blockchain. In the context of property tokenization, these features of Ethereum are particularly relevant. The ability to create dApps using Solidity allows for the development of smart contracts that can automate various aspects of property transactions, thereby eliminating intermediaries and enhancing transparency. Furthermore, Amazon Managed Blockchain facilitates the deployment of Ethereum Mainnet nodes on the PoS network, simplifying the setup and management of blockchain infrastructures necessary for property tokenization. By leveraging AWS's capabilities alongside Ethereum's robust framework, stakeholders can effectively implement tokenization strategies that streamline real estate transactions while ensuring security and compliance. This integration not only promotes trust through immutable records but also opens opportunities for fractional ownership and broader market participation in property investments. As such, the intersection of blockchain technology and property tokenization represents a transformative shift in how real estate assets are managed and exchanged.

5.4.4 A proposed technical architecture

NFTs are issued and managed through smart contracts, with the ERC-721 [80] token standard being one of the most widely recognized frameworks within the Ethereum ecosystem. The ERC-721 standard establishes a unique identifier for each token, linking it to a user's public address derived from their public key, and designates a storage location for the underlying media or metadata associated with that NFT. This standardization is particularly relevant in the realm of property tokenization, as it facilitates the representation of unique real estate assets on the blockchain. The proposed architecture for an NFT full-stack application consists of five tiers, each serving a distinct function in the overall system.

The first tier is a presentation layer that can be deployed on Amazon Simple Storage Service (Amazon S3) [81] and accessed via Amazon CloudFront [82]. Amazon S3 provides scalability, data availability, security, and performance, making it suitable for storing static assets related to property tokenization applications. Meanwhile, Amazon CloudFront acts as a content delivery network, ensuring global reach and high performance for users interacting with the application.

The second tier implements the business logic of the application, connecting the presentation layer to the blockchain network. This layer encompasses core functionalities such as a frontend web application that enables users to create, manage, and sell tokens, as well as an API that interfaces with the NFT smart contract running on Amazon Managed Blockchain. Utilizing AWS serverless compute services, such as Amazon Elastic Kubernetes Service Fargate [83], Amazon Elastic Container Service Fargate [84], or AWS Lambda [85], facilitates automatic scaling and high availability while optimizing costs.

The third tier consists of the smart contract that adheres to the NFT specification, operating on an Ethereum node managed by Amazon Managed Blockchain. This service simplifies node management by eliminating the need for manual hardware provisioning and allows for dynamic scaling based on application demands. Each node functions as a fully operational Ethereum node capable of executing smart contracts and performing all necessary Ethereum tasks.

The fourth tier serves as a data extraction layer that captures blockchain activities and ingests them into an Amazon S3 bucket via an Amazon Kinesis Data Firehose stream [86]. This layer is invoked by the main application to offload event data sent and received by the blockchain into a persistent storage solution for subsequent analytics. By leveraging a serverless streaming service like Amazon Kinesis Data Firehose, organizations can efficiently manage varying throughput levels while delegating infrastructure scaling to AWS.

Finally, the fifth tier comprises an analytics layer that enables data analysts to perform SQL-like queries on blockchain events. By utilizing services such as Amazon Athena [87], a serverless interactive analytics service, analysts can query transactions and tokens on the blockchain to gain insights into property tokenization activities.

In summary, the integration of NFTs through standards like ERC-721 within a cloud-based framework not only enhances the management of unique real estate assets but also facilitates transparency and efficiency in property transactions. By leveraging AWS's capabilities alongside Ethereum infrastructure, stakeholders can implement property tokenization strategies that facilitate access to real estate investments while ensuring secure ownership records (Figure 5.7).

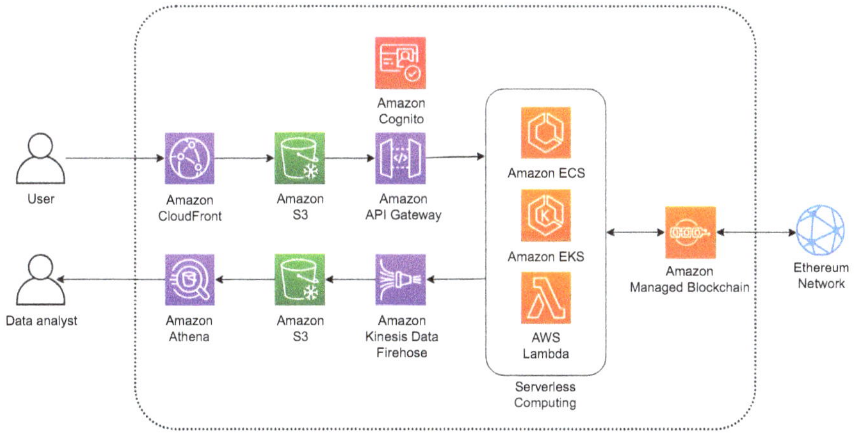

Figure 5.7 Scheme of the proposed possible architecture

The utilization of cloud computing offers numerous advantages, particularly in the context of property tokenization. By leveraging suitable services, organizations can deploy a comprehensive infrastructure with minimal effort, allowing them to concentrate on the core business of tokenization rather than the complexities of managing the underlying systems. The proposed architecture enables seamless scaling of resources, accommodating fluctuations in user demand; this means that during periods of high activity, resources can be scaled up efficiently, while during quieter times, they can be scaled down without the burden of overprovisioning or retaining unused assets.

Moreover, the entire infrastructure can be defined as code, facilitating easy duplication and implementation across different geographies or markets. This capability is particularly significant for property tokenization, as it allows businesses to replicate their tokenization solutions in various regions, thereby expanding their market reach and adapting to local regulations and market conditions.

In summary, the proposed architecture exemplifies how the integration of a provider such as AWS cloud services could streamline the operational aspects of property tokenization and enhance flexibility and scalability. This allows stakeholders to focus on innovation and service delivery while ensuring that the necessary infrastructure [88] can be efficiently deployed and managed across diverse environments. As a result, cloud computing plays a vital role in advancing property tokenization initiatives by providing a robust and adaptable framework for managing digital assets in real estate [89].

5.5 Conclusions, challenges, and future directions

This document discussed the foundational principles and impact of property tokenization and the potential digital infrastructure required for its implementation. It proposed a proposed framework for the tokenization of real estate assets supported by GIS, BIM, smart contract, blockchain, and tokens technologies. Using the Italian's real estate sector, recognized for its complexity and challenges, as a context, this chapter showed the solution's components, demonstrated how economic analysis for tokens can be executed, and advocated that the tokenization of properties in a blockchain environment will facilitate the needed flexibility in asset ownership arrangements. This plasticity, or the capacity to be shaped or modelled encompassing various contexts, includes but is not limited to the following options:

- fractional ownership of an asset with full rights;
- fractional ownership of an asset with limited ownership rights;
- time-bound ownership and time-share ownership of an asset.

The implementation of these arrangements is expected to enable new business models to support communities financially, raise funds for the retrofitting of the building stock in Europe, mitigate the existing housing challenges, and improve the current inclusivity in the asset market for underprivileged clusters. This will ultimately lead to more dynamic asset ownership and use modes catering for different needs and financial capabilities, and relieving municipalities and governments of pressures for low-carbon renovation processes or affordable housing policies, securing the property tokenization framework to sustainability as a key goal in the built environment.

The chapter discussed the key technological component involved in a property tokenization solution. GIS and BIM, when integrated and maintained concurrently in the tokenization solution, form the backbone of information needed at the urban and asset level. Their combined use improves the build-up of the information to be block-chained and controlled in this agreement to a finer detail. This is also necessary to provide reliable data for tokenization applications and render them futureproof from a technology perspective.

With the ongoing rapid digitalization in the built environment, the tokenization approach can be expanded to components other than buildings (e.g. parks, roads, and trees), where funds raised through tokens can be employed for environmental, economic, and social sustainability purposes. It is expected that the current level of technology backbone and asset value knowledge is adequate to operationalize and pilot the property tokenization system and the blockchain architecture depicted can effectively support the tokenization and registration processes.

The integration of blockchain technology into the real estate sector, particularly through property tokenization, could present a transformative opportunity to enhance liquidity, accessibility, and efficiency in real estate transactions. The current landscape reveals significant disparities in investment attractiveness across European countries, with regions like northern Europe attracting far more investments than Italy, primarily due to regulatory challenges and market dynamics. The concept of property tokenization allows for fractional ownership, enabling minor investors, such as young individuals, seniors, and those with limited financial means, to participate in real estate markets that were previously inaccessible.

This democratization of investment is crucial for fostering social inclusion and addressing the barriers that have historically excluded economically disadvantaged groups from property ownership. Moreover, the real estate sector's low digital maturity highlights an urgent need for innovation. Blockchain applications could mitigate inefficiencies related to fraud and corruption while streamlining transaction processes. The legal frameworks surrounding tokenization must evolve to provide clarity and consistency across jurisdictions, as regulatory uncertainty remains a significant barrier to widespread adoption. The potential for digital tokens to enhance investment effectiveness and liquidity is evident, particularly within crowdfunding platforms. However, challenges such as the lack of a centralized marketplace for trading real estate security tokens must be addressed to realize their full potential.

Overall, the exploration of property tokenization within this chapter underscores its capacity to redefine ownership structures and improve market fluidity while emphasizing the necessity for supportive regulatory environments and technological advancements to facilitate this shift.

In the chapter, the data structure of the system and an example of architecture for the implementation are presented to demonstrate the technological feasibility however regulations and current practitioners are far from this kind of evolved solution.

It is worth noticing that this type of asset tokenization on the blockchain is expected to deliver assurances against illegal and illicit activities in asset proprietorship (e.g. double sale of the same asset to two buyers). In time, a dynamic asset token-exchange market can be formed enabling spin-off businesses in this area. As with all disruptive arrangements, legal, social, technological, and management readiness are key concerns to be tackled before a real-life implementation of this arrangement.

Beyond economic sustainability, two additional scenarios related to social and environmental sustainability by tokenization are proposed. First, beyond creating new ways of ownership for the property market, it is possible to configure ultimately the use of these arrangements to mitigate homelessness or housing problems, or to create extra revenue for low-income families as part of social sustainability programs. Second, the token sale can be used to create funds for the refurbishment/retrofitting of aged assets, to reach the goal of low-carbon emission, which is one of the major concerns in Europe related to the built environment.

As demonstrated and discussed in the chapter, the technological backbone is not the real challenge in implementing tokenization models and opportunities. The technology infrastructure is evolving and available with increasing provisions. What is needed is making the currently static asset ownership laws and regulations, and asset recording and registry administration compliant with tokenization. It can be achieved in two ways: leadership by policymakers and governments, or public pressure with increased awareness of tokenization and blockchain-based applications.

In addition, several other limitations hinder the widespread adoption of property tokenization. One significant barrier is the lack of consumer awareness and understanding of blockchain technology and tokenization processes. Many potential investors may be hesitant to engage with tokenized assets due to a lack of familiarity with how these systems operate, which can lead to skepticism and reluctance to invest. Furthermore, the existing financial infrastructure may not be fully equipped to handle the nuances of tokenized assets, creating friction in transactions and potentially deterring participation.

Furthermore, persistent concerns about security and fraud persist in the digital domain are amplified by the growing prevalence of cyber threats that expose blockchain systems and tokenized assets to potential vulnerabilities. Regulatory uncertainty also stands as a critical challenge; disparate legal frameworks across jurisdictions complicate cross-border transactions and undermine investor confidence. In addition, the lack of a centralized marketplace for trading real estate security tokens restricts liquidity, posing significant hurdles for investors seeking to efficiently buy or sell their holdings. Collectively, these issues create a multifaceted landscape that must be carefully managed to unlock the transformative potential of property tokenization in the real estate sector.

References

[1] Demyanyk, Y., and Van Hemert, O., 2011. Understanding the subprime mortgage crisis, *The Review of Financial Studies*, 24(6), pp. 1848–1880.

[2] PWC, 2021. Urban Land Institute, Emerging Trends in Real Estate® An uncertain impact Europe, https://www.pwc.com/emergingtrendsrealestateeurope2021 (accessed 18 March 2025).
[3] Dahlan, N. H. M., 2014. *Corporate Insolvency Laws in Abandoned Housing Projects: Issues and Prospects (UUM Press)*. Sintok: UUM Press.
[4] Bischoff, O., and Maennig, W., 2011. Rental housing market segmentation in Germany according to ownership. *Journal of Property Research*, 28(2), 133–149.
[5] Li, H. X., Ma, Z., Liu, H., Wang, J., Al-Hussein, M., and Mills, A., 2020. Exploring and verifying BIM-based energy simulation for building operations. *Engineering, Construction and Architectural Management*, 27, 1679–1702.
[6] Norta, A., Fernandez, C., and Hickmott, S., 2018. Commercial Property Tokenizing with Smart Contracts. In *2018 International Joint Conference on Neural Networks (IJCNN)*, pp. 1–8.
[7] Wang, G., and Nixon, M., 2021. SoK: Tokenization on Blockchain. In *Proceedings of the 14th IEEE/ACM International Conference on Utility and Cloud Computing Companion*.
[8] Mistrangelo, P., Tagliabue, L. C., and Tezel, A., 2022. Property Tokenization Digital Framework for Inclusive and Sustainable Asset Markets Development. In *EC3 Conference 2022*, Vol. 3, pp. 53–60. University of Turin.
[9] Garcia-Teruel, R. M., and Simón-Moreno, H., 2021. The digital tokenization of property rights. A comparative perspective. *Computer Law & Security Review*, 41, 105543.
[10] Mendoza, R.U., and Thelen, N., 2008. Innovations to make markets more inclusive for the poor. *Development Policy Review*, 26(4), 427–458.
[11] Savills, 2016. Around the world in dollars and cents. SAVILLS WORLD RESEARCH [Internet]. Available from savills.com (accessed 18 March 2025).
[12] International Energy Agency, United Nations Environment Programme, 2022. Global Status Report – Towards a zero-emission, efficient and resilient buildings and construction sector [Internet]. Available from https://www.unep.org/resources/publication/2022-global-status-report-buildings-and-construction (accessed 18 March 2025).
[13] Calvino F., Criscuolo C., Marcolin L., and Squicciarini M., 2018. A taxonomy of digital intensive sectors [Internet]. OECD Science, Technology and Industry Working Papers; vol. 2018/14. Report No. 2018/14. Available from https://www.oecd-ilibrary.org/industry-and-services/a-taxonomy-of-digital-intensive-sectors_f404736a-en (accessed 18 March 2025).
[14] Manyika J., Ramaswamy S., Somesh K., *et al.*, 2015. Digital America: A Tale of Haves and Have-Mores [Internet]. McKinsey Global Institute. Available from https://www.mckinsey.com/industries/technology-media-and-telecommunications/our-insights/digital-america-a-tale-of-the-haves-and-have-mores.
[15] Saari A., Vimpari J., and Junnila S., 2022. Blockchain in real estate: Recent developments and empirical applications. *Land Use Policy*, 121, 106334.

[16] Wang, D., 2021. Tokenisation of corporate real estate on the blockchain: New strategies for corporate ownership and financial management. *Corporate Real Estate Journal*, 10(2), 168–180.

[17] DECRETO-LEGGE 14 dicembre, 2018, no. 135 Disposizioni urgenti in materia di sostegno e semplificazione per le imprese e per la pubblica amministrazione (18G00163) (GU Serie Generale n.290 del 14-12-2018).

[18] LEGGE 11 febbraio, 2019, no. 12 Conversione in legge, con modificazioni, del decreto-legge 14 dicembre 2018, no. 135, recante disposizioni urgenti in materia di sostegno e semplificazione per le imprese e per la pubblica amministrazione (19G00017) (GU Serie Generale n.36 del 12-02-2019).

[19] Creta F., and Tenca F., 2021. Tokenomics: A new opportunity in the real estate business? A qualitative approach to crowdfunding and blockchain interaction. *First Monday*, 26(10). Available from https://doi.org/10.5210/fm.v26i10.10699.

[20] Chow Y.L., and Tan K.K., 2022. Real estate insights: Is tokenization of real estate ready for lift off in APAC? *Journal of Property Investment & Finance*, 40(3), 284–290.

[21] Baum, A., 2021. Tokenization—The future of real estate investment? *The Journal of Portfolio Management*, 47(10), 41–61.

[22] Konashevych, O., 2020a. Constraints and benefits of the blockchain use for real estate and property rights. *Journal of Property, Planning and Environmental Law*, 12, 109–127.

[23] Bennett, R.M., Pickering, M., and Kara, A.K., 2021. Hybrid approaches for smart contracts in land administration: Lessons from three blockchain proofs-of-concept. *Land*, 10(2), 220.

[24] Rodima-Taylor, D., 2021. Digitalizing land administration: The geographies and temporalities of infrastructural promise. *Geoforum*, 122, 140–151.

[25] Adam, I., and Fazekas, M., 2021. Are emerging technologies helping win the fight against corruption? A review of the state of evidence. *Information Economics and Policy*, 57,100950.

[26] Li, J., and Kassem, M., 2021. Applications of distributed ledger technology (DLT) and blockchain-enabled smart contracts in construction. *Automation in Construction*, 132, 103955.

[27] Latifi, S., Zhang, Y. and Cheng, L. C., 2019. Blockchain-Based Real Estate Market: One Method for Applying Blockchain Technology in Commercial Real Estate Market. In *IEEE International Conference on Blockchain (Blockchain)*, pp. 528–535.

[28] Konashevych, O., 2020b. General concept of real estate tokenization on blockchain: The right to choose. *European Property Law Journal*, 9, 21–66.

[29] Kifokeris, D., Dounas, T., Tezel, A., and Moon, S. 2023. What is the potential value of tokens and token engineering for the architecture, engineering, and construction industry? A positional paper. In *Proceedings of the 2023 European Conference on Computing in Construction and the 40th International CIB W78 Conference*, http://www.doi.org/10.35490/EC3.2023.304.

[30] Naderi, H., Heydari, M. H., and Shojaei, A., 2024. Applications of Tokenization in the Architecture, Engineering, and Construction (AEC) Industry. In *Construction Research Congress 2024*, pp. 269–278.

[31] Farnaghi, M., and Mansourian, A., 2020. Blockchain, an enabling technology for transparent and accountable decentralized public participatory GIS. *Cities*, 105, 102850.

[32] Norta, A., Fernandez, C., and Hickmott, S., 2018. Commercial Property Tokenizing with Smart Contracts. In *2018 IEEE International Joint Conference on Neural Networks (IJCNN)*, pp. 1–8.

[33] Wang, Q., Li, R., Wang, Q., and Chen, S., 2021. Non-fungible token (NFT): Overview, evaluation, opportunities and challenges. *arXiv preprint arXiv*:2105.07447.

[34] MSCI, 2021. Market Size Report on Global Real Estate. Available from https://www.msci.com/www/research-report/real-estate-market-size-2021-22/03296053034, accessed 30 April 2023.

[35] Deloitte, 2020. Property Index Overview of European Residential Markets. Available from https://www2.deloitte.com/content/dam/Deloitte/at/Documents/presse/at-property-index-2022-final.pdf, accessed 30 April 2023.

[36] Scheinert C., 2016. Virtual currencies challenges following their introduction, European Parliamentary Research Service.

[37] European Union, Regulation (EU) 2022/858 of the European Parliament and of the Council of 30 May 2022 on a pilot regime for market infrastructures based on distributed ledger technology, and amending Regulations (EU) No. 600/2014 and (EU) No. 909/2014 and Directive 2014/65/EU.

[38] Saari, A., Junnila, S., and Vimpari, J., 2022a. Blockchain's grand promise for the real estate sector: A systematic review. *Applied Sciences*, 12(23), 11940.

[39] Benítez-Martínez, F. L., Romero-Frías, E., and Hurtado-Torres, M. V., 2023. Neural blockchain technology for a new anticorruption token: Towards a novel governance model. *Journal of Information Technology & Politics*, 20(1), 1–18.

[40] Darusalam, S., Janssen, M., Said, J., Omar, N., and Indra Saputra, M., 2023. Smart Contracts for Creating Transparent Transactions to Reduce Corruption. In *Proceedings of the 24th Annual International Conference on Digital Government Research*, pp. 355–361.

[41] Graglia, J. M., and Mellon, C., 2018. Blockchain and property in 2018: At the end of the beginning. *Innovations: Technology, Governance, Globalization*, 12(1–2), 90–116.

[42] Legge del 11/02/2019 no. 12 – Conversione in legge, con modificazioni, del decreto-legge 14 dicembre 2018, no. 135, recante disposizioni urgenti in materia di sostegno e semplificazione per le imprese e per la pubblica amministrazione. Pubblicato in Gazzetta Ufficiale no. 36 del 12 febbraio 2019.

[43] European Union, Regulation (EU) No. 910/2014 of the European Parliament and of the Council of 23 July 2014. On electronic identification and trust services for electronic transactions in the internal market and repealing Directive 1999/93/EC, Electronic Identification – eIDAS.

[44] Saygili, M., Mert, I. E., and Tokdemir, O. B. 2022, A decentralized structure to reduce and resolve construction disputes in a hybrid blockchain network. *Automation in Construction*, 134, 104056.

[45] Kalyuzhnova, N., 2018. Transformation of the real estate market on the basis of use of the blockchain technologies: opportunities and problems, In *MATEC Web of Conferences*, p. 06004, EDP Sciences.

[46] Mistrangelo, P., Tagliabue, L.C., and Tezel, A., 2023. Digital and Legal Environment Definition for Property Tokenization Uptake in Italy, In *2023 European Conference on Computing in Construction and the 40th International CIB W78 Conference*, Heraklion, Crete, Greece.

[47] Heiskanen, E., Hyvönen, K., Laakso, S., Laitila, P., Matschoss, K., and Mikkonen, I., 2017. Adoption and use of low-carbon technologies: Lessons from 100 Finnish pilot studies, field experiments and demonstrations. *Sustainability*, 9(5), 847.

[48] Noardo, F., Wu, T., Arroyo Ohori, K., Krijnen, T., Tezerdi, H. and Stoter, J., 2020. GeoBIM for Digital Building Permit Process: Learning from a Case Study in Rotterdam. In *ISPRS Annals of Photogrammetry. Remote Sensing and Spatial Information Sciences*, Copernicus GmbH, pp. 151–158.

[49] Wang, S., Ouyang, L., Yuan, Y., Ni, X., Han, X. and Wang, F. Y., 2019b. Blockchain-enabled smart contracts: Architecture, applications, and future trends. *IEEE Transactions on Systems, Man, and Cybernetics: Systems*, 49, 2266–2277.

[50] Nawari, N. O., and Ravindran, S., 2019. Blockchain and the built environment: Potentials and limitations. *Journal of Building Engineering*, 25, 100832.

[51] Chirtoaca, D., Ellul, J., and Azzopardi, G., 2020. A Framework for Creating Deployable Smart Contracts for Non-fungible Tokens on the Ethereum Blockchain. In *IEEE International Conference on Decentralized Applications and Infrastructures (DAPPS)*, pp. 100–105.

[52] Wang, H., Pan, Y., and Luo, X. 2019a. Integration of BIM and GIS in sustainable built environment: A review and bibliometric analysis. *Automation in Construction*, 103, 41–52.

[53] Centro Studi Confartigianato, 2015. Available from: https://www.confartigianato.it/2015/11/studi-oltre-due-milioni-gli-edifici-residenziali-in-cattivo-stato-di-conservazione-2-051-808-pari-al-168-e-il-288-consumi-di-energia-nel-settore-residenziale/ (accessed 18 March 2025).

[54] Agenzia delle Entrate, 2023. Osservatorio del mercato immobiliare (OMI) Non Residenziale, Statistiche IV trimestre 2022, Agenzia delle Entrate. Available from: https://www.agenziaentrate.gov.it/portale/documents/20143/4509119/StatisticheOMI_NON_RES_3_2022_20221203.pdf/33d09878-5374-55ad-9d7e-18bf272feab2 (accessed 18 March 2025).

[55] Ma, Z., and Ren, Y., 2017. Integrated application of BIM and GIS: An overview, *Procedia Engineering*, 196, 1072–1079.

[56] Nguyen, Q. K., 2016. Blockchain – A Financial Technology for Future Sustainable Development, *3rd International Conference on Green Technology and Sustainable Development (GTSD)*, pp. 51–54.

[57] Adams, R., Kewell, B. and Parry, G. 2018. Blockchain for Good? Digital Ledger Technology and Sustainable Development Goals. In: Leal Filho, W., Marans, R. W. and Callewaert, J. (eds.), *Handbook of Sustainability and Social Science Research*. Cham: Springer International Publishing, pp. 127–140.

[58] Li, J., Greenwood, D., and Kassem, M. 2019. Blockchain in the built environment and construction industry: A systematic review, conceptual models and practical use cases. *Automation in Construction*, 102, 288–307.

[59] Perera, S., Nanayakkara, S., Rodrigo, M. N. N., Senaratne, S., and Weinand, R., 2020. Blockchain technology: Is it hype or real in the construction industry? *Journal of Industrial Information Integration*, 17, 100125.

[60] Agenzia delle Entrate, 2020. Disposizioni di attuazione degli articoli 119 e 121 del decreto-legge 19 maggio 2020, no. 34, per l'esercizio delle opzioni relative alle detrazioni spettanti per gli interventi di ristrutturazione edilizia, recupero o restauro della facciata degli edifici, riqualificazione energetica, riduzione del rischio sismico, installazione di impianti solari fotovoltaici e infrastrutture per la ricarica di veicoli elettrici.

[61] Kasinger, J., Krahnen, J. P., Ongena, S., Pelizzon, L., Schmeling, M., and Wahrenburg, M., 2021. Non-performing Loans – New risks and policies? NPL resolution after COVID-19: Main differences to previous crises, *Study Requested by the ECON Committee, European Parliament, Economic Governance Support Unit (EGOV) Directorate-General for Internal Policies.*

[62] Consob – Commissione Nazionale per la Società e la Borsa, 2020, Report on financial investments of Italian households Behavioural attitudes and approaches. Available from: https://www.consob.it/documents/46180/46181/rf2020.pdf/ccfe7ad2-810f-4490-bd7e-413daa24c391 (accessed 20 November 2024).

[63] Global Property Guide Europe, 2021. Available from: https://www.globalpropertyguide.com/Europe/Italy (accessed 18 March 2025).

[64] Agenzia delle Entrate Testo unico del 31/10/1990 no. 346. Available from: https://www.agenziaentrate.gov.it/portale/documents/20143/244997/Modello+Testo+unico+del+31+ottobre+1990+n+346_Testo+unico+del+31_10_1990+n.+346+-+.pdf/b0d37750-c30f-23a6-7ba7-e87b75331b86 (accessed 18 March 2025).

[65] Animoca Brands, https://www.animocabrands.com/coverage-2021-jul-dec (accessed 18 March 2025).

[66] Tian, Y., Minchin, R. E., Chung, K., Woo, J., and Adriaens, P., 2022. Towards inclusive and sustainable infrastructure development through blockchain-enabled asset tokenization: An exploratory case study. *IOP Conference Series: Materials Science and Engineering*, 1218(1), 012040.

[67] Amazon Web Services, n.d. Blockchain [Internet]. Seattle: AWS. Available from https://aws.amazon.com/blockchain/ (accessed 18 March 2025).

[68] Amazon Web Services, n.d. Command Line Interface [Internet]. Seattle: AWS. Available from https://aws.amazon.com/cli/ (accessed 18 March 2025).

[69] Amazon Web Services, n.d. CloudFormation [Internet]. Seattle: AWS. Available from https://aws.amazon.com/cloudformation/ (accessed 18 March 2025).

[70] Amazon Web Services, n.d. Amazon CloudWatch [Internet]. Seattle: AWS. Available from https://aws.amazon.com/it/cloudwatch/ (accessed 18 March 2025).
[71] Ethereum Foundation, n.d. Ethereum [Internet]. Available from https://ethereum.org/en/ (accessed 18 March 2025).
[72] Solidity Team, n.d. Solidity [Internet]. Available from https://soliditylang.org/ (accessed 18 March 2025).
[73] Ethereum Foundation, n.d. Ethereum Virtual Machine (EVM) documentation [Internet]. Available from https://ethereum.org/en/developers/docs/evm/
[74] MetaMask, n.d. MetaMask [Internet]. Available from https://metamask.io/ (accessed 18 March 2025).
[75] Argent Team, n.d. Argent Wallet [Internet]. Available from https://www.argent.xyz/ (accessed 18 March 2025).
[76] Rainbow Team, n.d. Rainbow Wallet [Internet]. Available from https://rainbow.me/ (accessed 18 March 2025).
[77] Ethereum Foundation, n.d. Ethereum Upgrades [Internet]. Available from https://ethereum.org/en/upgrades/ (accessed 18 March 2025).
[78] Ethereum Foundation, n.d. The Merge Upgrade [Internet]. Available from https://ethereum.org/en/upgrades/merge/ (accessed 18 March 2025).
[79] Ethereum Foundation, n.d. The Merge and Energy Consumption [Internet]. Available from https://ethereum.org/en/upgrades/merge/#merge-and-energy (accessed 18 March 2025).
[80] ERC721 Team, n.d. ERC721 Standard [Internet]. Available from https://erc721.org/ (accessed 18 March 2025).
[81] Amazon Web Services, n.d. Amazon S3 Product Page [Internet]. Seattle: AWS. Available from https://aws.amazon.com/pm/serv-s3 (accessed 18 March 2025).
[82] Amazon Web Services, n.d. CloudFront [Internet]. Seattle: AWS. Available from https://aws.amazon.com/cloudfront/ (accessed 18 March 2025).
[83] Amazon Web Services, n.d. Elastic Kubernetes Service (EKS) [Internet]. Seattle: AWS. Available from https://aws.amazon.com/eks/ (accessed 18 March 2025).
[84] Amazon Web Services, n.d. Elastic Container Service (ECS) [Internet]. Seattle: AWS. Available from https://aws.amazon.com/ecs/ (accessed 18 March 2025).
[85] Amazon Web Services, n.d. Lambda [Internet]. Seattle: AWS. Available from https://aws.amazon.com/lambda/ (accessed 18 March 2025).
[86] Amazon Web Services, n.d. Kinesis Data Firehose [Internet]. Seattle: AWS. Available from https://aws.amazon.com/kinesis/data-firehose/ (accessed 18 March 2025).
[87] Amazon Web Services, n.d. Athena [Internet]. Seattle: AWS. Available from https://aws.amazon.com/athena/ (accessed 18 March 2025).
[88] Choubey, A., Choubey, S., Jaiswal, D., and Jaiswal, M., 2024, Integrating Blockchain in Cloud Computing for Enhanced Data Management and Security. In *2024 11th IEEE International Conference on Reliability, Infocom Technologies and Optimization (Trends and Future Directions)(ICRITO)*, pp. 1–7.
[89] Roy, S., Shovon, A. R., and Whaiduzzaman, M., 2017, Combined Approach of Tokenization and Mining to Secure and Optimize Big Data in Cloud Storage. In *2017 IEEE Region 10 Humanitarian Technology Conference (R10-HTC)*, pp. 83–88.

Chapter 6

Digital building logbooks on the blockchain: first conceptualisation and future research directions

Theodoros Dounas[*,†], *Daniel M. Hall*[‡], *Dimosthenis Kifokeris*[§], *David Christie*[||], *Jens Hunhevicz*[¶,**], *Firehiwot Kedir*[††], *Joseph Mante*[||], *Goran Sibenik*[‡‡], *Marijana Sreckovic*[§§], *Lavinia Chiara Tagliabue*[||||] and *Catherine De Wolf*[**]

The built environment and its industries lie at the core of the twin transitions of the European Union. Twin transitions aim at uniting ongoing green and digital transformations to address the climate crisis, reduce dependency on finite natural resources and ensure European companies remain competitive in the future. Furthermore, climate goals and the need for digitisation are interdependent – as noted in the Energy Efficiency Directive (EED) EU/2023/1791 [1]. Ongoing programmes such as the *European Green Deal* [2] and *A Europe Fit for the Digital Age* [3] reflect these aims as well. Such programmes ascertain that achieving a truly sustainable and circular built environment in Europe, requires gathering, maintaining and operationalising detailed data about buildings over long time horizons (European Commission 2022, 2023) [2,3]. Thus, there is a need for many types of building information including their history, physical characteristics, material composition, energy usage and operational performance.

[*]School of Textiles and Design, Heriot-Watt University, Scotland, UK
[†]Adventurous Systems Ltd, London, England, UK
[‡]Department of Management in the Built Environment, Faculty of Architecture and the Built Environment, Delft University of Technology (TU Delft), Netherlands
[§]Department of Architecture and Civil Engineering, Chalmers University of Technology, Sweden
[||]School of Law and Social Sciences, Robert Gordon University, UK
[¶]Urban Energy Systems Laboratory, Empa, Switzerland
[**]Department of Civil, Environmental and Geomatic Engineering, ETH Zurich, Switzerland
[††]Implenia AG, Glattpark (Opfikon), Switzerland
[‡‡]Singapore-ETH Centre, Future Cities Lab Global Programme, Singapore
[§§]Department of Integrated Planning and Industrial Building, Institute of Building and Industrial Construction, Vienna University of Technology (TU Wien), Austria
[||||]Department of Computer Science, University of Turin, Italy

Digital building logbooks (DBLs) is a recent concept proposed by the European Commission [4,5] that can act as the repository for most aforementioned information. A DBL is defined as a centralised, dynamic repository that captures and organises all relevant data throughout a building's lifecycle [6]. Widespread usage of DBLs can enhance transparency, facilitate informed decision-making and support data sharing across various stakeholders including building owners, occupants, financial institutions and public authorities [7]. Furthermore, users can use the DBL to document significant changes such as ownership transfers, usage modifications and updates from renovations or maintenance [8]. The DBL can include detailed records like administrative documents, building plans, descriptions of the property and its surroundings, technical systems, material traceability and performance data like energy consumption and indoor environmental quality. By linking to building certifications and utilising data from intelligent systems, the DBL not only helps manage a building efficiently but also supports sustainability and compliance with regulations [6,7].

However, the implementation of DBLs faces three significant barriers [4–7,9]:

1. **Incentives.** There is often a lack of motivation to update the logbook contents, a critical component that requires user engagement and awareness of the DBL's benefits. Users, such as facility managers or homeowners, often do not see the immediate benefits of updating the logbook, which can diminish their motivation to engage with the system regularly.
2. **Administration costs.** The administrative burden and high costs associated with maintaining DBLs can deter stakeholders, especially when these costs are passed on to homeowners or end-users. This issue is exacerbated by the absence of a clear business model for the operation and sustainability of DBLs.
3. **Data interoperability, privacy and security.** The challenges of interoperability, data privacy and security must be addressed to ensure seamless integration with other systems and compliance with data protection regulations (including General Data Protection Regulation (GDPR), see EU Document 32016R067). There is a need for systematic and standardised approaches to data collection, storage, processing and updating, while also acknowledging that there are many differences that exist across regions and contexts, adding another layer of complexity.

These barriers necessitate a strategic approach, including clear legislative frameworks and incentives, to encourage widespread adoption and effective utilisation of DBLs. However, a review of the current state-of-the-art finds that most ongoing efforts in research and industry cannot meet the long-term needs of DBLs. Instead, we find three main technological solutions proposed, each with their own limitations. The *open Building Information Modelling* (openBIM) approach relies on file-based exchange executed between silos of information on local machines or common servers [10]. However, there is currently no sufficient evidence that this necessarily provides a long-term immutable record or loss-free interactions within a complex network of participants, even though automatically generating Information Delivery Systems files could improve model verification [11]. Furthermore, building information modelling (BIM) systems do not address all stakeholders downstream without extensive retooling of the data. The *standard platform* approach offers a unified approach to logbook data within centralised platforms, but at times these can create lock-in effects to specific vendors or governmental systems [12]. The *linked data and semantic web approach* shows promise for interconnected data but could be augmented with approaches to further incentive users towards trusted data governance [13].

In the proceeding sections, we position a blockchain-based approach as the appropriate technical solution for the DBL. We first describe our motivations for this; we believe that the fundamental characteristics of blockchain – trust, immutability and decentralised peer-to-peer (P2P) transactions – are both a good fit for data in the built environment and the needs of the DBL. Second, we propose a conceptual architecture for a blockchain-based DBL technical system, across the data, logic and services, and user layers. Overall, this consists of a novel decentralised data architecture, a novel smart contract blockchain architecture, data bridges to legacy data (either found in databases or standalone) and a series of user-friendly interfaces for a multitude of stakeholders. Finally, we describe the implications and future research directions for such an approach to DBLs, including the development of novel business models, decentralised marketplaces and a new legal framework for DBLs.

6.1 Background

6.1.1 Implementation challenges for digital building logbooks

The design, construction, operation and decommissioning of a building require many stakeholders with various incentives to interact over long time horizons – especially in the current European context, where the drive for renovating the existing building stock to account for current energy and quality-of-life demands is extending such horizons even more [14]. The design and construction of a building can take years and the average lifespan of a building is over five decades – although this can vary depending on the building typology [15]. In this setting, it is difficult to assemble a complete information record such as a DBL. The European Union Framework [4,5] identifies nine gaps across finances, user needs, data and legal requirements that restrict the implementation of the DBL (see Table 6.1).

6.1.2 Possible technical approaches to digital building logbooks

In a review of current approaches to solve these challenges for DBL, we generally observe three current approaches. However, each approach has key limitations.

The first technical approach is the openBIM approach [16]. This approach advocates for the use of open BIM standards such as Industry Foundation Classes (IFC). Development attempts might look to standardise the parameters of DBLs within the IFC data schema [17]. However, the openBIM approach suffers from

Table 6.1 Identified gaps to implementing the DBL [4,5]

Gap	Description
1. Lack of sound funding model	There is a lack of a sound funding model, including sustainable financial strategies to support the initial development and ongoing maintenance of DBL systems.
2. Benefits not clear to all stakeholders	The benefits of DBLs are not clearly defined or communicated to stakeholders, resulting in lower engagement and adoption rates.
3. Inconsistent scope and purpose	The scope and purpose of DBLs are not always clear or aligned with the users' needs, which can lead to misalignment in expectations and usage.
4. Lack of user-friendliness	DBL systems often suffer from poor user interface designs, making them difficult to use and deterring stakeholders from regularly updating and using the logbook.
5. Barriers to updating the DBL	The process for updating DBLs is cumbersome, potentially requiring automation to ensure data remains current without excessive manual input.
6. Low interoperability of the repository	Lack of interoperability with other data systems and tools complicates data integration and reduces the effectiveness of DBLs.
7. Poor data governance	Issues surrounding data privacy, ownership and access control are not adequately addressed, raising concerns over data security and usage rights.
8. Lack of a defined legal framework	There is an absence of a comprehensive legal framework that supports the proper use and operation of DBLs, potentially leading to regulatory challenges.
9. Uncertain role for government (e.g. for the EU and Member States)	The roles and responsibilities of local, regional, national and international governments and policy makers in promoting and governing DBLs need clearer definition to enhance coordination and support.

lack of cross-silo, flexible and long-term data storage. Additionally, it does not address legacy data, that probably exists in two-dimensional, paper format. Because it is a file-based approach and not a data-based approach, openBIM approaches struggle to deal with data outside of the project-specific environment; there is a need to constantly revise data until the completion of a construction process, then deliver, archive and occasionally aggregate the data with additional sources thereafter. Even if all DBL can be successfully compiled in an IFC file, there is a risk that this file can be lost or damaged over the long-lasting lifecycle of the building. Furthermore, IFC files use a single schema about the building as an object and do not encapsulate the rich datasets that a building operation produces. openBIM is therefore not a permanent record. There is a need for a data infrastructure that is flexible at the outset but provides a secure record of DBL in the long term.

The second is the standard platform approach. Software-as-a-service vendors that store DBL data tend to have a centralised structure. The platforms provide an infrastructure for data exchange (e.g. material passports) by acting as a digital intermediary, creating a link between data providers and data buyers. In addition, the marketplace operator acts as an overview agent, brings together supply and demand and coordinates pricing. They keep the DBL data – in the form of databases or uploaded PDF documents – on their own servers. The platforms are also hopeful to develop service layers that can calculate key performance indexes such as circularity or carbon performance [12].

While these platforms are an important development step for the DBL, the *standard platform approach can be limited by the platform lock-in effect and permissioned nature found in centralised solutions* [18]. Closed metadata schemas can create a lock-in effect by making users dependent on a specific software or vendor, limiting flexibility and innovation due to the costs and risks involved in migrating data across incompatible systems [12,19]. Even when platforms are developed by local or regional governmental consortia, the platform lock-in effect makes it very difficult for users to leave the platform and very difficult for different platforms to be compatible with one another, across a market that is theoretically integrated like the one of the European Union. From that point of view, a single platform, while serving one or two countries, is unable to adapt and serve all the needs of the EU and may not be desirable as it does not allow for a multi-segment and company size market to develop.

The third approach is the *linked data* approach. The linked data approach shows a lot of promise using principles of the semantic web for sharing machine-readable interlinked data on the Web [20]. Nonetheless, there is still a need to provide governance models for authorisation of data provision (i.e. who fills in which parts of a DBL) and for validating data across multiple platforms [21]. Stakeholders need to be able to trust that data is correct and that someone has not manipulated the inputs, through the ability to view a verified record of transactions. Further, stakeholder benefits need to be considered through the design of an incentive system for the data, to overcome coordination challenges such as a lack of trust, poor information exchange and supply chain fragmentation. In general, the linked data approach is flawed in one respect. Even though they promise interoperability and vendor-neutral data access, most implementations tend to be technologically centralised [22]. This eliminates incentives for stakeholders to participate, as they are still highly dependent on a standard platform. In addition to the need for an open access, vendor-neutral and decentralised data foundation, a trusted incentive system must also be elaborated to motivate stakeholders participating in the DBL to complement the linked data approach.

We therefore need a data and logic architecture for the DBL that can accommodate a decentralised architecture and governance, embeds incentives for all stakeholders into the operation and data integrity of the DBL, allows for interoperability while thriving for decentralisation and integrating legacy data. This data and logic architecture must also allow for the creation of cross-purpose and cross-countries markets and uses of the data and integrate into one golden thread the data that a user might need, while retaining the resiliency of the data for decades. We outline in the following part of the chapter how this can be achieved using distributed data networks, blockchains and smart contracts.

6.1.3 Distributed ledger technologies, blockchains and smart contracts

Distributed ledger technologies (DLTs) refer to distributed databases where each participant holds a copy of the data (across sites and geographies), supported by a consensus mechanism to synchronise between these copies [23]. In DLTs, security is achieved by using public-private key cryptography and signatures to establish identities [23]. Blockchains are a common subset of distributed ledgers [24] and a decentralised version was first created in 2008–2009 to facilitate the idea behind digital cash, specifically, Bitcoin [25]. Earlier computing implementations or proposals brought forward the idea of Merkle trees in collapsing arrays of transactions into hashes or using cryptography to solve the problem of communication in a Mutually Suspicious group cryptographically. Cryptographer David Chaum first proposed a blockchain-like protocol in his 1982 dissertation *Computer Systems Established, Maintained, and Trusted by Mutually Suspicious Groups* [26]. Each block in a blockchain is connected immutably with the previous one by containing its cryptographic hash; this mechanism along with the incentivisation, secures the immutability of the blockchain, as users cannot go back and change the data, while incentives hold computing nodes true to the purpose and mission of the network which is the reliability and security of the transitions in a peer to peer fashion [24].

Blockchains can be either public (accessible by everyone) or private (closed, not publicly accessible), and permissioned (i.e. have restricted participation in the network) or permissionless (no restrictions and control to participate in the network) [27]. For the purposes of this chapter, we will focus on public, permissionless blockchains with the following features:

- An incentive for computing nodes to stay honest via the awarding of new crypto-tokens to the nodes,
- The collapse of all transactions at a given moment in a cryptographic hashing mechanism called a Merkle tree [28], where transactions are reduced to a single hash by progressively hashing them in pairs,
- A timestamp feature,
- A consensus mechanism using a proof of stake algorithm and
- A nonce, which is a number for each block used only once to prove authenticity, that is, the previous hash in a block of information which the computing nodes store on their ledger.

Beyond their cryptography, blockchains have been expanded to include the potential for running virtual software machines in a distributed fashion. Software executed on the blockchain is called a 'smart contract' as its immutable nature equates the concept of code execution with law

In this chapter, we refer to smart contracts as code[1] that get executed reliably on a blockchain, as a result of a data transaction. Smart legal contracts can also be used to define and perform legally binding and enforceable obligations [29]. Note that a smart contract does not necessarily have to constitute a valid binding agreement by law, but it must comply with a relevant legal framework.

Smart contracts can be explained as the computing code equivalent of automated vending machines. Deployed smart contracts act then as automatons, with the blockchain automatically executing their code when specific events trigger the computation. This creates an infrastructure automation layer where public permission blockchains can be used as global computing platforms. Blockchains are an excellent fit for processes that require the coordination of a sequence of events (e.g. supply chains or events of changes in a building), the coordination of trust of multiple stakeholders, privacy and security of data, and the operation of projects where not all actors need to trust each other.

6.1.4 Other key enabling technologies

6.1.4.1 Decentralised storage

Although blockchains can also be used as a data storage system, the cost of doing so on a public blockchain is often prohibitive [30,31]. As a complement to blockchain, decentralised data networks have emerged to enable storage and provision of data through decentralised systems [12]. For example, the InterPlanetary File System (IPFS) aims to solve the web data's typical availability problems, such as the limited lifetime of content, firewall blocks and out-of-business services [32]. It uses cryptographic checksums (i.e. digits representing the sum of the correct digits in stored or transmitted digital data, against which later comparisons can be made to detect data errors) of file contents to address and link data, which originates in GIT, BitTorrent and distributed hash tables. IPFS provides content not via the address where the content is located, but via what the content is. As such IPFS browsers locate content based on its hash, avoid as such duplication across containers.

While the development of IPFS is primarily aimed at file systems, so-called blockchain databases can transform conventional storage systems. The advantages of such blockchain-like technology are decentralised control, tamper resistance, value generation and transmission. IPFS and similar decentralised data storage approaches offer the opportunity to replace current cloud-based, highly centralised solutions [32]. Blockchain applications consider not only the record of the ledger but also the fundamental strategy of data organisation and its impact on data availability.

[1]Our definition of smart contracts is specific to the context of a blockchain-based DBL. Even so, instructions executed by smart contract code may have legal implications. Broadly, the readers should know that the debate about the status of smart contracts as code or legal contract is still in the process of being settled.

6.1.4.2 Cryptoeconomics and decentralised marketplaces

Any system built using blockchain smart contracts requires a systems-level thinking logic to govern digital transactions and interactions; in an institutional setting, this would for example reflect the transactional incentives of different actors adhering to the respective context's rules, norms, symbols and beliefs [33]. The term *cryptoeconomics* describes the design of P2P cryptographic systems that guarantee certain kinds of information security properties using incentives and/or penalties to regulate the distribution of efforts, goods and services in new digital economies [34]. Such blockchain-based mechanisms potentially integrate the increasingly available digital information with economic coordination (i.e. processes, contracts and finance) towards a more efficient construction industry.

Part of the epistemic and practical novelty of blockchains as crypto-economic systems is that trust, transactional infrastructure and incentives are encoded in the computing protocol of the network rather than decided by existing structures [33,35]. As such the implementation of cryptoeconomics has the potential to radically change the way various industries operate, for example in the development of markets for data in the built environment [33–35].

Although the sale of data has been practised for a long time, a data marketplace is a relatively new business model – for example as is reflected in the integration of transactional encryption and private AI in e-commerce [36]. A data marketplace refers to an intermediary service for data exchange. The key difference with the centralised marketplaces is the level of privacy and security provided by having two decentralised components in the web3 technology stack [37]. In other words, the data owners have secure and smooth access to their data, while maintaining privacy and confidentiality. This is because ownership never shifts to a central entity, meaning the data creator is paid for the data, not an intermediary. As a result, network participants are motivated to make their data available collaboratively while benefiting monetarily. Moreover, the decentralised architecture based on the web3 technology stack offers the advantage that the implementation of an incentive system is simpler due to the characteristics of the decentralised trust layer.

6.1.4.3 Cryptographically secure digital building twins

Blockchain technology can also be used to create a cryptographically secure digital building twin. This offers a transformative approach to managing building data [38]. The digital twin acts as a dynamic and tamper-proof replica of the building, encapsulating detailed records from the inception of the building through its entire lifecycle [39]. Because the data is secured on a blockchain, it remains immutable, meaning once information is entered, it cannot be altered or deleted without a trace. This level of security enhances both transparency and trust among all stakeholders, from builders and owners to regulators and occupants. The digital twin thus serves not only as a fully trusted source of information but also as a tool for enhancing decision-making and improving building management practices, fostering a more integrated and accountable approach to construction and facility management.

6.2 Conceptualisation

6.2.1 Motivation and objective

In this section, we describe our motivations and conceptualisation of the technical architecture and associated services for a blockchain-based DBL. In response to the above limitations, we find that a blockchain-based solution is the most appropriate technology to answer all the gaps identified for implementation of DBLs. We believe blockchain technologies can and should act as the backbone of the DBL. Blockchain records are immutable and will not be lost; this provides data and infrastructure resilience over the long-time horizon of a building lifecycle. Blockchain enables system-level trust by shifting the trust of managing records and data to the computer protocol of the blockchain. It can ensure native encryption and data security while featuring embedded incentives that encourage stakeholders to update the logbook. Blockchain offers decentralised, P2P transaction approach for the DBL that can decouple the business model logic from the data storage. This enables an opportunity to create novel governance, operating and business models for the DBL data and the buildings themselves.

Blockchain technologies are particularly suited for this type of data architecture as they decouple access and logic of the data from their storage, that is, data can be stored in legacy databases, decentralised storage architecture or in simple files and still be easy to reference them in the DBL blockchain. Further blockchain technologies allow the participation of multiple stakeholders in maintaining and accessing the data without compromising user experience, usability, the integrity of the data or the custody and ownership of data.

Other industries have seen this potential. For example, the shipping company Maersk, in collaboration with IBM, developed TradeLens, a blockchain-based platform aimed at enhancing transparency, efficiency and security in the global shipping industry. Launched in 2018, TradeLens utilised blockchain technology to digitise and streamline supply chain processes, reducing reliance on paper-based documentation and improving real-time visibility of cargo movements. The platform facilitated secure information sharing among various stakeholders, including shippers, freight forwarders, port operators and customs authorities, thereby minimising delays and errors in shipping operations [40]. However, TradeLens faced challenges in achieving widespread industry adoption. In November 2022, Maersk and IBM announced the discontinuation of the platform, citing its inability to meet the necessary level of commercial viability and industry collaboration. There are also other attempts at digital product passport traceability. For example, the company Circularise3 already works with manufacturing companies to digitise and trace materials across complex supply chains on a public blockchain without risking confidentiality.

To summarise, the characteristics of blockchain – trust, immutability and decentralised P2P transactions – are well aligned with the needs of the DBL and the loosely coupled, project-based built environment. Efforts to use blockchain has been done for product passports and supply chain tracing in other industries.

Through blockchain, there is opportunity to make the DBL more efficient, transparent and accountable between all involved participants.

6.2.2 Conceptual architecture

We envision a DBL standard that is constructed as a hybrid model, allowing the stakeholders and homeowners to use it either as a simple gateway to existing legacy data and databases but also, should the infrastructure and type of building allow it, as a container for decentralised data storage for the building. This is a deliberate architecture that allows the DBL to be future proof and resilient data architecture, that can adapt and evolve according to needs of the stakeholders and the industry.

There are three main layers to the system: data layer, logic layer (including services) and user layer (Figure 6.1). Within these three layers, we identify six key technical innovations of the blockchain-based DBL (see Table 6.2).

6.2.2.1 Data layer

The foundation of the data layer is a decentralised data network. Based on Table 6.2, we suggest that there is a need for a custom decentralised infrastructure that will be created for the purposes of hosting the various types of DBL data (see Table 6.3). There are many possible ways to organise this; the reader is referred to [41] for guidance on selecting a decentralised data network. One example approach would be to set up multiple IPFS nodes. Although it is possible to use public, permissionless blockchains for the infrastructure of the demonstrator, there are several issues related to speed of development, the possibility to test various consensus and the participation mechanisms that motivate this approach [42].

Within this infrastructure, users can integrate legacy data containers and existing databases alongside IPFS storage, resulting in a hybrid data storage system with both decentralised and centralised data. The IPFS nodes can run in cluster mode and as such will appear as one storage container where data is addressed by content identifiers (CID), ensuring that there is no unneeded duplication of data, that is, anyone uploading a second set of an existing data point/set will be informed that it already exists along with its CID.

In our conceptualisation, we suggest that user interfaces that will allow the stakeholders, data owners, users, data marketplace users and anyone downstream to access the DBLs when they have the appropriate access rights. The smart contracts architecture will be governed with a three-tiered structure of smart contracts: one that is the minimum viable product for a DBL, a second tier that deals with tokenisation and governance smart contacts and a third tier that establishes the smart contracts which deal with zero-knowledge proofs (protocols used between parties affiliated with a smart contract, where one party uses them to convince another that a given statement is true, without conveying any information beyond the fact of that statement's truth) a privacy enhancing tool of blockchains and a smart contract that manages data for a digital twin of the building. A web-layer will play the role of the user interface, that is served by visiting the address of the demonstrator. This layer

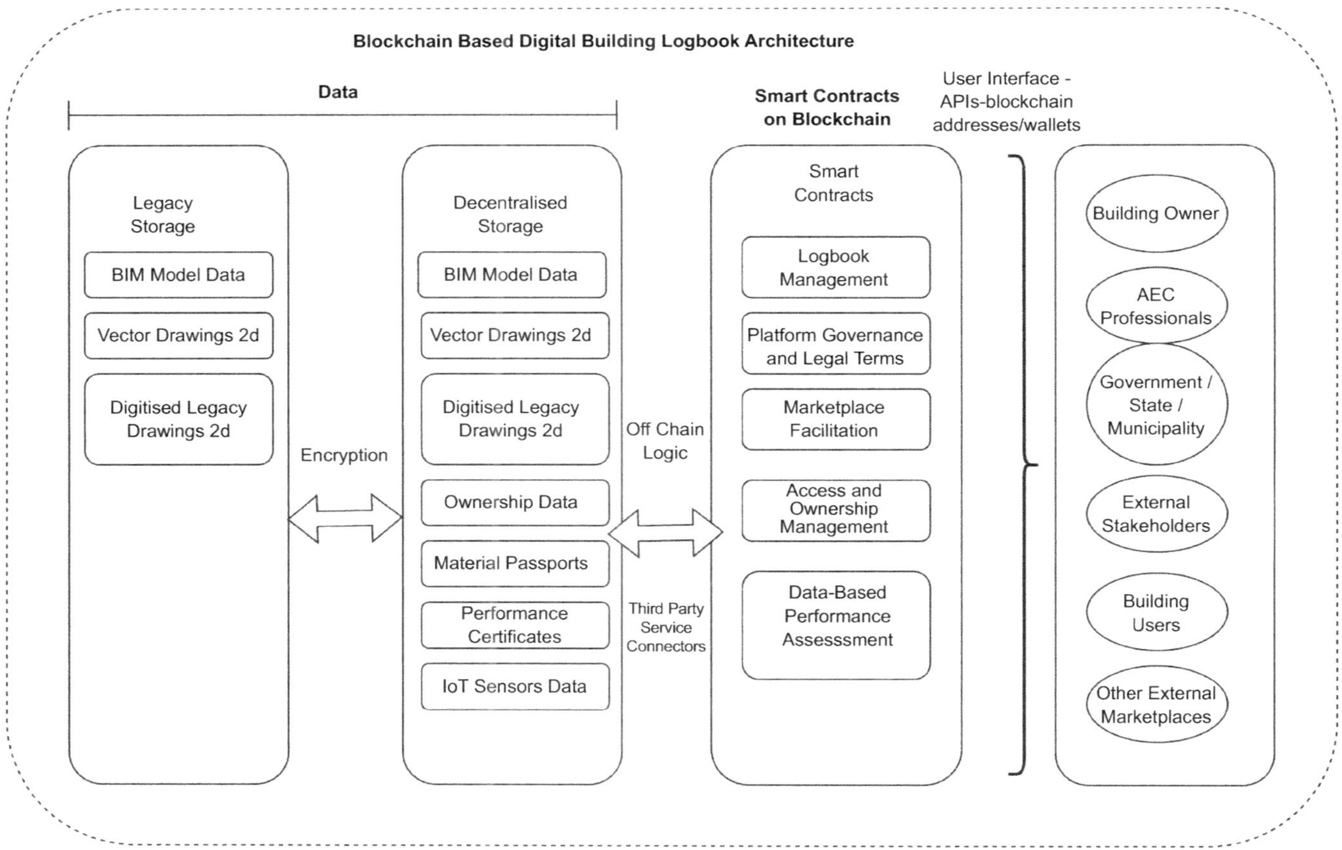

Figure 6.1 *Conceptual diagram illustrating the three pillars of the conceptualisation*

Table 6.2 The six key technical innovations identified for a blockchain-based DBL

Layer	Features	Description
Data layer	Decentralised data storage	Data is stored either on decentralised data networks or linked cryptographically to legacy databases. This hybrid approach ensures robustness and flexibility, allowing new buildings with IoT integration and older buildings under refurbishment to maintain a consistent and secure data architecture. This setup provides resilience against data loss and unauthorised access.
	Cryptographically secure digital building twin	Utilising blockchain creates a digital twin of the building that is secure and tamper-proof. This digital twin provides a detailed, immutable record of the building's data across its lifecycle, enhancing transparency and trust.
Logic and services layer	Smart contracts	The use of smart contracts automates the recording of changes to the DBL, including legal and ownership data. Operating on a proof of stake blockchain, these contracts provide a secure and efficient way to handle transactions and updates, minimising errors and enhancing processing speeds.
	Legacy data integration	Blockchain's architecture supports the integration of existing, older building data, ensuring that valuable historical information is preserved and utilised alongside new data. This integration is crucial for buildings under refurbishment, allowing for a seamless transition and maintenance of comprehensive data records.
User layer	Crypto-economic incentive system	Blockchain introduces incentives for various stakeholders to participate and update the DBL reliably. These incentives ensure active and honest participation, enhancing the data's integrity and reliability.
	Decentralisation of participation	Blockchain allows for decentralised control over the DBL, enabling various stakeholders such as owners, builders and regulators to participate without a central authority. This decentralisation increases engagement and accessibility while reducing the risk of data manipulation or control by a single entity.

will allow the registration of users, manage access to the rights, interact with both the decentralised storage clutter and the smart contract architecture and enable a streamlined experience for users.

Table 6.3 Data types that should be captured, stored and/or calculated by a blockchain-based DBL

Scale	Category	Data types	Primary source
Building	Contracts	Contract to design, contract to construct, other services contracts, build permit, asset ownership, tenancy agreement, maintenance service contract, warranty and guarantee, insurances, contract to demolish	PDF
	Building description	Unique building identifier, dimensions, floor area, room definition, fire safety plan, accessibility plan, water system design, heating system design, electrical system design	BIM, GIS
	Design and construction record	Architectural design record, structural design record, structural design calculations, mechanical design record, electrical design record, plumbing design record, cost data, schedule, information required by regulation (such as health and safety plans)	BIM
	Operational data	Renewable energy production, heating consumption, electricity consumption, hot water consumption, energy performance certificate, rainwater collection, weather, surface temperature	Digital building twin
	Performance data	Circularity indicator (e.g. by Madaster), building circularity index, life cycle analysis, material flow analysis, sum of embodied carbon, sum of operational carbon, whole lifecycle carbon, climate resilience potential, EPC rating, structural reliability index, residual structural service life	External assessment tools (e.g. Madaster)
Product	General information	Product name, purchase order, manufacturer's details, manufacture date, global trade item number, product owner, function, dimensions, weight, density, thermal resistance and conductivity	Material passport
	Material health and sustainability	Health and safety certifications, fire resistance class, embodied carbon, operational carbon, percentage of renewable materials, percentage of recycled material amount, material circularity indicator	Material passport, external assessment tools
	Production information	Quality test, waste, product certifications and labels, humidity, temperature	Material passport
	Construction log	Current geolocation, present status, assembly instruction, tracing mechanism, acceleration, rotation, humidity, temperature, date of installation	Digital construction twin
	Product performance data	Life cycle analysis, embodied carbon, construction carbon, target service life, compatibility with design, Product circularity index	External assessment tools
	Use, maintenance and end of life	Cleaning and maintenance instructions, warranty and guarantee, maintenance contractor details, maintenance service dates, storage requirements, service life and expected use times, availability of spare parts, proof of recycling	PDF

6.2.2.2 Logic and services layer

The logic layer of a blockchain-based DBL enables a transparent and efficient management system on top of the data layer. There are five categories of smart contracts that could be developed: logbook management, platform governance and legal terms, marketplace contracts, access and ownership management, and data-based performance assessment. Table 6.4 provides a summarised table capturing the five main categories of smart contracts for a blockchain-based DBL.

Table 6.4 A description of the five main categories of smart contracts for the DBL

Logic and services	Specific functions of smart contract	Further description
Logbook management	Recording	Automates the entry of new building data into the logbook.
	Verification	Validates the accuracy and authenticity of the data recorded.
	Updating	Manages updates to the logbook, ensuring data integrity and accurate history via timestamping.
	Event logging	Captures and records construction milestones and maintenance activities automatically.
Platform governance and legal compliance	Role definition	Specifies roles and responsibilities for each blockchain participant.
	Operational rules	Establishes the guidelines for how the blockchain operates.
	Legal alignment	Ensures that smart contract operations comply with relevant laws and regulations.
	Decision-making	Facilitates governance decisions through smart contracts based on predefined rules.
Access and ownership management	Access control	Regulates who can access or edit the DBL data, ensuring privacy and confidentiality.
	Data protection	Protects sensitive information by restricting access to authorised users only.
	Ownership rights	Clarifies and enforces data ownership rights on the blockchain.
	Legacy system integration	Manages the inclusion and integration of data from existing systems into the blockchain system.
	Data trading	Enables direct P2P exchange of data, eliminating the need for intermediaries.
	Privacy and anonymity	Ensures that participants can interact anonymously and securely in the marketplace.
	Transparency	Provides transparency in transactions, making all details visible and verifiable by participants.
	Ownership transfer	Manages the legal and technical aspects of transferring data ownership and rights during transactions.
Data-based performance assessment	Performance metrics	Establishes performance criteria such as energy efficiency and sustainability benchmarks for buildings.
	Real-time monitoring	Allows for continuous monitoring and evaluation of building performance against set criteria.
	Decision support	Provides actionable insights based on performance data to support maintenance, retrofitting or redevelopment decisions.
	Environmental impact	Evaluates and reports on the impact of buildings on the environment, promoting sustainability.

The first set of smart contracts for *logbook management* serve as the backbone of the DBL, enabling the recording, verification and updating of building data throughout its lifecycle. They ensure the integrity of the logbook entries, allowing for accurate tracking of a building's history, renovations and transactions. Through timestamping, these contracts would automate the logging of events, such as construction milestones or maintenance activities, ensuring that the logbook is an immutable and reliable source of the building's history.

A second set of smart contracts serve as the guidelines for *platform governance and legal compliance*. Together, these smart contracts implement a comprehensive governance framework that dictates how the blockchain operates, defines the roles and responsibilities of all participants and establishes the decision-making process. Such smart contracts could include a Non-Fungible Tokens Generator and a Fungible Tokens Generator. This model should align with the legal and regulatory requirements of the industry and the jurisdictions in which the system operates. They act as a framework within which all other smart contracts operate, thus serving as a legal and operational guideline for the DBL ecosystem. They also act as a point in which those regulatory requirements which relate to the collation and sharing of information can be met, stored and interacted with.

The third set of smart contracts are for *access and ownership management*. These contracts govern the permissions and rights associated with the DBL data, determining who can view or modify the logbook entries. They ensure that sensitive data is accessible only to authorised stakeholders and that ownership rights are clearly defined and protected. This category is crucial for managing the data inputs from legacy systems, as well as integrating new data streams in a controlled manner.

The fourth set of smart contracts comprise the *Decentralised data marketplace*. The decentralised data marketplace is a P2P system where participants can trade with each other and exchange information. The marketplace should be open so that anyone can join and add value to the network and community. At the same time, the marketplace will provide privacy, anonymity, transparency and access control. In practice, this means a data owner can be directly paid for the provision of data used by third-party services, instead of paying fees to an intermediary or the data being sold by a centralised platform. The smart contracts will encode the logic to access, modify and transfer data and their respective ownership (Figure 6.2).

The fifth set of smart contracts act as a service layer for *data-based performance assessment*. Smart contracts in this category would, for example, automate the evaluation of building performance against predefined criteria, such as energy efficiency benchmarks, sustainability indexes or circular economy standards. These could be connected to a blockchain-assured digital building twin. These contracts would enable real-time performance tracking, providing insights for optimisation and supporting decision-making for maintenance, retrofitting or redevelopment projects. They are essential for assessing the long-term value and impact of buildings on the environment and society.

The above two main components (data and trust layer) need to be connected to each other and legacy systems with what we call data bridges. The connection

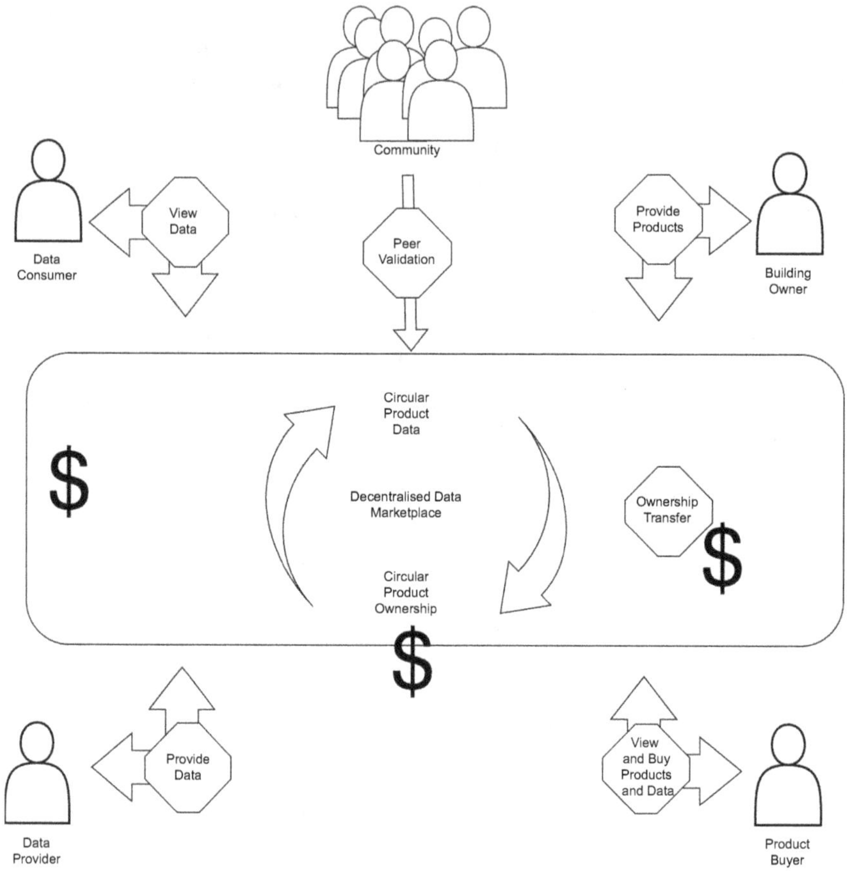

Figure 6.2 Example of a decentralised data marketplace set on top of the data layer (adapted from [43])

between the decentralised data storage and the blockchain will be enabled through third-party service connectors. In brief these servers will route data over APIs (e.g. the Web3.js or Web3.py libraries) between the two systems. In a similar way, the user interfaces needed for the stakeholders to interact with the data and logic will route data using third party servers and APIs between the web front end, the blockchain wallet, the blockchain smart contracts and the decentralised data storage. Finally, a series of processes and tools will be developed to ensure transfer of data between legacy data systems and the here proposed decentralised system. This ensures that stakeholders can input data from existing buildings in the most used data formats.

There is one important remaining point about the relationship between the logic layer and the data layer. The smart contracts can timestamp and record changes to the DBL legal on the decentralised data network. However, as a hybrid system is imagined, the smart contracts should also be designed to interact with legacy databases and data storages. As such, it is possible to use the same smart contracts to monitor a new building with embedded IoT sensors while maintaining the same legacy data of an existing older building that would be under refurbishment. In other words, blockchain allows a separation of the governance and economic layer from the data layer, so that multiple different types of data storage systems can be used underneath the same network of smart contracts.

6.2.2.3 User layer

In the blockchain-based DBL system, crypto economic incentives play a pivotal role in motivating stakeholders to share valuable data while ensuring their privacy and control. These incentives are designed to create a virtuous cycle where information exchange is not just encouraged but rewarded, balancing transparency with the stakeholders' need for anonymity. Such an incentive system can target data quality, seamless interoperability, availability, usability, classification, consistency, integrity, security and control models. For example, sharing high-quality data can be incentivised by a mix of monetary and non-monetary incentives. By aligning user actions with the overall health of the DBL, these incentives create a participatory ecosystem where each stakeholder is motivated to maintain the integrity and richness of the database.

For instance, a crowdsource incentive could refer to a user who contributes to the digital twin's accuracy by providing timely updates may receive tokens, which could in turn be used for services within the platform or gain access to advanced analytics that can inform better decision-making for future projects. This symbiotic relationship between users and the DBL enhances the system's value and ensures its sustainability, as each participant becomes a custodian of the shared digital infrastructure, empowered by the very act of contribution and incentivised through tangible rewards that reflect their effort and input.

To develop the crypto economic incentive system, the following future research steps are needed. First, there is a need to determine which actions within the DBL should be incentivised. These actions could include data entry, validation, updates, quality checks and the provision of resources for the platform. Where possible, these should reflect criteria which can be objectively determined to mitigate the risk of disputes as to whether these actions have been accurately carried out. The identification process should focus on activities that add value to the DBL's accuracy, completeness and reliability. Second, there is a need to create a system of rewards that will motivate users to perform the identified actions. This can include tokens, access rights or other benefits, as well as deciding on the form these incentives will take (e.g. cryptocurrency, service discounts, enhanced platform features), and ensuring they align with the users' values and needs. Third and finally, these rules can be encoded on the trust layer via smart contracts that can automate these processes transparently and without the need for a central authority.

A web-layer will play the role of the user interface, that is served by visiting the address of the demonstrator. This layer will allow the registration of users, manage access to the rights, interact with both the decentralised storage clutter and the smart contract architecture and enable a streamlined experience for users. The user interface should make it easy for users to understand how to earn and use their incentives. It should clearly display earned rewards, available actions that yield rewards, and provide a simple way to redeem or spend tokens.

6.2.3 Summary

As described above, we conceptualise three layers of architecture to the blockchain-based DBL.

At the core of the DBL's data layer lies a secure, accessible storage environment, allowing participants self-service access to information they can trust and understand. To enhance this trust, the DBL leverages cryptoeconomic incentives to encourage the sharing of high-quality data. These incentives are thoughtfully constructed to include monetary rewards, such as tokens for consistent contributors of verified data, and non-monetary benefits, like enhanced reputation within the platform or access to exclusive analytical tools. This multifaceted incentive system ensures that the data within the DBL is not only abundant but also accurate, reliable and valuable for all users.

The trust layer underpins the DBL's data marketplace, establishing the rules and logic that govern data exchange. Here, smart contracts play a critical role in automating the transfer of rights, such as allocating the future reuse of building materials, exemplified by the door trading scenario. This mechanism introduces a new dimension to the construction industry's marketplace, where not only goods and services but also potential future uses of a building's components can be commodified and exchanged. Such innovations pave the way for a sustainable and circular economy, rooted in the principles of transparency and conservation.

The integration of data bridges forms the final piece of the puzzle, linking the decentralised data storage and smart contracts with legacy systems. This connectivity ensures the platform's robustness, allowing a seamless flow of diverse data types, including real-time IoT feeds and historical records from conventional databases. Special attention is given to crafting a user-centric interface, where stakeholders interact through intuitive protocols, enabled by APIs that connect the blockchain layer with the user-facing web front. By prioritising ease of use and streamlined interaction, the DBL not only appeals to current users but is also positioned to adapt to future developments in data management and stakeholder engagement.

6.3 Potential impacts and future research directions

What would a blockchain-based DBL offer? Should the concept be implemented, we describe below seven potential impacts for the built environment that would come with such a new system. We further describe possible areas of future research for each of the areas. Table 6.5 summarises these areas.

Table 6.5 Impacts and future research directions for a blockchain-based digital building logbook

Potential impact	How it works	Future research
Enabling a new information infrastructure backbone	Acts as a consolidated platform for diverse building data, enhancing visibility and trustworthiness of information.	Investigating how decentralised networks support a circular economy and integrating fragmented industries like AEC. Exploring open-source smart contracts for integrated building management.
Accelerating the business case for renovation and reuse	Supports integration of existing and new buildings for sustainability goals, promoting circular economy concepts.	Developing business models such as renovation passports and servitisation in building management, tailored for energy efficiency and resource optimisation.
Enabling new P2P business models	Facilitates innovative digital business models and decentralised marketplaces across the building lifecycle.	Examining transformations in business models and governance through blockchain, focusing on P2P interactions and efficiency in the AEC industry's value chain.
Improving existing legal structures	Challenges existing legal frameworks with its novel approach, emphasising data security and multi-jurisdictional operations.	Creating toolkits for legal adaptation to blockchain technologies, enhancing data confidentiality and developing standard legal protocols for smart contract operations.
Fostering new governmental policies, incentives and regulations	Encourages adoption of sustainable practices and compliance through blockchain-enabled transparency and accountability.	Exploring how stakeholder engagement shapes policies for DBL adoption, analysing the effectiveness of governmental incentives for sustainable practices in the built environment.
Moving towards a more circular Europe	Promotes sustainability by enabling more efficient resource use and supporting circular practices in the building sector.	Investigating the impact of DBLs on European Green Deal goals, enhancing resource management and resilience in the AEC sector, assessing the environmental performance of blockchain technologies and addressing concerns about the energy consumption of blockchain technologies.
Enabling better security and privacy of data	Enhances data protection with robust cryptographic security and privacy-centric features, fostering user trust and engagement.	Advancing the integration of privacy-preserving technologies like zero-knowledge proofs in blockchain frameworks and assessing their impact on user confidence and system adoption.

6.3.1 Enabling a new information 'self-infrastructuring' backbone

The proposed blockchain-based DBL can act as a new information infrastructure backbone for the data of a building, consolidating disparate sources of information, from BIM to drawings, material passports and energy performance certificates. But how can this come about? We think the concept of Web3 as 'self-infrastructuring' is a key future research direction [44]. Self-infrastructuring occurs when relevant actors have the ability to place boundaries around their own actions in relation to shared purposes or goals, that are then expressed in technical and institutional infrastructure [44]. Future research can better identify how such actors can organise decentralised data networks can act as the information layer for a circular economy for the built environment and also can assist in integrating a fragmented industry such as the architecture, engineering and construction (AEC) industry. This will increase the visibility and trustworthiness of blockchain technologies as a decentralised information system for use in Architecture Engineering construction. The open-source nature of smart contracts on the logic layer can both educate and empower researchers and professionals to find integrated approaches to building management that emphasises carbon and energy performance.

6.3.2 Accelerating the business case for renovation and reuse

The AEC industry encompasses diverse market segments, including manufacturing, design, construction and software development such as BIM. These segments not only operate independently but also collaborate, sharing data, services and resources. As industry moves towards embracing a circular economy, the concept of treating the built environment as an urban mine becomes crucial. This involves using digital tools like material passports, energy certificates and digital twins throughout the lifecycle of buildings to support sustainability goals such as reducing resource extraction and achieving net-zero emissions.

The EU estimates that there are about 131 million buildings, with a significant portion built before the latest thermal standards [45]. This is also reflected in the fact that, across the EU, most residents still live in buildings that are at least half a century old – which are inadequate for today's energy efficiency demands [45]. The buildings' low energy efficiency can negatively impact their occupants' quality of life, while their suboptimal environmental microclimate can have a strong economic and ecological impact [14]. To address this, EU's Energy Performance of Buildings Directive (EPBD) EU/2010/31 and the new EED EU/2023/1791 are emphasising the renovation of existing buildings, as constructing new dwellings is getting increasingly more difficult, especially in densely populated areas: usable empty sites within cities are fewer, and demolishing and dismantling older buildings may have an even worse environmental impact [1,46]. A salient example of a national context affected by these EU demands could be found in Sweden: Nearly 933,000 (45%) of one- or two-dwelling buildings were built between 1961 and 1990, 20% of one- or two-dwelling buildings were built before 1931 and 51% of multi-dwelling buildings were built between 1951 and 1980 [47] –

which showcases a massive buildings stock in need of renovation. According to Sveriges Allmännytta [48], aligning the renovation needs of such an aging building stock with EPBD, along with the general goal of attaining a carbon-neutral construction sector in Sweden until 2045 [49], would be crucial but also very expensive – with early calculations pointing to a cost of many billion Euros.

Examples like that of Sweden and the general EU regulatory framework showcase a large opportunity for energy retrofits, which can be efficiently managed using a blockchain-based DBL. The proposed system supports the easy integration of both new and existing buildings into its system. For new constructions, it allows data to be entered at the design phase, while a 'legacy bridge' enables the inclusion of older buildings by updating and integrating their data. This process supports the creation of new business models like building renovation passports and servitisation, where companies manage installations and are incentivised to maintain them.

6.3.3 Enabling new peer-to-peer business models

The logic layer offers the opportunity to facilitate novel digital business models, considering the creation, maintenance and interoperability of DBL along the building's lifecycle. Those novel digital business models will comprise added value for the network (value constellation) and value offered to stakeholders or customers (value proposition) through the application of DBL. Both values will be combined through blockchain. One advantage of decoupling data storage from the logic layer, is that blockchain-based DBL remains flexible with regards to the types of logics proposed at the service layer. This enables local and regional markets to develop new business models that best fit the local markets and regulatory conditions. In other words, using our proposed conceptualisation, technical infrastructure and approach, various implementers can then create their own user interface and marketplace layers that are interoperable with our architecture. This allows the concept of DBL to scale while still accounting for local and regional building considerations.

The novelty here includes the necessary transformation of processes/activities and traditional (contractual and relational) governance mechanisms into new cooperation and coordination modes enabled by blockchain/smart contracts governance and the creation of new decentralised marketplaces and incentive mechanisms (cryptoeconomic; non-monetary) for participation in the DBL and exchange of information. Hence, future research of the expected business model transformation could investigate how changes in P2P transactions occur across the NICE framework [14] adapted from [50–52] *Novelty* (adoption of new activities, structures, approaches), change in *Lock-in effects* (attracting stakeholders to become business model participants), *Complementariness* (bundling of activities for the generation of added value, along the value chain) and *Efficiency* (reorganising and reshaping activities in order to reduce tacitness/transaction costs).

6.3.4 Improving existing legal structures

The implementation of blockchain-based DBLs confronts existing legal structures and frameworks that have yet to evolve to match the technology's novelty and data security demands. The challenges are amplified by blockchain's international and multi-jurisdictional nature, complicating the identification of applicable laws and enforcement of legal frameworks. The issue here is not the absence of a legal framework that can regulate the DBL environment but the adequacy and or adaptability of such framework to the changing socio-legal landscape. Legal systems (whether based on the common law or civil law) are often adaptable and are able to address challenges posed by technological innovations such as the DBL concept. Fundamentally, much of the processes outlined here will be underpinned by contract law in one form or the other and statutory provisions around data handling and privacy as reflected in the GDPR (see EU Document 32016R067). However, more work still remains in terms of adapting these existing frameworks to the DBL concept specifically. This necessitates a thorough investigation into current and prospective regulations concerning DBLs to address critical issues such as data confidentiality and to prepare for future legal developments. Future research could include the development of toolkits that outline consensus-building protocols and enhance data security are pivotal for navigating the legal landscape.

Furthermore, existing legislation could inadvertently impede the adoption of blockchain technology, posing significant barriers to implementation. Thus, there is an imperative for ongoing research into legal structures tailored to blockchain-based DBLs. Legal scholars are encouraged to develop toolkits that elucidate legal rights across jurisdictions, laying the groundwork for smart contract operations and the creation of standard protocols that will underpin internal governance structures. Such foundational work will clarify how the law interacts with blockchain's innovative incentive mechanisms, fostering a more profound understanding and smoother integration of blockchain within the AEC industry's regulatory environment.

Finally, however, DBLs, tailored appropriately, could provide a vehicle for the collation, distribution and analysis of the information which is required by regulatory bodies in various jurisdictions. Justified in the name of transparency, accountability and information sharing, different regulatory regimes - such as those which protect health and safety or the environment, require those involved in construction projects to provide specific information in terms of the activities being undertaken and with a view to meeting the particular regulatory requirements. Such regulations often also require there to be regular updates to these documents. The ability to record and update the necessary information in real time and to secure it on the blockchain, is therefore a potentially valuable avenue for the DBLs and one where the legal structures provide a positive additional use-case, assuming the DBLs are appropriately calibrated.

6.3.5 Fostering new governmental policies, incentives and regulations

If a blockchain-based DBL is shown to be technically feasible, future market adoption will depend on a series of orchestrated events such as adoption of standards by international commissions and national governments, as well as local regional and municipal authorities. Future research would need to look at how a proactive engagement with a broad spectrum of stakeholders – from large-scale building owners to municipal and national environmental authorities – could shape favourable conditions for the widespread adoption and effective implementation of DBLs.

A blockchain-based DBL enables third-party access to building logbooks across Europe. This could foster new approaches to governmental incentive systems that support a more circular and sustainable economy in the built environment. For example, city authorities can use blockchain to offer tax reductions to building owners who show that their buildings have a minimal environmental impact. These tax incentives can be managed and monitored through smart contracts, ensuring transparent and automatic enforcement of environmental standards.

Additionally, the DBL could allow for the integration of recycled and reused building materials during the design phase of new projects. Access to the DBLs makes it easier for city authorities and builders to verify that reclaimed materials are used correctly, adhering to sustainability practices.

6.3.6 Moving towards a more circular Europe

The adoption of DBLs is set to support the European Green Deal [45] by enhancing sustainability throughout the lifecycle of buildings. This improvement may help reduce net carbon emissions in the building sector by promoting more efficient resource use and lessening reliance on raw materials, which are core tenets of European Green Deal.

The use of DBLs will promote circular practices in design, construction and operation within the AEC industry. For instance, the transfer of ownership of building components can be simplified through Product-Ownership-Tokens on the blockchain, facilitating easier and more transparent transactions – even if the component is still in use. These transactions can occur directly or via the DBL marketplace using methods like auctions or set pricing, ensuring the new owner gains all associated rights and data.

Implementing digital logbooks more broadly can improve security by providing more reliable data, which makes buildings easier to update and adapt, increasing their resilience. It encourages sustainable practices such as designing for recycling and supports the use of environmentally friendly building materials. Moreover, it allows for the tracking of materials globally, enhancing the industry's ability to manage resources effectively and respond to crises with greater agility. This comprehensive approach not only reduces waste but also boosts the overall environmental performance of buildings, contributing to a more sustainable future for Europe's built environment.

As a short note, there are many rightful concerns about the current environmental performance of blockchains. While we share these concerns, we also want to note the ongoing momentum in blockchain development to solve this problem. Specifically, the current consensus mechanism for many blockchains (e.g. the bitcoin blockchain) is known as proof of work, which requires the solving of a cryptographic puzzle using energy-expensive algorithms. There are alternative approaches that can be used for a DBL, such as a proof of stake where consensus is ensured by participants risking some of their own stake. Proof of stake consumes much less energy. For example, the Ethereum blockchain reduced its energy footprint by 99.9% when it made the transition to proof of stake in September 2022 [53].

6.3.7 Enabling better security and privacy of data

Blockchain technology, recognised for its robust cryptographic security, plays a crucial role in enhancing data protection and privacy. Incorporating zero-knowledge proofs within blockchain frameworks, which allow the verification of data without exposing the actual data, further strengthens this approach. This combination ensures that personal data remains private even as it is used to verify transactions or interactions within a network.

By deploying these advanced security measures, blockchain technology supports the safe participation of stakeholders in shared data environments such as the decentralised data marketplace proposed above. This secure setup is critical for fostering trust among users, as it guarantees that their private information is not compromised while still allowing them to benefit from the system. The privacy-centric features of blockchain not only protect data but also enhance user confidence and willingness to engage with the platform.

The continued development and integration of these technologies are likely to accelerate the societal and commercial acceptance of blockchain-based systems. As stakeholders recognise the dual benefits of robust security coupled with strict privacy controls, the adoption of blockchain for managing sensitive information in a transparent yet secure manner is expected to increase. This will play a significant role in transforming how data is shared and utilised across industries, driving innovation while safeguarding user privacy. In this vein, it is encouraged that future empirical studies will be conducted to test the proposed impacts and validate the expounded conceptual model.

6.4 Conclusion

The exploration of a blockchain-based DBL as detailed in this chapter could represent a significant advancement in the AEC industry, particularly in fostering sustainable and efficient practices within Europe's built environment. By deploying blockchain technology for DBLs, it can be possible to enable a robust, secure and transparent framework for managing extensive data across a building's lifecycle. This approach not only supports the operational needs of building management but also aligns with strategic environmental goals such as those outlined in the European Green Deal. The integration of blockchain ensures that all changes, from construction through maintenance to demolition, are recorded immutably, enhancing trust and reliability in building documentation.

Moreover, the proposed architecture for DBLs incorporates decentralised data storage, smart contract functionality and the creation of a decentralised data marketplace, which together improve interoperability, reduce administrative burdens and enhance user engagement. These technological advancements facilitate more accurate tracking of material usage, energy consumption and overall building performance, contributing to the circular economy objectives. As the chapter concludes, it calls for ongoing research to refine these blockchain applications, ensuring that they can meet future demands and continue to evolve with the industry's technological landscape. This future-focused approach underscores the necessity for continuous improvement and adaptation in the implementation of blockchain technologies within the AEC sector.

References

[1] Directive (EU) 2023/1791 of the European Parliament and of the Council of 13 September 2023 on Energy Efficiency and Amending Regulation (EU) 2023/955 (recast) (Text with EEA relevance) [Internet]. Available from http://data.europa.eu/eli/dir/2023/1791/oj/eng, accessed 1 November 2024.

[2] Factsheet – The European Green Deal – Delivering the EU's 20 [Internet]. European Commission. Available from https://ec.europa.eu/commission/presscorner/detail/en/fs_23_4813, accessed 1 November 2024.

[3] European Commission. Joint Research Centre. A Europe Fit for the Digital Age: Tracking Europeans' Interest in EC Priorities Using Online Search Data. [Internet]. LU: Publications Office; 2022.

[4] European Commission. Executive Agency for Small and Medium-sized Enterprises, Volt J, Toth Z, Glicker J, *et al.* Definition of the digital building logbook – Report 1 of the study on the development of a European Union framework for buildings' digital logbook. Luxembourg: Publications Office of the European Union; 2020.

[5] European Commission. Executive Agency for Small and Medium-sized Enterprises, Carbonari G, Ricci M, Dourlens-Quaranta S, *et al.* Building logbook state of play – Report 2 of the study on the development of a European Union framework for buildings' digital logbook. Luxembourg: Publications Office of the European Union; 2020.

[6] Gómez-Gil M, Espinosa-Fernández A, and López-Mesa B. A new functionality for the Digital Building Logbook: Assessing the progress of decarbonisation of national building sectors. *Environ Impact Assess Rev*. 2024; 105:107393.

[7] Gómez-Gil M, Sesana MM, Salvalai G, Espinosa-Fernández A, and López-Mesa B. The Digital Building Logbook as a gateway linked to existing national data sources: The cases of Spain and Italy. *J Build Eng*. 2023; 63:105461.

[8] Al-Sadoon N, Fauth J, and Menzel K. Digital Twins as Enabler for Long Term Data Management Using Building Logbooks. In: Camarinha-Matos LM, Boucher X, and Ortiz A, (eds.), *Collaborative Networks in Digitalization and Society 5.0 [Internet]*. Cham: Springer Nature Switzerland; 2023.

pp. 673–86. (IFIP Advances in Information and Communication Technology; vol. 688).

[9] Gómez-Gil M, Espinosa-Fernández A, and López-Mesa B. Contribution of new digital technologies to the Digital Building Logbook. *Buildings*. 2022;12(12):2129.

[10] Khudhair A, Li H, and Ren G. Knowledge-based OpenBIM data exchange for building design. *Autom Constr*. 2023;156:105144.

[11] Owerko T, Wrzosek M, Rochel M, *et al.* Acceptance tests of BIM models using the IDS standards within openBIM. In: *28th International Conference on Methods and Models in Automation and Robotics (MMAR) [Internet]*; 2024. pp. 476–80. Available from https://ieeexplore.ieee.org/abstract/document/10680816?casa_token=73_Y5N8FO5gAAAAA:F_78zsnLS4NcFBOrDku-MOrl2TrnOXWJqtPsJ76rzuY1-MkpdNSo7qufHjmB6pdMQgnXghTpfmA, accessed 1 November 2024.

[12] Bucher DF, Hunhevicz JJ, Soman RK, Pauwels P, and Hall DM. From BIM to Web3: A critical interpretive synthesis of present and emerging data management approaches in construction informatics. *Adv Eng Inform*. 2024;62:102884.

[13] Reimann P, and Hachmann R. Linking design principles to context and evidence: A semantic web approach. In: Christensen IMF, Markauskaite L, Bonderup Dohn N, Ripley D, and Hachmann R (eds.), *Creating Design Knowledge in Educational Innovation*. London: Routledge; 2024.

[14] Lihtmaa L, and Kalamees T. Emerging renovation strategies and technical solutions for mass-construction of residential districts built after World War II in Europe. *Energy Strategy Rev*. 2024;51:101282.

[15] Andersen R, and Negendahl K. Lifespan prediction of existing building typologies. *J Build Eng*. 2023;65:105696.

[16] Azzalin M, Lauria M, Metastasio C, and Buglisi S. Digital Twin Approaches and BIM-Based Protocols for the Governance of the Built Environment. In: Battisti A, and Baiani S (eds.), *ETHICS: Endorse Technologies for Heritage Innovation: Cross-Disciplinary Strategies [Internet]*. Cham: Springer Nature Switzerland; 2024. pp. 171–85. Available from: https://doi.org/10.1007/978-3-031-50121-0_11.

[17] Daniotti B, Bolognesi CM, Lupica Spagnolo S, *et al.* An interoperable BIM-based toolkit for efficient renovation in buildings. *Buildings*. 2021;11(7):271.

[18] Amelio A, Giardino-Karlinger L, and Valletti T. Exclusionary pricing in two-sided markets. *Int J Ind Organ*. 2020;73:102592.

[19] Bucher D, and Hall D. Common data environment within the AEC ecosystem: Moving collaborative platforms beyond the open versus closed dichotomy; 2020.

[20] Mêda P, Hjelseth E, Calvetti D, and de Sousa H. Data templates and digital building logbooks boosting digital twin in construction; 2021.

[21] Mulhall D, Ayed AC, Schroeder J, Hansen K, and Wautelet T. The product circularity data sheet – A standardized digital fingerprint for circular economy data about products. *Energies*. 2022;15(9):3397.

[22] Valra A, Madeddu D, Chiappetti J, and Farina D. The BIM Management System: A Common Data Environment Using Linked Data to Support the Efficient Renovation in Buildings. In: *The 8th Annual International Sustainable Places Conference (SP2020) Proceedings [Internet]*. MDPI; 2021. p. 18.

[23] Li J, Greenwood D, and Kassem M. Blockchain in the built environment and construction industry: A systematic review, conceptual models and practical use cases. *Autom Constr*. 2019;102:288–307.

[24] Kifokeris D, and Tezel A. Blockchain and lean construction: An exploration of bidirectional synergies and interactions. *Archit Eng Des Manag*. 2025;25 (1):171–89.

[25] Nakamoto S. Bitcoin: A Peer-to-Peer Electronic Cash System; 2008. www.bitcoin.org, accessed 1 November 2024.

[26] Chaum DL. Computer Systems established, maintained and trusted by mutually suspicious groups [Internet]. Riverside, CA: Electronics Research Laboratory, University of California; 1979. Available from https://chaum.com/wp-content/uploads/2022/02/techrep.pdf, accessed 1 November 2024.

[27] Hunhevicz JJ, and Hall DM. Do you need a blockchain in construction? Use case categories and decision framework for DLT design options. *Adv Eng Inform*. 2020;45:101094.

[28] Kuznetsov O, Rusnak A, Yezhov A, Kuznetsova K, Kanonik D, and Domin O. Merkle trees in blockchain: A study of collision probability and security implications. *Internet Things*. 2024;26:101193.

[29] Ferreira A. Regulating smart contracts: Legal revolution or simply evolution? *Telecommun Policy*. 2021;45(2):102081.

[30] Hunhevicz JJ, Bucher DF, Hong K, Hall DM, and De Wolf C. Cheaper Smart Contracts for the Built Environment? Linking On-Chain and Off-Chain in a Blockchain-Governed Approach. In: *2024 European Conference on Computing in Construction [Internet]*. European Council on Computing in Construction; 2024. pp. 42–9.

[31] Hunhevicz JJ, Motie M, and Hall DM. Digital building twins and blockchain for performance-based (smart) contracts. *Autom Constr*. 2022;133:103981.

[32] Medina J, and Rojas-Cessa R. AMI-Chain: A scalable power-metering blockchain with IPFS storage for smart cities. *Internet Things Neth*. 2024;25:101097.

[33] Davidson S, and Potts J. Institutional Cryptoeconomics. In: Caton J (ed.), *The Economics of Blockchain and Cryptocurrency. [Internet]*. Cheltenham: Edward Elgar Publishing; 2022. pp. 1–12.

[34] Brekke JK, and Alsindi WZ. Cryptoeconomics. *Internet Policy Rev [Internet]*. 2021;10(2).

[35] Davidson S, De Filippi P, and Potts J. Blockchains and the economic institutions of capitalism. *J Institutional Econ*. 2018;14(4):639–58.

[36] Kuchipudi R, Prathima T, Palamakula RB, Satyanarayana Murthy T, and Gangadhara Rao K. Private AI in e-commerce. In: Maheswari VU and Aluvalu R (eds.), *Sustainable Development Using Private AI [Internet]*. 1st ed. Boca Raton, FL: CRC Press; 2024. pp. 232–9. Available from https://www.taylorfrancis.com/books/9781032716749/chapters/10.1201/9781032716749-12.

[37] Bucher D, and Hall D. New Ways of Data Governance for Construction? Decentralized Data Marketplaces as Web3 Concept just around the Corner. In: *Proceedings of the 29th EG-ICE International Workshop on Intelligent Computing in Engineering [Internet]*. Aarhus University; 2022 [cited 2024 Sep 4]. p. 1872. Available from: https://www.research-collection.ethz.ch/handle/20.500.11850/560904.

[38] Hadabi A, Qu Z, Elhabob R, *et al.* Proxy re-encryption with plaintext checkable encryption for integrating digital twins into IIoT. *Comput Electr Eng*. 2024;116:109164.

[39] Ghansah FA, and Lu W. Major opportunities of digital twins for smart buildings: A scientometric and content analysis. *Smart Sustain Built Environ*. 2024;13(1):63–84.

[40] Rukanova B, Ubacht J, van Engelenburg S, *et al.* Realizing Value from Voluntary Business-Government Information Sharing through Blockchain-Enabled Infrastructures: The Case of Importing Tires to the Netherlands Using TradeLens. In: *Proceedings of the 22nd Annual International Conference on Digital Government Research [Internet]*. New York: Association for Computing Machinery; 2021. pp. 505–14. Available from: https://dl.acm.org/doi/10.1145/3463677.3463704.

[41] Hunhevicz JJ, Hall DM, Brasey PA, Bonanomi MM, and Fischer M, Decentralized project delivery on the crypto commons: Conceptualization, governance mechanisms, and future research directions. *Project Leadership and Society*. 2024;5:100132.

[42] Hunhevicz JJ, Hall DM, Brasey PA, Bonanomi MMM, and Fischer M. Decentralized project delivery on the crypto commons: Conceptualization, governance mechanisms, and future research directions. *Proj Leadersh Soc*. 2024;5:100132.

[43] Wild N. *Material Product Passport for Blockchain Assured Ownerships*. Zurich: ETH Zurich; 2021.

[44] Web3 as ‘self-infrastructuring’: The challenge is how – Kelsie Nabben [Internet]; 2023. Available from https://journals-sagepub-com.tudelft.idm.oclc.org/doi/full/10.1177/20539517231159002.

[45] The European Green Deal – European Commission [Internet]; 2021. Available from https://commission.europa.eu/strategy-and-policy/priorities-2019-2024/european-green-deal_en.

[46] Directive 2010/31/EU of the European Parliament and of the Council of 19 May 2010 on the Energy Performance of Buildings (recast) [Internet]; 2010. Available from: http://data.europa.eu/eli/dir/2010/31/oj/eng, accessed 1 November 2024.

[47] Just over 4.9 million dwellings in Sweden [Internet]. Statistikmyndigheten SCB. Available from https://www.scb.se/en/finding-statistics/statistics-by-subject-area/housing-construction-and-building/housing-construction-and-conversion/dwelling-stock/pong/statistical-news/dwelling-stock-2018-12-31/, accessed 1 November 2024.

[48] Kommande energirenoveringskrav – En konsekvensanalys från Sveriges Allmännytta [Internet]. Sveriges Allmännytta; 2024 Available from https://www.sverigesallmannytta.se/trycksaker/kommande-energirenoveringskrav-en-konsekvensanalys-for-sveriges-allmannyttas-medlemmar/, accessed 1 November 2024.

[49] Hermansson C, Sandén B, Kjellström E, *et al.* Klimatpolitiska rådets rapport 2023; 2023. Available from https://lup.lub.lu.se/record/42e5a733-262c-4f05-959f-2c7b27d33dd7, accessed 1 November 2024.

[50] Zott C, and Amit R. Business model design: An activity system perspective. *Long Range Plann.* 2010;43(2):216–26.

[51] Wirtz BW. *Digital Business Models: Concepts, Models, and the Alphabet Case Study [Internet].* Cham: Springer International Publishing; 2019 (Progress in IS).

[52] Zott C, and Amit R. Business model design and the performance of entrepreneurial firms. *Organ Sci.* 2007;18(2):181–99.

[53] Kapengut E, and Mizrach B. An event study of the Ethereum transition to proof-of-stake. *Commodities.* 2023;2(2):96–110.

Chapter 7

Blockchain for energy management in seaport infrastructure: a microgrid case study

Alexander James Howe[*], *Ioan Petri*[*] *and Yacine Rezgui*[*]

This chapter investigates the application of blockchain technology in enhancing energy management within seaport infrastructure, focusing on a microgrid case study at the Port of Milford Haven in Wales. The chapter contextualizes the growing energy demands of seaports, driven by refrigeration, air conditioning, and other energy-intensive activities, which collectively contribute significantly to greenhouse gas (GHG) emissions. The authors examine blockchain's role in decentralizing energy management, particularly using Ethereum-based smart contracts.

The chapter introduces a blockchain-enabled microgrid framework that supports renewable energy integration, leveraging smart contracts for energy transfers and pricing while ensuring tamper-proof record-keeping. Using real-world data, two scenarios are analyzed to evaluate the scalability and economic feasibility of the proposed framework. Scenario 1 focuses on five buildings within the port utilizing locally produced renewable energy, while scenario 2 scales up to include 200 additional community households. The analysis highlights significant cost implications tied to Ethereum's gas prices and the computational requirements of smart contracts. Although the system demonstrates potential for reducing energy costs and enhancing local energy resilience, the volatility of gas and cryptocurrency prices presents challenges to widespread adoption.

The chapter discusses key barriers, including the need for user training, hardware upgrades, and ensuring the integrity of energy sources within the grid. It emphasizes the importance of further research into private blockchain and centralized alternatives to overcome cost and scalability challenges.

7.1 Introduction

As the global population continues to grow and urbanization becomes more prevalent, the energy demand is also increasing. Meeting this demand sustainably and

[*]School of Engineering, Cardiff University, UK

cost-effectively poses a significant challenge for the energy sector. Renewable energy sources (RES) have witnessed significant advancements in recent years and will play a crucial role in global efforts to reduce GHG emissions.

The climate has undergone significant changes in recent times due to the effects of global warming and the rising concentration of carbon dioxide in the atmosphere. Consequently, there is a major drive to reduce carbon emissions from all sectors of the built environment. This includes decarbonization throughout the whole life cycle of built environment projects, from the design phase through to end-of-life decommissioning. Infrastructure, as a subset of the built environment, is necessary to support the global supply chain, international trade and the distribution of water and electricity. Therefore, reducing industrial carbon emissions can play a crucial part in achieving climate goals.

The seaport industry represents an important sector that plays a significant role as an industry in the global economy. Within the seaport industry, energy consumption predominantly revolves around refrigeration and air conditioning systems, which are integral to the various processes involved [11]. It is worth noting that the expansion of ports has contributed to approximately 4% of the total global GHG emissions [24], with refrigeration and air conditioning accounting for around 15% of the global energy consumption, a significant portion of which is attributed to the seaports industry [12].

There are indeed several factors that can influence the energy consumption of seaport industries, including seasonal variations, weather conditions impacting annual fish processing volumes, and the number of processing lines and boats involved. In the fish processing sector, energy usage can be categorized into two main areas: direct consumption for lighting, heating, and washing machines, and indirect usage related to cooling cycles and freezing equipment. Similar energy consumption mechanisms extend throughout other types of infrastructure within the built environment, such as lighting and air conditioning through industry in general, or cooling and freezing within the food supply chain industry.

One of the key challenges faced by seaports is reducing their carbon footprint by transitioning to renewable and sustainable energy sources. Embracing renewable energy options like solar, wind, and hydropower can provide seaports with greater energy flexibility and optimized energy consumption. For instance, solar panels can be installed on large buildings within seaports to harness solar energy, converting sunlight into electricity. Similarly, wind turbines can be strategically placed, especially in coastal areas near seaports, to generate electricity and cover the energy demand of the port. By adopting renewable energy solutions, seaports can contribute to mitigating their environmental impact and moving toward a more sustainable future.

One way to enhance and implement locally sourced renewable energy is through the use of microgrids which enhance the integration of RES into the seaports industry's energy operation. Microgrids are modern small-scale electricity networks that operate independently from a main power grid such as the national grid that can distribute and store electricity in a localized area such as a neighborhood, community, or port. They use technologies such as sensors, automation, and communication systems to improve the reliability, efficiency, and sustainability of energy transfers within a community, especially benefiting from using local RES [13] allowing communities to reduce their carbon footprint and contribute to combatting climate change.

The combination of RES and microgrids has been studied widely [10] and it has been found that microgrids can help reduce system losses, and increase the reliability, security, and efficiency of electricity from renewable sources for producers, consumers, and prosumers [14].

Existing power grids were originally designed to operate in a one-way flow of electricity from the national grid to consumers, with limited means of monitoring and controlling the network. Adversely, microgrids are designed to facilitate a two-way flow of energy and reliable real-time information between producers and consumers, allowing for more dynamic and flexible management of the grid [12]. Microgrids can help to reduce energy consumption and costs for consumers by reducing reliance on the main grid and prioritizing the use of renewable energy.

Microgrid technology can enhance the integration of RES into the seaport industry and the communities nearby. Microgrids can optimize the distribution of energy using batteries to manage the intermittent and variable energy produced by renewable sources [21]. This can lead to lower costs for consumers involved in the smart grid after the initial investment of a battery. The increased use of smart grid hardware such as smart meters has allowed significant progress in moving toward more autonomous smart grid networks. The hardware can handle more complex information, using energy consumption patterns to optimize the price of energy for consumers [6,32]

Blockchain technology has the potential to enhance microgrid operations by enabling secure, transparent, and efficient transactions and data exchanges [5]. Blockchain technology is inherently secure due to its decentralized, distributed nature [31]. This means that there is no central point of failure, and the data is tamper-proof. The transparency of blockchain technology makes it ideal for tracking and verifying transactions in smart grids. It ensures all parties have access to the same information, which reduces the likelihood of fraud and errors [8]. Smart grids utilizing blockchain can automate the processes on the grid, streamlining operations and reducing the need for intermediaries [3,22].

Smart contracts are self-executing digital contracts that are stored on a blockchain and are programmed to automatically trigger actions or transactions when conditions are met. They behave exactly as programmed and cannot be changed [21]. When smart contracts are created and executed, it is recorded on the blockchain and become part of the permanent tamper-proof record.

In the context of smart grids, blockchain technology can be used to enable peer-to-peer energy trading on a secure transparent and decentralized system for tracking and managing energy transactions. Consumers and producers using the smart grid can buy and sell energy directly with one another utilizing smart contracts, avoiding going through a centralized utility company [12]. This could result in lower costs and increased efficiency for consumers, especially if the microgrid can utilize excess renewable energy stored in batteries [17].

The objective of this chapter is to evaluate the utilization of blockchain technology and smart contracts for managing energy flows in seaports. The analysis is applied to a case study at the Milford Haven port in the UK, aiming to assess the costs associated with implementing smart contracts for energy flows in a port industrial context. Additionally, this chapter investigates the energy usage and transaction times of smart contract transactions, helping to understand their impact on an energy microgrid scenario. The overarching goal is to determine if blockchain can provide a feasible solution for managing energy at the port level, including the integration of building energy systems and microgrids.

This chapter explores the use of smart contracts for energy flows in a site of buildings at the Milford Haven Port, the largest energy port in the UK, for which a blockchain model will be implemented and tested using real data and models. An additional scenario including 200 additional households within the local community will be investigated to explore how the scalability of blockchain transactions affects the cost of the microgrid. Section 7.2 will explore related work to renewable energy in ports, microgrids for renewable energy, and blockchain use in microgrids with a discussion around limitations in prior work. In Section 7.3, the theoretical framework for the investigation, the assumptions, the variables, and the limitations are presented. Section 7.4 contains the experimental framework with associated scenarios demonstrating the use of blockchain for seaport microgrids. Section 7.5 will discuss the results and their implications.

Overall, this chapter explains how blockchain can be leveraged to enable a smart contract to monitor the exchange energy flows for a set of five buildings from Milford Haven port in South Wales. The chapter provides an investigation of the effect of scalability and potential viability of using Ethereum smart contracts to enable a microgrid prioritizing trading renewable energy within seaports and expanding to the wider community.

7.2 Related works

Table 7.1 presents a list of key contributions from the literature in three critical areas relevant to this chapter: renewable energy applications in the port industry, microgrids for renewable energy integration, and the role of blockchain technology in managing microgrids. These studies provide an understanding of the interplay between sustainable energy systems and emerging digital technologies such as blockchain and distributed ledger technologies. Each of these categories will be reviewed in detail in one of the following sub-sections, offering a comprehensive analysis of the current state of research.

7.2.1 Renewable energy in industry

7.2.1.1 Renewable energy for the port industry

The utilization of microgrids to enhance the integration of renewable energy in ports has received significant attention in the literature. Sdoukopoulos *et al.* [29] examine the energy efficiency and prospects of seaports, highlighting the dual role

Table 7.1 Related studies

Author	**Renewable energy in the port industry**	**Microgrids for renewable energy**	**Blockchain for microgrids**
Sdoukopoulos *et al.* [29]	X		
Hentschel *et al.* [15]	X		
Misra *et al.* [21]	X	X	
Abusharkh *et al.* [1]		X	
Alzahrani *et al.* [6]	X	X	
Alzahrani *et al.* [7]	X	X	
Agung and Handayani [2]		X	X
Vieira and Zhang [32]		X	X
Buccafurri *et al.* [9]		X	X
Akhras *et al.* [4]		X	X
Mollah *et al.* [22]			X
Mengelkamp *et al.* [20]			X
Sabounchi and Wei [27]			X
Kim *et al.* [12]			X
Kermani *et al.* [17]	X	X	X
Petri *et al.* [25]	X	X	X

of ports as both substantial energy producers and consumers, and emphasizing the importance of addressing energy consumption as an environmental priority. Given their significant energy demands and production capacities, ports have a critical role to play in promoting and leveraging RES. The primary objective for ports is to transition toward the utilization of renewable energy, thereby reducing reliance on fossil fuels.

Hentschel *et al.* [15] investigate the potential of renewable energy cooperatives in facilitating the sustainability of the Port of Rotterdam through the adoption of RES. Renewable energy cooperatives comprise groups of consumers and producers who collaborate to develop and promote renewable energy solutions. These cooperatives operate democratically and transparently, diverging from the conventional practices of energy companies and ensuring fair distribution of renewable energy among its members. By investing in low-carbon energy and decentralizing the energy network, citizens and communities can attain greater self-sufficiency, and reduce reliance on the main grid. The exploration of renewable energy cooperatives in the context of ports is of utmost relevance, as ports are known for their high energy consumption and the flexibility of their energy demands. Electricity constitutes a significant cost factor for most port companies, making energy cost reduction measures crucial for improving their overall business profile. Moreover, the efficient distribution of renewable energy within a community not only reduces dependence on fossil fuels but also leads to a substantial reduction in carbon emissions.

7.2.1.2 Microgrids for renewable energy distribution in industry

Misra *et al.* [21] explore the potential of RES in reducing GHG emissions within the port industry. The study highlights the significant contribution of microgrids in facilitating the integration of local RES, enabling seaports to meet their electrical demands independently from a centralized grid dominated by non-renewable energy. Leveraging on-site renewable energy generation can lead to substantial reductions in pollution and energy expenses, making ports an ideal location for both land-based and offshore renewable energy installations. The authors present a case study of the Port of Chennai, India, wherein the installation of a battery with a capacity of 60% of the total daily energy requirements proves instrumental in harnessing renewable energy due to its intermittent nature.

In a study by Abusharkh *et al.* [1], the viability of microgrids in the UK's energy future is examined. The research highlights that the majority of electricity generated and consumed in the UK currently relies on a centralized power system, primarily relying on large-scale fossil fuel and nuclear power plants. The study suggests that microgrids have the potential to effectively integrate small-scale RES, thus enhancing the efficiency of meeting the UK's energy demands. The investigation reveals that there are no underlying technological barriers to the adoption of microgrids in the UK energy system, and economically, microgrids can be as viable as energy provided by the national grid.

Alzahrani *et al.* [6] conducted a study focusing on optimizing a smart energy trading system for the Milford Haven Port, aimed at reducing energy consumption from the grid and maximizing the utilization of on-site RES. Through simulations of optimized energy strategies, the study demonstrated the clear benefits of implementing a suitably sized battery system in terms of energy savings and cost reduction.

Building upon their previous research, Alzahrani *et al.* [7] further expanded their investigation by exploring and forecasting potential scenarios for the Milford Haven smart grid. These additional scenarios highlighted the possibility and benefits of integrating electric boats at the port to store electricity generated by on-site renewable sources. This integration would reduce the initial capital investment required for a separate battery system, as the electric boats already in use could serve this purpose. The electric boats would always maintain a charge level above a threshold of 50% to ensure their availability, with any excess energy being sold back to the grid. Additionally, a peer-to-peer energy-sharing strategy was proposed, allowing the local community of approximately 200 neighboring houses to buy and sell excess energy from the port. This strategy aims to increase the community's independence from the grid and promote the use of renewable energy within the local area. Featuring the local community within a renewable energy smart grid has the potential to spread the benefits of local renewable energy production to the whole community and will be further investigated within this chapter within the context of a blockchain microgrid solution.

7.2.2 Blockchain for microgrids

Agung and Handayani [2] demonstrate how blockchain technology, specifically through the use of smart contracts, enables secure, and authorized small-scale energy transfers. Blockchain technology ensures an immutable transaction history, facilitating audits, and resolving transaction disputes without the need for a centralized intermediary. The study presents a system wherein producers can sell energy in exchange for tokens, which can then be traded on the network at a predetermined price. Consumers have the option to purchase tokens in advance, deducting, and "burning" them upon energy consumption, thereby signifying the utilization of energy. It is noted that the use of Ethereum, a public blockchain, introduces variability, and it is suggested that the tokens generated by producers should be tied to a fiat currency. However, a potential limitation arises from the possibility of producers generating energy from non-renewable sources, thereby undermining the purpose of the microgrid concept.

Vieira and Zhang [32] conducted a study to investigate the functionality of smart contracts in utilizing blockchain technology in a smart grid. The research demonstrates the technical and economic feasibility of peer-to-peer energy trading using blockchain technology. Cost analyses were performed on different frameworks using real-world data, showing that the presented peer-to-peer trading system can effectively work with the current centralized energy grid, enhancing the utilization of local renewable energy. However, certain limitations exist, such as transmission and distribution costs, as well as battery inefficiencies. Moreover, as transaction volumes increase, not all transactions may be processed, resulting in market congestion and incomplete information. Additionally, the volatility of Ethereum introduces uncertainty to energy forecasting, leading to fluctuating costs in running a microgrid over time.

Buccafurri *et al.* [9] propose an Ethereum based solution leveraging public blockchain technology to enable a smart grid. A framework has been produced showing how the users of the smart grid interact to enable energy trading, a vital function of a smart grid. A Solidity energy trading smart contract has been demonstrated and tested to investigate the potential costs for users to bid for energy on the smart grid. The study introduces an additional user type, an energy retailer, who can purchase energy off the smart grid to sell its own consumers. The study also describes how an energy authority would be required to deploy the smart contracts and store verified retailers. The paper concludes that energy trading can be robustly and profitably enabled through a smart grid using Ethereum smart contracts.

A secure smart grid communication network has been proposed with a central utility company, which controls a source of renewable energy, to control and operate a smart grid communication network [4]. The utility can communicate with consumers using Ethereum smart contracts to send data between the utility and smart meters. Smart contract functions are proposed allowing energy consumers to join the grid, enable communication, and actively transmit data. The study produces a set of gas fees and hence real-life fees for each smart contract function required for users to join the grid and send data to the utility company.

Mollah *et al.* [22] conducted an insightful investigation into the diverse applications of blockchain technology in smart grids. The study highlights how blockchain-based smart grids offer secure and trustworthy systems, addressing the inherent challenges associated with centralized microgrids. However, the research also brings attention to the challenges and future directions related to blockchain technology and smart grids, particularly in terms of throughput. It is noted that public blockchains like Ethereum face limitations in throughput, which affects their ability to compete with existing financial systems. Private blockchains, however, can potentially offer higher throughputs but may compromise the complete decentralization provided by public blockchains.

In a similar vein, Mengelkamp *et al.* [20] introduced the concept of a decentralized blockchain-based energy microgrid. The authors evaluated the Brooklyn Microgrid project as a case study, observing its operation over a three-month test trial period. The study demonstrated the viability of blockchain technology in facilitating decentralized microgrid energy transfers. Notably, the example microgrid effectively integrated RES to manage the inherent uncertainty of renewable energy production. The project received positive public perception. However, the authors also acknowledged the limitations imposed by current regulations in energy markets, necessitating further progress and investigation in this aspect. The study calls for additional evaluation of the incentives required to encourage community participation in localized microgrid projects.

Sabounchi and Wei [27] proposed a peer-to-peer energy trading mechanism, utilizing smart contract transactions, to address critical challenges faced by the centralized energy grid. The study presents a peer-to-peer microgrid model where local energy producers and consumers can engage in trading through a decentralized blockchain system using Ethereum. The research demonstrates that a decentralized model can alleviate the burden on the existing grid, and smart contracts can be leveraged to enable fast, secure, and certifiable peer-to-peer energy trading.

Security and privacy are indeed important considerations when applying blockchain technology to smart grids. In the context of a blockchain-based smart grid, users would need to utilize public pseudonym addresses for anonymous transactions, ensuring their identities remain undisclosed. However, it is worth noting that transaction-related information such as senders, receivers, records, and values are publicly available on the blockchain. This raises the possibility of analyzing public information to unveil the real users behind the pseudonyms, compromising the privacy offered by blockchain smart grids. While cryptographic security options exist to protect this information, they often require the involvement of centralized entities, which contradicts the decentralized nature of blockchain technology.

7.2.2.1 Blockchain for microgrids in ports

A related study [25] investigated the Port of Milford Haven as a case study to explore the potential use of blockchain technology in supporting the formation of energy trading communities. The researchers proposed the implementation of Ethereum smart contracts to manage energy transactions and analyzed the costs associated with the necessary functions within the smart contract for facilitating peer-to-peer energy trading. The analysis considered various gas prices based on the 2020 Ethereum price. The results showed that blockchain technology can provide a secure and immutable record of all transactions within the energy community.

In a different study, Kermani *et al.* [17] investigated the application of blockchain technology in the context of the Port of Long Beach in California. The authors proposed a microgrid energy management system based on blockchain to reduce the port's reliance on the main grid and minimize power consumption. The proposal highlights the increasing significance of energy consumption as an environmental concern over the past two decades. Therefore, approaches that aim to minimize grid reliance and instead utilize local RES can bring significant environmental benefits, including improved air quality and reduced noise pollution. The authors demonstrate how a blockchain microgrid can provide the port with a secure and efficient energy management system. Their analysis suggests that ports are ideal locations for on-site local energy production, given their proximity to the sea, high wind levels, and ample roof areas for solar panel installation.

The next chapter aims to investigate the potential costs of a public Ethereum microgrid formed around five buildings at the Port of Milford Haven considering data and models from the EU INTERREG piSCES project around how the buildings consume and produce energy. The effect of Ethereum and the transaction gas prices on the functions within the smart contract has been explored and the average year-to-date Ethereum and gas prices were used to represent the costs of the energy transactions based on current prices.

7.3 The blockchain for energy transactions

The Ethereum network is a decentralized platform that has transitioned to a more secure and energy-efficient consensus algorithm called proof of stake (PoS) in 2022. This change reduces the computational complexity of transaction validation compared to the previous proof of work algorithm. In PoS, validators participate in the network by holding a stake or deposit of cryptocurrency such as Ethereum. The likelihood of a validator being chosen to validate a transaction is proportional to the amount of cryptocurrency they have staked. This stake serves as an incentive for validators to act honestly, as they risk losing their deposit if they engage in malicious or fraudulent behavior.

To validate and add transactions to the blockchain, several steps are involved. First, a transaction is initiated and broadcast to the network. Then, validators are selected based on their stake to authorize the transaction. The chosen validators utilize their computational power to validate and add the transaction to the blockchain. In return, validators receive rewards in the form of transaction fees and the

Ethereum cryptocurrency. Ethereum can be exchanged for fiat currencies like US dollars or Great British Pounds based on the prevailing market rates.

On the public Ethereum network, users need to allocate a certain amount of gas to complete their transactions. The amount of gas required depends on the computational power needed to verify and add the transaction to the blockchain. Validators offer their computational power to validate transactions, and users must pay a fee in Ethereum, known as the gas price, for each unit of gas used. Gas prices on the public Ethereum network can vary depending on network demand, with higher prices observed during periods of high transaction traffic. If users desire faster transaction processing, they have the option to overpay the current gas price, which incentivizes miners to prioritize their transactions. However, it's important to note that overpaying can lead to increased costs for users, particularly during times of high network congestion.

The equation for the cost of a transaction is shown below:

$$GasUsed * GasPrice = Cost\ (InEthereum).$$

The proposed public Ethereum microgrid introduces a new dimension to transaction times, as they can vary depending on the incentivization of chosen validators to include transactions in the next block. Overpaying the current gas price can expedite transaction processing. Real-time transaction time data has been analyzed to understand the impact of this strategy. Gas, measured in units called "Gwei," represents a fraction of one Ethereum. For instance, a transaction using 1,000 gas at a gas price of 10 will incur a cost of 10,000 Gwei, which is equivalent to 0.00001 Ethereum.

The fiat cost of a transaction is influenced by the exchange rate between fiat currency and Ethereum. Similar to gas prices, the price of Ethereum is highly volatile, as it is subject to market supply and demand dynamics. Increased demand for Ethereum leads to price appreciation. Currently, there are over 120 million Ethereum coins in circulation. However, there are no guarantees regarding the future price of Ethereum. Ethereum can experience significant price increases or, like many other cryptocurrencies, face a decline, and potential collapse.

To enable peer-to-peer trading, a microgrid can leverage smart contracts on the Ethereum network. These smart contracts are written in the Solidity programming language. Once deployed on the network, smart contracts become accessible to all participants within the microgrid. The rules embedded within smart contracts are immutable, meaning they cannot be modified once deployed. This feature, inherent to blockchain technology, ensures that all users have equal access, and abide by the same rules. Parties cannot alter the smart contract; any changes to the rules would require deploying a new version onto the blockchain. Smart contracts guarantee that both energy consumers and producers agree on the conditions of energy trade.

Smart contracts serve as a versatile tool for managing various aspects of the energy system. In the context of this chapter, a microgrid smart contract template (refer to Appendix B) has been employed to facilitate essential transactions for enabling peer-to-peer energy trading. This smart contract encompasses three functions specifically designed for microgrids. The framework for the smart contract to enable a renewable energy trading microgrid is presented in Figure 7.1.

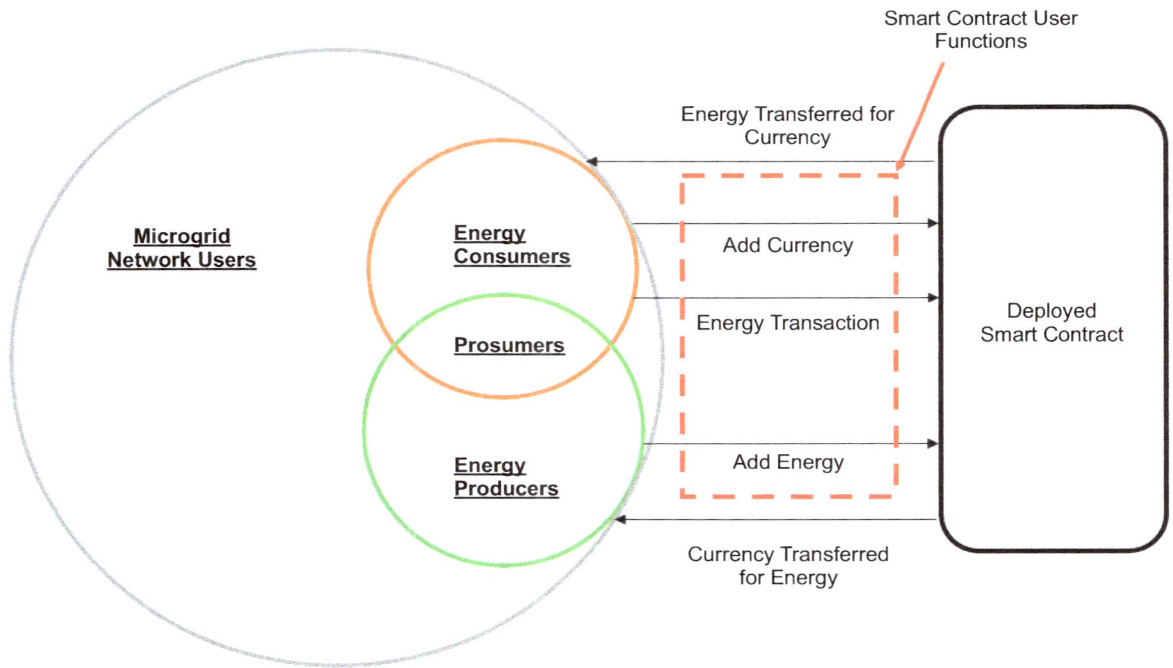

Figure 7.1 Microgrid smart contract framework

The first function, known as "Add Energy," empowers microgrid producers to sell surplus energy generated by renewable sources like solar panels or wind turbines. It provides producers with the means to contribute their excess energy to the grid.

The second function, named "Add Currency," enables consumers to add a virtual currency that can be utilized to purchase energy within the microgrid. This function allows consumers to participate in the energy trading process.

Lastly, the "Energy Transaction" function facilitates the transfer of energy from a producer to a consumer in exchange for the agreed-upon currency. It forms the core of the peer-to-peer energy trading mechanism within the microgrid.

In this proposed system, consumers can acquire the currency used for energy trading by purchasing it with fiat currency. Conversely, producers who earn the currency can trade it back for fiat currency. The smart contract provides the flexibility to accommodate the trading of four different types of renewable energy, with the option to adjust currency prices depending on the specific energy source. While the experiment does not consider the impact of energy type on gas usage, it demonstrates the practical application of the smart contract. The proposed energy trading system has benefits over an auction-based system. The system shown in Figure 7.1 requires less smart contract function uses for a single user to buy energy and can reduce the number of wasted functions by removing the situation of wasted bids. Reducing the number of smart contract functions can reduce the cost for users to use the microgrid, especially if gas prices are high.

To validate and record the functions of the smart contract on the blockchain, gas is utilized. The amount of gas consumed by each function depends on its complexity, and a gas price is paid per unit of gas. Additionally, gas is required for the deployment of the smart contract itself.

7.3.1 Scenarios of analysis

To investigate the potential costs of the transactions required to run a microgrid for the Port of Milford Haven, two scenarios have been considered. Milford Haven Port is one of the largest energy ports in the country with the capacity to supply almost 30% of the gas required in the UK and is regarded as the greatest handler of gas and oil. The port is situated in the western part of Wales and offers experience in a wide range of maritime activities, including freight, renewable energy, passenger ferries, fisheries, commercial property management, leisure, and retail.

According to the Milford Haven port's environmental report, around 1,600 MWh of electricity was consumed overall in 2012, and 790 tons of CO_2 were produced. PV panels are being installed throughout the port authority's facilities to meet energy demands and reduce carbon emissions. This will assist the port in meeting the UK's energy use strategy for 2050. Within its boundaries, the port put 2,500 panels on 25 different structures. The Packaway building has a PV solar system, lighting systems, smart meter, ice store freezer, and flake ice maker. Before fishing, the fishermen gather boxes and ice from this structure. They take out how much ice they'll need and fill each box with half full of fish and ice.

Nonetheless, the technology instantly applies power from the central grid system when the fishermen need it for the building's major appliances. The energy transmission firm calculates how much energy is produced and used at the building level during a known period based using smart meters.

The K Shed is an open hall that is mostly utilized by MHPA for storage. An external tenant has a sectioned area designated for them, and they also use it for storage, which includes a refrigerator and freezer. Additionally, there is a cold room on the site that is powered by solar panels. Since the office space is vacant, there is no need for heating within the building. The PV solar capacity of this building is 50 kW. Following the process, the fish is kept in the chiller storage room for an average of 12 h.

Several tenants currently occupy the M Shed building, which has different appliances and lighting systems installed. F Shed is a new build that has been separated into six units: Unit 1, Unit 2 Ground Floor, Unit 2 First Floor, Unit 3 Ground Floor, Unit 3 First Floor, and Unit 4. Fish processing facilities are available in the ground floor apartments, while light storage, such as fish boxes, is available in the two first-floor units. Except for group 4, which is occupied by a tenant, the building is currently empty.

J Shed is presently occupied by various tenants and has a highly intricate structure with numerous electrical systems. The structure is divided into three occupied apartments, one of which has a retail space with an area for fish processing and upstairs office space, the second unit is used for fish processing and storage, the third unit is a fish hawkers hall that is used by individuals for small processing and storage.

Figure 7.2 illustrates the Milford Haven Port site of analysis consisting of five main buildings, 2 of which can produce renewable energy due to having photovoltaic panels installed on their roofs. There is also a local solar farm with around a 5-MW power capacity.

As such, *scenario 1* develops around the formation of a microgrid community of five buildings with solar farm energy capability.

Scenario 2 considers the same five main buildings and solar farm but in addition, considers a higher scale of 200 houses within the local community. This scenario aims to investigate the scalability of an Ethereum microgrid when the number of users and transactions increases.

For both scenarios, the potential costs of the microgrid transactions per year have been calculated. This has been achieved by estimating the potential number of transactions of each function in the smart contract for each building throughout the year and relies on general assumptions.

It has been assumed that each building equipped with energy-producing capabilities, such as the two buildings with photovoltaic panels and the solar farm, will sell their surplus energy at the end of each day. This assumption necessitates that these buildings have sufficient battery storage to store the excess energy generated by the solar panels on high-demand days. Additionally, it is assumed that the smart contract only needs to be deployed once per year. Once the contract is deployed, all microgrid users can access it and participate in the energy transaction system.

If there is a need to modify the conditions of the smart contract, the contract must be redeployed, incurring additional gas fees. Therefore, it is essential to

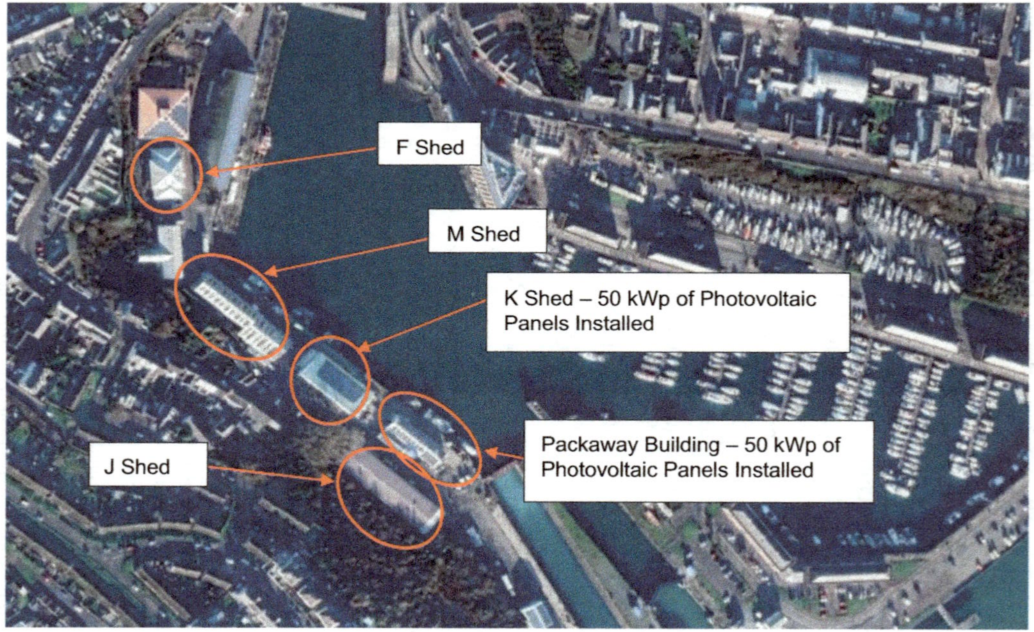

Figure 7.2 Milford Haven Port microgrid

minimize the number of deployments to reduce costs. Users of the microgrid are required to have an Ethereum wallet to cover the gas fees for their transactions. These wallets are anonymous and do not contain any personal user information.

Furthermore, it is assumed that consumers will utilize the "Add Currency" function of the smart contract every week. However, due to transaction costs, users are encouraged to minimize the frequency of their function usage.

7.3.2 Framework specification

It is assumed that all consumers and prosumers of the microgrid have a smart meter installed. These allow users to monitor their energy consumption and production habits and allow them to better plan transactions such as "Add Currency" to limit the number of transactions needed on the smart contract. The number of transactions being completed will be varied from 1 to 5 to investigate how increasing the number of transactions increases the cost of the microgrid due to the gas fees.

The implemented Solidity smart contract for the microgrid experiment, as shown in the box below, was developed using the Remix Integrated Development Environment (IDE). Remix IDE is a web-based platform that allows developers to write, compile, and test Solidity smart contracts. It provides essential tools for deploying and testing Ethereum smart contract functions, including the determination of gas requirements for various contract operations such as deployment and specific program features.

```
pragma solidity >=0.4.24 <0.6.0;
contract EnergyUse{
//mapped a supplier to an amount of energy and money from a subscriber
mapping(address => uint) public balanceOf;
event Transfer(address buyers, address sellers, uint amnt);
mapping(address=>uint)public sellAmountSollar;
mapping(address=>uint)public sellAmountWind;
mapping(address=>uint)public sellAmountHydro;
mapping(address=>uint)public sellAmountGas;
event Consumed(address seller,uint energy);
function seller(address sellers,string EngTyp, uint tmuse) {
uint amount;
function buyer( address buyers, uint money) {
balanceOf[buyers] += money;
}
}}
```

In this experiment, it is assumed that the gas amounts calculated by Remix IDE accurately reflect the gas requirements for real-world microgrid operations. These gas amounts play a crucial role in evaluating the economic viability of using public Ethereum smart contracts to enable a microgrid system of the scale required for a seaport. The experiment's results and associated costs will be compared to a similar study conducted by Vieira and Zhang [32], as well as the current national grid [23] averages and network costs.

Furthermore, the costs associated with scenario 2 will be compared to the expenses of the existing energy distribution system established by the national grid for the port.

To incentivize consumers in the Milford Haven community to adopt the microgrid and utilize locally generated renewable energy, the cost of energy provided by the microgrid should be lower than what they currently pay for energy supplied by the national grid, which may originate from GHG-emitting sources.

7.4 Evaluation and results

In this section, we evaluate two scenarios to showcase the application of blockchain in the establishment and operation of a seaport microgrid. The results and findings are presented to illustrate the effectiveness of utilizing blockchain technology in this context.

Table 7.2 provides an overview of the gas consumption associated with each smart contract function. These values were determined through the simulation of transactions using the Remix IDE. Gas consumption serves as a metric to measure the computational effort and resources required by each function within the smart contract.

Table 7.2 shows that the amount of gas required to deploy the smart contract onto the blockchain is much larger than the amount of gas needed for the other function. However, the impact of deployment costs on the annual operation of the microgrid is minimal, as the contract only needs to be deployed once. The primary factors influencing the cost of smart contract transactions are the gas price and the price of Ethereum. Table 7.3 provides an overview of the price variations for these factors.

Table 7.2 Gas prices of smart contract functions

Smart contract function	Amount of gas
Deploy	929,895
Add energy	28,035
Add currency	26,921
Energy transaction	42,188

Table 7.3 Variations of Ethereum and gas prices in 2023

Commodity	Max price 2023	Mean price 2023	MIN price 2023
Gas price (Gwei)	100	32	17.9
Ethereum (USD)	$2120.54	$1658.90	$994.98

The variation in gas price and Ethereum price, as illustrated in Table 7.3, highlights the potential for significant fluctuations in the cost of operating the microgrid on the public Ethereum network. Consequently, the overall cost of the microgrid can vary widely based on the prevailing prices at the time of conducting transactions. Since gas prices and Ethereum prices are linked to the supply and demand dynamics of the public market, they can be unpredictable and subject to substantial fluctuations over short periods.

To provide the best estimate of potential prices for the remainder of the year, the mean Ethereum and gas prices in 2023 have been calculated. These prices have been utilized in the microgrid's yearly cost scenarios to extrapolate the cost of operating the microgrid over an extended period. The significant variability in gas prices has implications for the cost of smart contract deployment, as depicted in Figure 7.2. This figure highlights how the cost of deploying the smart contract can vary depending on the prevailing gas price, further emphasizing the potential volatility in costs.

It is important to consider this variability when conducting experiments or making cost projections, as using a single value for gas price and Ethereum price at a particular time could result in volatile cost outcomes, heavily influenced by the values at the time of the experiment.

The variation of gas price on the cost of the deployment and functions of the smart contract that enable the microgrid are shown in Figures 7.3 and 7.4.

Table 7.4 shows the potential maximum and minimum costs of the smart contract processes assuming the mean Ethereum cost.

Table 7.4 shows that the cost of deploying the smart contract could vary from \$154.26 to \$27.61 dollars. This is a significant cost difference and highlights how susceptible the microgrid community is to the current gas price.

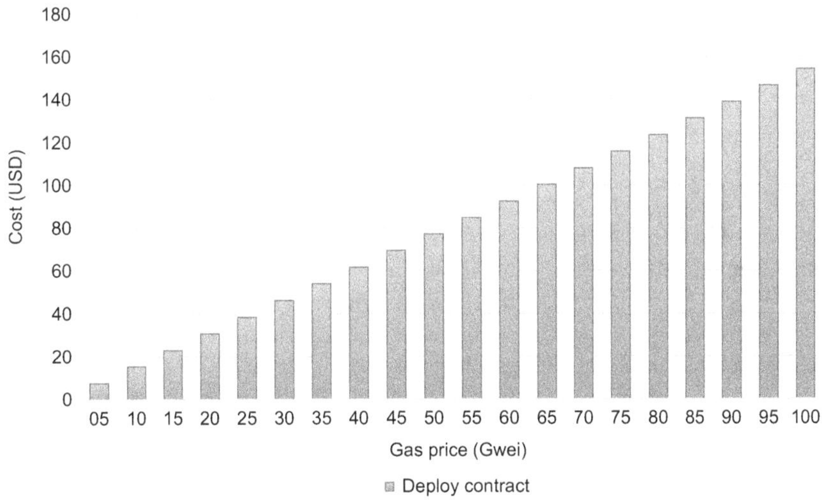

Figure 7.3 Graph showing effect of gas price on deployment cost

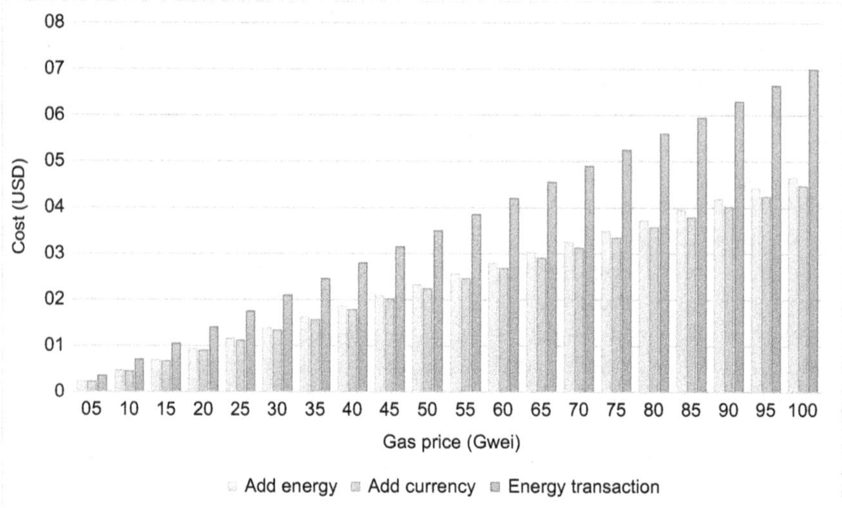

Figure 7.4 Graph showing effect of gas price on smart contract function cost

Table 7.4 Costs of smart contract functions

Function	Maximum gas price 2023	Mean gas price 2023	Minimum gas price 2023
Deployment	$154.26	$49.36	$27.61
Add energy	$4.65	$1.49	$0.83
Add currency	$4.47	$1.43	$0.80
Energy transaction	$7.00	$2.24	$1.25

For each scenario, the costs of running the blockchain microgrid for a year were calculated using the previous assumptions. The mean gas price and Ethereum price have been used to best represent the expected costs based on the year so far.

Assuming only one transaction is required per consumer per day, Figures 7.5 and 7.6 show that the cost of running the microgrid yearly for scenario 1 is $5,594 and $172,225 for scenario 2. This shows that increasing the number of consumers on the microgrid results in a great increase in transaction costs.

The cost per building per day has also been determined for each scenario to allow scenario comparison. However, Figure 7.7 allows the comparison of the scenarios per building per day and demonstrates that scenario 2 works out to be cheaper per building per day than scenario 1. This is because only one deployment of the smart contract is required in both scenarios.

Figures 7.5–7.7 demonstrate that increasing the number of energy transactions per day results in higher costs for the users of the microgrid. With the mean gas and

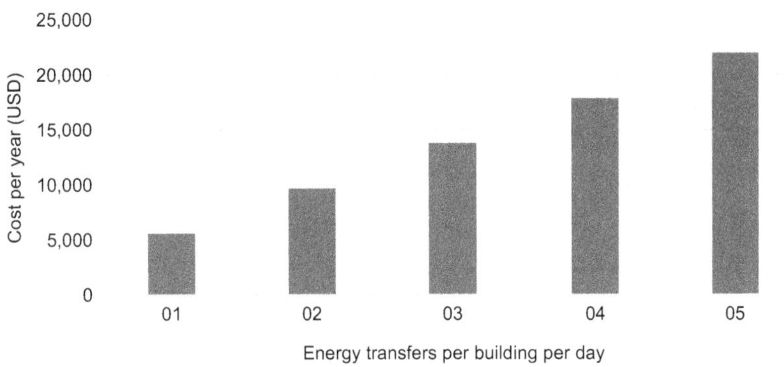

Figure 7.5 Graph showing the costs per year of running the scenario 1 microgrid

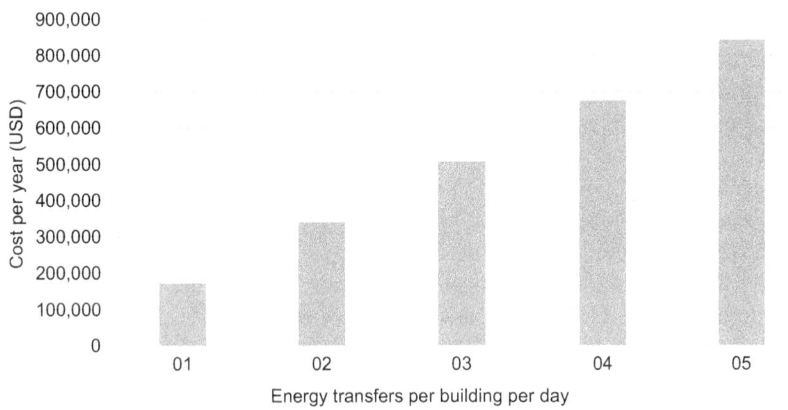

Figure 7.6 Graph showing the costs per year of running the scenario 2 microgrid

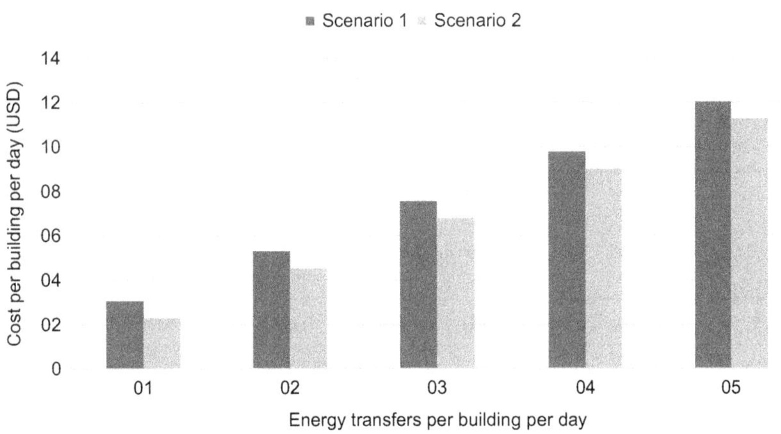

Figure 7.7 Graph comparing the cost per building per day of scenarios 1 and 2

Ethereum prices, the cost of a single energy transaction amounts to $2.24. However, if multiple transactions are conducted per day by consumers, the costs can escalate significantly. This indicates that it is advisable to minimize the number of energy transactions between consumers and producers on the grid to reduce overall expenses. Consumers can achieve this by actively monitoring their energy consumption or utilizing smart meters to identify usage patterns and make informed decisions.

It's important to note that the costs incurred by consumers and producers in the market differ because they need to perform distinct functions to complete the transaction steps. Energy producers are responsible for covering a portion of the deployment costs and utilizing the "Add Energy" function. However, consumers share the deployment costs and utilize both the "Add Currency" and "Energy Transaction" functions. The costs associated with each scenario can be found in Tables 7.5 and 7.6.

In the case of prosumers, they are categorized as producers since they generate enough energy to fully power themselves, eliminating the need to purchase energy from the microgrid.

Based on the yearly network costs of approximately $900 for consumers, as indicated in Table 7.5, it would be necessary to have an electricity price of 7p/kWh to achieve more cost-effective network expenses by utilizing the proposed microgrid instead of relying on the current national grid.

However, it is important to note that the current cost of electricity in 2023 stands at around 34p/kWh, as reported by EDF in 2022. This price is significantly higher than the required electricity cost to make the microgrid an efficient alternative. Given this disparity, public energy producers and prosumers currently have the option to sell their energy to the grid for up to 24p/kWh. Consequently, there is little incentive for them to sell their energy to the proposed microgrid, particularly considering the substantial fees associated with the "sell energy" function of the smart contract for each transaction.

Table 7.5 Yearly costs for producers and consumers

	Yearly cost (USD) scenario 1	Yearly cost (USD) scenario 2
Producer	$553.08	$543.45
Consumer	$901.62	$891.99

Table 7.6 Daily costs for producers and consumers

	Daily cost (USD) scenario 1	Daily cost (USD) scenario 2
Producer	$1.52	$1.49
Consumer	$2.47	$2.44

The breakdown of what costs come from each function within the smart contract is shown in Figures 7.8 and 7.9 for scenario 1 and Figures 7.10 and 7.11 for scenario 2.

The findings presented in Figures 7.8–7.11 reveal that the deployment cost of the contract constitutes only a small fraction of the overall expenses for both consumers and producers on the microgrid. However, it is crucial to consider the potential implications of redeploying the contract multiple times within a year due to necessary updates and adjustments in the conditions and rules. Such frequent redeployments could potentially result in higher costs for the grid users.

To provide further context, the cost per day per building for scenarios 1 and 2 has been compared to a similar experiment conducted by Vieira and Zhang [32] as well as the average household electricity costs reported by the national grid in 2021. For this comparison, it is assumed that £1 is equivalent to $1.25. Additionally, the proportion of the national grid bill attributed to network costs is also highlighted.

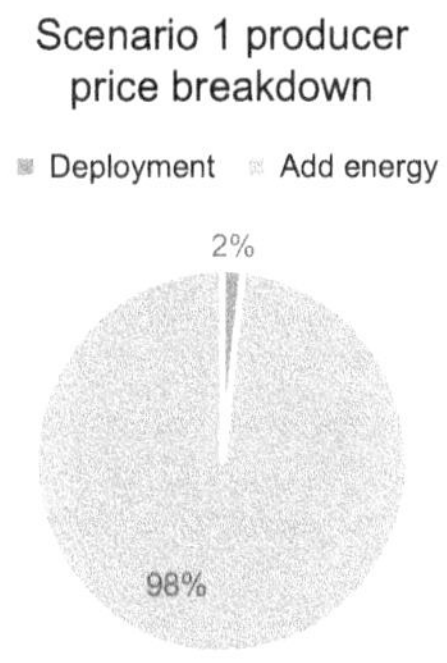

Figure 7.8 Scenario 1 producer cost breakdown

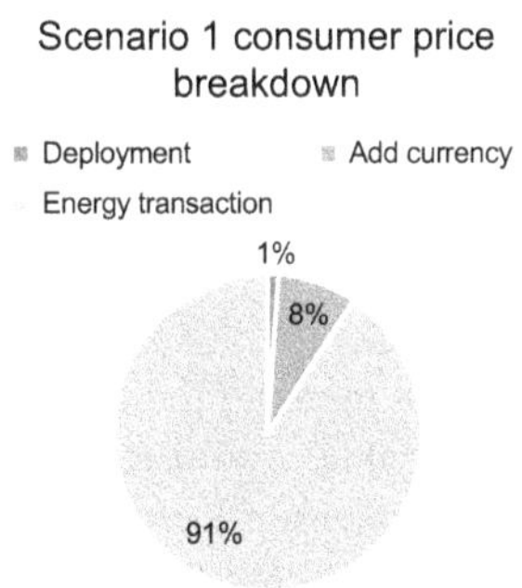

Figure 7.9 Scenario 1 consumer cost breakdown

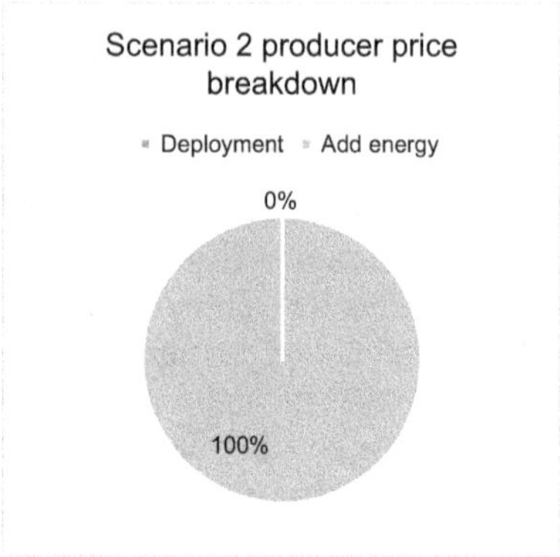

Figure 7.10 Scenario 2 producer cost breakdown

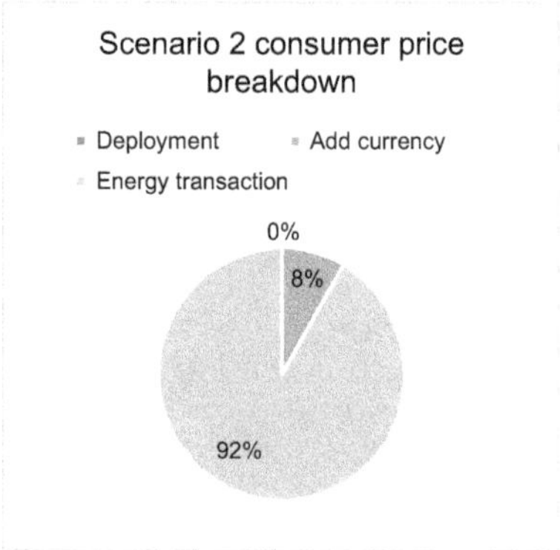

Figure 7.11 Scenario 2 consumer cost breakdown

Table 7.7 presents the outcomes of scenarios 1 and 2, considering one energy transaction per building per day. These results are compared to a similar experiment conducted by Vieira and Zhang as well as the average household electricity consumption price reported by the national grid. Upon comparing scenarios 1 and 2 to the findings of Vieira and Zhang, it becomes evident that the cost of operating a microgrid is considerably higher in this particular experiment. This disparity can be primarily attributed to the increase in the price of Ethereum, which differs from the assumed and utilized value in their study. Consequently, this demonstrates that even if a cost-effective public Ethereum microgrid were to be proposed, the uncontrollable nature of cost variables could lead to a significant increase in the framework's price, rendering the microgrid expensive for its users.

Furthermore, Table 7.7 highlights the cost difference between the Ethereum microgrid network and the existing national grid network costs for an average household. Specifically, scenario 1 incurs a cost of $3.07 per day, while scenario 2

Table 7.7 Yearly costs for producers and consumers

System	Cost per building per day (USD)	Ethereum price (USD)	Gas price (Gwei)
Scenario 1 (not including energy costs)	$3.07	$1658.9	32
Scenario 2 (not including energy costs)	$2.30	$1658.9	32
Viera and Zhang (2021) blockchain framework	$0.74	$250	20
National grid 2023 average household electricity cost	$3.96	N/A	N/A
National grid network cost proportion (3% of total bill)	$0.12	N/A	N/A

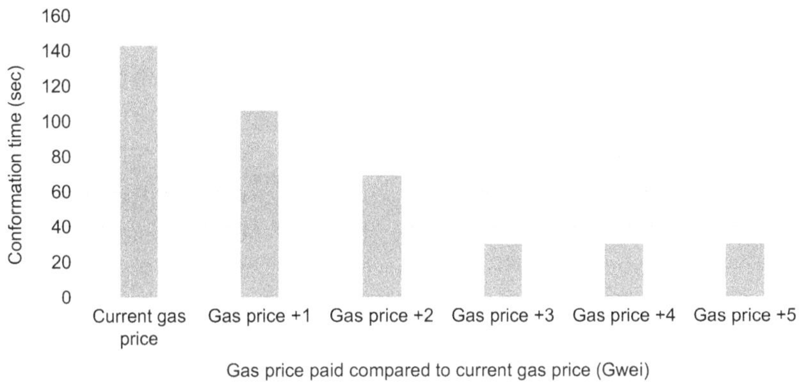

Figure 7.12 Influence of the increasing in gas price on transaction times

amounts to $2.30 per day. In contrast, the national grid network costs the average consumer household a mere $0.12 per day. This substantial difference underscores the expensive nature of operating a microgrid energy trading network using public Ethereum transactions. It's worth noting that the transaction time for public Ethereum transactions fluctuates based on the extent to which the user overpays the current gas price. The impact of overpaying the gas price has been examined and depicted in Figure 7.12, utilizing real-time Ethereum tracking.

Figure 7.12 illustrates the time taken for public Ethereum transactions to be completed. The data reveals that transactions can range from 142 seconds when paying the current gas price to a minimum of 30 seconds if users choose to overpay the gas price. These transaction times are considerably longer when compared to potential alternatives such as a private Ethereum blockchain or a centralized private database. For instance, Rouhani and Deters [26] demonstrated that private Ethereum transactions can be completed in under 0.2 seconds, while Lei *et al.* [19] showed that centralized private databases can achieve a throughput of hundreds of transactions per second.

Faster transaction times within a microgrid offer several benefits. They contribute to the efficient and swift distribution of energy, minimizing the risk of wastage or

Table 7.8 Yearly energy consumptions

Scenario	Total energy consumption per year (kWh)	Energy consumption per building per year (kWh)
Scenario 1	84.45	16.89
Scenario 2	2,340.45	11.42

shortages and enhancing the stability and reliability of the microgrid. Additionally, reducing transaction times enables users to respond promptly to changes in energy supply and demand, ensuring the efficient distribution of energy across the network.

Furthermore, it is estimated that with the new PoS consensus mechanism, a public Ethereum transaction consumes approximately 0.03 kWh of electricity [28]. Utilizing this estimation, the potential energy usage for both scenarios has been calculated and presented in Table 7.8.

Table 7.8 demonstrates that the energy consumption associated with utilizing the public Ethereum network for the Port of Milford Haven microgrid is below 20 kWh for both scenarios. In comparison, the average UK household is estimated to consume approximately 2,900 kWh of electricity per year [16]. This indicates that the additional energy usage resulting from using public Ethereum transactions represents only a small fraction of the total energy consumed annually.

Considering the relatively low energy consumption of the public Ethereum network in the context of the microgrid, it can be argued that the impact on overall energy consumption is minimal. However, it is important to acknowledge that even small energy increments can accumulate and have an impact when scaled up across numerous transactions and users. Therefore, it remains crucial to explore ways to optimize energy usage within the microgrid and identify potential improvements in energy efficiency.

7.5 Discussion

Blockchain has several potential advantages for the seaports industry and supply chains such as increased responsibility, openness, and trust between many parties involved in the supply chains. Smart contracts are efficacious instruments that enable energy transaction automation, real-time energy tracking, and monitoring with energy audit traces and transaction histories. Smart contracts for energy micro-grids in seaports represent a step forward in the ongoing climate change agenda whereby compliance, legitimacy, or certification is ensured as well as energy compliance and security standards.

7.5.1 Comparison to related studies

Related studies [32] report that the costs of the blockchain microgrid per household/ building per day can fluctuate based on the requirements of the pilot project and the difference in price of Ethereum used in the analysis. Although, the volatility of the Ethereum price can impact on the overall cost efficiency, blockchain technology can be applied to create a transparent, verifiable public energy trading network.

Buccafurri *et al.* [9] presented a similar framework using a public Ethereum smart contract, with individual smart contract functions available for users to purchase energy. The energy transaction framework proposed in the study involves consumers using an auction system to bid against each other to purchase energy. This could allow for more reactive costs for consumers where a new smart contract with new prices would be required to change the price of energy. The real-life cost to perform the energy transaction functions due to gas fees was found to be much higher as using a 2023 year-to-date price of Ethereum. The paper also proposed new smart grid users such as an energy retailer who could purchase energy from the smart grid and distribute it to its own customers, and highlights the need for an Energy Authority, needed to deploy the smart contract and verify retailers. The proposed framework is based on a scenario where the need for a central utility company to have control the smart grid is not necessary. Having a central utility company in charge of energy transactions defeats the decentralized nature of public blockchain technology. The proposed framework may be benefited by using a private blockchain, as the public Ethereum blockchain used in the study would cause the each of the smart grid functions to be more expensive.

Further exploration of a private blockchain microgrid or a private centralized database microgrid would be beneficial for assessing the current feasibility of these frameworks in implementing microgrids in such scenarios. Conducting investigations on the potential costs, energy consumption, and transaction durations associated with these frameworks could provide valuable insights into the viability of microgrids, while also determining the most suitable framework for adoption in seaports. The utilization of blockchain technology has the potential to bring communal advantages by facilitating transparent and secure transaction recording without relying on a single overseeing private entity, thereby encouraging the utilization of the microgrid by the 200 community households.

7.5.2 Challenges and barriers

While blockchain technology holds the potential of enhancing energy security and decentralization of microgrid systems, its practical application is not without substantial challenges.

The variable nature of Ethereum and gas prices creates a level of price uncertainty. This acts as a barrier in justifying the implementation of a public Ethereum smart contracts based microgrid economically. A potential price change in either direction could significantly alter the costs to use the energy trading functions of the smart contract, which could make the microgrid unusable if the price of Ethereum were to rise too high. As a result, this could deter investment and interest in the technology because of the inability to accurately predict operational use costs. This barrier could be reduced or removed by additional real life application case studies of public blockchain microgrids with an aim to show economic viability.

In order to access the functions of the smart contract to buy or sell energy on the microgrid, users would be required to own and know how to use an Ethereum wallet. This would require additional training for dedicated users. However, this would cause greater problem for implementation of a blockchain microgrid within the general public. Currently the technical knowledge of cryptocurrency wallets within the general public is low, which would be a barrier to extending the microgrid to include the general public as in *scenario 2*. To solve this problem, a solution such as an awareness campaign could be used to identify users who are interested in joining the microgrid.

Another potential issue with an anonymous blockchain microgrid is the potential for bad actors to sell unclean energy within the grid. A malicious user could attempt to sell energy from fossil fuel sources, such as diesel generators, on the grid as if they were from renewable sources. This would reduce the environmental benefits of implementing a microgrid within an energy community.

To set up an effective microgrid, possible hardware upgrades in the community may need to be installed to enable two-way data transfers, and energy storage, and to monitor energy transfers. Additional energy battery storage may need to be installed to store the excess renewable energy produced by the microgrid producers where the energy storage of the system can be optimized to fully utilize locally produced renewable energy. Additional smart meters and sensors are required to facilitate the smart energy transfers in the microgrid and additional hardware requirements to enable and enhance the energy trading on the microgrid would incur additional costs. Setup, installation, operation, repair, and maintenance costs that come with operating an energy distribution network need to be also considered when developing a blockchain-enabled microgrid system for seaports.

7.6 Conclusion

Energy-intensive infrastructures such as seaports face increasing pressure to enhance their sustainability by integrating local RES. This is particularly critical given the significant GHG emissions associated with port operations, including cooling, refrigeration, and various logistical activities. Seaports are ideal locations for RES such as offshore/onshore wind turbines and solar panels and these RES can be implemented to benefit communities effectively and efficiently by allowing the distribution of locally sourced energy between community members. The adoption of renewable energy not only helps to mitigate the environmental impact but also supports regulatory compliance and enhances the resilience of port operations against energy market volatility.

Microgrids serve as a highly efficient solution for integrating and harnessing renewable energy within port environments. They offer the potential to significantly reduce communities' dependency on the national grid, enabling them to monetize locally generated renewable energy instead of relying on grid-supplied energy. To incentivize the adoption of microgrids, they need to be more cost-effective than the current national grid, benefiting both consumers and producers.

This necessitates the incorporation of locally produced renewable energy within the port ecosystem.

In the realm of microgrids, blockchain technology, particularly Ethereum smart contracts, can play a pivotal role in decentralizing the infrastructure. By leveraging blockchain, transactions within microgrids can be conducted in a transparent and tamperproof manner. The public Ethereum blockchain, coupled with smart contracts, facilitates seamless transactions between microgrid users, ensuring a secure and accountable environment for all parties involved. Economic variability makes a public blockchain framework microgrid a risky investment due to the uncontrollable and unpredictable nature of Ethereum and the gas prices that could lead to the microgrid becoming extremely expensive for the microgrid users.

The energy consumption associated with the smart contract functions and the overall network is a crucial consideration when evaluating the effectiveness of the public Ethereum microgrid framework in facilitating the distribution of renewable energy. In the past, when Ethereum employed a proof-of-work consensus mechanism, there were concerns regarding high energy usage during transactions. However, it is worth noting that the energy consumed by adding new Ethereum transactions to the blockchain has significantly decreased over time and now represents only a fraction of the total energy costs per household per year.

To gain a comprehensive understanding of the potential costs, transaction times, and energy usage, further research could be conducted on the expenses associated with a potential private blockchain and private centralized network. This comparative analysis would provide valuable insights into the viability of these microgrid frameworks and assist in determining the most suitable model for implementation in ports and renewable energy communities, using the case study of Milford Haven Port as a reference.

The findings from this chapter can inform the scalability of blockchain-enabled microgrid frameworks, particularly in energy-intensive infrastructures like seaports. Scalability is a critical factor for widespread adoption, as the ability to handle increasing numbers of transactions, participants, and energy assets without compromising performance or efficiency is essential for large-scale deployment.

Further comprehensive studies are needed to evaluate the implications of deploying blockchain-based microgrids, including initial setup, operational, and maintenance aspects. Future research should investigate the legal and regulatory implications of deploying blockchain-based energy systems, focusing on compliance with energy and data protection regulations in various jurisdictions.

References

[1] Abusharkh, S., Arnold, R.J., Kohler, J., *et al.*, (2006), Can microgrids make a major contribution to UK energy supply? *Renewable and Sustainable Energy Reviews*, vol. 10, no. 2, pp. 78–127, www.sciencedirect.com/science/article/pii/S1364032104001194. Accessed 31 March 2023.

[2] Agung, A., and Handayani, R., (2022), Blockchain for smart grid. *Journal of King Saud University – Computer and Information Sciences*, vol. 34, no. 3, pp. 666–675, www.sciencedirect.com/science/article/pii/S1319157819309000. Accessed 28 March 2023.

[3] Ajomand, N., Ullah H., and Aslam, S., (2020), A Review of Blockchain-based Smart Grid: Applications, Opportunities, and Future Directions, https://stc.computer.org/blockchain/wp-content/uploads/sites/29/2020/11/November2020-1.pdf. Accessed 28 March 2023.

[4] Akhras, R., El-Hajj, W., Majdalani, M., Hajj, H., Jabr, R., and Khaled, S., (2020), Securing Smart Grid Communication Using Ethereum Smart Contracts, *2020 International Wireless Communications and Mobile Computing (IWCMC)*, pp. 1672–1678, https://ieeexplore.ieee.org/document/9148345. Accessed 22 September 2024.

[5] Alladi, T., Chamola, V., Rodrigues, J., and Kozlov, S., (2019), Blockchain in smart grids: A review on different use cases. *Sensors*, vol. 19, no. 22, p. 4862. www.mdpi.com/1424-8220/19/22/4862, Accessed 15 March 2023.

[6] Alzahrani, A., Petri, I., Rezgui, Y., and Ghoroghi, A., (2020), Developing smart energy communities around fishery ports: Toward zero-carbon fishery ports. *Energies*, vol. 13, no. 11, p. 2779, https://doi.org/10.3390/en13112779. Accessed 13 March 2023.

[7] Alzahrani, A., Petri, I., Rezgui, Y., and Ghoroghi, A., (2021), Optimal control-based price strategies for smart fishery ports micro-grids, *2021 IEEE International Conference on Engineering, Technology, and Innovation*, https://ieeexplore.ieee.org/document/9570267A. Accessed 13 March 2023.

[8] Andoni, M., Robu, V., Flynn, D., (2019), Blockchain technology in the energy sector: A systematic review of challenges and opportunities. *Renewable and Sustainable Energy Reviews*, vol. 100, pp. 143–174, https://doi.org/10.1016/j.rser.2018.10.014. Accessed 15 March 2023.

[9] Buccafurri, F., Lax, G., Musarella, L., and Russo, A., (2023), An Ethereum-based solution for energy trading in smart grids. *Digital Communications and Networks*, vol. 9, no. 1, pp.194–202, https://doi.org/10.1016/j.dcan.2021.12.004. Accessed 22 September 2024.

[10] Cecati, C., Citro, C., and Siano P., (2011), Combined operations of renewable energy systems and responsive demand in a smart grid. *IEEE Transactions on Sustainable Energy*, vol. 2, no. 4, pp. 468–476, ieeexplore.ieee.org/document/5951797. Accessed 14 March 2023.

[11] FAO, (2023), Publication Card | FAO | Food and Agriculture Organization of the United Nations. 2023, www.fao.org/publications/card/en/c/89457227-b55f-4ac6-95ad-cccd53a3dcb0/. Accessed 13 March 2023.

[12] Gellings, C., (2009), *The Smart Grid: Enabling Energy Efficiency and Demand Response*. New York: Rivers Publishing, https://doi.org/10.1201/9781003151524. Accessed 14 March 2023.

[13] Gungor, V., Lu, B., and Hancke G., (2010), Opportunities and challenges of wireless sensor networks in smart grid. *IEEE Transactions on Industrial*

Electronics, vol. 57, no. 10, pp. 3557–3564, https://doi.org/10.1109/tie.2009.2039455. Accessed 14 March 2023.

[14] Gungor, V., Sahin, D., Kocak, T., *et al.*, (2011), Smart grid technologies: Communication technologies and standards. *IEEE Transactions on Industrial Informatics*, vol. 7, no. 4, pp. 529–539, https://doi.org/10.1109/tii.2011.2166794. Accessed 14 March 2023.

[15] Hentschel, M., Ketter, W., and Collins J., (2018), Renewable energy cooperatives: Facilitating the energy transition at the Port of Rotterdam. *Energy Policy*, vol. 121, pp. 61–69, https://www.sciencedirect.com/science/article/pii/S0301421518304026. Accessed 18 March 2023.

[16] EDF Energy, Forrest, J., (2022), What Is Typical Energy Usage in the UK? | EDF. *EDF*, 27, www.edfenergy.com/energywise/what-is-the-average-energy-bill-in-the-uk. Accessed 8 May 2023.

[17] Kermani, M., Parise G., Shirdare, E., and Martirano L., (2020), Transactive Energy Solution in a Port's Microgrid Based on Blockchain Technology, *2020 IEEE International Conference on Environment and Electrical Engineering and 2020 IEEE Industrial and Commercial Power Systems Europe (EEEIC / I&CPS Europe)*, https://doi.org/10.1109/eeeic/icpseurope49358.2020.9160833. Accessed 4 April 2023.

[18] Lee, J, and Khan, V.M., (2020), Blockchain and smart contract for peer-to-peer energy trading platform: Legal obstacles and regulatory solutions. *SSRN Electronic Journal*, vol. 19, pp. 285, https://doi.org/10.2139/ssrn.3556260. Accessed 16 March 2023.

[19] Lei, Y., Ma, C., Mirza, N., Ren, Y., Narayan, S.W., and Chen, X., (2022), A renewable energy microgrids trading management platform based on permissioned blockchain. *Energy Economics*, vol. 115, p. 106375, https://doi.org/10.1016/j.eneco.2022.106375. Accessed 7 May 2023.

[20] Mengelkamp, E., Garttner, J., Rock, K., Kessler, S., Orsini, L., and Weinhardt, C., (2018), Designing microgrid energy markets: A case study: The Brooklyn Microgrid. *Applied Energy*, vol. 210, pp. 870–880, https://doi.org/10.1016/j.apenergy.2017.06.054. Accessed 18 March 2023.

[21] Misra, A., Venkataramani, G., Gowrishankar, S., Ayyasam, E., and Ramalingam, V., (2020), Renewable energy based smart microgrids—A pathway to green port development. *International Solutions to Sustainable Energy, Policies and Applications*. Denmark: River Publishing, pp. 321–333, https://doi.org/10.1201/9781003150978-18. Accessed 14 March 2023.

[22] Mollah, M.B., Zhao, J., Niyato, D., *et al.* (2021), Blockchain for future smart grid: A comprehensive survey. *IEEE Internet of Things Journal*, vol. 8, no. 1, pp. 18–43, https://doi.org/10.1109/jiot.2020.2993601. Accessed 16 March 2023.

[23] National Grid, (2021), Breaking Down Your Bill – National Grid Electricity Transmission, https://www.nationalgrid.com/electricity-transmission/who-we-are/breaking-down-your-bill. Accessed 8 April 2023.

[24] Parker, R., Blanchard, J., Gardner, C., *et al.*, (2018), Fuel use and greenhouse gas emissions of world fisheries. *Nature Climate Change*, vol. 8, no. 4, pp. 333–337, https://doi.org/10.1038/s41558-018-0117-x. Accessed 8 March 2023.

[25] Petri, I., Barati, M., Rezgui, Y., and Rana, F., (2020), Blockchain for energy sharing and trading in distributed prosumer communities. *Computers in Industry*, vol. 123, p. 103282, https://doi.org/10.1016/j.compind.2020.103282. Accessed 8 May 2023.

[26] Rouhani, S., and Deters, R., (2017), Performance Analysis of Ethereum Transactions in Private Blockchain, *2017 8th IEEE International Conference on Software Engineering and Service Science (ICSESS)*, https://doi.org/10.1109/icsess.2017.8342866. Accessed 7 May 2023.

[27] Sabounchi, M., and Wei, J., (2017), Towards Resilient Networked Microgrids: Blockchain-Enabled Peer-to-Peer Electricity Trading Mechanism, *2017 IEEE Conference on Energy Internet and Energy System Integration (EI2)*, https://doi.org/10.1109/ei2.2017.8245449. Accessed 28 March 2023.

[28] Sarkar, A., (2022), The Merge Brings down Ethereum's Network Power Consumption by over 99.9%. *Cointelegraph*, www.cointelegraph.com/news/the-merge-brings-down-ethereum-s-network-power-consumption-by-over-99-9. Accessed 8 May 2023.

[29] Sdoukopoulos, E., Boile, M., Tromaras, A., and Anastasiadis N., (2019), Energy efficiency in European ports: State-of-practice and insights on the way forward. *Sustainability*, vol. 11, no. 18, p. 4952, https://doi.org/10.3390/su11184952. Accessed 4 April 2023.

[30] Ethereum, (2023), Smart Contracts, *Ethereum.org*, 2023, ethereum.org/en/smart-contracts/. Accessed 16 March 2023.

[31] *The Economist*, (2015), The Trust Machine. *The Economist*, www.economist.com/leaders/2015/10/31/the-trust-machine. Accessed 15 March 2023.

[32] Vieira, G., and Zhang J., (2021), Peer-to-peer energy trading in a microgrid leveraged by smart contracts. *Renewable and Sustainable Energy Reviews*, vol. 143, p. 110900, https://doi.org/10.1016/j.rser.2021.110900. Accessed 28 March 2023.

Chapter 8

Decentralising architectural design through data governance

Theodoros Dounas[*,‡], *Davide Lombardi*[†] and *Hico McDonald*[‡]

This chapter explores the processes of decentralisation of architectural design through a series of design experiments and prototypes that develop an alternative governance model for architectural design, one routed into a new political economy for architectural design with a scope of transforming architecture into a non-extractive, allographic discipline.

8.1 Introduction and background – crises through development

The world is facing two concurrent crises, one of housing an ever-increasing global population and one of climate collapse, caused in part from unrestrained industrial growth. Within this growth, the architecture, engineering and construction (AEC) industry is responsible for providing solutions to the first crisis, with innovation, industrial solutions or radical other solutions, and responsible for 28% of the carbon equivalent for operating the built environment and 11% for constructing the built environment [1]. This inefficiency is partly due to the nature of the AEC industry in that it is fragmented, unproductive and adversarial, and more specifically, the nature of architectural design, as its current configuration is both autographic [2] and extractive [3]. The AEC industry is currently extractive towards the environment, in the sense that historically, the processes we use for building started with the extraction of raw materials. There are indications as well that certain forms of AEC development are also economically extractive; however, for the most part, the AEC industry provides vital infrastructure for the preservation of life and everyday activities. Instances of the AEC industry being economically extractive are probably defined locally in terms of particular ecosystems [3]. Subsequently, in this chapter,

[*]School of Textiles and Design, Heriot-Watt University, Scotland, UK
[†]Department of Architecture, Xi'an Jiaotong-Liverpool University, Suzhou, China
[‡]Adventurous Systems Ltd, London, England, UK

we explore how the utilisation of blockchain (BC) technologies and the concepts of decentralisation can turn architecture into an allographic discipline produced in a decentralised manner. While this also has implications for the discipline becoming non-extractive, we do not investigate this particular direction in depth.

This change is particularly relevant within the frame of the Fourth Industrial Revolution, which sees an accelerated convergence of biological and computational systems [4,5] alongside unrestrained automation of the industrial and building ones. Emerging information technologies such as BC, IoT and AI are then responsible for orchestrating their operations.

On the use of BC in the AEC industry, we have seen the rise of a series of prototypes in the past 5 years, after a series of speculative research papers on the potential of the technology. Within this chapter, we summarise our work on developing some of the first prototypes, including the development of a novel smart con and the creation of the first architecture decentralised autonomous organisation (DAO). Our motivation to developing the chapter is to capstone our efforts, towards taking our prototypes to the next impact level, that is, industry implementation.

8.1.1 The nature of design in the AEC industry and the current political economy

Nelson Goodman introduced the terms Allographic and Autographic for works of Art [6]. The initial definition is tied to the manner in which these works are produced. Music and novels are considered Allographic in the sense that every copy of each partiture, or every execution of the music or every printed copy of a novel is equally an instance of the work. An original painting is instead considered Autographic, as the author, the painter, has produced the work of art herself directly. In Autographic works, the history of the production is the important aspect, producing unique artefacts, while Allographic works use developed systems of notation to reproduce the work itself. In this context, architecture is considered by Goodman both Allographic and Autographic. Goodman mentions that the Allographic and the Autographic nature of architecture is diachronic and that a clear transformation of architecture to a fully allographic discipline will only take place once architecture develops a notation system that 'acquired full authority to divorce identity of work in all cases from particular production' [6]. The developments of the twenty-first century in computational representation of architecture are certainly moving towards that direction when one examines linked data and graph representations. We believe that transforming the architecture part of the AEC industry into an allographic discipline, will also help with turning it non-extractive.

8.2 Primitives

8.2.1 Blockchain, decentralised ledger technology and data storage

A BC is as a distributed computer network, with each node possessing a copy of a distributed ledger containing values and a history of transactions. In terms of computation, a BC serves as both a state machine capable of executing smart contracts – software code akin to an automatic vending machine – that change state and an unbroken, immutable chain constructed from discrete blocks of information. Each block encapsulates a cryptographic hash of the preceding block and encompasses a set of transactions or modifications to the distributed ledger in the form of a Merkle tree, that is, an encrypted device where transactions are hashed in pairs until you end up with a root encrypted hash. In the realm of architectural design, these may represent a sequence of synthetical actions executed in a bottom-up manner and encoded into a block [7].

At regular intervals, the BC network, facilitated by an incentivised participation system, selects the next block to be inscribed in the ledger/chain. Public, permissionless BCs, inherently by their nature, serve as trust mediums (trust machines) [8] between agents who may not be aligned in scope or acquainted. As such, trust shifts from other types of organisation unto the BCs. In this, BC act as a medium of decentralisation, in the sense that they are no gate keepers or central points of authority: everyone is free to participate and engage with the chain. Instead of a set of regulations, the computing protocol and the smart contracts become the rules and regulatory framework that is implicitly and explicitly accepted. They exhibit resilience, being able to withstand the loss of a substantial portion of the network portion, without destruction of the BC. The immutability arises from the fact that one cannot retroactively delete information; the cryptographic hashes of blocks are embedded into subsequent ones, creating an immutable chain. Operationally, they rely on cryptoeconomic incentives [9] – economic mechanisms motivating behaviours that upholds or enhances the system, not strictly in a monetary sense. Still, from an economic standpoint, a BC acts as a decentralised trust machine, enabling the establishment of peer-to-peer economies through smart contracts, tokens and their associated computer protocols.

BCs combine elements of cryptography, data immutability, information entropy and the simultaneous continuous and discrete nature of information, contributing to establishing trust.

Distinctive and promising BC applications include creating tokens (fungible and non-fungible), establishing DAOs [10,11] that operate through the BC medium and implementing decentralised finance (DeFi) [12]. These applications are built using smart contracts [13], with additional layers for interfaces and connectors that facilitate interactions between the BC and its external environment. Given the immutable nature of the BC as a record, ensuring the quality and truthfulness of data passing through and recorded on the BC is crucial. To address this, the concept of an oracle is introduced. Oracles are trusted entities situated outside the BC, incentivised and disincentivised, and are entrusted with feeding data into BCs. In a sense, the design of Oracles for BCs, and more so for AEC, is highly complex and in many cases requires the re-formulation of incentives and business model structures [14,15].

Parallel to BCs are distributed file systems designed for decentralised file storage, rather than data storage. One such file system is the Interplanetary File System (IPFS), which operates based on content rather than addressing. Within IPFS, the focus is on 'what' rather than 'where', unlike the World Wide Web:

Content on IPFS is cryptographically signed with a unique hash, ensuring its uniqueness and facilitating easy retrieval. On top of IPFS, one can develop further containers that use encryption, protecting the information stored on IPFS.

8.2.2 Decentralised autonomous organisations

BC shifts trust and responsibility to the computing protocol. As such, normal hierarchies, recede, giving rise to DAOs, that is, organisations that use BC/distributed ledger technologies and smart contracts to operate. Their infrastructure requires 'trustless' automation, shared control amongst the participating agents, incentives for participation and validation of decision-making frameworks. Theoretically, any sort of organisation could transition into a DAO, shifting its hierarchical relationship to the BC protocol and mirroring its structure there. Of course, we could envision a future where all collective governance minded organisations become some sort of DAO, in an ever-increasing effort to reduce corruption and single points of failure which are common in centrally organised organisations. The first aspect of governance for DAOs is the governance about the data of the organisation itself. As this chapter develops various forms of DAOs for the Architectural design discipline, it is helpful to note their main pros and cons.

In theory, these organisations might help alleviate some of the performance problems of the AEC space, while conversely, they are difficult to build and maintain; furthermore, they are requested to compete with centralised institutions in the AEC realm that have been operating for years if not in some cases for centuries. Consequently, the transition is not always easy or desirable. In other cases, we have examples of problems that do not require BC at all for effective governance [16]. Still, our expectation is that DAOs will start rising and become an effective tool in the AEC industry as we transform the industry from fragmented to integrated, from unproductive to productive and from adversarial to harmonious [17].

8.3 Blockchain is a governance technology (why decentralisation matters: decentralisation primitives)

The question of whether to use BC technology in architectural design reviles around the question of design governance. While at first glance, BC technologies are founded upon the idea of peer-to-peer transactions for monetary value [18], their inherently embedded decentralisation frames more questions of governance rather than monetary value. Governance, of course, is another high-level concept for commanding value generation. Within this context, decentralisation is not meant to directly signify geographical decentralisation, but rather governance power diffusion at the protocol level. This is also why BC technologies are a good fit for the architecture-engineering-construction ecosystem, as it is a fragmented ecosystem as many researchers have mentioned, in particular Hunhenvicz and Hall [16]. This does not preclude of course decentralisation in any other conceptual or organisational level.

Within our work, governance has been expressed in BC prototypes in many different manners, in a sense prodding the envelope to uncover possibilities and the true potential for BC technologies. In our prototypes, governance takes a collective form and is gradually explored from simple optimisation problems to qualitative and full wicked design problems, to high level governance of organisations. As such, we present in the following part of the chapter an interlude, that is, our first project on BC and architectural design and then a series of architectural design governance prototypes, which develop in ever-increasing complexity of the solution. In each of these cases, our computing prototype develops an ever increasing distance from the mechanics of the design problem and attention shifts to design agents solving the problem, that is, we shift into second-order cybernetic loops, where the design agent becomes part of the design problem [19]. Note that the prototypes presented are simple and crude with today's complex data landscape in BCs; however, the performance and shape grammar prototypes, when developed in 2017–2020, were sufficiently novel for the space. The ArchiDAO token prototype is still in active development and as such is presented as a snapshot of a living lab research project.

8.4 Interlude: a design competition as a driver for research in blockchain for architectural design

In 2017, we entered into a city design competition, where we would use computational design tools to do most of the design work. Our effort was research-driven, attempting to scale a previous computational design tool that was developed for increasing density in housing in the city of Suzhou, China, while still maintaining diversity of form, a problematic issue in China where a lot of cookie-cutter approaches prevail in architectural design.

The following paragraphs detail the development of the competition and our thinking in using BC nodes as a driver for architectural design.

The strategic use of urban development plays a pivotal role in China's governmental efforts to spur economic growth and alleviate poverty. Following decades of rapid urbanisation along the eastern coast of China, the 13th five-year plan emphasises the establishment of city clusters in the hinterland. A notable cluster is situated in central Guizhou, prompting the local government to initiate an international design competition for a new city within the core area of Qiushui Lake in the Yilong district of Guizhou [20].

Our decision to participate in the competition stemmed from the intention to test a set of algorithms we had previously developed for creating dense urban agglomerations [21]. These algorithms were appropriately configured to generate new cities rather than traditional urban clusters. The alignment of our strategies with the competition's theme became evident within the competition framework, which sought a juxtaposition and collaboration between a 'traditional' oriental landscape and a futuristic city. In the context of the term 'oriental', the competition description categorises the mountainous natural landscape of Guizhou, emphasising discussions on city life grounded in nature.

The primary emphasis within the competition description shifts towards establishing a novel connection between the typical Chinese landscape and the futuristic city development. Simultaneously, a key aspect highlighted in the competition aligns with the Chinese government's strategic focus on developing Guizhou's industries, particularly in the realm of Big Data. This includes the establishment of data centres for information processing and related service industries. Such an approach mirrors the Chinese government's overarching tactic for mainland development, involving the strategic relocation, establishment and concentration of vital industries in specific cities and city clusters.

The Yilong district emerges as a region in China undergoing significant transformations in urban design, providing an ideal context for conceptualising a forward-looking city. It grapples with common challenges encountered by contemporary cities, including issues such as traffic congestion, urban homogenisation and pollution. Additionally, the district is marked by distinctive natural concerns such as persistent water erosion and a unique network of underground water canals, rendering the area delicate with limited space for construction.

Notably, leading up to October 2017, expansive data centres engaged in BC mining operations were commonly found in specific areas of the Chinese countryside, especially in Inner Mongolia, where cold weather helped the formation of such high-tech power plants [22]. These mining centres generated substantial heat and consumed significant amounts of electricity. The decision to locate these computational mining facilities in these areas resulted from policies governing the placement of data centres, considering climate conditions and the low cost of electricity, often subsidised by the local governments. During the time of the competition, the computational requirements for Bitcoin generation were producing heat equivalent to sustaining a typical Chinese home's warmth for an entire week [24].

Recognising the potential benefits, we realised the advantage of situating these data centres, whether engaged in processing big data or running Bitcoin mining algorithms, within the city itself. This strategic deviation from the conventional Chinese policy of keeping such functions separate held logical merit. By doing so, we envisioned leveraging the excess heat generated by these centres to contribute to the heating requirements of the emerging city.

These explanations align with the objectives of our initial algorithm focusing on urban densities [21] – aiming to augment urban density while concurrently preserving building and urban footprints, thereby mitigating the erosion of agricultural land on the outskirts of Suzhou in Jiangsu province. Consequently, our approach to participating in the competition involved developing a revised iteration of the algorithm. This updated version incorporates features in line with the directives of the Chinese government and the competition's framework, emphasising the equilibrium between the natural landscape and the expedited trajectory of economic growth.

Simultaneously, any proposed urban intervention or avenue for growth contends with the potential erosion of the picturesque natural landscape inherent to the mountainous terrain of Guizhou. Despite these challenges, the competition outlines a directive to accommodate 800,000 residents in the new city – an amount considered moderate by Chinese standards but sufficiently intricate to introduce bottlenecks and contrasts in the development process.

The impartiality of data, coupled with their inherent mathematical nature, seamlessly aligns with the necessity to amalgamate various requests and constraints. This alignment led us to opt for an algorithmic language as the primary tool to contend with the extensive volume of information. The urban generation algorithm (Figure 8.1) unfolds through a dual-stage process: initially retrieving the centres of the new city and subsequently determining the distribution of buildings.

In the initial phase, a strategy is employed to minimise distances between the outer boundaries of the initial area. This strategy is complemented by a cellular automata (CA) algorithm, which is instrumental in defining the initial urban sprawl. The minimal path is derived through the digitisation of Frei Otto's well-known minimal path approach [23]. This involves dividing the opposite sides of the Yilong area into a specified number of points, interconnecting them to establish an initial rigid network. This net undergoes further high-density subdivision, generating crucial points where attraction forces are applied, transforming the rigid net into a minimal lattice path. Parameters controlling the strength of this physical process are meticulously configured based on the geometrical characteristics and extension of the area. Subsequently, the lattice geometry is assessed to pinpoint a position at each interconnection. These points, due to their vital location within the lattice, serve as the starting points where the CA algorithm will generate the new city. These become pivotal for applying the social condenser technique (Figure 8.2).

The second phase revolves around establishing a two-dimensional grid, serving as the masterplan for the three-dimensional development of the city. Initial calculations involve formulating the driving rules for the process and determining the starting grid where the first generation of CAs should emerge. Diverse tests have been conducted, applying varying time steps to the CA's evolving process. The quest for an optimal layout involves analysing the positioning of cells within specific areas of the city. Specific constraints are incorporated to prevent the algorithm from constructing within bodies of water or lakes, enabling considerable portions of the area to remain undeveloped, thereby allowing the natural landscape to flourish.

Once the initial layer of cells is defined, the algorithmic process proceeds to delineate the actual new city with buildings characterised by varying heights. The fundamental concept involves adapting the city's shape to harmonise with the surrounding landscape through three distinct organisational approaches: areas with densely built structures reaching the height of the hills, areas with buildings aligning with the hill's outline and areas featuring megastructures connecting the lower layer of the city with its highest points (Figure 8.3).

The three-dimensional progression of urban development relies on a CA algorithm, commencing from the centre identified in the two-dimensional study. To ensure heightened urban density, a lower value has been designated as the survival threshold for the CA. Similar to the preceding analysis, diverse timings have been established and tested to assess how varying CA distributions would react to the initial parameters. In the 3D design phase, the analysis of cell

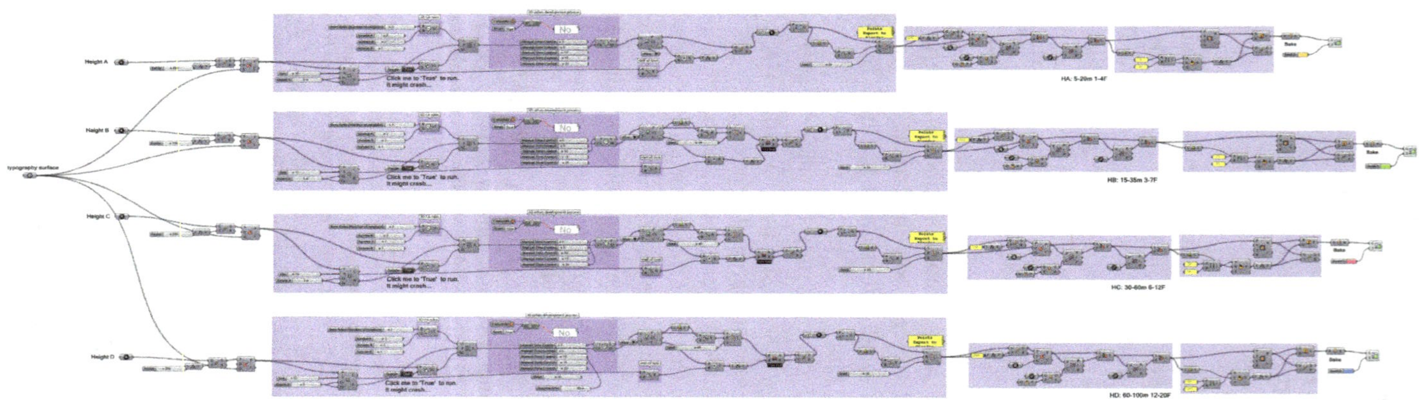

Figure 8.1 Execution of the algorithmic design process (captured and edited by Theodoros Dounas)

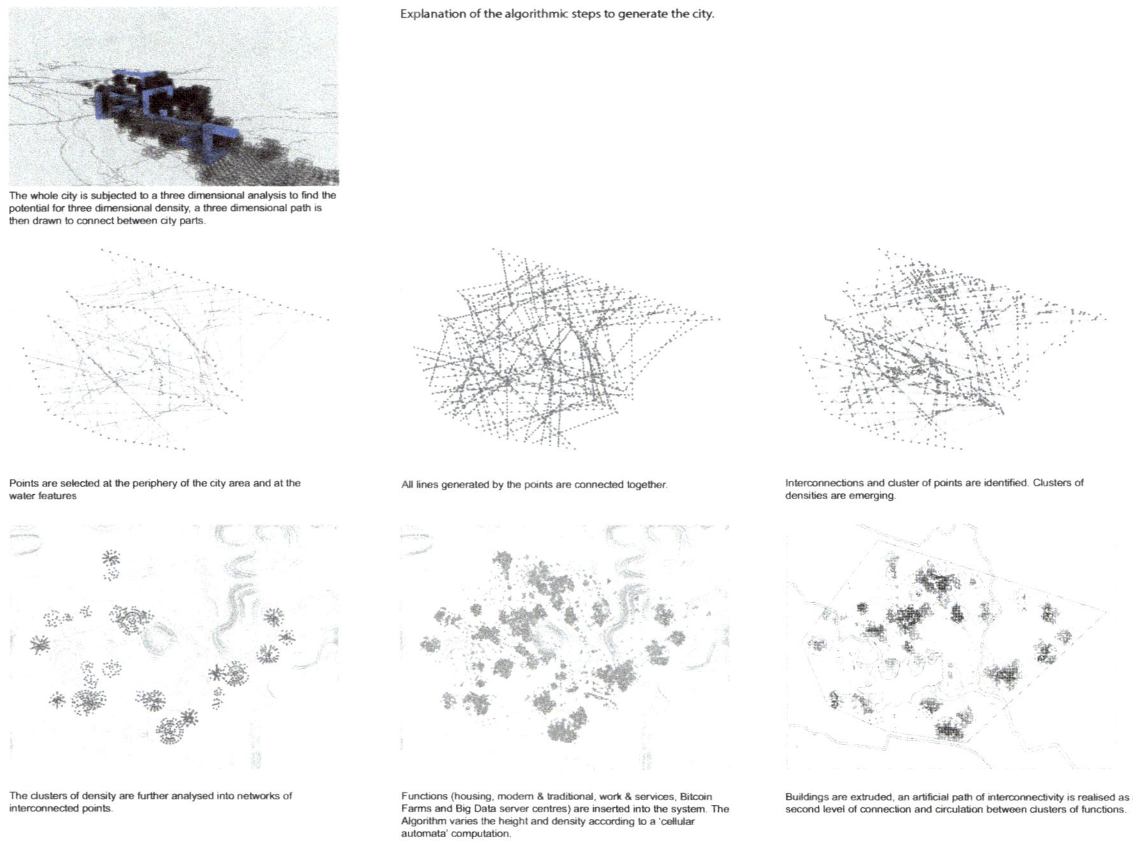

Figure 8.2 Stage 1 of the algorithmic process (authored and edited by Theodoros Dounas)

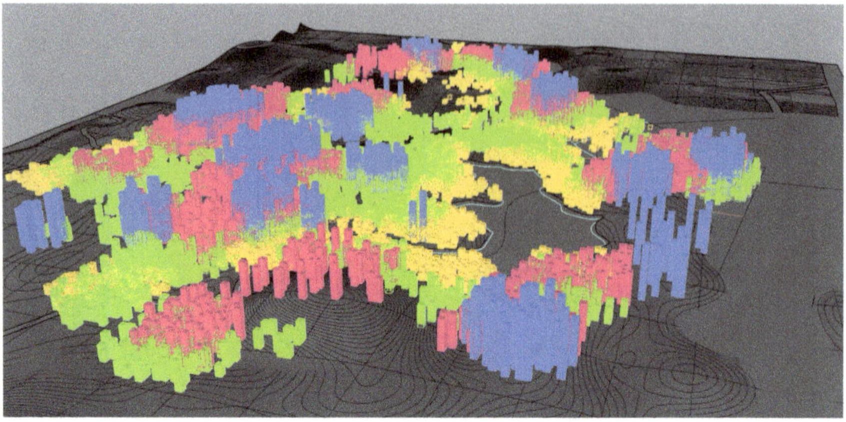

Figure 8.3 Schematic of the process (screenshot captured by Theodoros Dounas)

positions assumes a pivotal role, aiding in the selection and removal of elements that do not align with the functional plan and geographical characteristics of the site (Figure 8.3).

The ultimate outcome of the algorithm takes shape from this point onward. Buildings are vertically extended based on their designated functions, and seamless connectivity is established corresponding to the functional zones within the city. This connectivity is impartial to the type, encompassing options such as roads, urban trains or underground passages (Figure 8.4).

The city conceived by our algorithm stands as a paradigm of pure programming, density and organisation devoid of conventional forms. In a realm where architectural forms can be diverse, our emphasis lies in the interconnections, network structures and energy integration, defining the essence of architectural prowess. Our algorithm introduces a novel approach to creating grids, fostering a network of cities and orchestrating an interplay of functions. Consequently, we envision our algorithm as the precursor to a masterplan framework for the future, envisioning a landscape where architects operate within genuine 3D cities, with minimal reliance on traditional ground references.

Our entry into the competition won a third award out of 1048 entries, and the design became the nucleus with which we started to develop in 2018 our first questions on the affinity between BC technologies and computational design. The questions that arose from this revolved around whether BC could replace standard databases in building information modelling (BIM) systems, how could we connect computationally designed algorithms to BC systems, what the role of smart contracts would be and how these would operate.

In this, we were joined by other researchers who were investigating how non-manifold topology representation could be linked to BC technologies for a seamless workflow and database management [25].

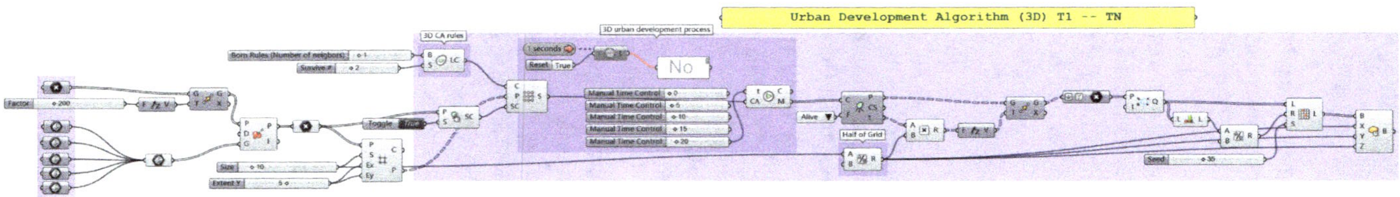

Figure 8.4 Stage 2 of the algorithmic process (captured and edited by Theodoros Dounas)

8.5 Decentralising architectural design: optimisation: creating architectural design decentralised autonomous organisations

The following examples are describing our speculative work on developing DAOs for architectural design. The next chapter describes a performance-based scenario, where trust and resolution of activities is shifted on to the BC protocol and handled by a series of smart contracts, in a crescendo of technological complexity, from the initial ideas on CAD–BC integration till a full scenario where the optimisation of design aspects is handled within the DAO itself [26]. This is a DAO in the sense that design agents compete and/or collaborate for resolving a design problem, set by another agent, but where there is no established trust between the designers and the problem setters.

8.5.1 Prototype 1

This study identified four levels of connection and integration between CAD/BIM and BC. The foundational level integrated a simple BC into the CAD/BIM software. The second level, feasible with contemporary technology, established a connection between the CAD/BIM client and a BC, specifically a test network built on Ethereum. The third and fourth levels entailed incorporating a computational node of Ethereum within the computational structure of the CAD/BIM software, with the highest level involving the operation of the entire CAD/BIM software database directly on the Ethereum BC, essentially implementing a CAD/BIM decentralised application (dApp) on Ethereum (Figure 8.5).

Commercial computational tools were selected for the implementation, but similar demonstrations could be conducted using BIM or CAD packages featuring a computer language interpreter and network communication capabilities. The first level of integration involved incorporating a basic BC written in Python within the Grasshopper environment on Rhinoceros. This basic BC leveraged the public network capacity of a typical BC, recording information in cryptographically hashed blocks.

In this implementation, a simple command was executed to copy a circle from one position to another. These actions were referenced inside Grasshopper, analysed into control points and subsequently converted into mathematical sets. The difference between the coordinates of the two groups of control points was calculated and fed into the Python BC. The BC operated autonomously within Grasshopper, devoid of connections to the external world. It comprised five elements: a definition of a block, a BC, a hashing algorithm, the first 'genesis' block and the executive function of the code. The number of blocks to be generated was specified for simplicity in the proof of concept. The full implementation would connect each generated block with the actions of the computational designer, creating one block for each action. The difference was fed into the variable 'diff', while a simple string was input into the first block of the chain. The BC then encrypted the first string with a hash, generated the first genesis block and continued producing the subsequent blocks, each incorporating the string representing the difference between the original and copied circle.

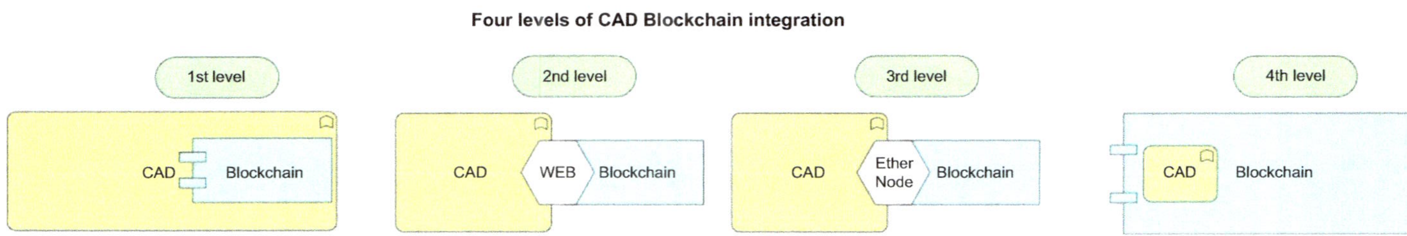

Figure 8.5 Levels of integration between CAD systems and blockchain (re-drawn by Theodoros Dounas, adapted from [26])

Due to its simplicity, this basic CAD/BC integration implementation lacked distributed and trustless verification of block contents. However, it underscored the significance of recording differences within the software, emphasising the transition from one state to another. Analogously, Grasshopper's origins lay in creating an explicit modelling history aligning conceptually with the understanding that this initial BC implementation created an immutable record, capturing the history of a difference between an original and a copy. The next development led to a proof-of-concept optimisation algorithm to identify the highest point on a curved surface. The algorithm endeavoured to find this optimal point and transmitted its corresponding optimisation value as a percentage over the internet. A smart contract on the Ethereum BC was configured to listen for these optimisation values, promptly disbursing issued tokens once a value above the required minimum threshold was received.

The smart contract on Ethereum was structured to receive optimisation values over the web, concurrently capturing the identity of the computational designer on another segment of the Smart Contract. In this scenario, the CAD system Rhino/Grasshopper functioned as an 'Oracle', acting as a node external to the Ethereum BC and providing essential information to the BC. Notably, identification was not mandatory in our implementation as the contract operated within a private BC. In a public Ethereum BC setting, identification would be imperative on the Rhino/Grasshopper side. This would necessitate each CAD node to possess its own ETH address on the network, aligning more closely with the third level of implementation.

In a prospective third and fourth level of integration, the system would have entailed constructing an Ethereum node within Gh. This BC node would have managed the communication and registration of the Gh definition with an address on the Ethereum network. Subsequently, the node could have documented the differences between successive versions of the file, encoding them as blocks. This process would have facilitated the creation of an immutable record, for instance. The fourth level of integration would have entailed executing a comprehensive CAD/BIM application directly on the BC, thereby transforming CAD/BIM into a dApp.

Those initial studies and results led us to move to the next level, where a DAO was adopted as a collaborative platform and demonstrated through a design scenario where BC technology was implemented for validation purposes. The envisioned scenario simulated designers proposing multiple solutions for a given task, utilising shape grammars, environmental analysis and regulations as design drivers. The proposed solutions were uploaded, stored, presented and evaluated within a DAO-controlled BC-enabled, decentralised storage system, validating the decision-making process through the reputation of the participants and its governance system [27].

This study established the groundwork and catalysed the development of a broader framework wherein design collaboration and competition were nurtured, and outcomes were secured, influencing design value and financial transactions.

8.5.2 Prototype 2

Optimisation problems in design do not capture the richness of qualitative decision making that a design team executes during architectural design. This is directly connected with the nature of the design problems that develop in Architecture: All design problems are essentially wicked problem [28], that means that they do not have a straightforward pathway to a singular answer, rather the production of a design is a composite effort of multiple factors that need to be properly balanced in decision making. As such any optimisation solution, including multiple objectives optimisations, rely on a series of previous steps to frame the design space, so that an optimisation question can be posed.

Our prototypes developing on the qualitative design decisions of DAOs use extensively the tools of shape grammars to represent designs, in an effort to capture the development of a design and change the basic encoding of grammars by using BCs as the underpinnings (Figure 8.6). Additionally, we use voting mechanisms to develop the prototypes. We chose shape grammars as the vehicle for this. Shape grammars are productive systems, that is, they have embedded in their notation the production or generation of a design, while capturing intermediate states of design development, an important feature that fertilises collaboration between architects, as our intention was to frame collaboration during the design process. Instead of thinking of one author as the author of the shape grammar to be developed, we place the DAO in its place, where the DAO consists of several shape grammarists (X number), sufficiently large number to make decision making complex enough so that the need for using a BC and automation of governance becomes apparent.

8.5.3 Shape grammars and blockchain

The common ground between shape grammars and BCs is larger than what it might appear in the first instance. Computationally, both are state machines, in essence well-polished versions of a universal Turing Machine, that is, computation machines that can emulate other Turing machines. In terms of their computational underpinnings, both use a rulebook that (most) of the time it is treated as sacrosanct. In BC terms, this means that the code is the law, in shape grammars terms this means that if you find a rule that was not there before, your grammar needs revising. As such, the people who work with both technologies think in terms of policies: bundles of rules that produce the desired outcome. In the same sense, both also need external triggers to halt computation, in the form of Shape grammars this takes the shape of a halting rule and in the form of smart contract enabled BCs takes the shape of needing gas to execute computation.

As such within this prototype we propose developing a shape grammar through the mechanisms of a smart contract and decentralised storage, operated by X shape grammarists. Within our prototype lies the nucleus of being able to develop designs in public, through a decentralised team that is not hierarchically structured apart from the governance mechanisms employed on-chain [11,29].

8.5.4 The blockchain infrastructure: DAOstack

Within a DAO, agents may compete with one another, as in any organisation. If structured around current economic models, a non-cooperative Nash equilibrium emerges in agent interactions – some agents act competitively to achieve a local optimum, not a global optimum. This poses a key challenge for DAOs in facilitating

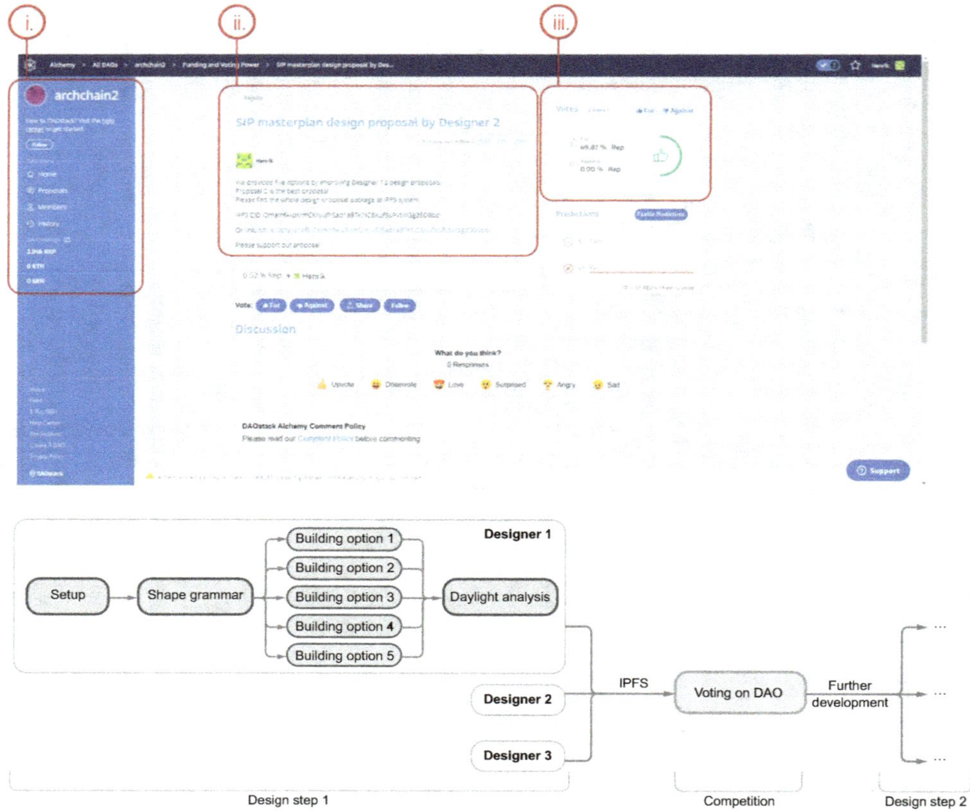

Figure 8.6 *Voting-based governance using shape grammars and optimisation (authored and edited by Theodoros Dounas and Davide Lombardi)*

positive decision-making from the bottom-up, avoiding both the non-cooperative Nash equilibrium and excessive concentration of voting power among tokens.

A structural solution for organisations trapped in non-cooperative Nash equilibriums is incentivising and rewarding shareable resources [30], nurturing growth and openness. This further aligns DAOs with Common Pool Resources Philosophies and Ostrom Principles [31]. We have used DAOstack in our prototype as it was the first DAO that would propose a solution to the trilemma between collective decision-making (have we taken everyone's needs and opinions into account), competence (is everyone capable of making the right decision?) and attention (do we all have time to provide governance input for every single decision making?). DAOstack created a dual system to handle this, the GEN token [32], which could be used for rewards and stakes and a reputation system that needed nurturing to be kept, that is, one had to be active to have a high reputation.

Regarding voting power, DAOstack allowed transferring GEN tokens but not reputation. Reputation represents 'influence power, forming a meritocratic governance system, where the most appreciated members, due to past contributions, have the most influence' [30]. In our system, reputation accrues through proposals, for example awarding reputation to a researcher for contributions. It can also derive algorithmically from citations, impact metrics, etc., translating research impact into decision-making power. Through reputation and GEN tokens, DAOstack aligns an empowered minority with the majority's interests, ensuring transparency, resilience (as reputation endures despite token transfers), scalability (via BC participation) and 'Holographic Consensus' – edge decisions reflecting consensus, like holograms containing the whole image.

The shape grammar:

As shape grammars are generative systems for design. A shape grammar consists of an initial shape set A1 and a set of shape transformation rules R1 which, when applied to A1, generate new shapes described on the right-hand side of each rule. A final 'halt' rule K terminates computation when activated, since shape grammars being universal Turing machines have no manner of stopping unless. Halting rules is included [33].

A two-dimensional shape grammar has:

$$A1 = \{a1, b2, c3, \ldots, xn\}$$

$$R1 = \{r1, r2, r3, \ldots, rn\}$$

Typically, the shape grammarist retains full authority in all aspects of the grammar. Sometimes they emulate a known style, innovating new examples computationally within the same language. Relevant issues include rule selection/modification, shape recognition/emergence and shape ambiguity.

We describe the standard process of creating and applying a layout grammar, with the implication that the grammarist receives a ready-made grammar. The grammarist selects shape a1 and rule r1, moving a1 one meter right. Next b2 is selected and rotated 45 degrees by r2. C3 is selected and doubled in size by r3. A rule labels the new spaces, which are checked against constraints [34], terminating the grammar after N rule applications upon recognising shape K.

Algebraically:

$$\begin{aligned} r1 &: a1 -> a1' \\ r2 &: b2 -> b2' \\ r3 &: c3 -> c3' \\ &\dots \\ rn - 1 &: K' -> K \\ rn &: \text{halt} \end{aligned}$$

Consider as such a set Gx of grammarists collaborating to create the most variety of compact apartment layouts by minimising the apartment's global integration/floor area ratio, with minimum shape sizes for liveability. What possibilities exist for internet collaboration with a record of all steps? How does one negotiate rule selection?

With A1, R1, Gx {F, E, D, K, S, …, G883}, and validation V = global integration/floor area, with constraints on S1:

While F, E, D, K, S should guide Gx, innovative ideas from all members are needed. Addressing X members at each step creates excessive communication overhead.

To address this issue, we propose a BC Grammar, the first grammar engine on a DAO stack, enabling human–machine collaboration. The BC Grammar provides token distribution, reputation assignment based on research impact, collective data curation through grammar creation/application and governance upgrading. Reputation decays over time without ongoing academic contributions. Initially, all receive equal tokens, but reputation stems from publications, h-index and c-index in shape grammars. Voting power Vt = Tokens + Papers + h-index + c-index. The BC Grammar runs on DAOstack, using Ethereum for decentralisation. At each step, all can propose rule applications. With 30 proposals, all 888 votes using reputation and tokens. Senior members should have more influence, but innovative solutions can also pass. A minority, like K and S, could pass a proposal despite equal tokens, if their high reputation overrides other voters.

8.5.5 Prototype 3

In order to validate the study, we presented a design exercise and a DAO implemented to lead a decision-making process for a design problem. The design exercise engaged with the goal of maximising the light conditions in a series of apartment towers in China via generative design applied to shape grammar rules and with the support of a plugin for the evaluation of lighting conditions and analysis [29]. The computational modelling environment utilised Rhinoceros alongside Grasshopper plugins to execute shape grammar transformations and daylight simulations. In generating design variants, shape grammars served as a robust tool, encapsulating design variations and generative options. The application of shape grammars harnessed their generative power to produce diverse design alternatives akin to those that would emerge in a tangible design setting. Given the objective of exploring design project governance, shape grammars proved to be an excellent instrument for creating the full spectrum of design alternatives.

The workflow encompassed four main phases:

1. Preparation of the overall test and display environment.
2. Designers generated alternatives using shape grammars, transforming the final shapes into buildings for simulation purposes.
3. Daylight simulations were conducted in accordance with predefined regulations and specifications.
4. Data extraction for the DAO voting process was carried out.

The described workflow served as the foundation for simulating the collaborative efforts of three designers engaged in the same design task within a competitive scenario. Following the execution of the simulation test, which generated proposals from the three designers, a total of five alternatives were produced, each accompanied by daylight hour simulation results.

In this simulated context, Designer 1 initiated the process by submitting design proposals, along with the modelling file and simulation data, to the DAO system through the IPFS system. Subsequent to the participants' voting, the proposal was rejected due to inadequate daylight performance that failed to meet regulatory requirements. Notably, Designer 1 lacked design options wherein over 90% of the main façade adhered to the minimum regulatory stipulations. Following this outcome, Designers 2 and 3 presented their proposals, building upon the initial design file provided by Designer 1. Both designers submitted options aligning with daylight regulations, and participants cast their votes after considering the quality of the designs. Following a detailed comparison of the building designs, Designer 3's proposal emerged as the favoured option, garnering the most votes for further development. In this example, we developed essentially a cooperation system, with BC at the foundational level, where data, information and governance are using the BC as the substrate of information. While with a small number of designers, one could claim that the BC is not really needed, we needed to develop the proof of concept to analyse the mechanics of how a voting-based BC governance could be developed for architectural design.

8.5.6 Decentralised architectural design

The above experiences were further fed into a further prototype that integrates BIM with the Ethereum BC to address issues of consensus, trust, transparency, responsibility and incentivisation among design agents in a distributed design environment. The prototype allows interoperability between digital tools that communicate with the Ethereum BC, enabling the simulation of multiple design agents working within their respective software platforms. The system is designed to scale to accommodate thousands of agents, regardless of whether they are human designers or machine learning algorithms [35].

The presented scenario involves three design agents attempting to solve a structural design problem initiated by a DAO as a smart contract on the Ethereum BC. Notably, there is no need for formal legal contracts or established trust between the agents and the DAO. The DAO serves as the stakeholder setting the design problem, and any Ethereum address holder can participate by staking tokens towards desired outcomes. This inclusive approach allows non-expert participants to contribute to the governance of the design project without diminishing the expertise of professionals.

The DAO, acting as the problem setter, utilises Ethereum smart contracts to make strategic decisions and allocate funds for incentivisation. Designers have the flexibility to reformulate the problem, fostering an environment that accommodates 'wicked' problems inherent in design. The prototype supports both competitive and collaborative modes, and large data storage is facilitated through the IPFS, a decentralised file system that ensures data distribution without relying on centralised servers.

The structural design optimisation problem involves finite element analysis using the Karamba3D plugin within the Grasshopper environment. Multiple smart contracts on the Ethereum BC serve specific purposes, including BIMManager, IPFSHashHolder, Problem, Solution and Solutionise. These contracts manage the connection between BIM software and the Ethereum BC, handle storage of IPFS hashes and represent the design problem, solution and the awarding of incentives.

The BIMManager contract connects the BIM software to the Ethereum BC, defining parameters such as the problem ID, cryptographic hash of the problem file on IPFS, desired inputs/outputs, minimal required performance, reward amount and expiry date. Design agents provide their Ethereum address, username ID and payout ETH address. When an agent solves the design problem, they upload the solution to IPFS and a plugin submits relevant data parameters to the BIMManager smart contract. The smart contracts facilitate comparison of solutions, synchronisation among agents and the achievement of design consensus.

The prototype employs a federated organisation of the DAO, allowing experts and non-experts to participate in decisions. The IPFS ensures secure and decentralised file storage, while Ethereum smart contracts manage the flow of data and ETH payments. The design consensus is reached through synchronised interactions between agents and the BC. Overall, the system aims to revolutionise collaborative design processes by leveraging BC technology to enhance trust, transparency and incentivisation.

8.5.7 DAOs and shape grammars further development

These processes would allow us to create a public repository of designs which we can use for multiple purposes (for example training machine learning algorithms for architecture) but incentivise architects to create exciting designs and constructs that are re-usable - as we could come up with an incentive structure on BC terms, where excellent design is rewarded, creating an excellent example of a global common pool of designs. Now this commonality means that we could essentially start running shape grammars on BCs, which would allow one designer to launch a design grammar on a BC, represented as a set of smart contracts for policies, a set of graph grammars and shapes, stored unto decentralised storage. then another one, without asking permission, to change it, modify it to purpose and re-launch it. There are obvious constraints in such a system, for example it is difficult to prevent people from being extractive to the system, that is, cheating by buying the designs and then copying them off-chain. A potential solution to this is to enforce a data structure such as linked data where there is no longer a file to be copied.

8.6 Building a decentralised autonomous organisation as a living lab

8.6.1 The architecture decentralised autonomous organisation 'archiDAO'

Inspired by the developments of BC in AEC, in December 2021, a group of architects and scholars of architecture came together to realise the first architecture DAO. ArchiDAO is a 'decentralised architecture studio, for design, research and learning' thus uses BCs as the infrastructure for most of its activities, with a flat hierarchical structure. Immediately the issues that arose is the lack of tooling to build a DAO that focuses on architecture, but also in general a DAO that is essentially a decentralised version of entities that already exist, such as architecture studios. It became important then to document all efforts and perceive the DAO as a living lab, setting initially its mission and values. As a pre-ample of the mission and values, archiDAO framed the AEC industry as fragmented, unproductive and resistant to innovation mentioning our earlier prototypes as the basis of how we would explore how archiDAO would operate, setting the vision 'that blockchain-based tools and modes of operation will transcend the whole lifecycle of the AEC industry, from design to construction, operations and decommissioning'. This led to mission and values declaration of the DAO.

The mission of archiDAO was declared thus (quoted from the mission and values document of archiDAO):

> **Our mission** is to establish, expand and continuously develop our Web3 framework for architecture by onboarding collaborators to the DAO, scaling a broader community of architects, researchers, and engineers. As part of our mission, we will design and develop the architecture and spatial experiences of the future, whether on the Metaverse or in Physical reality. This takes place through commissions and competitions and prototypes.
>
> We will scale by developing learning pathways and products, Blockchain and smart contracts infrastructure for architectural design, but also offer consulting services to other AEC entities to help them adopt **web3 technologies**. The three pillars of learning, projects and services are interconnected and supported all by our community [36].

Followed by its values:

> Blockchain technologies and web3 will bring transformational change not just to the AEC industry, but all people who experience spatial structures – whether **physical or virtual**. For example, coupling blockchain primitives with architectural design could facilitate new models of collective authorship for buildings. Our values are driven by **stigmergic work** rather

> than capital, a vision for the commons rather than an over-financialization of assets. We wish to embed desirable economic behaviours into spatial design to transform the use and maintenance of space. We should also incentivize **sustainable development** in the industry: $ tokens could incentivize design strategies that reduce carbon emissions, or non-fungible tokens for building components could enable a new circular economy. (Buildings are responsible for over **40%** of global carbon emissions, after all.) We envision an equitable future for all, including supporting basic rights such as housing and shelter for all.
>
> Our work is collective, open and designed around **our community.** Through a series of **workshops, learning pathways and engagements** we help develop the members' skillset and understanding of web3. We believe that learning is a public good and thus we strive to offer **learning opportunities** to all members of the DAO and the community' [36].

To develop the operational capacity of ArchiDAO we embarked on writing the first draft of a white paper, attempting to examine stigmergic and collective authorship principles to operate with. A key issue that arose in terms of governance was the distinction between governance of a project and governance of the organisation itself. This emerged as it became clear that governance of the organisation had to do more with key mission, value and allocation of resources, while governing a project had to focus more on expertise and operational actions rather than resources allocation.

While the narrative of the discipline of architecture often focuses on sole authorship, contemporary architectural practice involves shared effort across disciplines. The design and construction process necessitates collaborative work organised around collective labour, expertise and authorship. This has driven research into communication protocols, remote collaboration and other mechanisms for cooperative creation. To establish the dimensions of cooperative action we embarked on a systematic review of papers through CUMINCAD on collective authorship in architectural design, with 37 results and with a further filtering we narrowed the field down to work that discussed specifically architectural co-authorship (Table 8.1).

Key themes emerging from this literature relate to remote collaboration, communication protocols and knowledge sharing. A significant portion of papers focus on virtual coordination and its implications for pedagogy and design. However, few explicitly involve computer protocols for communication and collaboration. Additionally, projects in the public realm more frequently incorporate shared commons, understandable given their larger scale. To relate these concepts to stigmergic principles, articles were assessed on dimensions of communication, feedback loops, decentralisation and use of a shared substrate. Communication between members and the design artefact scored highly. Decentralised authority was less prominent, suggesting an opportunity for further exploration.

In summary, architectural collaboration increasingly recognises distributed authorship and collective effort facilitated by networked technologies. Stigmergic

Table 8.1 CuminCAD Co authorship publications as per [38]

No	Authors	Year	P1	P2	P3	P4	P5	P6	P7	P8
1	Merrick, Gu	2011	2	3	3	3	3	1	1	1
2	Fioravanti, Loffreda	2009	3	3	2	3	3	3	0	1
3	Dortheimer, Neuman, Milo	2020	3	3	3	3	2	3	1	1
4	Vallejo *et al.*	2017	3	2	2	3	2	1	0	1
5	Eleutheriou *et al.*	2015	3	2	1	0	0	2	1	2
6	Angulo, Fillwalk, Velasco	2009	3	3	1	3	3	2	1	1
7	Yaniv Ophir	2009	3	3	3	0	0	1	0	1
8	Jose Sanchez	2015	1	3	3	1	3	3	2	2
9	Hsieh Chih-Wen	2005	3	3	1	3	2	0	2	1
10	Velasco	2007	3	2	1	3	2	3	1	1
11	Abdelmohsen, El-Khouly	2004	3	1	3	3	1	1	0	0
12	Dorta *et al.*	2019	3	3	3	2	2	2	1	1
13	Schnabel, Ham	2013	3	3	3	1	2	2	1	2
14	Briscoe, Hadilou	2013	2	3	3	3	3	2	1	2
15	Adamantidis, Kidao, Tsiliakos	2013	2	3	2	0	3	1	0	2
16	Merrick *et al.*	2011	3	3	2	3	2	2	0	1
17	Paulini, Maher, Murty	2011	3	3	2	1	2	1	1	2
	PX: Score		46	46	38	35	35	30	13	22

principles offer a model for decentralised, indirect coordination without hierarchical oversight. This review highlights relevant precedents while identifying areas for additional research at the intersection of stigmergy, co-authorship and computational design.

8.6.2 Stigmergy

Stigmergy represents a key conceptual framework underpinning this chapter, referring to a decentralised, self-organising mechanism observed in biological systems like insect colonies. The term was first coined by French entomologist

Pierre-Paul Grassé to describe the indirect coordination achieved in social insects through environmental signals left from previous work that stimulate future action. The etymology combines the Greek words stigma (mark or sign) and ergon (work or action). Stigmergic principles enable simple behavioural rules at the individual level to elicit complex collective outcomes. A salient example is pheromone trails in ant colonies which facilitate colony-wide coordination without any individual possessing a global perspective. Similarly, honeybee colonies utilise chemical signalling to regulate division of labour and organisation of the hive according to basic stigmergic principles. The key insight is that ignorance at the individual level does not impede sophisticated group-level coordination. This chapter utilises stigmergy as a model for decentralised, self-governing systems that can engage in complex collaborative endeavours through simple signalling mechanisms. The aim is to translate these biomimetic principles into guidelines for a DAO focused on architectural design and construction [38].

The viable system model (VSM) provides a conceptual framework for decentralised organisations that align with the stigmergic principles explored in this chapter. Developed by Stafford Beer, the VSM offers tools to steer interactions towards effective structural mechanisms for the collective endeavour. It is organised around three pillars: cybernetics, requisite variety and recursion. Cybernetics frames organisations as adaptive systems, while requisite variety refers to matching internal diversity to external complexity. Recursion enables nested autonomous units. According to Espejo [39], the VSM helps groups self-organise through synergistic relationships without pre-existing hierarchical control. Ibrahim *et al.* demonstrate the VSM's utility for capturing iterative features in architectural design. The model promotes decentralised governance through recursive subsystems, reflecting stigmergic principles. Autonomous units can self-regulate and communicate to provide services to the wider organisation without top-down oversight.

Parallels can be drawn to the Chilean Cybersyn model for economic self-governance. The VSM offers a generalisable framework for decentralised coordination applicable to a DAO focused on architectural collaboration. It conceptualises organisations as networks of recursive, adaptable units with sufficient diversity to solve complex problems. By emphasising autonomy, emergence and bottom-up self-regulation, the VSM aligns with the biomimetic principles of stigmergy identified in this chapter. It provides a precedent for decentralised organisational design without centralised authority.

8.6.3 Smart contract and token designs for stigmergy – the ArchiDAO non-fungible token

A key insight emerging from this analysis is the role of predefined communication modes in enabling decentralised organisation through feedback loops, aligning with stigmergic principles. Rather than top-down oversight, indirect coordination arises through information transfer that is bidirectional and recursive. In the VSM, strategic intelligence represents feedback between a system and its environment. In collective design, iterative improvement involves collaborative feedback loops.

And in DAOs, governance protocols embed feedback mechanisms on-chain and off-chain. While specific mechanisms vary, common principles facilitate stigmergic self-organisation:

- Feedback loops, encoded or not, achieve collective action through second-order coupling.
- Clear communication channels distribute information.
- Boundary conditions constrain interactions.
- Transparent influence generation and accrual prevent centralisation.
- Explicit limitations on influence utilisation further distribute authority [37,40].

In summary, decentralised coordination relies on protocols for recursive information flow within defined constraints. By formalising feedback, communication, boundaries and influence, stigmergic systems can self-organise without hierarchical control. These principles offer guidance for designing distributed governance, as demonstrated across organisational theory, design collaboration and DAOs. The key insight is that collective outcomes emerge not from top-down oversight but from simple, decentralised rules for flexible interaction.

8.6.4 The ArchiDAO proposed structures

Taking these principles into account, within the ever-alive archiDAO community, we developed a series of on-chain specifications and prototypes on-code that would enable us to realise architectural design projects, or revolving around architectural design activities such as education or culture. Within this chapter we present the key instrument for governing the organisation, a proposed 'soulbound' Non-Fungible Token [41] that would represent a member's history, skills and activity, essentially a system of decentralised identity and credentials of professional capacity. We envisioned this as a vehicle for scaling the organisation in a decentralised manner, while we kept initial decisions about resource allocation, that is, spending from the treasury unto simpler decentralised structures, namely voting mechanisms through the snapshot protocol, where certain Ethereum addresses, belonging to core and long-term members were allowed to vote.

8.6.5 The ArchiDAO soulbound token [42]

The code for the development of the ArchiDAO soulbound non-fungible token that acts as an identification and governance-gated scheme follows basic NFT templates and connects the token with a series of skills and a personal identity.

Two issues emerged through the design of the token, namely the issue of recording personal information on chain, or similar to on-chain mechanisms for example, through IPFS, and the recording of skills for a particular token that needed to be immutable and on-chain so that people that own the tokens can participate in archiDAO projects accordingly, that is, they would be selected based on the kind and levels of the skills recorded. Additionally, the skills posed another classification problem in the sense that the DAO needed, of course architects, but also other designers and engineers with a wide variety of skills and capacities. Thus, we needed a manner in selecting and consistently representing the members' skills. Part of other soulbound tokens followed solutions that resembled the systems for recording experience in Role Playing Games, however we felt that we needed a system that would recognise experience in the manner we have in more traditional professional guilds, such as the Architects Registration Board. As such, classification schemes that we examined to develop the skills part of the NFT included the UNESCO International Standard of Classification of Education, the International Standard for the Classification of Occupations and the EU multilingual classification of skills, competencies and occupations. Out of these schemes we ended up with a list of 10 core skills that would be an a-priori classification, where the archiDAO member could add more types of skills on their own.

This posed an additional challenge for the creation of the NFT as, as changing skills and their rating would change the content identity of the metadata of the NFT thus needing a re-minting. We also needed a way to secure and make private the skill levels of the tokens and only expose them to those other archiDAO members that would need access, for example, project managers. The solution was to develop the metadata of the soulbound token unto a zero-knowledge proof layer 2 chain, namely Polybase [43], which is a privacy focused zero-knowledge proof layer 2, following the following example (Figure 8.7).

This would allow the saving on-chain of the metadata, but it would only be revealed to either people within the organisation or addresses that would possess proper access. It also allows for the creation of a sophisticated, privacy defending decentralised credentials system, that would allow the smart contracts that would run the organisation to recognise its members and act accordingly. Additionally, the use of Polybase allows also transactions using the USDC stablecoin with almost zero fees, and preservation of the privacy of the users and the transactions executed, a unique feature among layer 2 BCs. At the time of writing, the emergence of account0based abstraction on the Ethereum ecosystem, would also allow a seamless on boarding of users on the DAO, but we have not had yet the opportunity to test this at scale.

8.7 Discussion – contribution and conclusion

We have explored a multitude of processes with which architects can create DAOs to create a more equitable, non-extractive discipline, in the sense that DAO bypasses traditional and entrenched hierarchies, to the cost, however, of making coordination much more difficult, as stigmergic systems need to attain balance to work effectively. This we have experienced empirically through archiDAO; however, we believe that there is the space and need for many more DAO experiments so that the AEC community understands which the design dimensions that are critical for success and which parameters are trivial. In the process of building the DAO examples we have proposed, we have run into the problem of tooling, that is, developing custom tools for our purposes, a common feature amongst early stage BC in AEC researchers. The space has improved massively since 2017, no less due to the

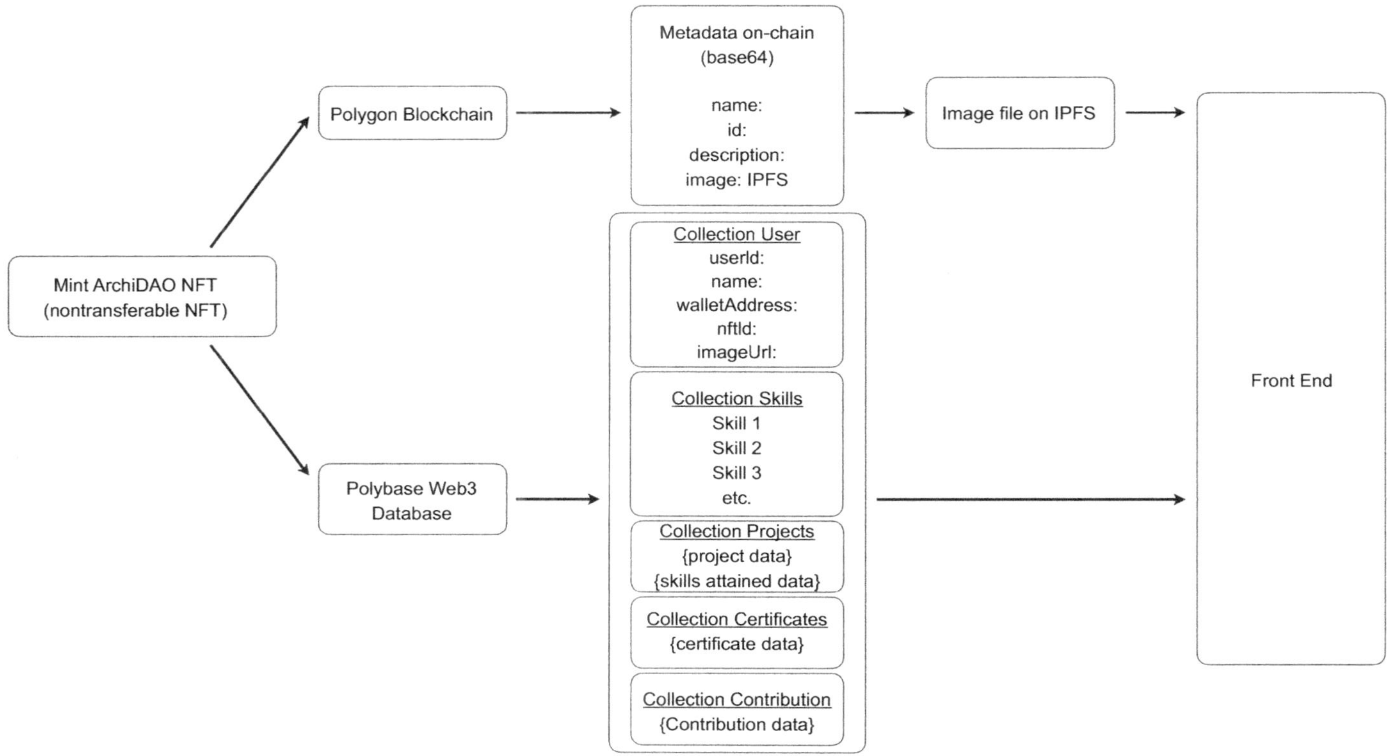

Figure 8.7 Polybase for soulbound NFT Process (authored by Hico McDonald and redrawn by Davide Lombardi)

collective early research. We still believe, however, that there is room to develop even more sophisticated tools, but also many more DAOs in the AEC space.

Acknowledgements

We would like to thank Professor Wassim Jabi for collaborating with us on many early prototypes. We would also like to thank the archiDAO community. TD and DL have the right to mention the chapter as the first authors. TD and DL created the structure, TD wrote the introduction, Yilong competition part, the grammars and the archiDAO part while DL wrote the optimisation prototype. All figures were redrawn by TD with input from TD and DL, except from Figure 8.7, which was created by HmcD and redrawn by TD. HmcD wrote the code along with TD on the archiDAO token and conceptualised its structure in using polybase as zero-knowledge layer.

References

[1] World Green Building Council [Internet]. Embodied Carbon. Available from https://worldgbc.org/advancing-net-zero/embodied-carbon/, accessed 30 September 2024.

[2] Routledge and CRC Press [Internet]. Practice: Architecture, Technique and Representation. Available from https://www.routledge.com/Practice-Architecture-Technique-and-Representation/Allen/p/book/9780415776257, accessed 30 September 2024.

[3] Crypto: Towards a New Political Economy in Architecture – Prospectives [Internet]. Available from https://journal.b-pro.org/article/crypto-towards-a-new-political-economy-in-architecture/, accessed 30 September 2024.

[4] Kimani D, Adams K, Attah-Boakye R, Ullah S, Frecknall-Hughes J, and Kim J. Blockchain, business and the Fourth Industrial Revolution: Whence, whither, wherefore and how? *Technol Forecast Soc Change*. 2020;161:120254.

[5] Ferrari TG. Design and the Fourth Industrial Revolution. Dangers and opportunities for a mutating discipline. *Des J*. 2017;20(sup1):S2625–33.

[6] Werning RC. Nelson Goodman's autographic–allographic distinction in architecture: The case of Mies van der Rohe's Barcelona Pavilion. In: Ernst G, Steinbrenner J, and Scholz OR (eds.), *From Logic to Art* [Internet]. Berlin: De Gruyter; 2013. pp. 269–92. Available from https://www.degruyter.com/document/doi/10.1515/9783110327199.269/html?lang=en.

[7] ethereum.org [Internet]. Ethereum whitepaper. Available from https://ethereum.org, accessed 30 September 2024.

[8] Brekke CJE. Disassembling the trust machine: Three cuts on the political matter of blockchain. PhD Thesis, Durham University. Available from: http://etheses.dur.ac.uk/13174/.

[9] Brekke JK. Hacker-engineers and their economies: The political economy of decentralised networks and 'cryptoeconomics'. *New Polit Econ*. 2021;26(4):646–59.

[10] Hunhevicz JJ, Wang H, Hess L, and Hall DM. no1s1 – A blockchain-based DAO prototype for autonomous space. In *ETH*; 2021 pp. 27–33 (Computing in Construction; vol. 2). Available from https://ec-3.org/publications/conference/paper/?id=EC32021_185, accessed 30 September 2024.

[11] Dounas T., and Lombardi D. Blockchain grammars – Designing with DAOs – The blockchain as a design platform for shape grammarists' decentralised collaboration. In: M Haeusler, M A Schnabel, and T Fukuda (eds.), *Intelligent & Informed – Proceedings of the 24th CAADRIA Conference – Volume 2*, Victoria University of Wellington, Wellington, New Zealand, 2019, pp. 293–302 [Internet]. CUMINCAD; 2019. Available from http://papers.cumincad.org/cgi-bin/works/paper/caadria2019_136, accessed 30 September 2024.

[12] Tolmach P, Li Y, Lin SW, and Liu Y. Formal analysis of composable DeFi protocols. In: Bernhard M, Bracciali A, Gudgeon L, *et al.* (eds.), *Financial Cryptography and Data Security FC 2021 International Workshops*. Berlin: Springer; 2021, pp. 149–61 (Lecture Notes in Computer Science).

[13] Dounas T., Jabi W., and Lombardi D. Smart contracts for decentralised building information modelling. In: Werner, L and Koering, D (eds.), *Anthropologic: Architecture and Fabrication in the Cognitive Age – Proceedings of the 38th eCAADe Conference – Volume 2*, TU Berlin, Berlin, Germany; 2020, pp. 565–74 [Internet]. CUMINCAD; 2020 [cited 17 January 2022]. Available from http://papers.cumincad.org/cgi-bin/works/paper/ecaade2020_190, accessed 30 September 2024.

[14] Dounas T, Hunhevicz J, and Byers B. Design dimensions for blockchain oracles in the AEC Industry. 2023. Available from https://ec-3.org/publications/conference/paper/?id=EC32023_297, accessed 30 September 2024.

[15] Lu W, Li X, Xue F, Zhao R, Wu L, and Yeh AGO. Exploring smart construction objects as blockchain oracles in construction supply chain management. *Autom Constr*. 2021;129:103816.

[16] Hunhevicz JJ, and Hall DM. Do you need a blockchain in construction? Use case categories and decision framework for DLT design options. *Adv Eng Inform*. 2020;45:101094.

[17] Hunhevicz J, Dounas T, and Hall DM. The promise of blockchain for the construction industry: A governance lens. In: Dounas T, and Lombardi D (eds.), *Blockchain for Construction [Internet]*. Singapore: Springer Nature; 2022. pp. 5–33 (Blockchain Technologies). Available from https://doi.org/10.1007/978-981-19-3759-0_2, accessed 30 September 2024.

[18] Nakamoto S. *Bitcoin: A Peer-to-Peer Electronic Cash System*. Available from: https://bitcoin.org/bitcoin.pdf.

[19] Heylighen F, and Joslyn C. Cybernetics and second-order cybernetics. *Encycl Phys Sci Technol*. 2001;4:155–70.

[20] The 13th Five-Year Plan, for Economic and Social Development of the People's Republic of China (2016–2020) [Internet]. Communist Party of China; 2016. Available from https://en.ndrc.gov.cn/policies/202105/P020210527785800103339.pdf, accessed 30 September 2024.

[21] Dounas T, Spaeth AB, Wu H, and Zhang C. Dense urban typologies and the game of life: Evolving cellular automata. In: Çagdas, G, Özkar, M, Gül, LF, and Gürer, E (eds.). *Future Trajectories of Computation in Design*. Istanbul, Turkey: CAAD Futures; 2017. pp. 648–66.

[22] Li C. China, a major bitcoin source, considers moving against it. *The New York Times [Internet]*. 2019. Available from https://www.nytimes.com/2019/04/09/business/bitcoin-china-ban.html, accessed 30 September 2024.

[23] Lopes J, Paio A, and Sousa JP. Parametric urban models based on Frei Otto's generative form-finding processes. *Proceedings of the 19th International Conference on Computer-Aided Architectural Design Research in Asia*, 2014, pp. 595–604.

[24] O'Dwyer KJ and Malone D. Bitcoin mining and its energy footprint. *25th IET Irish Signals & Systems Conference 2014 and 2014 China-Ireland International Conference on Information and Communications Technologies (ISSC 2014/CIICT 2014)*, 2014. pp. 280–5.

[25] Jabi W, and Chatzivasileiadi A. Topologic: Exploring spatial reasoning through geometry, topology, and semantics. In: Eloy S, Leite Viana D, Morais F, Vieira Vaz J (eds.), *Formal Methods in Architecture*. Cham: Springer International Publishing; 2021. pp. 277–85 (Advances in Science, Technology & Innovation).

[26] Dounas T., and Lombardi D. A CAD-blockchain integration strategy for distributed validated digital design – Connecting the blockchain. In: Kepczynska-Walczak A, Bialkowski S (eds.), *Computing for a Better Tomorrow – Proceedings of the 36th eCAADe Conference – Volume 1,* Lodz, Poland: Lodz University of Technology; 2018. pp. 223–30 [Internet]. CUMINCAD. Available from http://papers.cumincad.org/cgi-bin/works/paper/ecaade2018_226, accessed 30 September 2024.

[27] Dounas T, and Lombardi D. Designing with DAOs – The blockchain as a design platform for shape grammarists' decentralised collaboration. In CAADRIA 2019. Available from http://papers.cumincad.org/cgi-bin/works/paper/caadria2019_136, accessed 30 September 2024.

[28] Wortmann T. Surveying fitness landscapes with performance explorer – Supporting the design of a better tomorrow with interactive visualizations. In: Kepczynska-Walczak A, Bialkowski S (eds.), *Computing for a Better Tomorrow – Proceedings of the 36th eCAADe Conference – Volume 1*, Lodz, Poland:Lodz University of Technology; 2018, pp. 621–30 [Internet]. CUMINCAD; 2018. Available from https://papers.cumincad.org/cgi-bin/works/paper/ecaade2018_279, accessed 30 September 2024.

[29] Lombardi D., Dounas T., Cheung L.H., and Jabi W. Blockchain grammars for validating the design process. In: *Blucher Design Proceedings*. São Paulo: Editora Blucher; 2020. pp. 406–11.

[30] DAOstack whitepaper [Internet]. *The Whitepaper Database*. 2018. Available from https://www.allcryptowhitepapers.com/daostack-whitepaper/, accessed 30 September 2024.

[31] Hunhevicz JJ, Hall DM, Brasey P-A, Bonanomi MMM, and Fischer M. Decentralized project delivery on the crypto commons: Conceptualization, governance mechanisms, and future research directions. *Project Leadership and Society*. 2024;5100132.
[32] Home – DAOstack Arc Docs [Internet]. Available from https://daostack.github.io/arc/README/, accessed 30 September 2024.
[33] Stiny G. *Shape: Talking about Seeing and Doing*. Cambridge, MA: The MIT Press; 2006.
[34] Duarte J. A discursive grammar for customizing mass housing – The case of Siza's houses at Malagueira. In: *Digital Design 21st eCAADe Conference Proceedings*, Graz, Austria: eCAADe; 2003. pp. 665–74.
[35] Dounas T, Lombardi D, and Jabi W. Framework for decentralised architectural design BIM and blockchain integration. *Int J Archit Comput*. 2021; 19(2):157–73.
[36] Notion [Internet]. ArchiDAO mission and values. Available from https://www.notion.so, accessed 30 September 2024.
[37] Dounas T. The architecture decentralised autonomous organisation – A stigmergic exploration in architectural collaboration. In: Pak, B, Wurzer, G and Stouffs, R (eds.), *Co-creating the Future: Inclusion in and through Design – Proceedings of the 40th Conference on Education and Research in Computer Aided Architectural Design in Europe (eCAADe 2022) – Volume 1*, Ghent:CUMINCAD; 2022. pp. 567–76 [Internet]. Available from https://papers.cumincad.org/cgi-bin/works/paper/ecaade2022_246, accessed 30 September 2024.
[38] Theraulaz G, and Bonabeau E. A brief history of stigmergy. *Artif Life*. 1999; 5:97–116.
[39] Espejo R, and Reyes A. Organizational Systems: Managing Complexity with the Viable System Model. Springer Science & Business Media; 2011.
[40] Lombardi D, and Dounas T. Decentralised autonomous organisations for the AEC and design industries. In: Dounas T, and Lombardi D (eds.), *Blockchain for Construction [Internet]*. Singapore: Springer Nature; 2022. pp. 35–45. Available from https://doi.org/10.1007/978-981-19-3759-0_3.
[41] Ohlhaver P, Weyl EG, and Buterin V. *Decentralized Society: Finding Web3's Soul [Internet]*. Rochester, NY; 2022. Available from https://papers.ssrn.com/abstract=4105763, accessed 30 September 2024.
[42] GitHub [Internet]. ArchiDAONFT/contracts/ArchiDAO.sol at main ArchiDAO/ArchiDAONFT. Available from https://github.com/ArchiDAO/ArchiDAONFT/blob/main/contracts/ArchiDAO.sol, accessed 30 September 2024.
[43] Polybase Labs [Internet]. Available from https://ancient-head-267506.framer.app/, accessed 30 September 2024.

Chapter 9

Blockchain in construction supply chain management

Liupengfei Wu[*], Jinying Xu[†] and Weisheng Lu[‡]

9.1 Introduction

A supply chain is an interconnected system of organisations, resources, activities and information involved in producing and delivering a product from initial raw materials to final consumers via a carefully managed process of information, physical movement and financial transactions [1]. Supply chain management (SCM) refers to the coordination of the flow of goods, data and finances of a product or service, from raw materials procurement to final product delivery [2]. While supply chains and logistics are often equated, that is, logistics and SCM, logistics management is generally considered a subset of SCM.

Definitions of logistics, supply chains and SCM vary according to the context in which they are discussed. Nonetheless, the consensus is that the importance of SCM can never be overemphasised in contemporary economic development. SCM has existed ever since there were products to be manufactured, transacted and delivered. With the advent of industrialisation, it has become more sophisticated [3], evolving, for example, from tangible products to intangible services [4], from a simple party to a web of stakeholders [5] and from goods, services and money to information management [6]. With this sophistication comes enormous challenges. Bartman [7] reports labour shortages, equipment availability and the ripple effect of global bottlenecks as key challenges in global SCM, while Santhi and Muthuswamy [8] identify risk management, customer expectations, fluctuating prices, unanticipated delays, adaptability to changes, digitalisation and port congestion. Jabbar *et al.* [9] attribute these SCM challenges to transparency and traceability issues; a lack of trust amongst stakeholders at the root.

While the challenges faced by SCM and the construction industry share similarities, such as the need for coordination, risk management and efficiency, they

[*]Department of Real Estate and Construction, Faculty of Architecture, The University of Hong Kong, Hong Kong
[†]Department of Engineering, University of Cambridge, UK
[‡]Department of Real Estate and Construction, Faculty of Architecture, The University of Hong Kong, Hong Kong

also exhibit distinct characteristics due to the inherent differences in their operational contexts and sectoral heterogeneity. For example, the fact that the cost of materials, machinery and works accounts for a large percentage (e.g. up to ~70%–80%) of the overall construction project cost [10] underscores the critical role of SCM in the construction sector. Given the material- and labour-intensive nature of the industry, effective SCM is essential for optimising resource utilisation, controlling costs and ensuring timely project delivery.

Contrary to the traditional stereotype that construction predominantly relies on local materials and labour, contemporary construction practices Unlike the stereotype that construction often uses local materials and labour, in modern society, construction clients tend to source these key resources from various geographically dispersed places [11]. The renaissance of modular integrated construction (MIC) has further altered the global landscape of construction supply chain management (CSCM), in which managers must comply with heightened requirements related to carbon emissions [12], fair trade [13], environmental, social and corporate governance [14].

Inefficient communication between multiple parties leads to the sharing of outdated and inaccurate information. There may be constant design changes and errors in drawings, prolonged processes of error detection and correction, and an inability to meet quality specifications in manufacturing and delivering a specific product [15]. Establishing trust in this context poses a significant challenge, potentially leading to numerous well-documented issues throughout the global construction supply chain (CSC) such as delays, cost overruns and quality defects [16]. Technologies play a key role in addressing challenges in the general economy and the construction industry specifically. Computers, the Internet and advancements in information and communication technologies have helped to enhance information flow and communication and rebuild trust, improving global SCM efficiency. Nevertheless, new challenges to the global SCM constantly emerge such as the COVID-19 pandemic [17], the Russia–Ukraine War [18] or the changing geopolitics [19]. In this context, Industry 4.0 technology seems to provide an unprecedented opportunity to enhance CSCM. While there is no consensus on the definition of Industry 4.0 and derivative terms (e.g. Fourth Industrial Revolution, 4IR), it generally refers to the evolving paradigms of industries, businesses and social patterns driven by technologies such as the Internet of Things (IoTs), cloud computing, smart sensors, advanced robotics, big data analytics and 3D printing. In particular, blockchain technology is considered a game changer in streamlining information flow, facilitating transactions and building trust within CSCM [20]. However, this potential has not been adequately documented in the literature or applied in practice.

This chapter aims to investigate the potential of the potential of blockchain technology in enhancing CSCM focussing on challenges related to improved information sharing, transparency and traceability, and better security and risk management, especially in cross-border supply chain transactions. First, the chapter explores the benefits of blockchain in the CSCM domain. Second, it presents a blockchain-based CSCM system demonstrating its design, architecture and workflows. Third, the chapter illustrates the implementation of the system through a real-life case study of cross-border CSCM in offsite modular construction, focusing on the high-traffic Hong Kong–Mainland China corridor. Fourth, based on the understandings from the case study, the chapter discusses the prospects and challenges of blockchain in CSCM. Finally, the chapter summarises key takeaways and outlines directions for future research and development.

9.2 Benefits of using blockchain in construction supply chain management

Blockchain is an innovative technology that facilitates peer-to-peer value transactions within a secure and trustworthy environment [21]. It does so by utilising a decentralised and immutable digital ledger, composed of interconnected data blocks containing transaction information [22]. Each block's data is encrypted using a hash function, resulting in unique combinations of letters and numbers known as hash values which establish the links between consecutive blocks in the blockchain. This key characteristic prevents tampering with information, as any alteration would immediately change the corresponding hash value. By ensuring a tamper-proof framework, the blockchain guarantees the authenticity and immutability of transaction data [20]. All transaction data within the blockchain are replicated and shared across the network of participants, allowing them to actively monitor and verify the values exchanged in real time [23]. This combination of immutability and transparency fosters trust in the peer-to-peer exchange, eliminating the need for intermediaries. A detailed coverage of the characteristics of blockchain and smart contracts can be found in Chapter 1 of this book.

Blockchain technology can revolutionise CSCM information systems by enhancing trust, transparency, auditability, accountability, security, robustness, resilience, performance and equality [21]. Its unique features can help to establish a reliable and efficient digital infrastructure, resolving key CSCM problems and improving information sharing and stakeholder collaboration, traceability and transparency, security and risk management, and sustainability and operational performance. Each of these benefits is discussed in one of the subsequent subsections.

9.2.1 Enhanced information sharing and stakeholder collaboration

Current approaches of CSCM rely on fragmented and sequential methods of information sharing. There is a pervasive lack of trust among stakeholders, compounded by the absence of a single source of truth for information [24]. To address these challenges, researchers have proposed the utilisation of an IT-based communication platform specifically blockchain in conjunction with other technologies such as building information modelling (BIM), geographic information systems (GIS) and SCM systems (e.g. Enterprise Resource Planning software) as shown in Figure 9.1. BIM enables the storage and management of construction data, including construction processes and material types, in a 3D model. GIS allows

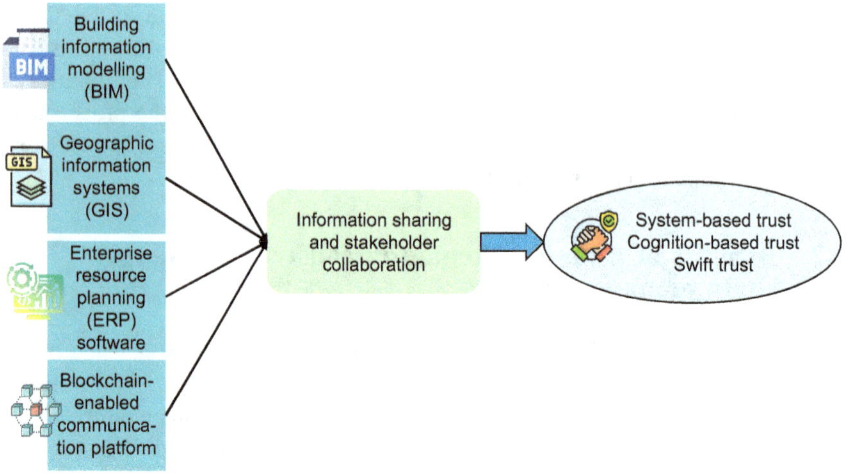

Figure 9.1 Solutions presented in literature for enhancing information sharing and stakeholder collaboration in CSCM

suppliers to gather logistical data, including delivery schedules, while SCM software enables the analysis of complex production scenarios and synchronised simulations with geographic information and demand and inventory data. If the data from these technologies can be shared on a blockchain-enabled communication platform, it becomes tamper-proof and traceable.

The advantages of such information sharing to enhance collaboration among stakeholders within the supply chain. The literature (e.g. [24–26]) consistently highlights how blockchain technology fosters trust in data authenticity, integrity and traceability, thereby promoting collaboration among stakeholders. This trust encompasses system-based and cognition-based trust [24] as well as swift trust [25]. Notably, Qian [24] concluded that trust in supply chains encourages collaboration by minimising opportunistic behaviours, while Longo *et al.* [26] demonstrate that trust-based collaboration positively impacts operational performance within the supply chain. Other studies indicate that blockchain simplifies supply chain processes and eliminates intermediaries, facilitating collaboration [27].

Several studies have explored the use of blockchain for transparent and efficient information sharing. Betti *et al.* [28] and Zhang *et al.* [29] demonstrate that blockchain improves information sharing in supply chains, leading to faster and more accurate transactions. Zheng *et al.* [30] found that adopting blockchain technology increased overall profits in the supply chain by enabling stakeholders to make optimal decisions on order quantity. These findings highlight the potential of blockchain as a communication platform to enhance operational performance within the CSC [31].

9.2.2 Improved transparency and traceability

Transparency refers to the availability of high-level information about the entire supply chain, such as product components, facility locations and supplier names [32].

This includes to the extent to which all stakeholders have access to critical and accurate information regarding various processes, transactions and decisions within the supply chain linked to sourcing, production, transportation and distribution activities. Data is central to establishing transparency and trust in supply chain, and blockchain technology has been utilised as a key enabler in this domain. It enables the exchange of transactional data among supply chain partners, ensures data immutability and promotes the sharing of a single version of the transactional database. Transparency requires visibility and disclosure, which necessitate data collection from supply chain suppliers and dissemination to internal and external stakeholders. Traceability involves the ability to track specific elements in the supply chain, establish communication criteria, collect accurate records, store traceability data and exchange data on a platform [32]. Through improved traceability and transparency, blockchain can improve visibility, facilitate provenance tracking, eliminate trade disputes and fraud, decrease error rates and facilitate product recall in a time-efficient way [33].

Blockchain and related technologies have been applied in different industries to improve traceability. For instance, Walmart reduced the time required to trace mangoes from seven days to just 2 s [34]. The utilisation of sensors, Raspberry Pi units, RFID/NFC tags and source-verified goods platforms further enhances the effectiveness and reliability of traceability systems, as exemplified in the collaboration between Maersk and IBM to increase information transparency and sharing in cross-border supply chains [35]. Dujak and Sajter [36] show that information sharing via the blockchain optimises material flow and fosters trust among stakeholders, and Xu *et al.* [37] have developed a construction material provenance tracking system using the IoT and blockchain. These examples demonstrate the potential of blockchain technology in enhancing transparency and traceability across various industries and supply chains, including CSC.

9.2.3 Enhanced security and risk management

The data used to assess suppliers, including information on product quality, logistics and transportation, manufacturing status and operational capabilities, is prone to fraud and tampering and suppliers may exaggerate their performance and capabilities to secure more job orders [38]. To mitigate fraudulent activities and claims, several studies (e.g. [21, 33, 38]) have proposed the utilisation of blockchain technology. Implementing a shared and immutable ledger with predefined rules, can greatly reduce or even eliminate reliance on internal systems and business processes for conducting audits, eliminating the need for intermediaries [21]. In addition, participants can remain anonymous or provide proof of their identity; a feature that nurtures trust and a sense of security among participants. All stakeholders involved can access the information stored on the blockchain and no single entity has the power to modify it unilaterally, thereby mitigating the risk of fraud.

Smart contracts have been widely explored for financial security by reducing payment fraud and delay risks. Smart contracts are automated contracts or sets of rules where the terms of the agreement between a buyer and a seller are written directly into lines of code. These contracts exist on a distributed and decentralised blockchain network. CSC stakeholders typically come together temporarily to deliver one-off projects. The introduction of smart contracts empowers them to establish digital rules and contracts using the data stored within the blockchain network [39]. The utilisation of smart contracts addresses the lack of governance or a system that provides appropriate rewards and penalties to facilitate information sharing and promote collaboration. In the CSC, stakeholders often lack trust in one another, and this leads to isolated pursuit of individual profits and limited collaboration within the supply chain. Contractors and suppliers, being independent entities in construction projects, may not feel obligated to engage in knowledge sharing and collaboration without proper incentives or penalties. In this context, owners can leverage smart contracts to implement a system of financial incentives and penalties for contractors and suppliers based on their operational performance. The smart contract can prescribe the required performance criteria as conditions to enforce rewards or punishments. Moreover, smart contract-powered solutions can significantly mitigate payment and related financial issues in the construction industry, including partial payments, non-payments, cost of finance, long payment cycle, retention and security of payments [39].

Smart contracts can also be employed within CSCs to regulate and enforce the sustainable performance of stakeholders. A notable example demonstrated by Manupati *et al.* [33] involves the use of smart contracts to minimise carbon emissions in the supply chain. By establishing upper and lower limits for emission rates within the terms and conditions of the smart contract, their proposed system ensures that when the emission rate exceeds the upper limit or falls below the lower limit, an automatic notification is issued. This enables managers to optimise control over emissions. Another work by Kouhizadeh and Sarkis [21] introduces a specific framework for a green supply chain, where all environment-related data is stored in the blockchain and evaluated based on the terms and conditions of the smart contract. Within this framework, suppliers are expected to reduce emissions or waste, as the acceptable amount is specified in the terms and conditions of the smart contract.

In the current SCM practice, goods typically flow in one direction. Enabling a reverse flow of products and transactions is a challenge when handling product faults or returns, and customers are often left with product faults and limited options for addressing any consequences. With IoT technologies to collect identification, trajectory and process data, blockchain can record and trace construction material provenance data along the CSC, capturing all the processes the materials undergo from their rawest state to the final product [37]. In the event of a quality issue arising in a reinforced concrete beam, for example, inspectors can examine the relevant external provenance data and analyse the potential causes behind the quality issue. Afterwards, inspectors can trace back the individuals and entities involved, by referencing the main blockchain for external provenance; furthermore, the responsible organisation can utilise their side blockchain to track the internal provenance, enabling them to pinpoint the specific procedure and individual accountable for the issue [37].

9.2.4 Improved sustainability and operational performance

The careful selection and evaluation of suppliers is crucial in enhancing CSC sustainability [33]. Prioritising suppliers and products that adhere to sustainable practices, organisations can reduce environmental impact, promote ethical labour standards and ensure resource efficiency throughout the supply chain. A dedicated blockchain-based data storage system for CSCM can help securely store and share supplier data and information related to sustainable construction materials. The immutability and traceability of data within the blockchain network contribute to improving trust in its authenticity and accuracy, and can empower stakeholders to evaluate and select reputable suppliers based on their track record of sustainable performance and to verify the sustainability performance of construction materials. Blockchain technology addresses a critical challenge in the industry by providing a tamper-proof and traceable platform for sharing data related to environmental sustainability performance within the supply chain. This capability enhances transparency and accountability, ultimately contributing to the overall sustainability of supply chain operations [21,38]. This has significant implications to green SCM in the construction sector by linking economic benefits derived from collaboration in emission reduction with the enhancement of sustainable processes within the CSC. Stakeholders can effectively manage energy consumption and waste in supply chains by closely monitoring environment-related data such as emissions, recycling processes and energy usage [33]. For instance, the terms and conditions of smart contracts can effectively regulate the distribution of economic benefits generated by emission reduction to suppliers based on their sustainability performance. With emission reduction data stored in the tamper-proof and traceable blockchain network, suppliers can trust that incentives or penalties will be fairly applied based on their efforts to reduce emissions during their operational processes [38]. By ensuring accountability and incentivising sustainable practices, blockchain is expected to drive the commitment among suppliers to enhancing sustainability within the CSC.

9.3 Implementing blockchain in construction supply chain management

Acknowledging the benefits of blockchain for CSCM, there are increasing efforts to explore the implementation of blockchain in CSCM. However, a comprehensive implementation, covering both the development and demonstration of the system, are still lacking in the literature. Using a pilot project, the chapter's authors have successfully implemented blockchain for CSCM in Hong Kong in a modular construction project. Based on this experience, the authors will summarise, in this section, the key considerations and steps of implementing blockchain in CSCM.

9.3.1 Key considerations for implementation

Before implementing blockchain in CSCM, it is essential to consider six key factors: stakeholder management, cost, human resource readiness, interoperability, scalability and legal compliance.

9.3.1.1 Stakeholder management

Consideration of stakeholder management is crucial to implementing blockchain in CSCM [20]. First, stakeholders who can initiate the use of a blockchain platform should be identified based on actual CSCM cases [40]. For example, where a governmental supervision unit (GSU) intends to use blockchain to regulate CSCM participants, the

GSU could initiate blockchain implementation [41]. Where a blockchain platform is built to track materials in a construction project, the project client might be the initiator [20]. Second, it is crucial to determine which stakeholders will participate in the blockchain network. Stakeholders in CSCM may include material suppliers, logistics providers, contractors, subcontractors, clients and GSUs [42]. It is important to collect specific requirements from the relevant stakeholders identified [43].

9.3.1.2 Human resource readiness

Human resource readiness needs to be assessed before implementing blockchain in CSCM. First, CSCM experts should be involved, as their in-depth domain knowledge allows them to identify and prioritise the CSCM challenges and barriers that blockchain needs to address [44]. Additionally, users should evaluate whether blockchain technology companies have sufficient and experienced consultants, system designers, developers and maintainers [45]. Where human resource readiness is ensured, users will understand blockchain properties and contextualise their potential uses in CSCM, while blockchain companies will be able to develop customised blockchain solutions.

9.3.1.3 Cost–benefit balance

Before implementation, the production-scale blockchain solution cost for an organisation in CSCM must be accurately forecasted [46]. The cost depends on various factors, such as specific user scenarios, complexity, type of blockchain, blockchain platform and other technology stacks [47]. Implementing a blockchain generally includes consulting, design, development, quality assurance, third-party services and maintenance costs [46]. Table 9.1 shows examples of cost percentages for implementing a blockchain solution. Given that a key advantage of blockchain is that it can facilitate collaboration among CSCM stakeholders, users should consider which stakeholders need to pay for the cost of blockchain implementation and in what proportion. Any cost forecasting model will rely on assumptions and benchmarking data until enterprise blockchain data is more readily available, applications are standardised and pricing becomes consistent [47].

Table 9.1 An example of cost percentages for implementing a blockchain solution (adapted from [46])

Cost item	Examples	Percentage
Consulting	Business process analysis, user requirement collection, general consultation	10%
Designing	System blueprint, user interface/experience design, low-fidelity designs with operation flow	15%
Development	Coding and testing	40%
Quality assurance	Issue detection and solving	10%
Deployment and third-party services	Deployment on cloud platforms, hosting, storage, notification system	5%
Maintenance cost	Maintenance, new features upgrading	20%

9.3.1.4 Interoperability

The interoperability of the blockchain solution should also be considered carefully [48], in two main ways. On the one hand, the developed blockchain solution should be able to communicate with other blockchains [49]. To meet the growing ecosystem demand for blockchain interoperability solutions, Chainlink is developing the cross-chain interoperability protocol [50], a new open-source standard for cross-chain communication that establishes universal connections between different blockchain networks through a single, simple interface. On the other hand, blockchain interoperability should support communication with existing systems or platforms in an organisation [51]. For example, Li *et al.* [52] point out that it should connect easily with the existing IoT-enabled BIM platform in CSCM.

9.3.1.5 Scalability

The scalability of blockchain is the ability of that platform to increase or decrease performance in response to changes in CSCM applications [53]. CSCM participants are often established and dispersed for different projects, and the human resources allocated to them constantly change [54]. For example, the construction of a two-tower student hostel at the University of Hong Kong involved about 20 suppliers in the supply chain [16], but the university's future construction projects may involve more or fewer suppliers. In addition, as the project progresses, there may be more and more large files, such as CSCM-related videos and BIM models. Thus, the developed blockchain solution should support an increasing or decreasing load of data and the number of peer nodes in the network.

9.3.1.6 Legal compliance

Legal compliance is critical for implementing blockchain [55]. Blockchain technology in the construction industry has not yet formed a mature legal environment with relevant regulations, laws, policies and standards [42]. Since regulatory uncertainty may make CSCM stakeholders hesitant to adopt blockchain, legal counsel should be sought to ensure that the developed blockchain solution complies with local laws [55]. The legal considerations surrounding blockchain adoption in the construction industry are discussed in depth in Chapter 12 of this book.

Only after thorough consideration of these factors can the project stakeholders proceed to the implementation of the blockchain system. While we are in the nascent stages of addressing various challenges, blockchain technology presents a promising opportunity for all stakeholders in the CSC.

9.3.2 Steps for successful implementation with a real-life case study

9.3.2.1 Case descriptions

This section illustrates the steps for successful blockchain implementation using a real case study in Hong Kong. Over the years, the Hong Kong Special Administrative Region of the People's Republic of China has moved its MIC factories to mainland China's Pearl River Delta (PRD) which offers more space and lower-cost labour and materials [56].

A key challenge in the MIC supply chain is low transparency. Since all modules are produced in off-site factories, supply chain processes are expected to be transparent to all involved parties so that there is trust in product quality. Another major challenge is the immutability of data. In contemporary supply chain systems, even digital records, in addition to traditional paper-based documentation, are susceptible to intentional or unintentional data modifications. Additionally, low traceability of data can obscure product accountability should issues arise. Consequently, project stakeholders are actively seeking innovative solutions to mitigate these challenges.

The case in this study is a MIC student residence project with two 17-floor tower buildings atop a three-story podium structure, as shown in Figure 9.2(a). More than 900 modules (e.g. rooms and toilets) are manufactured off-site in the PRD, and inspected and qualified modular products are then transported to Hong Kong for onsite installation, as presented in Figure 9.2(b). According to Buildings

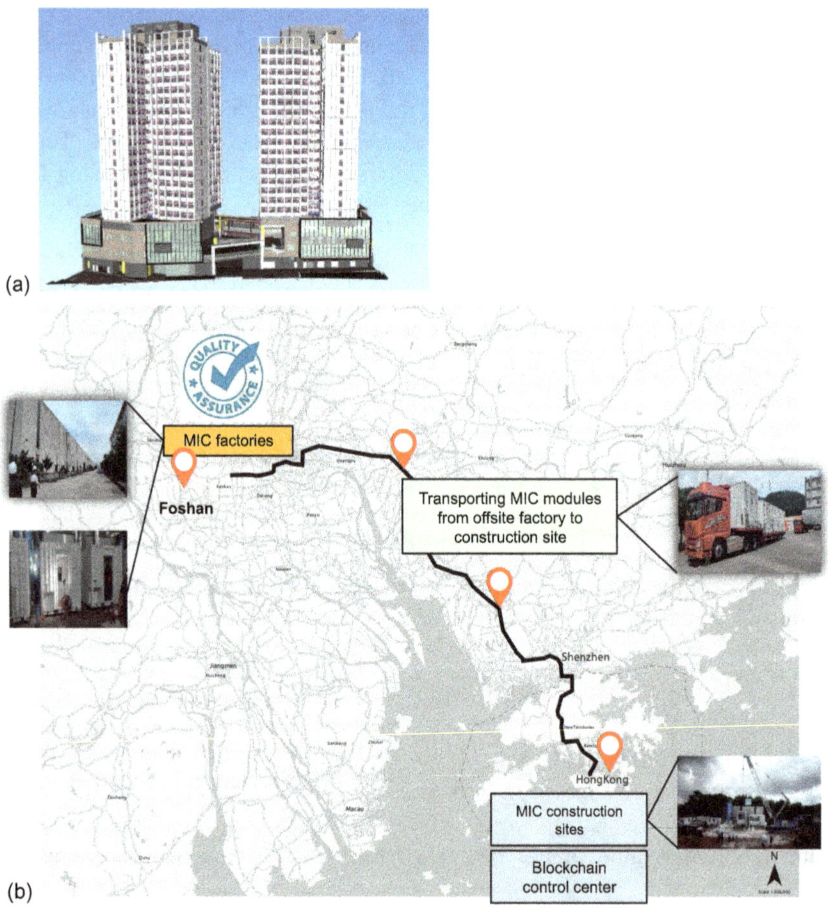

Figure 9.2 Case project overview: (a) prefabricated modules in the case project and (b) the supply chain of the case project

Department quality regulations, these modules must be quality-assured in the PRD. Blockchain is proposed to meet this requirement due to its transparency, immutability and traceability. The distributed database can enhance the transparency of the supply chain processes by sharing the relevant data with network participants. The cryptographic algorithm of the blockchain can safeguard the recorded data from manipulation, and each data record is timestamped with the blockchain, ensuring data traceability when an issue occurs. With these advantages understood, the research team and case project stakeholders have engaged with each other to explore blockchain solutions.

Prior to the implementation of the blockchain system, project stakeholders must evaluate the six aforementioned factors. Following this assessment, the implementation should be executed in accordance with the five primary steps depicted in Figure 9.3.

Step 1 – Feasibility of the adoption of the blockchain system

In the case project, the research team and project client identified a business case for the blockchain solution, the potential return on investment (ROI), the stakeholders (potential users) and the key performance indicators (KPIs) that they were seeking to improve and associated challenges. Several focus group meetings were held to discuss the application scenario. As a result of these meetings, the project stakeholders and the research team agreed to apply blockchain to oversee the off-site MIC production processes by endorsing and securing the records of each production step of the modules. Regarding ROI, this his concept extends beyond mere financial benefits. In this context, ROI is defined as achieving reduced complexity, thereby decreasing risk and enhancing collaboration through increased trust. The project stakeholders participating in the MIC production and supervision processes were identified as the potential users (e.g. the manufacturer, client representative (CL), authorised personnel (AP), building service inspector (BSI), registered structural engineer (RSE) and main contractor). KPIs defined by the project client with the research team included 100% checking in the production line, 1–2% rectification work saving, reduced resources for supervision and zero data records manipulation. Potential technical (e.g. scalability, interoperability), organisational (e.g. resistance to change) and environmental (e.g. lack of collaborative culture) challenges of blockchain implementation in the case project were discussed during the group meetings. During several research workshops, stakeholders were asked to learn as much as possible about blockchain technology and the proposed application scenario.

Step 2 – Workflow analyses

In developing the business case, all project stakeholders are required to map out the current workflow and illustrate where and how a blockchain solution can advance them. Here, the research team and project stakeholders conducted a business process analysis for off-site MIC production. They defined 40+ work steps that require the MIC manufacturer to upload production data (e.g. photos, videos and comments) to the blockchain for data transparency, immutability and traceability purposes, as shown in Figure 9.4. Next, project stakeholders identified governance rules for the blockchain system, and the project client was given the right to control

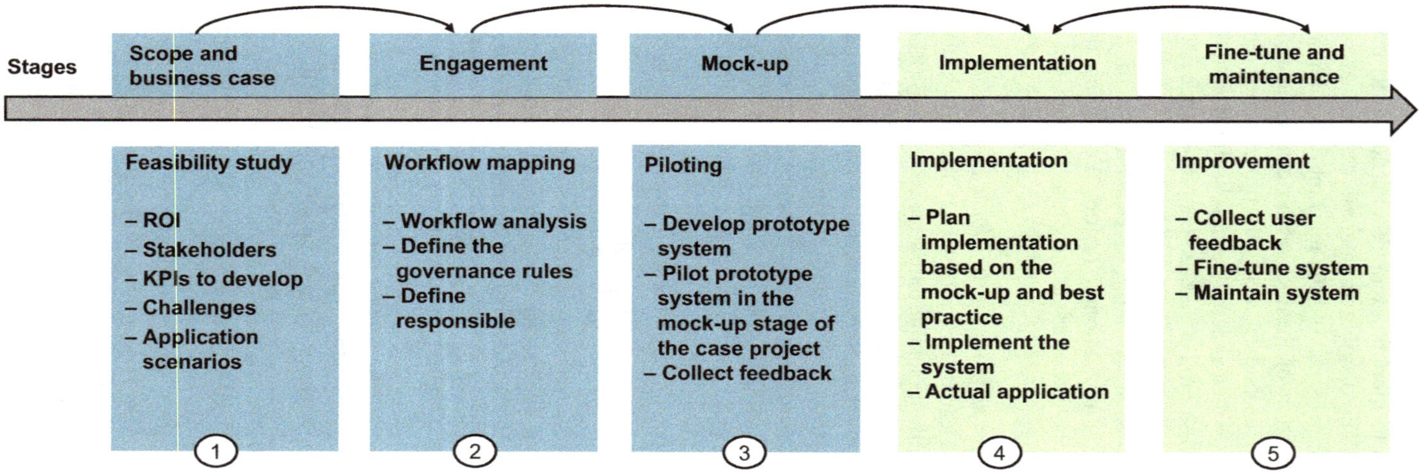

Figure 9.3 Steps for the implementation of a blockchain solution

Item		Description	Working Day		Inspection/ Test Method	Frequency	Responsibility: Paul Y.	RSE	AP	BSI	Client's REP	N/C Follow-up
	structure	Steel material on factory 鋼材到工廠			Rebar and corner cast sampling for test	As per BD requiement	H	C				Re-test
1		Board, profile pretreatment and punching 板、型材料處理及沖壓	Day	1	Material Grade and Size Spot Check	Routine	H					Re-inspection
			Day	2	Visual Inspection	Routine	H					Re-inspection
			Day	3	Visual Inspection	Routine	H					Re-inspection
2		2D Panel Assembly (Butt weld and Fillet Weld included) 二維部件組裝	Day	4	Visual Inspection	Routine	C					Re-inspection
			Day	5	Visual Inspection	Routine	C					Re-inspection
			Day	6	Visual Inspection	Routine	C					
3		2D Panel Weld Test 二維部件驗焊	Day	7	Visual Inspection (Butt Weld and Fillet Weld)	100%	H	C				Re-test
					Butt Weld (MPI & U/S)	100%	H	C				Re-test
					Fillet Weld (MPI & U/S)	10%	H	C				Re-test
4		2D Panel Hot Dip Galvanizing Treament 二維部件熱鍍鋅處理	Day	8								
			Day	9	Hot Dip Galvanized Thickness (RSE spot check)	Routine	H	C				Re-inspection
5		3D assembly including (welding work for bondek, Shear Stud, bracket, wall & ceiling sheet) 總裝焊（鋼板、栓釘、子結構及外掛板結構件）	Day	10	Cladding Sub Frame Alignment Check		H	C				Re-inspection
			Day	11	Dimension Check (AP spot check)	Routine	H	C				Re-inspection
			Day	12	Visual Inspection (Butt Weld and Fillet Weld)	100%	H	C				Re-test
					Butt Weld (MPI & U/S)	100%	H	C				Re-test
					Fillet Weld (MPI & U/S)	10%	H	C				Re-test
					Shear Stud (bend and weld)	5%	H	C				Re-test
			Day	13	Water Test for chassis enclosure (AP spot check)	100%	H		C		H	Re-inspection
6		Touch-up galvanized paint (Zinc Rich Primer) 用富鋅油漆修補熱鍍鋅層	Day	14	Visual Inspection	Routine	H	C				Re-inspection
7		Apply fire paint at strucutral member 結構塗防火油漆	Day	14								
			Day	15								
			Day	16								
			Day	17								
			Day	18								
			Day	19								
			Day	20	Thickness checking (AP RSE spot check)	Routine	H	C				Re-inspection
8		Rebar Fixing and Pouring Concrete 鋼筋固定及澆灌混凝土	Day	21	Rebar fixing Visual Inspection, Cleaning	100%	H	C				Re-inspection
					Concrete sampling for cube test	As per BD requirement	H	C				Re-test
					Concrete slump test	Every Truck	H	C				Re-test
9		Concrete Curing 混凝土養護	Day	22	Visual Inspection	Routine						Re-inspection
			Day	23	Visual Inspection	Routine	C	C				Re-inspection
10	Door/window	Door and window frame installation 門安裝、窗框安裝	Day	24	Dimension Check of Window Size & Setting Out	Routine	C					Re-inspection
11		Water Spray Test 窗滲漏測試	Day	24	Water Spray Test and Electrical continuity Test		H		C		H	Re-inspection
12	wall	Ceiling stud, rockwool in-fill and fire board installation at ceiling surface 天花防火板鋪設	Day	25		Routine	H		C		H	Re-inspection
			Day	26	Visual Inspection (AP-Spot Check)							
13		Fire board, Wall stud installation 結構第一層防火板鋪設(板及龍骨)	Day	27		Routine	H		C		H	Re-inspection
			Day	28	Visual Inspection (AP-Spot Check)							
14		Rockwool in-fill and 2 layer of fire board installation at wall surface 牆面第二層防火板鋪設	Day	29	Visual Inspection (AP-Spot Check)	Routine	H		C		H	Re-inspection
15		Installation of cement board wall partition at toilet 洗手間內隔牆水泥板安裝	Day	30	Visual Inspection and Internal Dimension Checking	Routine	H				H	Re-inspection

Figure 9.4 Workflow mapping

user registration. Also, for each defined work step, before the manufacturer proceeds to the next production step, there is a set of defined parties that need to verify the quality of the uploaded production data through the consensus mechanism of the blockchain. For example, after the manufacturer completes step 1 and uploads the data, the main contractor, RSE and CL verify the corresponding production data on the blockchain system; in step 2, the main contractor, AP and BSI verify the corresponding production data after the manufacturer completes step 2 and uploads the data. The user workflow diagram for quality chain in the CSCM is presented in Figure 9.5.

Step 3 – Developing and piloting the prototype system

A blockchain prototype system should be developed based on the mapped workflow and defined governance rules. The diagram in Figure 9.6 illustrates the system architecture of the proposed blockchain prototype system. It consists of three key layers. Starting from the bottom, these layers include Infrastructure as a Service

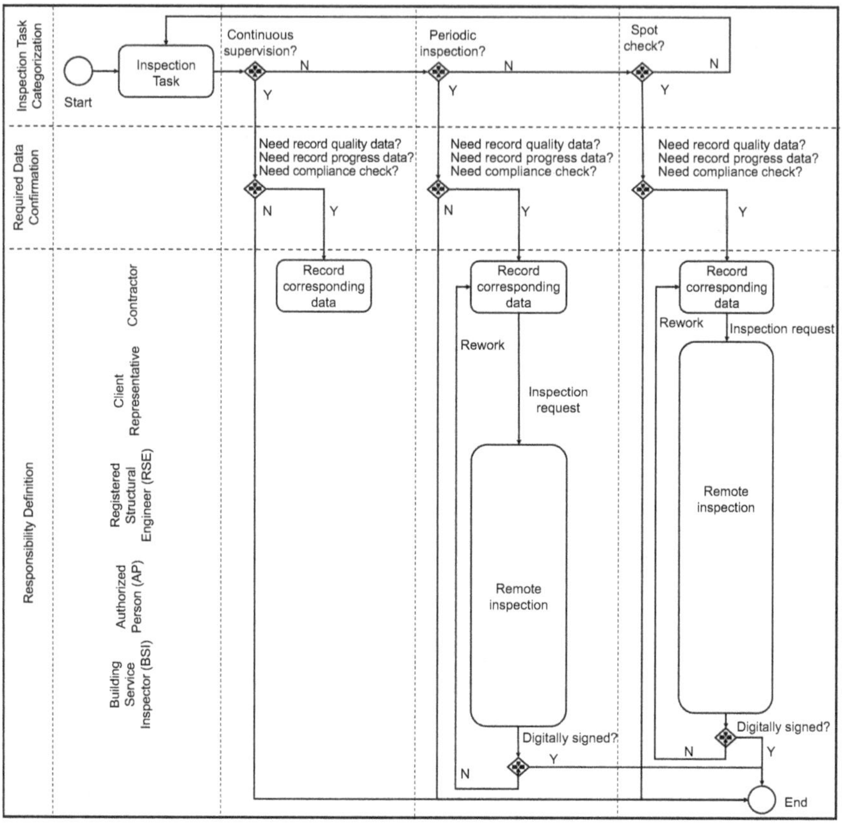

Figure 9.5 The user workflow diagram (adapted from [56])

(IaaS) of the system, which incorporates a mobile inspection application (A1), a remote video application (A2) and web-based dashboards (A3). A1 facilitates tasks such as user registration for all inspection participants, task generation by the contractor based on project plans and building regulations, and confirmation of tasks by other parties. After a construction phase is finished, the contractor can use A1 to input construction data, along with comments and photos. A2 allows CL, RSEs, AP and BSI to conduct remote real-time video supervision when necessary, while the contractor's inspectors recheck specific items at each hold point. Once the smart contract acknowledges receipt of data from the data interface, it transmits the information to the blockchain network for validation. Furthermore, A3 permits inspection participants to create inspection reports and track past inspection records.

The BaaS component of the system includes a core functional module responsible for integrating information related to inspection transactions. Within the BaaS framework, Hyperledger Fabric is utilised and encompasses various services such as a blockchain-based network, smart contracts, cryptography and ledger functionalities. The network is comprised of four peer nodes representing different

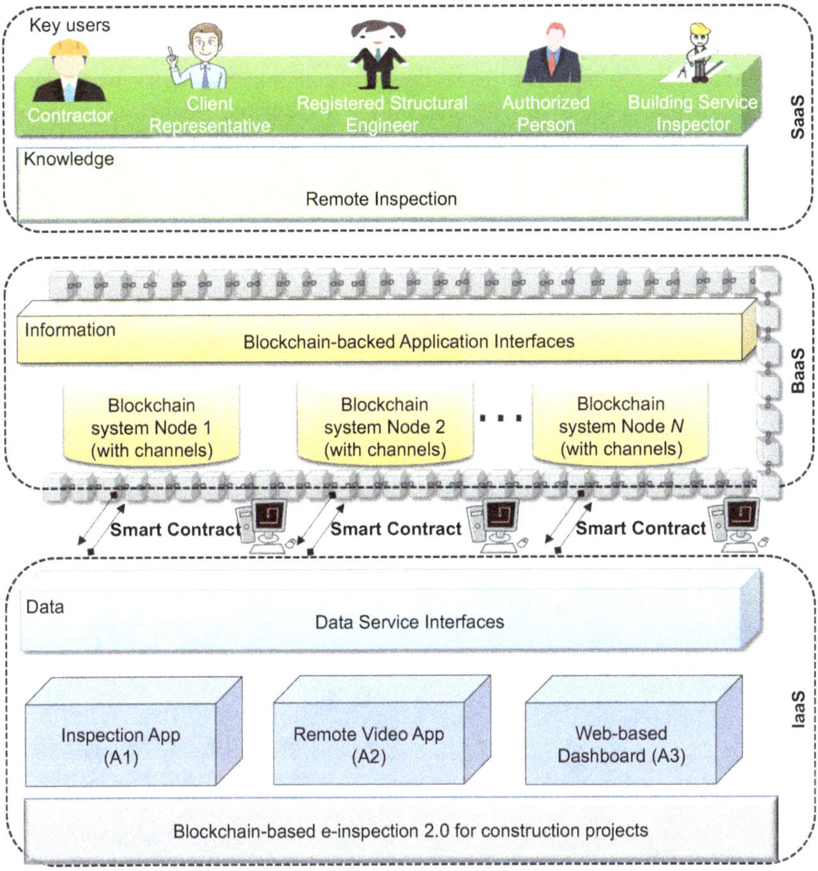

Figure 9.6 System architecture of the blockchain-based system (adapted from [56])

organisations involved in the inspection process, including the contractor, RSE, AP and BSI. Smart contracts play a crucial role in transferring inspection data from the data service interface to the blockchain network. Cryptography plays a key role with the main subcomponent being the hash algorithm (SHA-256), which ensures the security and immutability of the inspection transactions recorded. Additionally, inspection transactions are endorsed, digitally signed, timestamped and organised into blocks. Each block includes the hash of the previous block, ensuring that any alterations to a block would also affect all subsequent blocks. Finally, these blocks are stored in the ledgers of peer nodes, allowing for transparent sharing of information among the participants involved in the inspection process.

The SaaS layer of the system incorporates a consensus mechanism. According to the user workflow in Figure 9.5, the system employs a customised consensus mechanism known as 'proof of stakeholders'. This means that relevant stakeholders are required to approve and digitally sign the inspection transactions. For instance, during the 3D assembly checkpoint, the CL, RSE, AP and BSI must remotely endorse and digitally sign the inspection transactions. Similarly, in the case of lighting installations, the RSE, AP and BSI are responsible for digitally signing the inspection transactions. The system leverages Hyperledger Fabric as the foundational blockchain infrastructure in the BaaS layer, enabling the implementation of a pluggable consensus mechanism to accommodate various applications.

All network users should then test the application for an appropriate duration on a defined small scale, where performance can be easily determined. In the case project, the research team first transformed the 40+ production work steps into a mobile application named e-InStar (see Figure 9.7(a)). Then the research team developed the

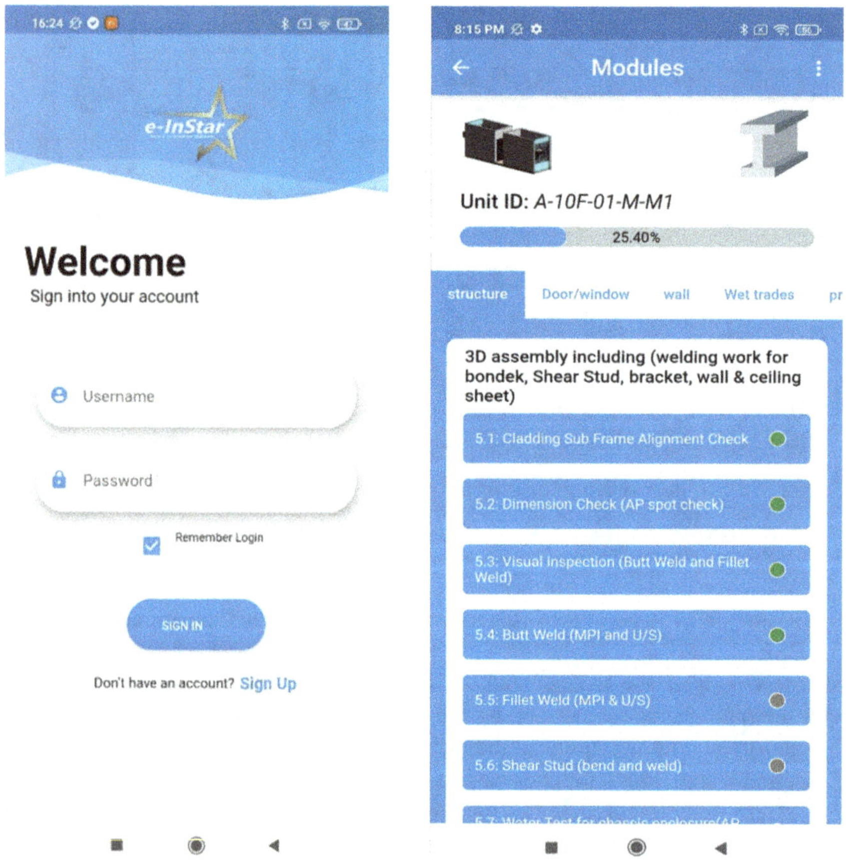

(a)

Figure 9.7 The developed blockchain-based prototype system: (a) e-InStar mobile application, (b) configured blockchain network, and (c) blockchain dashboard

```
# ---------------------------------------------------------------------------
# "PeerOrgs" - Definition of organizations managing peer nodes
# ---------------------------------------------------------------------------
PeerOrgs:
  - Name: Contractor
    Domain: Contractor.com
    EnableNodeOUs: true

    Template:
      Count:
      SANS:
        - localhost
    Users:
      Count:
  - Name: Registered Structural Engineer
    Domain: RSE.com
    EnableNodeOUs: true

    Template:
      Count:
      SANS:
        - localhost
    Users:
      Count:
  - Name: Authorized person
    Domain: AP.com
    EnableNodeOUs: true

    Template:
      Count:
      SANS:
        - localhost
    Users:
      Count:
  - Name: Building Service Inspector
    Domain: BSI.com
    EnableNodeOUs: true

    Template:
      Count:
      SANS:
        - localhost
    Users:
      Count:
```

(b)

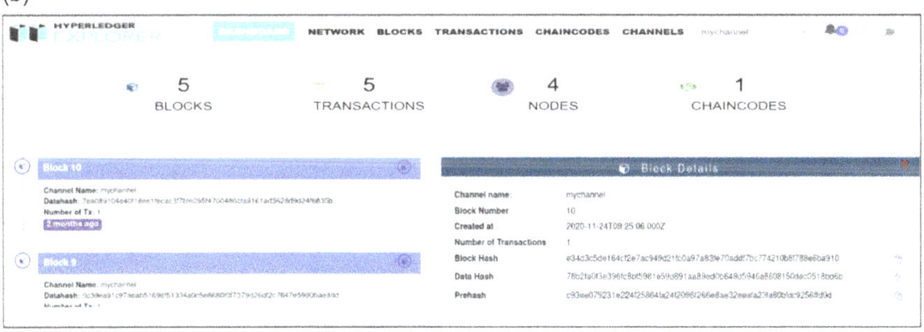

(c)

Figure 9.7 (Continued)

blockchain prototype system by configuring a blockchain network with defined stakeholders (see Figure 9.7(b)). A dashboard was also adopted enabling users to browse production data uploaded to the blockchain (see Figure 9.7(c)). The prototype system was piloted during the mock-up production stage of modules. So, the MIC manufacturer uploaded the production data (e.g. quality data) of eight modules to the blockchain system for endorsement via the e-InStar, and defined parties verified the production data via the consensus mechanism of the blockchain through the dashboard. User feedback was collected for adjustment and improvement in the next step.

Step 4 – Implementing the blockchain system

According to the implementation plan developed by the research team and stakeholders for the case project, the fully developed blockchain system was applied to all 952 modules during the mass production stage of MIC. The evaluation results of the system, included in Table 9.2, has helped the research team to improve the

Table 9.2 Evaluation of the proposed system (adapted from [56])

Interviewee	Response	Category
Question 1: What are the advantages of the prototype system?		Feedback categories
CL1	Enables a system for verifiable quality inspection The system improves the traceability of inspection records The system allows us to conduct quality inspections remotely during the COVID-19 pandemic at a safe distance	1 2 6
CL2	Recorded data is difficult to be manipulated due to the blockchain Remote inspections can be performed in a trustable manner	3 6
CL3	Only authorised parities can use the system Recorded data can be encrypted and only available for AP The system allows collaborative remote inspection	1 5 6
MC1	Getting quality inspection data to be timely shared among all parties will increase the feasibility of remote inspections during the pandemic	4, 6
MC2	When recorded quality inspection data is traceable and immutable, the remote inspection could happen	2, 3, 6
RES	This system enables the BD to trust digital inspections because all recorded data is traceable and immutable The system allows inspection records duly and remotely signed by all responsible parties at hold points	2, 3 6
AP	Would minimise data tampering and allow digital and remote inspections	3, 6
BSI	This system enables the BD to trust digital and remote inspections because all data is traceable and immutable recorded in the blockchain	2, 3, 6
Question 2: What are the disadvantages of the prototype system?		
CL1	The system may incur additional development and maintenance costs for all parties	7
CL2	Without a mature legal infrastructure and policies, I think the system cannot be used in current modular construction projects	8
CL3	If there are loopholes in the smart contract, the security of the blockchain network cannot be guaranteed	9
MC1	Compared with the existing process, the system takes extra time for relevant parties to endorse the information through digital signatures	10

(Continues)

Table 9.2 (*Continued*)

Interviewee	Response	Category
MC2	I think we have to bear the cost of the system	7
RES	The system will bring additional training costs to all parties	7
AP	Due to the unstable network, the video sometimes lags behind	11
BSI	If the cost is too high, I don't think the participants will accept it	7
Question 3: Would you use the prototype system? Why/Why not?		
CL1	Yes, as it could improve the verifiability and traceability of quality data and provide accessibility for remote inspections	Yes
CL2	I am not sure, maybe after the policies for the system are ready	Conditionally
CL3	The system allows me to conduct inspection at a safe distance, and the quality record is verifiable	Yes
MC1	No. The process is time-consuming	No
MC2	If the cost is known, I would use the system	Conditionally
RES	Yes, it may convince all participants due to the traceability and immutability of blockchains	Yes
AP	Definitely, yes. It secures the data and enables digital inspection.	Yes
BSI	All stakeholders can trust the remote inspection due to the introduction of the blockchain technology	Yes
Question 4: Do you have other recommendations to improve the prototype system?		
CL1	The system can be linked with BIM to visualise the inspection progress	12
CL2	I recommend testing the system in more projects to ensure its scalability	13
CL3	The total cost of the system should be determined before its actual implementation	7
MC1	The system should be able to send text messages to participants to remind them that they can proceed to the next inspection	14
MC2	I suggest a rough cost estimate for the system	7
RES	The system should be tested in different projects to determine cost.	7
AP	The system should be tested by a professional security audit company	15
BSI	I do not have any suggestions	–

Note: CL=Client, MC=Main contractor, RSE=Registered structural engineer, AP=Authorised personnel, BSI= Building service inspector.

1. Data verifiability, 2. Data traceability, 3. Data immutability, 4. Data transparency, 5. Data privacy, 6. Trust, 7. Additional cost, 8. Lack of legal infrastructure, 9. Security risk of smart contracts, 10. Extra time, 11. Unstable network, 12. Integration with BIM, 13. Scalability test, 14. Proactive reminders, 15. Professional security audit.

system. Participants (i.e. the contractor, CL, RSE, AP, BSI who used the system in the case project) were required to respond to the four interview questions outlined in Table 9.2 following their utilisation of the prototype system during the evaluation phase. The feedback received is summarised in Table 9.2. In relation to Question 1, interviewees noted that the proposed solution could address stakeholders' concerns, with strong support from the RSE, AP, BSI interviewed. The system is utilised daily in the project; however, in response to Question 2, interviewees highlighted various drawbacks, including the absence of cost evaluation, the inadequate legal framework and policy backing, potential security risks associated with smart contracts, increased time consumption and unreliable network connections. These issues are significant and should be resolved in future iterations. During the evaluation phase, participants are granted free access to the system, hence cost concerns are not applicable at this stage. The use of the blockchain system necessitates contractors to upload inspection videos and photos to a server in Hong Kong in real time to prevent tampering, potentially causing resistance from clients due to the lack of legal infrastructure and policy support. Previously, users could print out and submit photos and videos to the Buildings Department, leading to notable resistance, especially following network disruptions during the upload process.

When addressing Question 3, five interviewees indicated their willingness to use the system unconditionally. Two mentioned they would use it with certain conditions, while one rejected its use due to concerns about increased time compared to the traditional paper-based workflow. The main contractor expressed a neutral stance, stating they would adopt the system if it is accepted by the clients and Buildings Department. It is advisable to prioritise the responses of the client, RSE, AP and BSI in evaluating the system's acceptance. Regarding Question 4, interviewees suggested evaluating system costs, integrating BIM, conducting scalability tests, incorporating task reminder features and auditing system security. These recommendations present valuable avenues for further research and development.

After considering user feedback, the system was improved by adjusting the design layout of e-InStar and developing a new blockchain dashboard (see Figure 9.8(a) and (b)). Finally, the fully developed blockchain system was applied in mass production for full-scale use by the stakeholders, as shown in Figure 9.8(c) and (d).

Step 5 – Fine-tuning and maintaining the system

The final step is to continually fine-tune and maintain the system based on user feedback. This iterative process ensures that the system remains responsive to user needs, improves its functionality over time and adapts to evolving requirements. In the case project, the research team maintained both the front and back ends of the system during the application and engaged with the stakeholders to adjust the system continuously based on their needs.

The implementation of a blockchain system necessitates both internal and external collaboration. The successful adoption of blockchain technology goes beyond simply implementing new software; it requires embracing advanced business technologies and driving process innovation. For blockchain to deliver its full potential, the organisational environment must be conducive to change, supporting collaboration, innovation and technological integration. For example, organisational structures and the skills of construction practitioners will need to change to unlock the full potential of new blockchain-based supply chain business models. In this study, the developed blockchain system was used as a shadow system operating in parallel to the existing production system. This approach was adopted primarily due to the absence of a mature policy environment encompassing relevant regulations, standards, codes and laws for blockchain technology.

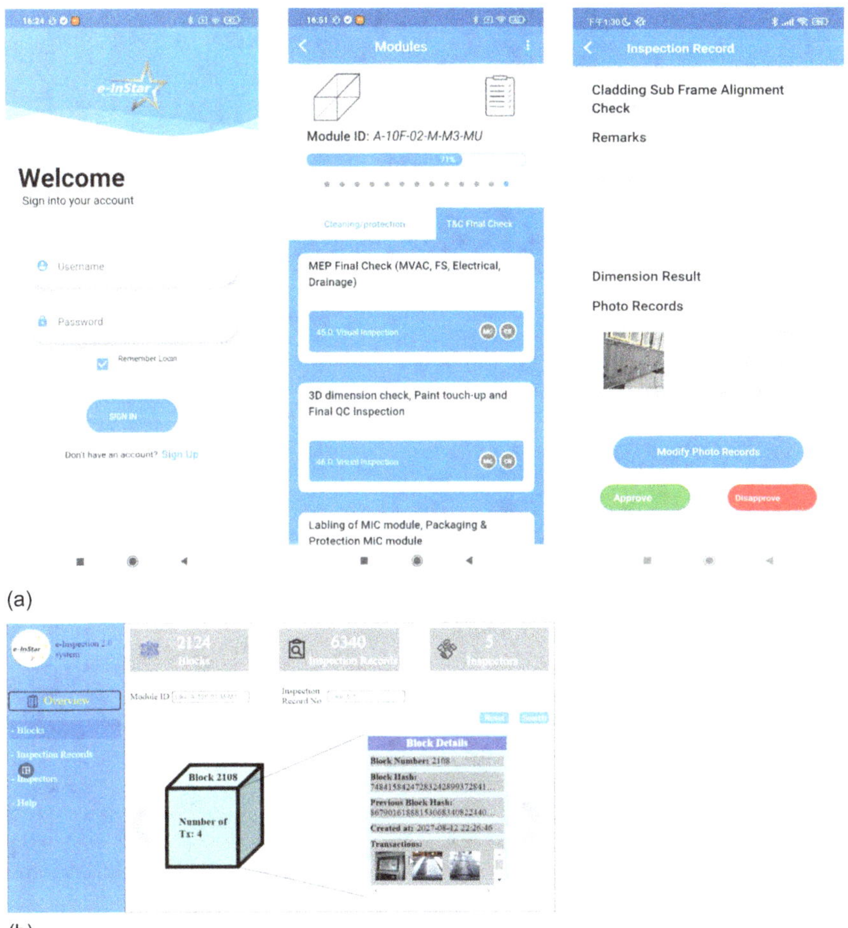

Figure 9.8 Fully developed blockchain system: (a) improved e-InStar mobile application, (b) newly developed blockchain dashboard, (c) application Scenario A (production data upload in the MIC factory), and (d) application Scenario B (data endorsement in the office)

(c)

(d)

Figure 9.8 (Continued)

9.4 Prospects and challenges of blockchain in construction supply chain management

The CSC is inherently complex, involving multiple stakeholders, extensive documentation and a high degree of interdependence. Blockchain technology has emerged as a promising solution to address many of the inefficiencies and risks in this sector, offering prospects for improved transparency, traceability and operational efficiency. The CSC is inherently complex, involving multiple stakeholders, extensive documentation and a high degree of interdependence. Blockchain technology has emerged as a promising solution to address many of the inefficiencies and risks in this sector. However, its adoption is not without challenges. Understanding these prospects and challenges is crucial for assessing the viability of the implementation of blockchain in this CSCM.

9.4.1 Prospects

As demonstrated in the case study, blockchain technology offers a range of distinctive and advanced features that have the potential to significantly improve the trust, transparency and security of information systems in the CSC sector. By providing a transparent and immutable ledger, stakeholders can ensure the integrity of data and transactions. The decentralised nature of blockchain allows stakeholders to verify and track the history of CSC information with ease, while the auditability and accountability provided by blockchain technology offer increased confidence in the accuracy and reliability of CSC information. The security measures inherent in blockchain, such as cryptographic encryption and consensus mechanisms, provide robust protection against unauthorised access and tampering, enhancing the overall security of CSC information systems. The most outstanding prospects of blockchain in CSCM are discussed in this subsection.

9.4.1.1 Integration with IoT and AI

A key prospect for blockchain applications in CSCM is its capability for integration with IoT and artificial intelligence (AI) to track and trace logistics. For a blockchain system to function seamlessly, a careful and systematic approach is required for collecting and storing data relating to supply chain processes, transaction flow, material flow and information flow. CSC stakeholders can benefit if they integrate blockchain technology into their recording material recording and data processing systems [15]. The integration of blockchain, BIM, GIS and other supply chain technologies can enhance the traceability and trustworthiness of CSC data. The resultant system also supports claims, auditing, forensic analysis and dispute resolution in CSCM. However, such integration also introduces a set of challenges. These include the complexity of system interoperability, high implementation costs, the need for substantial data standardisation and the demand for upskilling the workforce to effectively operate and manage the integrated technologies. Additionally, aligning these technologies within the existing regulatory and organisational frameworks can further complicate adoption and operationalisation [57].

Several studies have explored the integration of blockchain with BIM and the IoT. Lu *et al.* [16] demonstrated that blockchain can establish trust in distributed IoT networks through a sidechain, enhancing confidence in CSCM. Lu *et al.* [16] explored smart construction objects (which is an integrated and robust IoT modules with sensing, processing and communicating capacities to facilitate information exchange among various construction resources) as blockchain oracles, bridging the blockchain system and real-life physical project worlds, enhancing the transparency, traceability and immutability of construction data, and enabling collaboration and trust throughout the CSC. Li *et al.* [52] proposed a blockchain-enabled IoT-BIM platform for CSCM that can save storage costs to achieve a single point of truth in BIM and avoid a single point of failure for IoT networks with privacy and security-preserving mechanisms. Following up on these pilot studies, technical solutions to data fusion and update among IoT, blockchain and information systems await exploration with real implementation, as does the application of AI to the integrated secure information for decision-making.

9.4.1.2 Digital identity management

CSCM faces numerous challenges when it comes to digital identity management. The complex nature of the industry, involving multiple stakeholders, contractors, suppliers and subcontractors, makes it difficult to establish and maintain trust in digital identities. Blockchain technology offers a decentralised and tamper-resistant platform and therefore holds great promise in addressing these challenges and revolutionising digital identity management within the CSCM ecosystem.

Blockchain provides a decentralised and tamper-resistant ledger where identities can be securely stored, verified and managed. By leveraging blockchain in digital identity management, CSCM can be transformed into a more secure, transparent, efficient and user-centric process. The use of cryptographic techniques reduces the risk of identity fraud and unauthorised access to sensitive information, therefore ensuring the integrity and privacy of sensitive CSC data. This empowers multiple individuals and organisations along the supply chain to manage and authenticate their identities independently, streamlining the identity verification process across the supply chain. Overall, the integration of blockchain technology in digital identity management within the CSCM industry holds significant potential to enhance security, streamline processes and foster trust among stakeholders. Xu *et al.* [22] have explored role-based information access for occupational safety and health monitoring in the construction industry, but digital identity management in CSCM remains an area requiring further investigation.

9.4.1.3 Decentralised autonomous organisations

Decentralised autonomous organisations (DAOs) are built on the principles of decentralised decision-making and transparent governance, and blockchain serves as a foundational technology to enable these principles. In CSCM, blockchain and smart contracts have the potential to support the development of DAO in areas such as procurement and financing in through secure and transparent decentralised operations. This is through the automation of agreements execution when predefined conditions are met without the need for intermediaries, while simultaneously maintaining a tamper-proof record of all transactions and decisions. This is expected to reduce the reliance on intermediaries, decrease transaction costs and enhance trust among stakeholders. Furthermore, the immutable nature of blockchain ensures that all transactions and decisions are auditable, promoting accountability and fostering a more efficient and collaborative supply chain ecosystem. However, these anticipated benefits are yet to be fully investigated in literature.

The DAO literature in construction domain remains very limited. In public procurement, blockchain is able to provide a secure and transparent platform for managing transactions between DAOs and their suppliers [58]. The use of smart contracts on the blockchain can automate and enforce procurement agreements, ensuring that contractual terms are met and eliminating the need for intermediaries [58].

Additionally, blockchain-based CSCM within DAOs can improve transparency and traceability by recording every transaction and movement of construction materials, components and machinery on the blockchain [59]. This enhances the efficiency of CSC operations and helps maintain a clear audit trail, reducing the risk of fraud and improving overall accountability [59]. Nevertheless, more studies in the area of DAO for CSCM are needed. There have been some early studies in related areas such as the operation and maintenance of buildings [38], information sharing during construction site operation [60] application and self-owning houses equipped with their own IoT and treasury on the blockchain [61]. However, these are either conceptual or experimental studies confined to small scale or experimental prototypes. This area holds significant potential but necessitates extensive research in the context of CSCM.

9.4.2 Challenges

While the prospects of blockchain are widely recognised, there are also many challenges. Data privacy, cost, change management, regulation and the lack of practical applications to validate usability are all barriers that hinder CSCM from fully realising its prospects in real-life projects [62].

9.4.2.1 System security and data privacy

Although one of its main advantages is to enhance security, the blockchain-integrated CSCM system can face numerous security threats, with different types of attacks targeting specific components and layers within the blockchain infrastructure. These attacks can exploit vulnerabilities in transaction verification, consensus algorithms, network protocols and smart contracts at any node within the complicated network of organisations involved the CSC. One type is the majority or 51% attack, where an attacker possesses the majority of the hash rate in a network, gains significant control and can carry out various malicious actions. These include preventing the verification of transactions or blocks, which allows them to reverse transactions, isolating network participants temporarily and splitting the network to cause complete instability [15]. Although this risk is primarily associated with public, permissionless blockchain networks, it is largely mitigated in permissioned blockchains, where access is restricted and consensus mechanisms such as Practical Byzantine Fault Tolerance or Proof of Authority provide enhanced security and resilience against such attacks. The CSC usually operates within a permissioned environment meaning the network is not public and participation is restricted to authorised stakeholders. However, the number of supply chain members can still be quite large posing a significant security concerns that require robust security measures such as role-based access control, encryption and consensus mechanisms tailored to permissioned networks.

Additionally, the inherent transparency of blockchain raises data privacy concerns. While blockchain ensures that all transactions are recorded and visible to participants in the network, this openness can potentially expose sensitive information (e.g. construction material/component price, logistics location information, etc.) that CSC stakeholders may wish to keep confidential. The challenge lies in finding the right balance between transparency and privacy for CSCM applications through a combination of technical (e.g. encryption techniques, pseudonymisation) and governance measures (e.g. privacy rules coded in smart contracts).

9.4.2.2 Feasibility and scalability issues

The development and maintenance of blockchain-based software solutions present significant challenges. Hamida *et al.* [63] have highlighted a potential drawback in terms of poor recoverability or resilience. This aspect should be carefully considered when evaluating the feasibility and scalability of a blockchain solution in CSCM.

Especially, there exist several technical limitations that hinder the scalable application of blockchain technology in real-life CSCM projects. The first limitation is throughput, with the current maximum number of transactions per second being limited to seven, which may fall short for CSCM which involves an intensive amount of transactions at any stage in the CSC, especially during the construction phase. The second major limitation is latency. In most blockchain networks, such as Bitcoin, each block takes approximately 10 min to process, resulting in a delay for transaction confirmation. While this may not be as critical for CSCM – where contracts and processes tend to be long-term- the delay could still hinder real-time decision-making in high-pressure scenarios. The third significant technical limitation relates to size and bandwidth. Downloading the entire blockchain requires a significant amount of time, contributing to widespread time lags, which can be an issue in some critical checkpoints of the CSCM. To date, blockchain developers have not yet provided industry-tested blockchain applications that fully address these limitations for CSCM [15].

9.4.2.3 Resistance to change

Despite the willingness of 40–50% of construction companies to embrace digital supply chains and blockchain technology, risk aversion and the conservative nature of the industry mean there is little investment in emerging technologies. Construction companies are accustomed to traditional supply chains and establishing the necessary level of trust for successful project completion is challenging and often elusive [64]. These challenges further contribute to a hesitance to invest in new and unproven technologies, as construction companies seek reliability and effectiveness before committing to significant changes.

The complexity of the CSC, which involves numerous stakeholders, regulatory requirements and interdependencies, adds to the challenge. Implementing blockchain technology requires collaboration and coordination among various entities, which can be time-consuming and costly. Alleviating resistance to change requires demonstrating the value proposition of the technology, showcasing successful case studies (as demonstrated in this chapter) and exploring cost-sharing models. Additionally, industry-wide collaboration and knowledge-sharing platforms can help in addressing this challenge by sharing best practices and facilitating transformation. As the benefits of blockchain technology become more evident and its adoption in other industries grows, the construction sector may gradually embrace the transformative potential of blockchain in CSCM.

9.4.2.4 Significant investment

Implementing blockchain solutions requires substantial financial resources, including the development of custom software, infrastructure upgrades and ongoing maintenance and support. The construction industry, known for its low-profit margins and cost-sensitive nature, may face difficulties in allocating funds for such investments. Decision-makers within construction companies often prioritise immediate project needs and may be hesitant to invest in relatively untested blockchain solutions, requiring convincing evidence of the ROI and long-term benefits before committing significant financial resources. Actually, from the case study presented in Section 9.3, we are convinced that the investment in hardware and software development is relatively low compared to the benefits the solution can bring. With the diffusion of this technology, the availability and training of professional engineers who can develop blockchain solutions talents will decrease. Therefore, it is likely that blockchain investment in CSCM will drop significantly in the coming years.

9.4.2.5 Regulatory uncertainty

The construction industry operates within a complex web of regulations, compliance requirements and legal frameworks that govern various aspects of project execution and data management. When incorporating blockchain solutions into existing processes, legal barriers and uncertainty regarding compliance with industry-specific regulations may be encountered. Blockchain's decentralised nature and the immutability of data stored on the ledger can raise questions about data privacy, security and ownership rights. There is often a lack of clarity regarding how existing regulations apply to blockchain systems, particularly about commercial and sensitive data protection. Without clear guidance from regulatory authorities, CSCM stakeholders may hesitate to invest in blockchain technology due to the fear of potential legal and compliance risks.

The dynamic and evolving nature of regulatory environments further exacerbates the challenge. As blockchain technology advances and its applications in SCM expand, regulatory frameworks may struggle to keep pace with these rapid developments. While blockchain enjoyed both popularity and criticism, the lack of clear guidelines and regulatory frameworks creates a sense of uncertainty among CSCM stakeholders, inhibiting their willingness to adopt blockchain solutions, which can further hurdle information sharing among fragmented CSCM value chain.

9.5 Conclusion

This chapter explored the potential of blockchain technology in enhancing CSCM. Through an in-depth examination of its benefits, the development of a blockchain-based CSCM system and the application of this system in a real-life case involving cross-border modular construction between Hong Kong and Mainland China, the chapter highlights blockchain's transformative impact on the sector. The case study demonstrates how blockchain can improve transparency, efficiency and trust in complex, multi-stakeholder supply chains, offering valuable insights into its practical implementation and benefits.

Blockchain, with its core properties of decentralisation, immutability, transparency and security, offers important benefits in CSCM. These include enhanced information sharing, improved transparency and traceability, better security and risk management. Despite the excitement surrounding blockchain, challenges remain such as interoperability, scalability, investments and regulatory alignment. Our real-life case demonstrates the tangible benefits of blockchain integration with technologies like BIM and IoT for CSCM, though much potential remains unrealised. Future opportunities lie in IoT/AI integration, digital identity management and DAOs, but these must be navigated alongside significant challenges like system security, resistance to change and regulatory concerns.

Stakeholders must consider a series of factors when implementing blockchain technology in CSCM. For instance, they need to reconsider potential changes in stakeholder relationships as a result of blockchain adoption and assess how new arrangements can be effectively managed. arise regarding the readiness of talent to implement and maintain blockchain systems, as well as the feasibility of the associated operational and maintenance costs. From a technological standpoint, the challenges of interoperability and scalability of blockchain-based systems are significant, especially when integrating with existing infrastructure. To address the challenge of regulatory uncertainty, research collaboration between CSCM value chain participants, regulatory bodies and policymakers to discuss the legal and regulatory implications of blockchain adoption can help clarify the compliance requirements and identify potential areas for regulatory adaptation or amendment. Ultimately, blockchain's potential to transform operational workflows, reshape social and business structures and impact the foundations of trust presents both significant opportunities and complex challenges. This chapter aimed to explore these aspects specifically within the domain of CSCM.

References

[1] James C, John B, and Michael S. *American Production and Inventory Control Society [APICS] Dictionary*. New York: McKinsey & Company; 2013. Available from: https://books.google.com.hk/books?id=41ZaAAAAYAAJ

[2] Ramakrishna Y, Mamata Y, Wahab SN, and Wanganoo L. Achieving supply chain sustainability through effective talent management strategies. In: Ramakrishna Y, Wahab SN, (eds.). *Handbook of Research on Designing Sustainable Supply Chains to Achieve a Circular Economy*. Hershey, PA: IGI Global; 2023. pp. 573–93.

[3] Banzekulivaho J. The supply chain management: Challenges and solutions. In: *IV International Scientific and Practical Online Conference*; 2020. pp. 818–21.

[4] Giannakis M. Conceptualizing and managing service supply chains. *Serv Ind J*. 2011;31(11):1809–23.

[5] Litke A, Anagnostopoulos D, and Varvarigou T. Blockchains for supply chain management: Architectural elements and challenges towards a global scale deployment. *Logistics*. 2019;3(1):5.

[6] Gunasekaran A, and Ngai EW. Information systems in supply chain integration and management. *Eur J Oper Res*. 2004;159(2):269–95.

[7] Bartman T. *Overcoming Global Supply Chain Challenges*. New York: McKinsey & Company; 2022. Available from: https://www.mckinsey.com/industries/travel-logistics-and-infrastructure/our-insights/overcoming-global-supply-chain-challenges [accessed 25 August 2024].

[8] Santhi AR, and Muthuswamy P. Pandemic, war, natural calamities, and sustainability: Industry 4.0 technologies to overcome traditional and contemporary supply chain challenges. *Logistics*. 2022;6(4):81.

[9] Jabbar S, Lloyd H, Hammoudeh M, Adebisi B, and Raza U. Blockchain-enabled supply chain: Analysis, challenges, and future directions. *Multimed Syst*. 2021;27:787–806.

[10] Kaiser MJ, and Snyder B. *Offshore Wind Energy Cost Modelling: Installation and Decommissioning*. Vol. 85. Berlin: Springer Science & Business Media; 2012.

[11] Lu W, Ye K, Flanagan R, and Jewell C. Nexus between contracting and construction professional service businesses: Empirical evidence from international market. *J Constr Eng Manag*. 2014;140(2):04013049.

[12] Lee KH. Integrating carbon footprint into supply chain management: The case of Hyundai Motor Company (HMC) in the automobile industry. *J Clean Prod*. 2011;19(11):1216–23.

[13] Macdonald K. Globalising justice within coffee supply chains? Fair Trade, Starbucks and the transformation of supply chain governance. *Third World Q*. 2007;28(4):793–812.

[14] Tate WL, Ellram LM, and Kirchoff JF. Corporate social responsibility reports: A thematic analysis related to supply chain management. *J Supply Chain Manag*. 2010;46(1):19–44.

[15] Shemov G, Garcia de Soto B, and Alkhzaimi H. Blockchain applied to the CSC: A case study with threat model. *Front Eng Manag*. 2020;7:564–77.
[16] Lu W, Li X, Xue F, Zhao R, Wu L, and Yeh AG. Exploring smart construction objects as blockchain oracles in construction SCM. *Autom Constr*. 2021;129:103816.
[17] Chowdhury P, Paul SK, Kaisar S, and Moktadir MA. COVID-19 pandemic related supply chain studies: A systematic review. *Transp Res E: Logist Transp Rev*. 2021;148:102271.
[18] Mbah RE, and Wasum DF. Russian-Ukraine 2022 War: A review of the economic impact of Russian-Ukraine crisis on the USA, UK, Canada, and Europe. *Adv Soc Sci Res J*. 2022;9(3):144–53.
[19] Sodhi MS, and Tang CS. Supply chain management for extreme conditions: Research opportunities. *J Supply Chain Manag*. 2021;57(1):7–16.
[20] Wu L, Li X, Zhao R, Lu W, Xu J, and Xue F. A blockchain-based model with an incentive mechanism for cross-border logistics supervision and data sharing in modular construction. *J Clean Prod*. 2022;375:133460.
[21] Kouhizadeh M, and Sarkis J. Blockchain practices, potentials, and perspectives in greening supply chains. *Sustainability*. 2018;10(10):3652.
[22] Xu J, Lu W, Wu L, Li X, and Lou J. Balancing privacy and occupational safety and health in construction: A blockchain-enabled P-OSH deployment framework. *Safety Sci*. 2022;154:105860.
[23] Xu Y, Chong HY, and Chi M. Blockchain in the AECO industry: Current status, key topics, and future research agenda. *Autom Constr*. 2022;134:104101.
[24] Qian X, and Papadonikolaki E. Shifting trust in CSCs through blockchain technology. *Eng Constr Archit Manag*. 2020;28(2):584–602.
[25] Dubey R, Gunasekaran A, Bryde DJ, Dwivedi YK, and Papadopoulos T. Blockchain technology for enhancing swift-trust, collaboration and resilience within a humanitarian supply chain setting. *Int J Prod Res*. 2020;58(11):3381–98.
[26] Longo F, Nicoletti L, Padovano A, d'Atri G, and Forte M. Blockchain-enabled supply chain: An experimental study. *Comput Ind Eng*. 2019;136:57–69.
[27] Philipp R, Prause G, and Gerlitz L. Blockchain and smart contracts for entrepreneurial collaboration in maritime supply chains. *Transp Telecommun J*. 2019;20(4):365–78.
[28] Betti Q, Khoury R, Hallé S, and Montreuil B. Improving hyperconnected logistics with blockchains and smart contracts. *IT Prof*. 2019;21(4):25–32.
[29] Zhang X, Sun P, Xu J, *et al.* Blockchain-based safety management system for the grain supply chain. *IEEE Access*. 2020;8:36398–410.
[30] Zheng K, Zhang Z, Chen Y, and Wu J. Blockchain adoption for information sharing: Risk decision-making in spacecraft supply chain. *Enterp Inf Syst*. 2021;15(8):1070–91.
[31] Kim JS, and Shin N. The impact of blockchain technology application on supply chain partnership and performance. *Sustainability*. 2019;11(21):6181.
[32] Musamih A, Salah K, Jayaraman R, *et al.* A blockchain-based approach for drug traceability in healthcare supply chain. *IEEE Access*. 2021;9:9728–43.

[33] Manupati VK, Schoenherr T, Ramkumar M, Wagner SM, Pabba SK, and Inder Raj Singh R. A blockchain-based approach for a multi-echelon sustainable supply chain. *Int J Prod Res*. 2020;58(7):2222–41.
[34] Yiannas F. A new era of food transparency powered by blockchain. *Innovations*. 2018;12(1–2):46–56.
[35] Wong S, Yeung JKW, Lau YY, and Kawasaki T. A case study of how Maersk adopts cloud-based blockchain integrated with machine learning for sustainable practices. *Sustainability*. 2023;15(9):7305.
[36] Dujak D, and Sajter D. Blockchain applications in supply chain. In: Kawa A, and Maryniak A, (eds.). *SMART Supply Network*. Cham: Springer International Publishing AG; 2019. pp. 21–46.
[37] Xu J, Lou J, Lu W, Wu L, and Chen C. Ensuring construction material provenance using Internet of Things and blockchain: Learning from the food industry. *J Ind Inf Integr*. 2023;33:100455.
[38] Li Z, Guo H, Barenji AV, Wang WM, Guan Y, and Huang GQ. A sustainable production capability evaluation mechanism based on blockchain, LSTM, analytic hierarchy process for supply chain network. *Int J Prod Res*. 2020;58(24):7399–419.
[39] Wu L, Lu W, and Xu J. Blockchain-based smart contract for smart payment in construction: A focus on the payment freezing and disbursement cycle. *Front Eng Manag*. 2022;9(2):177–95.
[40] Shahbaz MS, Chandio AF, Oad M, Ahmed A, and Ullah R. Stakeholders' management approaches in CSC: A new perspective of Stakeholder's theory. *Int J Sustain Constr Eng Technol*. 2018;9(2):16–25.
[41] Lu W, Wu L, Zhao R, Li X, and Xue F. Blockchain technology for governmental supervision of construction work: Learning from digital currency electronic payment systems. *J Constr Eng Manag*. 2021;147(10):04021122.
[42] Wu H, Zhong B, Li H, Guo J, and Wang Y. On-site construction quality inspection using blockchain and smart contracts. *J Manag Eng*. 2021;37(6): 04021065.
[43] Amend J, Kaiser J, Uhlig L, Urbach N, and Völter F. What do we really need? A systematic literature review of the requirements for blockchain-based e-government services. In: Ahlemann F, Schütte R, and Stieglitz S, (eds.). *Innovation through Information Systems*. WI (eds.) 2021. Lecture Notes in Information Systems and Organisation, Vol. 46. Cham: Springer; 2021.
[44] Banihashemi SA, Khalilzadeh M, Antucheviciene J, and Edalatpanah SA. Identifying and prioritizing the challenges and obstacles of the green SCM in the construction industry using the fuzzy BWM method. *Buildings*. 2023; 13(1):38.
[45] Wang Y, Singgih M, Wang J, and Rit M. Making sense of blockchain technology: How will it transform supply chains? *Int J Prod Econ*. 2019; 211:221–36.
[46] Takyar A. How to Determine the Cost of Blockchain Implementation? LeewayHertz; 2023. Available from: https://www.leewayhertz.com/cost-of-blockchain-implementation/ [Accessed 22 April 2023].

[47] Brody P, Holmes A, Wolfsohn E, and Frechette J. *Total Cost of Ownership for Blockchain Solutions*. London: Ernst & Young LLP; 2019. Available from: https://theblockchaintest.com/uploads/resources/EY%20-%20Total%20cost%20of%20ownership%20for%20blockchain%20solutions%20-%202019%20-%20Apr.pdf [accessed 25 April 2024].

[48] Xue F, and Lu W. A semantic differential transaction approach to minimizing information redundancy for BIM and blockchain integration. *Autom Constr*. 2020;118:103270.

[49] Lafourcade P, and Lombard-Platet M. About blockchain interoperability. *Inf Process Lett*. 2020;161:105976.

[50] Chainlink. Build secure cross-chain applications quickly. 2023. Available from: https://chain.link/cross-chain#cross-chain-interoperability-protocol [accessed 25 April 2024].

[51] Celik Y, Petri I, and Barati M. Blockchain supported BIM data provenance for construction projects. *Comput Ind*. 2023;144:103768.

[52] Li X, Lu W, Xue F, *et al.* Blockchain-enabled IoT-BIM platform for SCM in modular construction. *J Constr Eng Manag*. 2022;148(2):04021195.

[53] Sanka AI, and Cheung RC. A systematic review of blockchain scalability: Issues, solutions, analysis and future research. *J Netw Comput Appl*. 2021;195:103232.

[54] Ekanayake EMAC, Shen GQ, Kumaraswamy MM, Owusu EK, and Saka AB. Modeling supply chain resilience in industrialized construction: A Hong Kong case. *J Constr Eng Manag*. 2021;147(11):05021009.

[55] Charles W, Marler N, Long L, and Manion S. Blockchain compliance by design: Regulatory considerations for blockchain in clinical research. *Front Blockchain*. 2019;2:18.

[56] Lu W, Wu L, Xu J, and Lou J. Construction E-inspection 2.0 in the COVID-19 pandemic era: A blockchain-based technical solution. *J Manag Eng*. 2022;38(4):04022032.

[57] Ali O, Jaradat A, Kulakli A, and Abuhalimeh A. A comparative study: Blockchain technology utilization benefits, challenges and functionalities. *IEEE Access*. 2021;9:12730–49.

[58] Monteiro F, and Correia M. Decentralised autonomous organisations for public procurement. In: *Proceedings of the 27th International Conference on Evaluation and Assessment in Software Engineering*; 2023. pp. 378–85.

[59] Borgogno O, and Martino E. Decentralised autonomous organisations: targeting the potential beyond the hype. *Law Innov Technol*. 2024;16(2):392–431.

[60] Lee D, Lee SH, Masoud N, Krishnan MS, and Li VC. Integrated digital twin and blockchain framework to support accountable information sharing in construction projects. *Autom Constr*. 2021;127:103688.

[61] Wang H. No1s1: An Autonomous Self-Owning House Goes to the World Economic Forum. ETH Ambassadors; 2022. Available from: https://ethambassadors.ethz.ch/2022/06/09/no1s1-an-autonomous-self-owning-house-goes-to-the-world-economic-forum/ [Accessed 19 November 2024].

[62] Hijazi AA, Perera S, Alashwal A, and Calheiros RN. Blockchain adoption in CSC: A review of studies across multiple sectors. In: *CIB World Building Congress 2019;* 2019; Hong Kong. Vol. 2021.

[63] Hamida EB, Brousmiche KL, Levard H, and Thea E. Blockchain for enterprise: Overview, opportunities and challenges. *13th International Conference on Wireless and Mobile Communications (ICWMC 2017)*; 2017; Nice, France.

[64] Li J, Kassem M, and Watson R. A blockchain and smart contract-based framework to increase traceability of built assets. In: *Proceedings of the 37th CIB W78 Information Technology for Construction Conference (CIB W78)*; 2020. pp. 347–62.

Chapter 10

Blockchain for data observability: integrating BIM, digital twins and enterprise CDEs

Abel Maciel[*] *and Klaudia Jaskula*[†]

This chapter discusses a main digital inflection point in a significant infrastructure initiative under development in the UK. Developed as a micro-ethnography and participatory observation by the main author, it describes some of the consideration in the selection of data capabilities needs for the wider enterprise common data environment (CDE), how the advent of building information modelling (BIM) and digital twins (DTs) play a role in this selection, giving emphasis to the observability of data in in respect to systems integration.

A literature review was conducted, focusing on infrastructure and construction industries' data lifecycle and examining the challenges of designing extensive CDEs for varied and fragmented data. It discusses data fragmentation challenges at the enterprise level, including data traceability issues. This broader context of enterprise-level data management is considered when discussing blockchain technology and the data architectures of DTs and BIM. Learning from this literature review, we build up on the subject of data observability discussion using a domain description on the participatory observation. This is followed by the application of triadic semantic questions, incorporating Function, Sequence and Attribution. The chapter thoroughly examines data architecture and the application of blockchain in integrated enterprise CDEs (E-CDEs) for infrastructure projects.

All participatory observation is anonymised and only CDE and data architecture concepts are discussed. No sensitive information has been included.

10.1 Introduction

This chapter examines the role of BIM as a crucial, yet singular component within the broader framework of large-scale and mega infrastructure projects. It underscores the importance of integrating BIM not just as an isolated set of tools and processes, but as a key element of a comprehensive enterprise-level strategy. In

[*]Bartlett Faculty of the Built Environment, University College London, UK
[†]The Bartlett School of Sustainable Construction, University College London, UK

these extensive projects, BIM forms one part of a vast ecosystem that includes a multitude of data sources and architectures.

The complexity of large-scale construction projects demands a nuanced and layered approach to data management [1]. While BIM plays a foundational role in the visualisation and management of construction processes, it represents only one facet of the overall enterprise data landscape. These projects entail a diverse array of data inputs beyond BIM such as supply chain logistics, workforce management, land use, environmental impact data and stakeholder communication channels. The integration of this data is critical for the successful realisation of the project, as well as for maintaining a coherent and consistent information flow [2].

In addressing this complexity, a holistic approach to data integration across projects, particularly in the context of developing an Enterprise DT is required. The concept of the DT extends beyond the physical construction aspects, encompassing the entire operational spectrum of the enterprise [3]. This involves harmonising BIM data with other enterprise data sources, ensuring a seamless and comprehensive data environment. Such integration is essential for creating a dynamic and interactive model of the project, offering real-time insights and decision-making capabilities across all facets of the enterprise [4].

The transition from traditional data management methods to a more integrated approach signifies a significant shift in the construction industry. This chapter explores the implementation of Blockchain, BlockDAG[1] and other distributed ledger technologies (DLT) within BIM, extending these innovations to the broader E-CDE context. The adoption of blockchain and DLT not only facilitates the secure and decentralised tracking of BIM data but also enables the integration of diverse data sources within a unified, transparent and accountable framework.

Data lineage and the concept of golden records[2] [5] represent two fundamental aspects of data engineering that stand to gain significantly from the integration of blockchain technology [6,7]. Data lineage, which traces the origin, movement and transformation of data through its lifecycle, is crucial for ensuring transparency and accountability in data handling. Blockchain, with its inherent characteristics of immutability and traceability, provides an ideal foundation for creating a tamper-proof and transparent record of data lineage. This immutable ledger capability ensures that every change, access or transfer of data is securely logged, enabling a clear and verifiable audit trail. However, the notion of 'golden records' – the creation and maintenance of a single, authoritative version of a data record – is equally vital in maintaining data integrity and consistency and fundamental for enterprise data governance.

[1]A block directed acyclic graph (BlockDAG) is an advanced structure used in blockchain technology. Unlike traditional blockchains that add blocks in a linear sequence, BlockDAGs allow multiple blocks to be added simultaneously in a networked pattern. This parallel processing enhances transaction speed, scalability and security.

[2]In informatics, a golden record is the valid version of a data element (record) in a single source of truth system. It may refer to a database, specific table or data field or any unit of information used. A golden copy is a consolidated data set.

This chapter thus presents a perspective for emerging architectures that transcends traditional BIM CDE. It highlights the need for an all-encompassing data integration approach that aligns with the broader objectives of the enterprise of mega projects. It discusses BIM as a synergistic component within the wider enterprise data ecosystem. By embracing this comprehensive vision, the construction industry can then manage the complex interplay of data in large-scale projects, setting a foundation for future developments of construction digital solutions and enterprise-wide information management.

10.2 The trajectory of CDEs

CDEs have emerged as a pivotal tool in the field of information management, particularly in sectors like construction and engineering. The concept of CDEs originated from the growing need for centralised data management and collaboration across complex projects involving multiple stakeholders. Prior to CDEs, the management of project information was often fragmented, leading to inefficiencies and errors [8]. Cloud-based CDE systems have fundamentally changed how built environment project information is managed, shared and stored. Throughout the course of the project, a CDE can assist stakeholders in information management and productive work [9]. A CDE might increase direct labour productivity by up to £6 for every £1 invested in information management, per KPMG research from 2021 [10]. Furthermore, since access to a CDE speeds up the process of locating and comparing crucial information, employing CDEs may make resolving construction disputes easier [9].

The idea of a CDE for BIM first appeared in PAS 1192:2007 [11] and PAS 1192-2:2013 [12] expanded on it. 2019 marked the release of ISO 19650 [13], an international standard for the CDE-based information management workflow [14]. Many software companies have since created an increasing variety of products that can be used as BIM CDEs with varying degrees of conformance to the ISO standard. In practice, a BIM CDE is typically a cloud-based repository that allows project data to be stored and accessed by all parties involved [15].

Unlike traditional FTP servers and other file-sharing tools, CDEs offer a more structured and integrated approach to data management [16]. While FTP servers provide basic functionalities for file storage and transfer, they lack advanced features such as version control, metadata management and workflow automation that are integral to CDEs [16]. Furthermore, CDEs are designed to support collaboration and data sharing in a more controlled environment, ensuring that all stakeholders have access to the latest, most accurate project information [15].

CDEs are distinct in their ability to maintain a single source of truth for all project data, which is crucial in complex projects with numerous interconnected elements [17]. This centralised approach reduces the risk of data silos and inconsistencies, promoting better decision-making and efficiency.

E-CDEs, in particular, bring unique value to large-scale organisations. They extend beyond the management of project-specific data, encompassing enterprise-wide data integration and management [18]. Unlike project-specific CDEs, E-CDEs deal with a broader spectrum of data types and are designed to cater to the diverse needs of an entire organisation. This includes integration with other enterprise systems, such as enterprise resource planning (ERP) and customer relationship management (CRM) systems, to provide a comprehensive view of the organisation's operations.

CDEs in general represent a significant advancement over traditional file-sharing methods. E-CDEs offer an even greater level of integration and functionality, aligning data management with the strategic objectives of large-scale organisations.

10.2.1 The particulars of BIM CDEs

The scope and functionalities of an E-CDE, as presented above, greatly differ from those of a BIM-focused CDE. For BIM-based construction projects, ISO19650 states that a CDE should be used to coordinate and share information [19]. It should do so by precisely specifying operational processes and, whenever possible, using open standards to enable a consistent exchange of information between all participating organisations. The standard defines a CDE as 'an agreed source of information for any given project or asset for gathering, organising and sharing each information container via a controlled procedure' [15, p. 5]. According to the BIM Dictionary [20], a CDE is made up of a 'CDE solution' and a 'CDE workflow' that coordinate the information flow over the course of an asset's lifecycle across four information container stages. Work in progress, shared, published or archived are the possible states for each file; changing between these states should require authorisation and approval procedures [19].

According to BSI [19], project development information layer progresses from stage 1 (structured and unstructured data) to stage 3 (object-based servers) via federated information models. Three phases of development for analogue and digital information management are distinguished by the ISO 19650 Standard, which covers four layers: business, information, technology and standards. In terms of technology, a CDE is anticipated as a query-based solution in stage 3, while in stages 1 and 2, it is still considered a file-based solution. To supplement ISO 19650 and provide more detailed specifications for the functional levels of CDEs, another standard, the DIN SPEC 91391, was released in 2019 (European Standards 2019).

An E-CDE transcends the scope of BIM CDE by encompassing a broader range of data functionalities and integration requirements that align with the comprehensive objectives of an enterprise, often lasting longer and counting many construction projects [17].

While a BIM CDE is primarily focused on the architectural, engineering and construction aspects of a project, facilitating the collaboration and sharing of BIM data, an E-CDE extends its functionality to cater to the wider operational needs of an organisation [17]. This includes the integration of diverse data types beyond the realm of BIM such as financial operations, human resources data, supply chain logistics and customer interactions. The capability needs of an E-CDE are therefore more expansive, requiring robust systems that can handle a larger variety and volume of data, ensuring seamless interoperability across different business units and external entities [21].

Because of the multifaceted nature of the E-CDE, the functional requirements are more complex, demanding advanced data management features such as enhanced security protocols, comprehensive data analytics tools and sophisticated data governance mechanisms [18]. These features are crucial in supporting the strategic decision-making processes and operational efficiency of the enterprise as a whole.

Thus, while both types of CDEs serve the crucial function of data management and collaboration, an E-CDE has to address a larger set of systems of interest, accommodating a wider array of data types and supporting a broader spectrum of organisational functions beyond the confines of BIM-focused activities. It should then leverage all BIM data and processes not only for design development and delivery and the DT, but also for the overarching enterprise services that need the BIM project data.

10.2.2 Providing for digital twins in the CDE

As CDEs continue to evolve in the realm of construction and infrastructure, their role in facilitating the deployment of DTs has become increasingly significant. DTs are virtual representations of physical assets or systems that are linked together to enable bi-directional exchange of information., offering immense potential for enhancing asset management, predictive maintenance and operational efficiency.

Incorporating DTs within the CDE framework presents unique challenges and opportunities. A key consideration is ensuring seamless integration to enable the exchange of data and insights throughout the asset lifecycle. Data orchestration and Data Hubs serve as central points for data integration and distribution within an enterprise [22]. They facilitate the aggregation, standardisation and dissemination of data from various sources, ensuring a unified and accessible data ecosystem for the enterprise.

Data orchestration involves the coordination of data flows across disparate systems, applications and databases. It ensures that data is collected, transformed and delivered to the appropriate destinations in a timely and efficient manner. This process is essential for supporting DTs, as it enables the synchronisation of data between the CDE and DT platforms, allowing for real-time updates and insights [23]. Data Hubs play a crucial role in data orchestration by serving as central repositories for storing and managing enterprise data. They provide a single source of truth for all data assets, allowing users to access, analyse and share information seamlessly. Data Hubs also support data standardisation efforts by enforcing consistent data formats, schemas and metadata across the enterprise [24].

At its core, a DT relies on accurate and up-to-date data to effectively simulate real-world scenarios and provide actionable insights [25]. Therefore, the CDE serves as the foundation for sourcing, storing and managing the data required to populate and maintain the DT.

By effectively addressing these considerations, the CDE can be a foundational platform for the deployment and operation of DTs; however, it is distinct in scope and functionality when compared to the definition of a BIM CDE. The integration of both concepts of CDE enables organisations to harness the full potential of these digitally transformed processes in design and operations to optimise asset performance, improve decision-making and drive innovation across the asset lifecycle.

10.3 The enterprise total lifecycle data governance

The enterprise data lifecycle encapsulates the comprehensive journey of data within an organisation, from its initial creation or acquisition to its eventual archiving or deletion. This lifecycle is marked by several key stages: data generation or acquisition, storage, usage, sharing and finally its archiving or disposal. Each stage demands careful data engineering considerations to ensure the data's integrity, security and usefulness throughout its lifecycle.

In contrast to project-specific data environments, the CDE for enterprises encompasses a broader spectrum and perspective. It integrates various data types from different business units, necessitating advanced solutions for data consolidation, management and governance to cater to the diverse needs of the enterprise. E-CDEs represent a distributed approach to data management and allows for decentralised control while maintaining a unified framework for data exchange and integration across the enterprise, enhancing collaboration and data consistency [26].

In the realm of construction and design, BIM serves as a key component of the data strategy. BIM facilitates detailed representation and management of the physical and functional characteristics of a facility, playing a crucial role in the data lifecycle of construction projects. Today, BIM integrates almost seamlessly with geographical information systems (GIS) [27]. GIS is pivotal in managing location-based data, providing spatial context and insights for planning, analysis and decision-making processes within an enterprise.

Enterprise content management (ECM) systems manage an organisation's unstructured data such as documents, emails and multimedia content. ECMs play a vital role in organising, storing and retrieving this content efficiently. As a separated data environment, information management systems (IMS) underpin the strategic application of technology, people and processes in the management of enterprise governance and host the governance, legal and financial documents.

Asset information management systems (AIMS) focus on the lifecycle management of enterprise assets. They encompass data related to the acquisition, operation, maintenance and disposal of assets, ensuring effective asset utilisation and management [28]. AIMS frequently inherits data from BIM objects and is the place that outcomes for Asset Information Requirements (via the asset information models) develop as part of BIM processes.

One of the features of heterogeneous but interconnected CDEs is the Data Hub orchestrating the various data streams across data sources. It is in effect a centralised platform or repository that serves as a comprehensive and integrated data management solution within an E-CDE. It collects, stores, organises and manages data from various sources and formats across the organisation.

A data hub architecture plays a central role in an E-CDE, as a seamless centre of data exchange, enabling integration and interoperability across different databases and systems. It acts as a unifying layer that connects disparate data sources, such as relational databases, data warehouses, data lakes, cloud storage and external APIs [29]. Organisations in the enterprise can then aggregate and harmonise data from diverse sources, ensuring consistency, accuracy and accessibility. This is central for enabling functionalities for data cleaning, transformation and enrichment, allowing users to prepare data for analysis, reporting and decision-making.

10.3.1 Enterprise digital twins: what is required?

CDEs for enterprise DTs and CDEs for BIM are designed to serve different, albeit complementary, purposes when considering the capability needs for each segment of the technology architecture. Their differences primarily lie in scope, functionality and data architecture, encompassing the entire data lifecycle of an organisation's assets to integrate and manage a wide variety of data types in the enterprise data catalogue [30]. They have to include operational, financial, environmental and maintenance data – across the entire organisation. The focus is on creating a dynamic and holistic representation of the organisation's physical assets and their performance in real time or, more realistically, in the 'right time' as required by business intelligence (BI). This broader scope facilitates strategic decision-making and long-term asset management.

BIM CDEs are more narrowly focused on the design, construction and initial lifecycle stages of a building or infrastructure project as a target state. As previously discussed, the scope primarily handles architectural, engineering and construction data, supporting collaboration and information sharing for the building process. The data managed within BIM CDEs is more detailed in terms of the physical characteristics of the building, such as geometry, materials, spatial relationships and construction sequencing.

Conversely, the data architecture for these Enterprise DT CDEs is preoccupied with complex and multifaceted intensive data streams, often involving integration with various enterprise systems like ERP, CRM and IoT platforms. They require robust data processing capabilities for real-time analytics and must handle large volumes of diverse, fast-changing data. The architecture is typically distributed and scalable to accommodate the expansive and evolving data needs like the ones of IoT swarms [29]. However, the architecture of BIM CDEs is typically more centralised and project design development. It is optimised for handling detailed design and construction data, with strong capabilities in version control and document management.

BIM CDEs are often considered as a foundational platform or part of DT CDEs [31]. The detailed design and construction data managed in BIM CDEs feed into the DT environment, providing the necessary granularity and accuracy for the initial creation of the DT. As the asset moves from the construction phase into operation, the DT CDE takes over, continuously updating and utilising data from the BIM CDE as well as other sources. This relationship between BIM CDEs to DT CDEs represents a shift from project-specific data management to ongoing, enterprise-wide asset management and optimisation.

10.4 The evolution of data architectures

The evolution of data architecture has been a journey marked by continuous innovation and adaptation, responding to the ever-increasing complexity and volume of data in the digital age. This evolution has distinct milestones compounding concepts and functionalities.

The concept of databases emerged in the1960s–1970s, with the development of the first hierarchical and network database models. These were significant for their ability to store and retrieve data efficiently [32]. The introduction of relational databases in the 1980s, notably SQL (structured query language), revolutionised data storage and querying. It provided a more flexible and user-friendly approach to data management [33].

The 1990s saw the rise of data warehousing and online analytical processing [34]. These technologies allowed for the aggregation and analysis of large datasets, aiding in BI. By the early 2000s, the explosion of the internet led to the era of Big Data. NoSQL databases gained popularity for their ability to handle unstructured data and scale horizontally.

In the early 2010s, Cloud computing transformed data architecture by providing scalable, on-demand data storage and computing resources. Data lakes emerged as a solution for storing vast amounts of raw data in its native format [35]. Subsequently, by the late 2010s, the integration of artificial intelligence (AI) and machine learning (ML) with data architecture allowed for advanced data analytics and predictive modelling. Real-time data processing has become increasingly important for instant insights. To bring reliability, scalability and performance enhancements to data lakes, Delta Lakes[3] are used to ensure data integrity by enabling atomicity, consistency, isolation, durability (ACID) transactions on big data workloads, facilitating structured streaming and simplifying data management. This technology allows companies to handle vast amounts of data efficiently, perform real-time updates and maintain a historical audit trail, which is crucial for analytics, reporting and decision-making processes [36,37].

Current developments and new capabilities in data architecture allow for innovation in how data grouping for Distributed Microservices, enabling more agile and flexible data management and processing [38]. As an extension of cloud computing, Edge Computing brings data processing closer to the source of data, reducing latency and improving efficiency, especially advantageous when thinking about localised microservices. This brings the advantage of Embed Analytics [39] happening closer to the customer and separately from BI. In practice, embedded analytics has to deliver a higher request rate for reports, and the corresponding burden on analytics systems goes up dramatically; access control is significantly more complicated and critical. Businesses may be serving separate analytics and data to thousands or more customers [29]. The surface for cyberattacks is much larger.

[3]A Delta Lake is an open-source storage layer designed to run on top of an existing data lake and improve its reliability, security and performance.

Blockchain technology is increasingly being explored for data security, particularly in ensuring data integrity and traceability. Unlike a blockchain, which is made up of blocks, a directed acyclic graph (DAG) uses vertices and edges, with crypto transactions being represented as vertices. These transactions are layered on top of each other. Similar to blockchain, transactions in a DAG are submitted through nodes, which must complete proof-of-work tasks to submit a transaction. While a blockchain resembles a chain, a DAG looks more like a graph. Due to its efficiency in data storage and transaction processing, the DAG model is considered a potential future alternative to traditional blockchains [40]. Building on this, the BlockDAG is a specialised type of DAG used in blockchain technology, where multiple blocks can reference several previous blocks. This structure enhances transaction throughput and scalability, offering advantages over traditional single-chain blockchains. In a further horizon, Quantum Computing, although still in its nascent stages, promises to revolutionise data processing capabilities, offering unprecedented speed and power [41].

The current landscape of data architecture is characterised by a shift towards more distributed, secure and real-time data processing capabilities [42]. This evolution reflects the ongoing challenges and opportunities presented by the vast amounts of data generated in our digital world.

Akin to this, the evolution of traditional computer-aided design (CAD) to advanced BIM mirrors the broader evolution of data architectures. Initially, CAD systems relied heavily on centralised databases, where data was stored and managed in a single location. This model, while straightforward, had limitations in scalability and flexibility. With the advent of BIM, the need for more dynamic data management became evident, leading to the adoption of more distributed data architectures. These allow for enhanced collaboration, better version control and more efficient handling of the complex data sets typical in BIM projects.

The shift from centralised to fully distributed systems in BIM and DTs has been driven by the need for greater scalability and real-time collaboration across geographically dispersed teams. Distributed systems enable multiple users to access and modify data simultaneously, often via a consensus mechanism [43], without bottlenecking at a central server, crucial for large-scale construction projects involving numerous stakeholders.

The evolution of data storage solutions from traditional data warehouses to data lakes and more recently, to data lakehouses, reflects the growing diversity and volume of data. Data lakes allow for the storage of vast amounts of raw data in its native format, which is particularly useful for BIM and DTs, where data comes in varied formats and from multiple sources. Data lakehouses combine the best features of data lakes and warehouses, offering both the flexibility of data lakes and the structured querying capabilities of data warehouses, supporting ACID transactions [29]. However, there's a risk of these repositories becoming data swamps if not managed properly, where data becomes inaccessible and unusable due to poor organisation and lack of governance.

Data fabrics and data meshes represent the next frontier in data architecture, with Delta Lake quickly becoming popular to manage regulatory compliance like

General Data Protection Regulation (GDPR) [44] and California Consumer Privacy Act of 2018 (CCPA) [45]. They provide a more holistic approach to data management, emphasising interoperability and accessibility. In the context of BIM and DTs, these architectures facilitate the integration of disparate data sources, ensuring seamless data flow and accessibility across different platforms and tools.

AI in BIM and DTs primarily feeds on large datasets comprising structured and unstructured data for predictive analytics, pattern recognition and decision-making processes. Blockchain technology enhances this by providing a secure and immutable record of data transactions, ensuring data integrity and trust, especially crucial in shared BIM environments. The role of analytics in BIM and DTs is becoming increasingly sophisticated. Real-time analytics allows for the proactive management of buildings and infrastructure, leveraging data from IoT devices and sensors integrated within the DT. This enables predictive maintenance, energy efficiency optimisation and enhanced operational performance.

Understanding the origin, movement and transformation of data – is vital in BIM and DTs for ensuring data integrity and accountability. Effective data lineage combined with interoperability – the seamless exchange of data between different systems and platforms – is crucial for the accuracy and reliability of BIM models and DT simulations. As data architectures evolve, ensuring robust data lineage and interoperability becomes increasingly important in managing complex construction projects and the lifecycle of built assets.

This research explores data architectures that are pivotal in bridging BIM processes with comprehensive enterprise-grade DTs. By examining the shift from centralised systems to distributed and interconnected frameworks, we aim to understand how these evolving architectures facilitate a seamless integration of BIM into the wider context of DTs. The focus will be on investigating the role of DLT and blockchain, as well as blockDAGs (directed acyclic graphs), in enhancing data integrity, security and traceability. The outcome of this research is expected to provide valuable insights into the development of more resilient, efficient and integrated data architectures, ultimately contributing to understanding of capability needs and functional requirements for broader application of DTs in the enterprise realm.

10.5 Methodology

Participant observation is a data collection method used by practitioners and scholars alike, typically in qualitative research and ethnography. It sets the premises of the first-hand and holistic immersion of the participant in the social situation observed with the dual purpose of engaging in the situation observed as well as observing the actors (agents), activities and physical aspects of the situation [46].

Typically, this observation includes the development of a domain description. In this particular study, this matrix focuses on the E-CDE and a domain and sets an alignment of data types along the *Y*-axis with the various dimensions of interest along the *X*-axis. The structured matrix is presented in Table 10.1, illustrating the relationships and questions at each intersection. Each cell in the matrix represents a unique intersection of these elements, leading to specific domain description questions.

Table 10.1 Enterprise CDE domain description

	Asset	Actor	Space	Time	Utility
Enterprise data	How does enterprise data inform asset management strategies?	How is enterprise data used to identify and manage actors within the organisation?	In what ways does enterprise data map and optimise the use of space?	How does enterprise data inform and predict time-based trends and patterns?	How is enterprise data utilised to enhance the utility and efficiency of resources?
BIM data	What role does BIM data play in the lifecycle of assets?	How does BIM data represent various actors involved in the construction process?	How does BIM data contribute to the planning and organisation of space?	How is BIM data used to track and plan timelines for construction projects?	How does BIM data optimise the utility of construction materials and processes?
DT data	How does digital twin data integrate with asset management and predictive maintenance?	How does digital twin data interact with and represent the various actors in a system?	What insights does digital twin data provide in the usage and optimisation of space?	How does digital twin data utilise time series analytics for predictive modelling?	What is the role of digital twin data in enhancing the utility of systems and processes?

Further on the development of this observation, we develop the progression of the domain description as a semantics analysis. Ethnographic semantics focuses on understanding the culturally significant aspects of language. Primarily used in anthropology, especially in studies of kinship and folk science, it includes various analytical methods like contrast-level mapping, componential analysis, programmed specification and semantic rules application. Contrast-level mapping organises native words into hierarchical levels of generality, while componential analysis breaks down terms into distinctive features for differentiation. Semantic rules are used to generate culturally acceptable statements or to simplify terms to their core meanings, providing insights into semantics and ethnology.

This approach, emphasising the importance of cultural context of the social situation or community of practice in hand, avoiding pitfalls like homonymy. This has become increasingly significant in social anthropology since the 1960s, with a growing focus on the semantic description of communicative codes in specific speech communities. We selected this particular method in addition to the domain description matrix, as it is employed as a linguistic device to distinguish the qualitative insights of the requirements engineering processes in the case study being observed in this micro-ethnographic exercise.

Because of the time and space constraints of this research, we restrained ourselves to only one semantic question using a triadic combination:

- Function-Sequence-Attribution (X is used for Y, and Y is a step in Z, and Z is a characteristic of A)

'How can blockchain (X) be used for ensuring data observability in BIM processes (Y), and why is this step crucial in the overall data architecture of Digital Twin development (Z), given that ensuring observability is a fundamental characteristic of effective enterprise CDEs (A)?'

This question incorporates the three semantic aspects as follows:

- Function (X for Y): It examines how blockchain integration (X) is utilised to ensure data observability in BIM processes (Y). This aspect explores the functional role of lineage tracking in maintaining high-quality data within BIM.
- Sequence (Y in Z): It considers why maintaining data observability (Y) in BIM processes is a crucial step in the broader context of developing DTs (Z). This reflects on the sequential importance of accurate BIM data as a foundation for reliable DT development.
- Attribution (Z is a characteristic of A): It seeks to understand how the emphasis on data architecture in DT development (Z) aligns with the fundamental characteristic of effective E-CDEs (A), which is to maintain high data quality and reliability across all processes and systems by leveraging observability.

This triadic semantic question explores the effect of blockchain in the interconnectedness of data lineage, BIM processes and DT development within the context of an E-CDE's overall objective to maintain data accuracy and integrity.

Both the description matrix and semantic analysis are integrated in one translation. This has annotations to aid on the levels of translations and are further references with relevant literature asserting the description in the particular translation.

10.5.1 Levels and annotation

For the ethnographic translation process, we adopt the following level of description:

- Level 2: cross-cultural descriptive statement
- Level 3: general statement about a social or cultural group
- Level 4: general statement about a specific cultural scene
- Level 5: specific statement about a cultural domain

Other levels above and below the levels described here are left out of this research to maintain focus but also anonymise in this research. At the time of this publication, any translation below level 5 may be commercially sensitive (i.e. Level 6: specific incident statement, is not used). Labels indicating the level of translation are occasionally used to indicate the level of description and are distinguished in a superscripted {Label} using italic braces on the left-hand margin of the document. For example, a level of translation 2 is represented as:

{LoT2}Text translated at Level 2.

The organisation of this participatory observation and semantic analysis can be greatly expanded with numerous strategic semantics 'mining' the complications of E-CDEs for value structures and setting out an ontological approach for further research.

10.6 Observing provenance in enterprise CDEs: accounting from BIM and DTs

{LoT2} The project under study is a substantial railway initiative operating in the UK, with the primary aim of establishing crucial transport links for communities between Oxford and Cambridge. It encompasses various key components, including upgrading existing railway sections, restoring disused segments and constructing new railway infrastructure.

This initiative addresses the transportation needs of communities along the route, which are known for their picturesque landscapes, cultural heritage and significant contributions to the national economy, amounting to approximately £110 bn annually. However, inadequate transport links pose challenges for residents and businesses, hindering their access to employment, education and leisure opportunities.

The project aims to support economic growth and job creation in the region by improving rail connections between intervening towns. By creating new connections and stations along the route and reducing journey times, the project seeks to make the region more attractive to businesses and housing development, thereby supporting growth and job creation.

Funding and oversight for the project are provided by the UK Department for Transport (DfT), which established an arm's-length body to plan and deliver the rail project. Additionally, Homes England, an arm's-length body of the Department for Levelling Up, Housing and Communities, advises on housing growth potential along the route.

In assessing how the project can support economic growth, DfT and its partners identified constraints such as a lack of space for commercial and laboratory facilities in Cambridge and Oxford, a limited labour market and high living costs affecting businesses' ability to attract staff and relocate. To address these issues, a long-term housing plan was announced in July 2023, focusing on the expansion of Cambridge to accommodate housing and businesses, thereby supporting local economic growth.

{LoT4}As of March 2023, the project's estimated cost ranges circa £6.0 bn, with circa £1.0 bn already spent. The majority of this expenditure went towards completing connection stage 1, with funds allocated for planning stages 2 and 3.

Once operational, the railway project will play a pivotal role in transforming transportation infrastructure, facilitating seamless travel across the region. Concepts such as infrastructure as a service and the ambition of piloting emerging technologies relevant to the industry are set to foster this project as an example of digital transformation across the national rail infrastructure in the UK.

The primary author has played a direct role in this project as the system integrator and leading the BIM and DT effort. Therefore, developing the digital engineering strategy, producing the DT roadmap and managing all BIM development, including the development of the information architecture of the BIM CDE.

10.6.1 Enterprise data lineage and observability

{LoT3} The orchestration of enterprise data lineage and observability stands at the forefront of revolutionising asset management. This integration harnesses real-time data capture and sophisticated data lineage tracking, thereby enabling a proactive approach to asset maintenance and decision-making.

{LoT4} When describing in detail the ways observable and traceable enterprise data integrates concepts of observable and traceable asset management in the E-CDE, Observable and traceable enterprise data integrates into asset management in the CDE through various means. It can utilise IoT sensors for real-time monitoring of asset conditions, enabling instant data capture and analysis. This data is used for predictive maintenance, identifying potential issues before they occur. Data lineage tracking ensures that every change in asset data is recorded and traceable, enhancing accountability and audit trails. Asset performance analytics derived from this data help in making informed decisions about asset utilisation, maintenance schedules and lifecycle management. The integration of GIS data provides spatial context to assets, aiding in their effective management across different locations.

However, the reliance on IoT sensors for real-time monitoring can lead to data overload in formed enterprise data architecture, where the sheer volume of information becomes challenging to manage effectively. Predictive maintenance, while useful, can lead to over-dependence on technology, potentially undermining the value of human expertise and intuition in asset care. Data lineage tracking, although enhancing accountability, can also increase the complexity of data management and may not always capture the nuances of asset performance. The integration of GIS data, while providing spatial context, might not always accurately reflect real-world complexities and dynamic changes in asset locations.

{LoT4} When describing in detail all the ways observable and traceable enterprise data interacts with and represents the various actors in the CDE, enterprise data interacts with various actors in the CDE by providing role-specific access and insights. For example, maintenance teams receive alerts and detailed performance data, enabling proactive asset care. Management can access high-level performance dashboards for strategic decision-making. The data also facilitates collaboration, ensuring that all stakeholders have a consistent view of asset status. Training modules can be developed using this data to enhance the skills of operators and technicians. Compliance officers utilise this data to ensure regulatory adherence and customer service teams use it to provide informed support to clients.

The interaction of enterprise data with various actors in the CDE, while facilitating role-specific insights, might also lead to information silos where data is not effectively shared across different roles. This could hinder holistic decision-making. Role-specific data provision may overlook the need for a comprehensive understanding of asset management across different levels. Training modules based on this data might not address the practical, hands-on experience required in certain roles. Over-reliance on data for compliance and customer service can lead to a lack of personalisation and human touch in interactions.

{LoT4} When elucidating the diverse spatial concepts for observable and traceable enterprise data within the CDE, the application of Phenomenological Space to observable and traceable data is pivotal. By mapping assets within their physical environment using 3D modelling and GIS data, enterprise systems visualise assets in their actual locations – a practice foundational for both planning and operational management. This approach mirrors Phenomenological Space, where space is comprehended through human interaction and experience, making it crucial for understanding how assets physically coexist and interact within their environments.

The use of *a priori* intuition is apparent in how spatial utilisation analytics are derived. These analytics do not just represent physical measurements but also an a priori framework that structures and informs facility layout and usage optimisation. This aligns with Kant's view of space as a necessary representation underlying all external experiences, where understanding space utilisation goes beyond observable metrics to include intuitions about how space should be optimally used.

Integrating environmental monitoring data into the spatial context reflects the concept of relational space [47]. This integration does not view space as an absolute container but as a set of relations affected by environmental conditions. It emphasises that the maintenance and performance optimisation of assets are contingent upon understanding the dynamic interrelations between space and environmental conditions.

However, the challenges associated with these spatial applications in the enterprise context also reflect the limitations of non-Euclidean geometry. Just as non-Euclidean geometry proposes that the true nature of space may not conform to traditional geometric intuitions (e.g. space may be curved), relying solely on 3D modelling and GIS data might not account for the non-linear and unpredictable changes in physical spaces. These tools may miss the complexities and dynamics that a flat or overly simplified model of space cannot capture.

Furthermore, potential inaccuracies in space utilisation analytics and environmental monitoring data suggest the applicability of concepts like Simultaneity[4] [48] or quantum field theory (QFT) and the Void [49]. In Einstein's framework, events perceived as simultaneous from one perspective may not be regarded as such from another, highlighting the subjective nature of observational data. Similarly, QFT posits that space is permeated with fluctuating energies, which can variably influence observations and measurements, much like how environmental conditions unpredictably affect data accuracy and timeliness. This is analogous to how real-time changes in physical conditions or human interactions with space might not be consistently observed or understood across different monitoring systems or analytical models. These observations underscore the need for a nuanced and relativistic approach to interpreting spatial and environmental data within enterprises, ensuring that such data is evaluated within its broader and dynamically changing context.

{LoT4} In the context of the CDE, understanding the nuanced concepts of time is crucial for effective management of enterprise data. Observable and traceable data within the CDE includes real-time monitoring, which is analogous to the concept of Presentism, where only the present moment is considered real. This approach allows for immediate decision-making and swift responses to current conditions. Historical data analysis, however, aligns with eternalism, suggesting that past, present and future data are equally significant, providing a comprehensive view that is essential for understanding long-term trends and asset lifecycles. This principle reflects the holistic approach of blockchain to historical data, where past, present and future transactions are treated with equal integrity, providing a comprehensive and unalterable ledger.

Time-stamped data represents discrete time, where data entries are distinct and separate events logged into the system, creating a chronological record of asset performance and maintenance activities. This method ensures that each data point is quantised and independent, facilitating clear timeline tracking and accountability. Such data is crucial for lifecycle analysis, where the growing Block universe model can be applied, considering the past and present as fixed while the future remains open and indeterminate, thus predicting future maintenance needs and optimising asset replacement schedules.

Furthermore, the integration of data for scenario planning and forecasting introduces the concept of the specious present, where decision-making processes consider a 'present' that extends slightly into the past and future, allowing for strategic planning that encompasses more than the instantaneous moment. This broader perspective helps mitigate the risk of overemphasis on immediate and past trends, which might otherwise lead to neglecting forward-looking strategies.

However, the reliance on time-stamped data and historical analysis might suffer from the relativity of simultaneity effect, where the same set of data might not be interpreted similarly by different stakeholders due to varying perspectives or information needs. This discrepancy can sometimes lead to inaccurate forecasts,

[4]Simultaneity is relative to the observer's frame of reference. Events that appear simultaneous to one observer may not be simultaneous to another observer in a different frame of reference.

particularly in rapidly evolving industries or in environments where future conditions are unpredictable.

To consider these philosophical concepts of time, an enterprise data management within the CDE can be more robust, ensuring that data not only informs about the past and present but is also used effectively to plan, thus balancing immediate needs with long-term strategic goals.

{LoT4} When describing in detail all concepts of utility in observable and traceable enterprise data as it is employed to account for systems and processes in the CDE, we note that observable and traceable enterprise data contributes to utility in systems and processes by enabling efficient resource management. It provides insights into energy consumption patterns, leading to optimisation strategies for energy use. This approach of marginal utility is the most aligned concept because each incremental improvement in energy efficiency derived from insights into consumption patterns represents additional utility gained from consuming one less unit of energy.

Predictive analytics derived from this data reduce downtime and improve operational efficiency. Expected utility is the fitting concept here as decisions to perform maintenance are made under uncertainty, with predictive analytics providing a way to calculate the expected benefits of such actions, hence optimising operational decisions based on the probable outcomes.

The data also supports process automation, where systems can adapt based on real-time data inputs, enhancing productivity and reducing errors. The aggregate increase in satisfaction from enhanced productivity and reduced errors due to automation reflects the total utility gained from implementing observable and traceable data in CDE. This data informs about the environmental impact of operations, aiding in sustainable resource utilisation and compliance with environmental regulations. Quantifying the satisfaction or utility from sustainable practices can be complex, but the reduction in environmental impact can be seen as a measurable improvement in utility, aligning with cardinal utility where satisfaction is quantified.

While observable and traceable data can enhance resource management and efficiency, it may lead to an over-reliance on automated systems, potentially reducing the scope for human discretion and adaptability. This perspective aligns with 'negative utilitarianism', which emphasises minimising suffering or negative outcomes – in this case, the potential downsides of over-reliance on technology, such as reduced human discretion.

Predictive analytics, though beneficial in reducing downtime, might not always be accurate, leading to unnecessary maintenance activities or missed issues. The concept of 'Law of Diminishing Marginal Utility' is applicable as the utility gained from predictive analytics may decrease with each successive application, especially if inaccuracies lead to diminishing returns on maintenance actions.

The focus on optimising energy consumption and operational efficiency might overshadow other important factors like employee well-being or broader environmental impacts. This potential 'preference utilitarianism' reflects the idea that optimising for one set of preferences (efficiency and energy savings) should not overshadow broader considerations such as employee well-being and environmental sustainability, emphasising a balance that leads to the greatest overall utility.

The complexity and cost of implementing such data-driven systems could be prohibitive for smaller organisations or those with limited resources. The concept of the 'utility monster' can be metaphorically applied here, suggesting that while large entities may derive vast benefits from these systems, smaller organisations might struggle, pointing to an imbalance in the distribution of utility gains across the broader community of practice.

10.6.2 BIM data lineage and observability

{LoT3} BIM data lineage and observability within the CDE are foundational for asset management across a building's lifecycle. By centralising and organising asset information, BIM supports lifecycle management, ensuring that all stakeholders have consistent data, thereby improving decision-making, reducing costs and enhancing overall asset sustainability.

{LoT4} When describe in detail the ways BIM data integrate concepts of asset management in the E-CDE, BIM integrates concepts of asset management in the BIM-CDE in several ways to enhance the lifecycle management of built assets:

BIM ensures consistent data standardisation, with structures and formats in the CDE, fostering consistency and connectivity throughout the asset lifetime. Because of this standardisation, asset management systems can be seamlessly integrated, making it simple to exchange and utilise data at each stage of an asset's lifecycle [50].

BIM enables information exchange across various asset lifecycle stakeholders, such as owners, designers, contractors and facility managers. BIM facilitates stakeholders' sharing of essential asset information, such as specifications, maintenance schedules, warranty details and performance statistics, by offering a centralised platform for data interchange within the CDE [51]. This ensures that relevant asset information is accessible to all stakeholders throughout the asset lifecycle, enhancing collaboration and decision-making [52]. BIM makes it possible to reduce the total cost and timing of the implementation of investment and construction projects and at the same time improve the quality of design and construction and installation work [53].

BIM extends beyond the design and construction phases to incorporate asset information modelling (AIM) for managing built assets throughout their lifecycle [54]. AIM involves the creation and management of digital representations of physical assets, including information related to their components, attributes, relationships and behaviour. By capturing comprehensive asset information within the BIM model, stakeholders can effectively manage assets from procurement and installation to operation and decommissioning [55].

BIM supports lifecycle data management within the BIM CDE by providing a platform for storing, organising and accessing asset information throughout its lifecycle [56]. This is however segregated from the E-CDE. The lifecycle data includes data related to asset specifications, maintenance history, performance metrics, regulatory compliance and sustainability requirements. By centralising lifecycle data within the CDE, stakeholders can track and manage assets more effectively, making informed decisions to optimise performance, reduce costs and mitigate risks [57].

BIM enables the tagging and identification of assets within the CDE using unique identifiers and metadata. Asset tagging allows stakeholders to associate digital information with physical assets, such as equipment manuals, maintenance records and inspection reports. By linking digital information to physical assets through BIM, stakeholders can easily locate and access relevant information within the CDE, streamlining asset management processes and improving operational efficiency [57].

BIM provides integrations with asset management systems to facilitate seamless data exchange and interoperability between the CDE and external asset management platforms [58]. Integration allows stakeholders to leverage BIM data within their existing asset management workflows, ensuring consistency and accuracy of asset information across systems. This enables stakeholders to optimise asset performance, streamline maintenance processes and improve decision-making based on real-time data and insights [59].

Overall, BIM greatly enhances asset management in the E-CDE by providing a framework and production pipeline for storing, managing and exchanging design information for assets throughout the asset lifecycle. The challenge then is to define what object types and classification hierarchies will be adopted, as this varies from industry to industry and asset data schemas are not sufficiently templated in BIM production solutions. Once a project asset data schema is accepted, standardising data formats within the integrated design development can be implemented to produce and transform data to be consumed by the AIM systems. BIM helps stakeholders optimise asset performance, reduce costs and enhance sustainability across the built environment [60] and it is foreseeable that this will be significantly improved as data schema production becomes more adaptive and enterprise focus.

{LoT5} When detailing the myriad ways BIM data interfaces with and embodies the diverse actors within the CDE, it becomes evident that BIM data interacts with and represents various participants engaged in the project lifecycle. These include architects, engineers, contractors, facility managers, owners and other stakeholders. Within the CDE, BIM data engages and represents each participant uniquely:

BIM enables Architects and Designers to create digital models of buildings, incorporating architectural, structural and MEP information. It aids in visualising design concepts, iterating designs and analysing building performance [61]. It also supports various Engineers (structural, mechanical, electrical, civil) in the design and analysis of building systems such as HVAC, electrical layouts and site infrastructure. BIM data represents components like beams, columns, ductwork and drainage systems.

BIM helps contractors to schedule, plan and coordinate construction activities. It visualises construction sequences, generates quantity take-offs and ensures spatial coordination. Contractors rely on BIM data for installation requirements and tolerances. Similarly, BIM supports facility managers throughout the building lifecycle. It enables space planning, asset tracking and maintenance scheduling. Facility managers rely on BIM to access digital models of buildings, infrastructure and assets. BIM data represents facility components such as equipment, systems and spaces, providing critical information about maintenance schedules, warranties and service history [62].

BIM empowers owners and developers to make informed decisions across the project lifecycle. From planning and design to construction and operation, owners leverage BIM for cost estimation, risk analysis and lifecycle planning. BIM data represents project information, including schedules, budgets and performance metrics, enabling effective monitoring and risk management [63]. BIM data may also interact with other external stakeholders such as regulatory authorities, consultants, suppliers and tenants. Regulatory authorities may review BIM data for compliance with building codes, standards and regulations. Consultants may provide specialised expertise in areas such as sustainability, accessibility and safety, leveraging BIM data for analysis and recommendations. Suppliers may contribute product information and specifications to BIM models, ensuring accurate representation of building materials and components. Tenants may use BIM data for space planning, interior design and fit-out projects, optimising the use of leased spaces and amenities [62,64].

{LoT5} When delineating the various spatial concepts for BIM data within the CDE, it's essential to recognise the pivotal role that the CDE plays in managing spatial information within digital models. Within the CDE, several distinct space concepts for BIM data can be identified including Absolute, Kantian, Phenomenological and most notably relations concepts.

Geometrical modelling of buildings and infrastructures in BIM can be seen as an application of Absolute Space. In this context, the BIM environment acts as a fixed, immovable framework within which the models of buildings are placed. This is similar to how Newton described absolute space as a container [65] – it is the digital 'container' where all BIM elements reside. The fixed coordinate system in a BIM software where a building model is positioned and the unchanging spatial boundaries defined in the model.

Conceptual modelling in BIM may be influenced by Kantian Space, where the space within which projects are designed is not merely a reflection of physical dimensions but also of the designers' intuitive structuring of space. The initial design sketches in BIM are guided by an architect's intuitive sense of spatial organisation before detailed measurements and empirical data are applied.

BIM spatial organisations (such as rooms, corridors and lobbies) are based on their function. Geometric representations define these spaces – either 3D volumes or 2D polygons. Associated attributes (such as name, area and volume) aid in space planning [66]. The definition of spaces through geometric representations and associated attributes emphasises the importance of the relationships among spaces, akin to Leibniz's view of space as a collection of relations among objects [47].

BIM establishes a hierarchical structure for spaces within a project. This hierarchy includes levels like floors, rooms and zones. Users navigate spatial information at different granularity levels, facilitating interaction from project-wide views to individual areas [66]. This hierarchical arrangement mirrors the relational space concept, where space is understood in terms of the positions of objects relative to one another.

BIM represents spatial relationships by connecting between spaces, including adjacency and containment. These relationships enable analysis of interactions, circulation patterns and accessibility [66]. BIM captures detailed information about space attributes characteristics, properties and requirements. Attributes cover:

- Physical properties: dimensions, area and volume.
- Functional properties: room type, occupancy load and usage.
- Environmental properties: lighting, acoustics, ventilation.
- Regulatory requirements: compliance with standards, fire safety codes.

Physical, functional and environmental properties, along with regulatory requirements, all pertain to how spaces are defined and understood in relation to their purpose and context, reinforcing the concept of space as a network of relations.

{LoT5} Time-related concepts are key for managing the lifecycle of a building or infrastructure project effectively as a BIM within the CDE. Different time concepts for BIM data in the CDE are mentioned below.

BIM data includes a comprehensive project timeline, outlining milestones and deliverables from inception to completion. It utilises tools like project schedules and Gantt charts to track progress [67]. This representation of discrete time involves distinct, separate moments (such as milestones) that are clearly demarcated and independent, akin to frames in a film reel.

The project lifecycle is divided into phases and stages, each with specific tasks, milestones and DataDrops. These include design, construction and operational phases, with stages marking key progress points [68]. This aligns with the Growing Block Universe concept where, as each stage completes, it adds to the 'block' of the project's history, while the future stages remain open and indeterminate.

BIM time-based information contains data for construction schedules and procurement deadlines, aiding in monitoring and decision-making to mitigate delays. This specious present involves a broader present that incorporates immediate past data and near-future expectations to facilitate current decision-making.

BIM supports visualising and simulating construction sequences, optimising task coordination to minimise conflicts and delays. 4D models integrate geometry with schedules for visualisation [67]. This exemplifies the continuous time concept because it portrays time as a smooth flow, essential for synchronising tasks seamlessly across the construction process.

BIM data covers the entire lifecycle phases, including design, construction, operation, maintenance and decommissioning, enabling holistic asset management. BIM data embodies eternalism. All phases, past, present and future, are considered within the model, providing a comprehensive, time-inclusive framework.

BIM enables temporal analysing of project performance, identifying schedule impacts and optimising schedules through techniques like critical path analysis and scenario testing [69]. Scenario testing aligns with A-theory of time that sees the passage of time as an objective feature of reality, crucial for assessing and optimising the sequence of project events.

The articulation of time concepts in CDE-orchestrated BIM production provides a common set of references for managing the temporal aspects of data and information beyond the immediate design of the building. By incorporating project timelines, phases, time-based information, construction sequencing, lifecycle phases and temporal analysis into BIM data, stakeholders can plan, execute and optimise projects to meet schedule objectives that integrate in the much wider context of the enterprise business process and time cycles.

{LoT5} When elaborating on the comprehensive concepts of utility, we refer to how BIM data operates within the CDE. This encompasses numerous facets of BIM data management and utilisation aimed at facilitating the effective operation and maintenance of built assets. Essential utility concepts of BIM data within the CDE comprise:

BIM data interoperability ensures seamless exchange between systems and stakeholders through standardised formats and protocols, facilitating collaboration [70]. The concept of ordinal utility may be relevant here because it involves ranking preferences – in this case, the preference for systems that can integrate smoothly over those that cannot, without quantifying the exact utility differences [71].

BIM systems integration produces data classifications and schemas that are aligned with external and parallel systems like asset and facility management, enabling streamlined workflows such as asset tracking and maintenance planning. Total utility is applicable as the aggregate satisfaction derived from the combined functionalities provided by integrated systems.

The data produced in BIM processes is used for asset lifecycle management, aiding informed decision-making. Expected utility, which calculates the probable outcomes to optimise decision-making, mirrors the predictive nature of lifecycle management strategies in asset optimisation.

BIM data allows stakeholders to visualise and simulate building systems and processes, aiding in performance assessment and design evaluation. This develops as a cardinal utility because the precise improvements in decision-making effectiveness can be quantified and measured against simulation outcomes.

BIM software allows for analytical capabilities to produce data on building system performance and efficiency. Expected utility is fitting as it involves calculating the probable utility outcomes from using BIM data in different scenarios, optimising decision-making under uncertainty.

Standardised frameworks for BIM collaboration [17,72] are data management and accessibility principles for robust planning and management to ensure integrity, security and accessibility. The cardinality or utility function concepts maps how different levels of data management and accessibility contribute to the utility for the users, helping to model choices and predict behaviours in data usage.

BIM workflows for data automation and optimisation of processes like clash detection and schedule optimisation, improving accuracy and decision-making. Marginal utility is relevant as each incremental improvement in process automation potentially adds additional utility to project management.

BIM data serves as a foundation for decision support, providing stakeholders with relevant information for project delivery and operational performance optimisation [67]. Expected utility again is key, as it guides the decision-making process by weighing the utility of various outcomes based on the data provided.

Overall, there is a degree of difficulty in assigning utility concepts in BIM data as they are involved in many processes within the CDE and influence the total lifecycle of building systems. Data and processes for interoperability, system integration and decision support leverage these concepts to improve project outcomes and asset performance. One example is the better development of sustainable built environments.

10.6.3 Digital twins data lineage and observability

{LoT3} The potential for DTs to enhance data lineage and observability by leveraging blockchain and AI technologies is an epic vision that will dramatically transform the production and operations of the built environment. The early days of this integration demonstrates the plausible transparent, tamper-proof tracking of data from origin through modifications, vital for auditability and regulatory compliance in asset management within an E-CDE well suited for GovTech.

{LoT4} When discussing the integration of observable and traceable data for asset management within the DT-CDE, numerous aspects come into play. DTs leverage IoT sensors placed on physical assets to gather real-time data about their status, performance and environment. This data stream can be continuous, providing a live feed of information that is observable and measurable [73]. To visualise these data streams, DTs employ advanced 3D modelling to create a virtual graphic representation of physical assets. This visualisation not only provides a detailed view of the assets but also integrates data from various sources, making it easier to observe and interact with complex information [74]. By analysing the data collected from sensors and other sources, DTs enable real-time analytics at the cloud and edge of the computing architecture. This allows for immediate observation and response to changing conditions, facilitating proactive asset management, maintenance and resilience [75].

DT has capabilities to store historical data, enabling the analysis of long-term trends and patterns. This historical perspective is vital for traceability and understanding the lifecycle of an asset, including its usage, wear and maintenance history [76]. By using ML algorithms and predictive analytics, DTs can forecast potential issues before they occur. This predictive approach relies on observable data trends to make informed predictions, enhancing asset management strategies with predictive maintenance and forecasting [77].

Blockchain brings traceability and security in edge and cloud data architectures for DTs. This enhances by providing an immutable record at the source of data capture. The interactions with the asset ensures that data production and transformations are transparent and tamper-proof. This capability of tracking the lineage of data, capturing where it originates, how it is produced, transformed and consumed, becomes foundational for auditability and accountability in asset management, ensuring that all decisions and changes are traceable back to their source. The DT becomes then centric to facilitating collaboration among various stakeholders. By providing a shared, observable platform, they ensure that all actors have access to the right data and insights. This is particularly powerful as DTs become federated and part of larger systems of systems [3].

{LoT4} When delineating the interactions between DT observable and traceable data and the diverse actors within the CDE, one crucial aspect is real-time or right-time monitoring of assets through data collected from IoT sensors. Addressing functional requirements for a particular data is particularly invaluable for actors such as maintenance personnel, operators and especially for the client. The engineering of the requirement for that data, when well implemented as a solution, mitigates problems of information asymmetry [78]. This is possible by leveraging AI and ML for parametric information distribution within the system where the many humans and machine agents can dynamically complement each other's work.

DT systems can do so by actively facilitating role-based access to data. For example, engineers may receive detailed technical data, whereas management might view high-level performance metrics. This parametric data presentation ensures that each actor interacts with data that is most relevant and comprehensible to their role [76]. Actors such as designers and architects interact with DT through detailed 3D models and simulations. These tools allow them to visualise and simulate and interact with assets in a virtual environment, experiment with different scenarios and understand the potential impact of their decisions [74]. Evidence confirms that maintenance teams benefit from predictive maintenance alerts generated by DT, allowing these teams to proactively schedule repairs and avoid downtime [77]. DTs provide a data-driven basis for decision-making grounded in traceable history and performance analytics of assets, therefore helping in strategic planning, budgeting and resource allocation [79].

Conversely, the data exchanges of the DTs and the wider E-CDE enhance adaptability of the business architecture by improving communication and collaboration [23]. The E-CDE, nesting the DT, presents and transforms DT data for various actors normally not having access to the DT. This provides a wider platform for data sharing that is transformed for consumption by stakeholders external to the DT.

The DT also became an invaluable asset for effective education and training. By simulating real-world conditions and scenarios, DTs provide a practical and risk-free environment for training, helping actors to develop necessary skills and knowledge in a virtual environment identical to the production environment. Because of this capability, DTs also play a role in engaging external stakeholders, such as customers or investors [80]. By presenting a transparent view of asset performance and management, they help in building trust and providing value-added services. For actors responsible for regulatory compliance, DTs offer a comprehensive view of asset compliance status. They can track whether operations adhere to industry standards and regulations, helping in audit preparations and ensuring regulatory compliance [81].

Incorporating AI agents such as Google Gemini [82], Microsoft Copilot [83] and ChatGPT [84] into the E-CDE further enhances the dynamic capabilities of DTs. These AI agents, by leveraging advanced ML algorithms and vast data sets for long context queries [85], offer predictive analytics and decision support, automating complex processes and providing insightful analytics. For example, Google Gemini might optimise operational sequences in real time, Microsoft Copilot could assist in predictive maintenance scheduling and ChatGPT could be used for natural language processing to interpret and generate reports or interact with users for troubleshooting and support. The integration of these AI technologies not only streamlines workflow but also enriches the interactive experience for all stakeholders, enhancing strategic decision-making and operational efficiency across the enterprise DT ecosystem [86].

{LoT4} Spatial concepts play a central role in the engineering of CDEs for DT observable and traceable data within that CDE. It's essential to acknowledge that DT data encompasses a comprehensive approach to spatial concepts. These concepts are particularly significant for understanding and managing the physical and operational characteristics of assets within a CDE.

DTs employ advanced 3D modelling to create detailed virtual representations of physical spaces, such as buildings, infrastructure and environments. This spatial visualisation allows for an immersive and interactive view of assets, providing a realistic context for how assets exist and operate within their physical space. This process integrates geospatial data, which is vital for understanding the spatial relationships and environmental context of assets. It can be used for various purposes such as planning, navigation and optimising asset location [87].

By analysing data from sensors and other sources, DTs facilitate the examination of how space is being utilised. This analysis can identify underused areas, optimise layout for efficiency and improve overall space management. DTs can simulate and run environmental monitoring within a space, such as temperature, humidity or air quality. This is essential for maintaining optimal conditions, especially in spaces like data centres, warehouses or manufacturing facilities [88]. Within this monitoring and simulations, there is a spatial subset of the place for Interaction with Physical Elements, encompassing the captured relationships of interaction of assets with other physical elements. This includes how machinery operates within a factory floor, how foot traffic flows through a building or how environmental changes impact asset performance.

DTs enable real-time spatial tracking of assets within a space. For movable assets, this can mean tracking their location, movement and operational status, which is critical for logistics, inventory management and security. This differs from spatial planning and simulating changes to the physical space, usually involving the testing of different layouts, ensuring that spaces meet accessibility and safety compliance and planning new construction, or simulating emergency scenarios to assess response strategies. The analysis of historical spatial data helps in understanding how the usage and effectiveness of a space have evolved over time and often inform future improvements and planning [89].

{LoT4} When detailing the time concepts for DT observable and traceable data within the CDE, it's crucial to recognise their multifaceted integration and significance for comprehensive asset management and analysis.

DTs utilise real-time monitoring, primarily through IoT sensors and other monitoring systems. This capability allows for the immediate observation and analysis of asset conditions, operations and environmental factors [88]. The concept of continuous time, where time flows smoothly without discrete breaks, best describes this ongoing, seamless data stream, reflecting the uninterrupted and ongoing data collection and updates in real-time monitoring.

Historical data analysis involves archiving and analysing past data trends and the evolution of asset performance over time. Eternalism, which posits that past, present and future events are equally real, provides a suitable framework for understanding how DTs integrate and use historical data alongside present data, treating all historical states of the asset as being just as real and accessible as current data. However, when the DT is used for predictive analytics and forecasting, the past and present are used to predict future states of assets. The Growing Block Universe, where the past and present are real but the future is not yet determined, mirrors this aspect of DTs. As new data comes in, it 'grows' the 'block' of known information used to forecast future conditions and behaviours.

Every data point in a DT being time-stamped resolves as a concept of discrete time, where time is made up of distinct moments. This allows for precise tracking of changes and conditions at any given instant, segmenting time into quantifiable units essential for diagnostics, data observability engineering and compliance.

Simulation and Scenario Planning in DTs, involving simulations based on discrete time events (what-if scenarios), reflects A-theory where time's passage is a distinct feature of reality. Each scenario is a potential timeline with a clear before and after, emphasising changes over distinct, measurable intervals.

As an externalism, DTs integration extends to the entire total lifecycle management of an asset, from inception and design to operation and eventual decommissioning. DTs provide a continuous thread of data across these stages, offering insights into long-term asset management and optimisation strategies as a block universe [79].

DTs ensure that operations across various systems are synchronised in real-time. This Synchronisation of Operations reflects Presentism and the view that only the present is real, aligning with the necessity in DTs to have current, up-to-date data reflecting the immediate state of assets for operational coordination.

Maintenance scheduling based on historical and predictive data reflects B-theory of time, where all points in time are equally real and the passage of time is seen as an illusion. This approach in DTs treats past, present and future data with equal importance, allowing for effective planning based on the full spectrum of time-indexed data without privileging the present.

{LoT5} When detailing the utility concepts employed by DTs that use observable and traceable data within the CDE, it's evident that the utility of DT data in a CDE is substantially bolstered by its observable and traceable characteristics. These aspects are harnessed to create utility across various dimensions:

DTs optimise operational efficiency by providing real-time data on asset performance. This enables timely decision-making and adjustments, leading to improved productivity and reduced operational costs. For example, in a manufacturing context, a DT can identify bottlenecks in production lines and suggest improvements [77]. Total utility here is maximised as the overall satisfaction from the use of DTs is increased through enhanced productivity and cost savings.

Using traceable historical data and real-time analytics, DTs facilitate predictive maintenance. This proactive approach prevents failures and extends the lifespan of assets by scheduling maintenance based on actual asset condition rather than predetermined schedules [77]. Expected utility happens as maintenance decisions are made based on probabilistic outcomes, optimising for the best expected condition of the assets. Similarly, the DTs enhance health and safety measures by simulating different scenarios and identifying potential hazards. This is particularly useful in high-risk environments, where safety is paramount. The concept of expected utility is relevant as it involves calculating the best outcomes under various risk scenarios, maximising safety [90].

However, DTs facilitate the efficient utilisation of resources such as energy or raw materials by analysing usage patterns and recommending optimisations. This approach can result in substantial cost savings and environmental advantages. The concept of marginal utility is pertinent here, as each incremental optimisation of resources brings additional savings and benefits, thus reflecting the extra utility derived from each unit of resource conserved. Similarly, in building management, DTs can optimise energy consumption by monitoring and regulating heating, ventilation, air conditioning and lighting systems based on real-time environmental and usage data [3]. Here, marginal utility illustrates the additional utility obtained with each incremental improvement in energy efficiency. Furthermore, the insights derived from DT data can propel innovation in product and process development. By comprehending how products are utilised and how they perform under actual conditions, companies can enhance their innovation efforts. Marginal utility again becomes relevant as each new innovation potentially increases the utility of the product or process.

In manufacturing and production environments, DTs contribute to quality control and assurance by continuously monitoring and analysing the output. They can quickly identify deviations from desired specifications, allowing for immediate corrective actions [91]. The utility function here represents the relationship between different quality levels and their associated utility, guiding decisions on quality adjustments. Similarly, the observable and traceable data in DTs assist in ensuring compliance with industry regulations and standards. By maintaining a detailed record of operations and changes, organisations can easily demonstrate compliance during audits and inspections. The utility function quantifies compliance levels and the utility derived from meeting various regulatory standards.

Through the integration of time and usage data, DTs support the entire lifecycle management of assets, from design and construction to operation and decommissioning [79]. This holistic view helps in making informed decisions about asset utilisation and retirement. Total utility here encompasses all the utility gained throughout the asset's lifecycle, from inception to decommissioning.

DTs provide a platform for engaging with customers and other stakeholders by visualising and demonstrating how products and systems function. This transparency can build trust and facilitate collaborative development. Ordinal utility is suitable here as it involves ranking different engagement strategies based on their perceived utility without specifying exact differences in satisfaction levels.

10.6.4 Summary of results

A summary of enterprise, BIM and DT data lineage by levels of translation is provided in Table 10.2.

Table 10.2 Summary of findings

Translation	Enterprise data lineage	BIM data lineage	Digital twin data lineage
LoT2	Enterprise data in large infrastructure projects, such as the UK railway initiative, reflects the global challenge of integrating multiple data sources, systems and stakeholders. This cross-functional data integration varies across different cultural contexts, requiring tailored solutions for data provenance that handle regulatory, social and operational diversity.	BIM is universally utilised in construction projects and the continuity of data provenance across design, construction and operational phases is also a universal challenge. Regional differences in regulations and standards can complicate cross-border data management, especially in countries with varying levels of digital maturity and cultural expectations around data transparency.	Digital twins, as a transformative technology, face global cultural challenges in implementation. The expectations and risks surrounding real-time data usage, privacy and security differ. In some contexts, there is broad acceptance, while in others, scepticism about digital data's reliability and ethical implications can hinder full-scale adoption of digital twins.
LoT3	In large-scale enterprise environments, such as infrastructure projects, stakeholders from diverse cultural and social backgrounds bring different expectations and approaches to data provenance. Some groups may prioritise transparency and traceability, while others may focus more on operational efficiency, leading to potential misalignments in data governance strategies.	BIM adoption and data lineage management are affected by the cultural norms within architectural, engineering and construction industries. In cultures that emphasise collaboration and openness, data flows are smoother, enhancing provenance tracking. In contrast, more hierarchical or traditional cultures may experience slower data management processes, which can lead to gaps in accountability.	Workplace cultures in manufacturing, logistics or asset management sectors, where efficiency is key, are more inclined to adopt digital twins for data provenance. Conversely, industries emphasising traditional methods may view digital twins with suspicion, fearing that real-time data systems could erode the value of hands-on experience and human oversight, which can complicate their acceptance.

(Continues)

LoT4	In UK infrastructure projects, particularly in transport, enterprise data lineage is tightly integrated with the cultural emphasis on public transparency and accountability. The UK's regulatory frameworks require rigorous data provenance systems, ensuring that all data related to public works can be traced and audited, reflecting a strong focus on maintaining public trust.	In technologically advanced nations like the UK, the construction industry's adoption of BIM as part of its digital transformation initiatives highlights the importance of data-driven decision-making. However, traditional construction practices can create tension, as full digital integration for data provenance may be resisted by sectors of the industry that value manual methods or conservative project management approaches.	In smart cities projects, digital twins are proposed as a system to manage infrastructure by real-time data integration for efficient urban operations. However, ensuring that data lineage aligns with local ethical standards and privacy, can present significant challenges. The seamless, traceable nature of digital twin data must be balanced against public concerns around data usage and security.
LoT5	In the UK transport sector, enterprise data is subject to strict legal and regulatory controls, ensuring clear and traceable data lineage. These frameworks mandate comprehensive audit trails for all infrastructure projects, allowing most decision to be linked back to specific data points. This aims to ensure transparency and accountability across all levels of data management.	In architecture and construction, BIM systems are integrated with UK regulations and standards, which demands data provenance. Compliance with these standards ensures that every stage of a project, from design through operation, maintains full data traceability, supporting regulatory compliance and project accountability.	In the energy and utilities sector, digital twins are adopted to optimise critical infrastructure, such as smart grids. In this domain, data provenance is crucial for real-time asset monitoring and the use of blockchain to ensure data integrity is becoming increasingly common. Real-time data management and predictive maintenance requires detailed provenance tracking to meet regulatory and operational standards.

10.7 Prospecting for enterprise digital twin data architectures

This exploration of the triadic semantic framework – function, sequence and attribution – examines how blockchain enhances data observability within the data architectures of DT, BIM and, more broadly, E-CDEs.

10.7.1 Functions

In the context of enterprise data management functions, data observability is crucial not only for BIM but also for DTs and CDEs. Numerous solutions adjacent to BIM and DT development such as ECM, IMS for enterprise governance, Requirements Engineering CDEs, GIS, land access and requirements management systems (LARMS), technical and solution libraries and BI data sources, also play significant roles. This practice extends beyond merely capturing and monitoring diverse data types; it encompasses the management and maintenance of data quality, availability and reliability. Ensuring the health and integrity of interoperable data across these various sources, platforms, systems and processes is vital for E-CDEs. Conversely, these adjacent solutions can greatly enhance the applicability and value of BIM and consequently DTs.

Blockchain technology plays a transformative role in this complex environment. Its application across various platforms aligns with the fundamental need for accurate, transparent and reliable data. Blockchain's ability to create a sovereignless, verifiable, immutable record of data transactions significantly bolsters the overarching goals of E-CDEs: maintaining data integrity, facilitating compliance and enabling effective data governance. This is particularly relevant in government mega-projects, as discussed in this chapter.

In BIM processes, blockchain ensures that every modification and transaction is recorded and traceable through various documentation processes, including journaling, version control via synchronisations of local models and the exchange and application of BIM objects. This feature is invaluable in complex project data management, where tracking data provenance, design revisions and accountability chains are critical. Blockchain's decentralised nature enhances the security and transparency of data, making it a key aspect of observability. Instead of relying on a central database, data is distributed across a network, mitigating risks of tampering and bolstering data reliability.

10.7.2 Sequences

Enterprises and projects should ideally begin with effective requirements development to structure the information architectures for various data sources and orchestrate them from the enterprise level to specific design solutions.

To maximise the utility of digital engineering, data observability in BIM must be meticulously considered and traceable back to the initial requirements. This development should not only ensure compliance with standards such as ISO 19650 but also anticipate the needs for DT development and enterprise BI. Accurate configuration of the total digital lifecycle ensures effective simulation, analysis and maintenance for the business and its components. This integrated, reliable and effective process supports decision-making that is predictive, preventative and precise.

In developing the E-CDE, elements such as ECM, LARMS, BI, IMS, Requirements and CRM are considered at the enterprise level to equip the engineering team with the capabilities needed to develop BIM, GIS and Technical Libraries. These elements, in turn, inform the data architecture of DT development.

This process can be orchestrated as a data hub, which must consider the sequencing of data production and transformation at the network edge, offloading this data to shore bases (e.g. in railways and other critical infrastructure projects) for local download and processing at locations where some data collection may occur offline. The shore base then consolidates this data in the cloud. Blockchain technologies, particularly BlockDAGs, can manage both online and offline consensus, using DLT as an overlay for DT decentralised systems to ensure immutable observability of these multiple layers of data transactions.

In sequencing the data architecture, several strategic steps are essential for data observability in BIM and DT contexts. These include:

- Defining key metrics: Tailored to the complexities of data pipelines to monitor crucial aspects like spatial accuracy and real-time data synchronisation.
- Selecting compatible tools: Capable of handling the multi-dimensional data typical in BIM and DTs.
- Standardising libraries: Creating a unified infrastructure for API and data management to facilitate seamless team integration and communication.
- Instrumenting the data pipeline: Embedding data collection tools at each stage to ensure comprehensive data capture and monitoring.
- Establishing scalable storage solutions: Accommodating extensive datasets generated by BIM and DT technologies.
- Utilising advanced data analysis tools: For in-depth analysis and intuitive visualisations.
- Configuring alerts and notifications: For swift issue detection and resolution, integrated with incident management platforms to enhance response efficiency.
- Regular reviews and updates: Continuously adapting the observability framework to match the evolving nature of BIM and DT technologies.

These steps ensure that the data architecture supports the goals of DT development, maintaining high standards of data quality, availability and reliability across diverse systems and processes.

10.7.3 Attributions

Data observability in BIM and DTs can improve data quality, expedited troubleshooting, enhanced collaborative efforts, by improving current attributions of the CDE architecture but also enabling new attributions in this architecture.

It is underpinned by five pillars – freshness, distribution, volume, schema and lineage – each integral to understanding the quality and reliability of complex data systems. While data observability bolsters data quality and supports governance, it is distinct from ensuring good data quality. It forms a critical component in the hierarchy of data observability, providing a comprehensive view of data's operational and qualitative status within the spheres of BIM and DT development.

The potential for blockchain to revolutionise data management can be further realised through the application of smart contracts computed in the blockchain. This integration can automate the consumption of observable data, leading to higher levels of data orchestration.

Smart contracts can be programmed to execute specific actions when predetermined data conditions are met. For an instance of this conditional data handling, in a BIM process, a smart contract could automatically trigger a review process if the data shows deviations from the standard architectural specifications.

Implementing smart contracts for data lineage allows for automatic tracking of data changes over time. This feature is essential in complex systems like DTs, where understanding the history and evolution of data is key to maintaining system integrity.

Smart contracts can facilitate secure and automated data sharing between different stakeholders, streamlining collaboration in multi-dimensional data environments. This enhanced alignment and collaboration ensure that data sharing adheres to predefined rules, enhancing collaboration efficiency while maintaining data security.

Decentralised autonomous organisations (DAOs), powered by blockchain technology, offer a robust solution for automating data governance protocols [92]. By embedding data governance rules into smart contracts, DAOs ensure strict adherence to these rules without centralised oversight. This mechanism maintains data quality and compliance across various platforms, mitigating the risks associated with manual governance processes [92,93].

Facilitating real-time data management: In dynamic environments such as BIM and DTs, real-time data management is paramount. DAOs enable automated decision-making processes based on real-time data inputs, thereby significantly reducing response times and enhancing operational efficiency. By leveraging blockchain's immutable ledger, DAOs ensure the integrity and reliability of real-time data, fostering trust among stakeholders.

Enabling decentralised data verification: Leveraging blockchain's decentralised architecture, DAOs facilitate robust data verification across multiple nodes. This decentralised approach enhances data observability by ensuring higher reliability and transparency, thereby mitigating the risk of tampering and inaccuracies. Through consensus mechanisms inherent in blockchain technology, DAOs verify data integrity autonomously, eliminating the need for centralised verification authorities.

Integrating blockchain with advanced data analysis: By integrating blockchain with sophisticated data analysis tools, organisations can unlock new capabilities for automated data analysis and insights generation. This integration enables the development of advanced analytics models that continuously learn and improve from the data stored on the blockchain. As a result, organisations can derive deeper insights, identify patterns and make data-driven decisions with greater precision and agility.

Data mesh architecture is a more recent advent and a decentralised approach to data management that emphasises domain-oriented data ownership and architecture. It aims to address the scalability and agility challenges faced by traditional monolithic data architectures.

Domain-oriented decentralisation allows domain-specific teams to manage their data, reducing bottlenecks and enhancing efficiency. This decentralisation leverages the concept of marginal utility by optimising resource usage and reducing inefficiencies, which incrementally improves operational efficiency across the enterprise.

Self-serve data infrastructure gives users the freedom to access and manage data on their own, boosting productivity and satisfaction. This feature embodies consumer surplus utility concepts, as it allows users to extract more value from the systems than what they invest, enhancing collaborative aspects in the CDE.

Product thinking for data focuses on treating data as a product that meets high standards of quality and usability. Emphasising total utility, this approach ensures that all users derive the maximum benefit from using the data, which is crucial in projects where data integrity directly affects outcomes.

Federated computational governance implements flexible yet standardised governance policies across different domains. This strategy maximises expected utility by balancing control with flexibility, ensuring compliance and operational efficiency across complex projects.

Unfortunately, as edge devices proliferate – including mobile handsets and IoT sensors – new attack vectors are emerging that take advantage of the proliferation of endpoints. The use of Edge-Shore-Cloud data architectures for enterprise DTs created more surfaces of vulnerability. They also offer the ability to move security elements closer to the originating source of attack, enable higher performance security applications and increase the number of layers that help defend the core against breaches and risk.

Given that ensuring observability is a fundamental characteristic of effective E-CDEs, the adoptions of these integrated functions, how they are laid as a system of systems and adopt new attributions like the ones required for the functioning of smart contracts (including their legal form) as well as DAOs.

10.8 Conclusion

The E-CDE is undergoing a significant transformation, evolving into a sophisticated ecosystem characterised by diverse and varied data sources acted upon by increasingly diverse and specialised human and machine agents. This ecosystem consists of numerous interconnected components working together to streamline organisational processes and optimise resources, ultimately enhancing the delivery of services and products with greater efficiency, cost-effectiveness and sustainability. Within this dynamic framework, some components engage in coordinated data interactions, while others operate independently.

In the presented case study and micro-ethnography, we discuss the critical data capabilities inherent in E-CDEs, particularly within the context of a major infrastructure project in the UK. Recognising the pivotal role of the CDE and its endogenous systems like BIM and DTs of attaining diverse utility to in maximising value across the lifecycle of an organisation's assets, it is essential to understand and effectively address these as enterprise requirements. This understanding is crucial for refining data architectures, which in turn facilitate value realisation across various stages such as architectural development, transition planning, governance and capability improvement.

At the heart of this discussion lies the holistic understanding of data architecture throughout the enterprise lifecycle. This understanding necessitates the integration of key considerations into BIM processes during project design, delivery and operational phases, aligning them with the data architecture principles of DTs. Concepts such as data lineage and observability emerge as critical for enhancing BI, as it does open standards for big BIM development.

Both BIM and DT development, as driven by enterprises, require this observability for establishing responsibility and accountability chains. Blockchain and adjacent technologies emerge as a cornerstone in enhancing data observability within BIM processes and, consequently, within the domain of DTs. This enhancement aligns with the core objectives of E-CDEs, which aim to maintain stringent standards of data transparency and integrity.

Blockchain technology provides a unified, observable data layer that seamlessly integrates with various CDEs, including ECM, IMS and GIS. This integration promotes a comprehensive and holistic view of information across the enterprise, essential for informed decision-making and strategic planning. Incorporating blockchain into the data architecture goes beyond mere additive functionality; it becomes a critical component reinforcing the principles of data provenance and accountability. This integration underscores the commitment to maintaining a transparent, reliable and efficient data management infrastructure within the E-CDE framework.

References

[1] Ghaleb H, Alhajlah HH, Bin Abdullah AA, Kassem MA, and Al-Sharafi MA (2022) A scientometric analysis and systematic literature review for construction project complexity. *Buildings* 12:482.

[2] Parsamehr M, Perera US, Dodanwala TC, Perera P, and Ruparathna R (2023) A review of construction management challenges and BIM-based solutions: Perspectives from the schedule, cost, quality, and safety management. *Asian Journal of Civil Engineering* 24:353–389.

[3] Moretti N, Xie X, Merino Garcia J, Chang J, and Kumar Parlikad A (2023) Federated data modeling for built environment digital twins. *Journal of Computing in Civil Engineering* 37:04023013.

[4] Yitmen I, and Alizadehsalehi S (2021) Towards a digital twin-based smart built environment. In: Yitmen I (ed.), *BIM-enabled Cognitive Computing for Smart*

Built Environment: Potential, Requirements, and Implementation. (1st ed.). Boca Raton, FL: CRC Press.

[5] Wayland J (2023) What is a golden record in MDM? Informatica. https://www.informatica.com/blogs/golden-record.html. Accessed 30 April 2024.

[6] Nakamoto S (2008) Bitcoin: A peer-to-peer electronic cash system. https://bitcoin.org/bitcoin.pdf. Accessed 7 July 2020.

[7] Tapscott D (2016) *Blockchain Revolution: How the Technology Behind Bitcoin Is Changing Money, Business, and the World*. Portfolio, New York.

[8] Kim EW, Park MS, Kim K, and Kim KJ (2022) Blockchain-based automatic tracking and extracting construction document for claim and dispute support. *KSCE Journal of Civil Engineering* 26:3707–3724.

[9] Keyte S (2023) Common data environments' impact on dispute resolution. BIM+. https://www.bimplus.co.uk/common-data-environments-impact-on-dispute-resolution/. Accessed 12 June 2023.

[10] KPMG (2021) The value of information management in the construction and infrastructure sector. https://kpmg.com/uk/en/home/insights/2021/06/the-value-of-information-management-in-the-construction-and-infr.html. Accessed 26 April 2024.

[11] BSI (2007) BSI PAS-1192-1:2007 Collaborative production of architectural, engineering and construction information. Code of practice.

[12] BSI (2013) BS PAS 1192-2:2013 Specification for information management for the capital/delivery phase of construction projects using building information modelling.

[13] BSI (2019) BSI ISO 19650 Webinar.

[14] AEChub (2022) Common data environment (CDE). https://aechub.io/common-data-environment-cde/. Accessed 10 October 2022.

[15] Turk Ž, Sonkor MS, and Klinc R (2022) Cybersecurity assessment of BIM/CDE design environment using cyber assessment framework. *Journal of Civil Engineering and Management* 28:349–364.

[16] Das M, Tao X, and Cheng JCP (2021) BIM security: A critical review and recommendations using encryption strategy and blockchain. *Automation in Construction*. 126:103682. https://doi.org/10.1016/j.autcon.2021.103682.

[17] Godager B, Mohn K, Merschbrock C, Klakegg OJ, and Huang L (2022) Towards an improved framework for enterprise BIM: The role of ISO 19650. *Journal of Information Technology in Construction* 27:1075–1103.

[18] Nanayakkara S, Rodrigo MNN, Perera S, Weerasuriya GT, and Hijazi AA (2021) A methodology for selection of a blockchain platform to develop an enterprise system. *Journal of Industrial Information Integration* 23:100215. https://doi.org/10.1016/j.jii.2021.100215.

[19] BSI (2021) Organisation and digitisation of information about buildings and civil engineering works, including building information modelling (BIM). Information management using building information modelling. Part 1: Concepts and principles. British Standards Institution.

[20] BIM Dictionary (2020) Common data environment (CDE). https://bimdictionary.com/en/common-data-environment/2. Accessed 12 December 2022.

[21] Nanayakkara S, Perera S, Senaratne S, Weerasuriya GT, and Bandara HMND (2021) Blockchain and smart contracts: A solution for payment issues in construction supply chains. *Informatics* 8(2):36. https://doi.org/10.3390/informatics8020036.

[22] Bhardwaj A, Karger D, Subramanyam H, *et al.* (2015) Collaborative data analytics with DataHub. *Proceedings of the VLDB Endowment* 8:1916–1919.

[23] Haße H, van der Valk H, Möller F, and Otto B (2022) Design principles for shared digital twins in distributed systems. *Business & Information Systems Engineering* 64:751–772.

[24] Gadepally V, and Kepner J (2020) Technical report: Developing a working data hub. https://doi.org/10.48550/arXiv.2004.00190.

[25] Javaid M, Haleem A, and Suman R (2023) Digital twin applications toward Industry 4.0: A review. *Cognitive Robotics* 3:71–92.

[26] Kiu MS, Lai KW, Chia FC, and Wong PF (2024) Blockchain integration into electronic document management (EDM) system in construction common data environment. *Smart and Sustainable Built Environment* 13(1):117–132. https://doi.org/10.1108/SASBE-12-2021-0231.

[27] Esri (2017) Esri: GIS mapping software, spatial data analytics & location platform. http://www.esri.com/. Accessed 23 August 2017.

[28] Maletič D, Marques de Almeida N, Gomišček B, and Maletič M (2023) Understanding motives for and barriers to implementing asset management system: an empirical study for engineered physical assets. *Production Planning & Control* 34:1497–1512.

[29] Reis J, and Housley M (2022) *Fundamentals of Data Engineering: Plan and Build Robust Data Systems.* O'Reilly Media, Sebastopol, CA.

[30] Oracle (2024) What is a data catalog? https://www.oracle.com/uk/big-data/data-catalog/what-is-a-data-catalog/. Accessed 11 May 2024.

[31] Walmsley K (2022) The past and future of digital twins at autodesk: From project dasher to autodesk tandem and beyond. In: *Design Computation Input/Output 2021.* https://doi.org/10.47330/DCIO.2021.ENAB9070.

[32] Huawei Technologies Co. Ltd (ed.) (2023) Introduction to databases. In: *Database Principles and Technologies – Based on Huawei GaussDB.* Springer Nature, Singapore, pp. 1–39.

[33] Kaufmann M, and Meier A (2023) Database management. In: Kaufmann M, and Meier A (eds.) *SQL and NoSQL Databases: Modeling, Languages, Security and Architectures for Big Data Management.* Springer Nature, Cham, pp. 1–24.

[34] Budianto G (2022) Data warehouse modeling using online analytical processing approach. *Jurnal Ilmiah Informatika dan Ilmu Komputer (JIMA-ILKOM)* 1:7–13.

[35] Zagan E, and Danubianu M (2023) Data lake architecture for storing and transforming web server access log files. *IEEE Access* 11:40916–40929.

[36] Haelen B, and Davis D (2023) *Delta Lake: Up and Running.* O'Reilly Media, Inc., Sebastopol, CA.

[37] Kamalakkannan S, Yasmin A, R Arunkumar, and Kavitha P (2023) A model for the analytical performance of data lake in stock market analysis with

databricks delta lake. In*: 2023 International Conference on Self Sustainable Artificial Intelligence Systems (ICSSAS)*. pp. 1065–1071.

[38] Velepucha V, and Flores P (2023) A survey on microservices architecture: Principles, patterns and migration challenges. *IEEE Access* 11:88339–88358.

[39] Ghosh AM, and Grolinger K (2021) Edge-cloud computing for internet of things data analytics: Embedding intelligence in the edge with deep learning. *IEEE Transactions on Industrial Informatics* 17:2191–2200.

[40] Deer M (2021) What is a directed acyclic graph in cryptocurrency? How does DAG work? Cointelegraph. https://cointelegraph.com/explained/what-is-a-directed-acyclic-graph-in-cryptocurrency-how-does-dag-work. Accessed 17 October 2024.

[41] Gill SS, Kumar A, Singh H, *et al.* (2022) Quantum computing: A taxonomy, systematic review and future directions. *Software: Practice and Experience* 52:66–114.

[42] Atitallah SB, Driss M, and Ghzela HB (2022) Microservices for data analytics in IoT applications: Current solutions, open challenges, and future research directions. *Procedia Computer Science* 207:3938–3947.

[43] Xu Y, Tao X, Das M, *et al.* (2023) Suitability analysis of consensus protocols for blockchain-based applications in the construction industry. *Automation in Construction* 145:104638.

[44] EUGDPR (2019) EUGDPR – Information portal. https://www.gdpreu.org/. Accessed 3 June 2019.

[45] CLI (2018) California Consumer Privacy Act of 2018 (CCPA) – AB-375 privacy: Personal information: Businesses. https://leginfo.legislature.ca.gov/faces/billTextClient.xhtml?bill_id=201720180AB375. Accessed 11 May 2024.

[46] Spradley JP (1980) *Participant Observation*, 1st ed. Holt, Rinehart & Winston, New York.

[47] Leibniz GW, and Loemker LE (1970) *Philosophical Papers and Letters*. D. Reidel, Dordrecht.

[48] Smith Q (2011) *Einstein, Relativity and Absolute Simultaneity*, 1st ed. Routledge, London.

[49] Yourgrau W (2003) *Variational Principles in Dynamics and Quantum Theory*, 3rd ed. Dover Publications Inc., New York.

[50] Paskaleva G, Mazak-Huemer A, Sint S, and Bednar T (2022) Standardized data integration in the AEC domains – What does it take to succeed? *IOP Conference Series: Earth and Environmental Science* 1101:082034.

[51] Lou J, Lu W, and Xue F (2021) A review of BIM data exchange method in BIM collaboration. In: Lu X, Zhang Z, Lu W, Peng Y (eds.) *Proceedings of the 25th International Symposium on Advancement of Construction Management and Real Estate*. Springer, Singapore, pp. 1329–1338.

[52] Kassem M (2015) BIM in facilities management applications: A case study of a large university complex. *Built Environment Project and Asset Management* 5(3):261–277. https://doi.org/10.1108/BEPAM-02-2014-0011.

[53] Peacock D (2022) 10 ways BIM can reduce the costs of your construction projects. TÜV SÜD. https://www.tuvsud.com/en-gb/resource-centre/stories/10-ways-bim-can-reduce-the-costs-of-your-construction-projects. Accessed 14 February 2024.

[54] ICE (2024) BIM and asset management factsheet. Institution of Civil Engineers (ICE). https://www.ice.org.uk/engineering-resources/briefing-sheets/bim-and-asset-management-factsheet/. Accessed 14 February 2024.

[55] NBS (2024) What is the asset information model (AIM)? NBS. https://www.thenbs.com/knowledge/what-is-the-asset-information-model-aim. Accessed 14 February 2024.

[56] Valra A, Madeddu D, Chiappetti J, and Farina D (2021) The BIM management system: A common data environment using linked data to support the efficient renovation in buildings. *Proceedings of the 8th Annual International Sustainable Places Conference* 65:18.

[57] Chang K-M, Dzeng R-J, and Wu Y-J (2018) An automated IoT visualization BIM platform for decision support in facilities management. *Applied Sciences* 8:1086.

[58] Condotta M, and Scanagatta C (2023) BIM-based method to inform operation and maintenance phases through a simplified procedure. *Journal of Building Engineering* 65:105730.

[59] Wanigarathna N, Jones K, Bell A, and Kapogiannis G (2019) Building information modelling to support maintenance management of healthcare built assets. *Facilities* 37:415–434.

[60] Wijeratne PU, Gunarathna C, Yang RJ, Wu P, Hampson K, and Shemery A (2024) BIM enabler for facilities management: A review of 33 cases. *International Journal of Construction Management* 24:251–260.

[61] Bejtullahu F, Nushi V, and Jakupi E (2015) Architects role in change influenced by technology and increasing urban data (BIM, GIS and CODES) from global and local perspective. *UBT International Conference*. pp. 46–52. doi:10.33107/ubt-ic.2015.59.

[62] Patacas J, Dawood N, and Kassem M (2020) BIM for facilities management: A framework and a common data environment using open standards. *Automation in Construction* 120:103366.

[63] Tsay GS, Staub-French S, and Poirier É (2022) BIM for facilities management: An investigation into the asset information delivery process and the associated challenges. *Applied Sciences* 12:9542.

[64] Yalcinkaya M, and Singh V (2014) Building information modeling (BIM) for facilities management – Literature review and future needs. In: Fukuda S, Bernard A, Gurumoorthy B, and Bouras A (eds.) *Product Lifecycle Management for a Global Market*. Springer, Berlin, pp. 1–10.

[65] Janiak A (2021) *Newton's Philosophy. The Stanford Encyclopedia of Philosophy*. Stanford, CA: Metaphysics Research Lab, Stanford University.

[66] Lorek S (2022) *What Is BIM (Building Information Modeling)*. Trimble Construction. https://constructible.trimble.com/construction-industry/what-is-bim-building-information-modeling. Accessed 20 Feb 2024.

[67] Borrmann A, König M, Koch C, and Beetz J (eds.) (2018) *Building Information Modeling: Technology Foundations and Industry Practice*. Cham: Springer. https://doi.org/10.1007/978-3-319-92862-3.

[68] RIBA (2020) RIBA plan of work 2020 overview. https://www.architecture.com/knowledge-and-resources/resources-landing-page/riba-plan-of-work. Accessed 18 January 2021.

[69] Karasu T, Aaltonen K, and Haapasalo H (2022) The interplay of IPD and BIM: A systematic literature review. *Construction Innovation* 23: 640–664.

[70] Bucher D, and Hall D (2020) Common data environment within the AEC ecosystem: Moving collaborative platforms beyond the open versus closed dichotomy. In: *EG-ICE 2020 Proceedings: Workshop on Intelligent Computing in Engineering*, pp. 491–500. https://doi.org/10.3929/ETHZ-B-000447240.

[71] Baccelli J (2024) Ordinal utility differences. *Social Choice and Welfare* 62:275–287.

[72] Churcher D, Davidson S, and Kemp A (2019) Information management according to BS EN ISO 19650: Guidance Part 1: Concepts.

[73] Jedermann R, Singh K, Lang W, and Mahajan P (2022) Digital twins for flexible linking of live sensor data with real-time models. In: *Sensors and Measuring Systems; 21st ITG/GMA-Symposium*, pp. 1–7.

[74] Zheng H, Liu T, Liu J, and Bao J (2024) Visual analytics for digital twins: A conceptual framework and case study. *Journal of Intelligent Manufacturing* 35:1671–1686.

[75] Akroyd J (2022) CReDo Technical Paper 1: Building a cross-sector digital twin.

[76] Correia JB, Abel M, and Becker K (2023) Data management in digital twins: A systematic literature review. *Knowledge and Information Systems* 65:3165–3196.

[77] van Dinter R, Tekinerdogan B, and Catal C (2022) Predictive maintenance using digital twins: A systematic literature review. *Information and Software Technology* 151:107008.

[78] Pernagallo G (2024) Overcoming asymmetric information: A data-driven approach. In: Raban DR, and Włodarczyk J (eds.), *The Elgar Companion to Information Economics*. Edward Elgar Publishing, Northampton, MA, pp. 135–153.

[79] Sacks R, Brilakis I, Pikas E, Xie H, and Girolami M (2020) Construction with digital twin information systems. *Data-Centric Engineering* 1:26.

[80] Fan Y, Yang J, Chen J, *et al.* (2021) A digital-twin visualized architecture for flexible manufacturing system. *Journal of Manufacturing Systems* 60:176–201.

[81] Papic D, and Cerovsek T (2019) Digital built environment maturity model: Digital twins advancing smart infrastructure asset management. In: *European Conference on Computing in Construction*, pp. 387–394.

[82] Google (2024) Gemini – Chat to supercharge your ideas. Gemini. https://gemini.google.com. Accessed 23 April 2024.

[83] Microsoft (2024) Copilot for Microsoft 365 – Business plans | Microsoft 365. https://www.microsoft.com/en-gb/microsoft-365/business/copilot-for-microsoft-365. Accessed 23 April 2024.

[84] OpenAI (2024) Introducing ChatGPT. https://openai.com/blog/chatgpt. Accessed 23 April 2024.

[85] Google (2024) What is a long context window? Google. https://blog.google/technology/ai/long-context-window-ai-models/. Accessed 23 April 2024.

[86] Sadri H, Yitmen I, Tagliabue LC, *et al.* (2023) Integration of blockchain and digital twins in the smart built environment adopting disruptive technologies – A systematic review. *Sustainability* 15:3713.

[87] La Guardia M, and Koeva M (2023) Towards digital twinning on the web: Heterogeneous 3D data fusion based on open-source structure. *Remote Sensing* 15:721.

[88] Pinmas P, Panyadee P, Kaewmoracharoen M, Khueankhan S, and Wannachai A (2022) Towards digital twin data center using building information modeling and real-time data sensing. In: *2022 37th International Technical Conference on Circuits/Systems, Computers and Communications (ITC-CSCC)*. pp. 908–911.

[89] Hossein Motlagh N, Zaidan MA, Lovén L, *et al.* (2024) Digital twins for smart spaces – Beyond IoT analytics. *IEEE Internet of Things Journal* 11:573–583.

[90] Kong L, Chen C, Zhao R, *et al.* (2022) When permissioned blockchain meets IoT oracles: An on-chain quality assurance system for off-shore modular construction manufacture. In: *2022 IEEE 1st Global Emerging Technology Blockchain Forum: Blockchain & Beyond (iGETblockchain*). pp. 1–6.

[91] Zheng X, Psarommatis F, Petrali P, Turrin C, Lu J, and Kiritsis D (2020) A quality-oriented digital twin modelling method for manufacturing processes based on a multi-agent architecture. *Procedia Manufacturing* 51:309–315.

[92] Faqir Y, Arroyo J, and Hassan S (2020) An overview of decentralized autonomous organizations on the blockchain. In: *16th International Symposium on Open Collaboration*, pp. 1–8.

[93] Reiff N (2019) What is a DAO? Investopedia. https://www.investopedia.com/tech/what-dao/. Accessed 23 May 2021.

Chapter 11

Blockchain and contract administration in construction

Mahir Msawil, David Greenwood* and Mohamad Kassem†*

11.1 Introduction

Construction contract administration (CCA) defines the processes for fulfilling the obligations and protecting the rights of the contracting parties through the duration of the contract. The effectiveness of CCA is indispensable for achieving project success for all contracting parties. However, poor CCA performance remains one of the most significant challenges in the industry, leading to delays, cost overruns and frequent disputes. There is abundant evidence of CCA failure in the industry and recent reports by Arcadis[1] and HKA[2] show that poor CCA continues to be one of the top ranked causes of disputes. Poor CCA can arise as a result of the individual or collective misinterpretation, misapplication, negligence or refusal to operate the contractual provisions and their associated processes, as well as incomplete information and contractual provisions. This can result in adverse effects (e.g. legal disputes) and ripple effects (e.g. bankruptcy of organisations involved). Hence, it is not difficult to appreciate the need for a solution to the problem of poor CCA while recognising that such a solution should provide mutual benefits to all involved organisations. As long ago as 1979 the eminent civil engineering lawyer Max Abrahamson[3] advocated the use of a digital solution for such contractual challenges [1]. An emerging digital innovation that could confront these challenges is blockchain technology.

*Department of Architecture and Built Environment, Northumbria University, UK
†Department of Engineering, Newcastle University, UK

[1]https://www.arcadis.com/en-us/knowledge-hub/blog/united-states/benjamin-eiss/2024/disputes-in-the-digital-age-findings-of-the-2024-construction-disputes-report

[2]https://www.hka.com/crux/

[3]Max Abrahamson (29 October 1932–7 October 2018) was one of the most internationally recognised experts in construction law. He is most known for the Abrahamson Principles often used in drafting construction contracts.

The advent of blockchain technology and its associated smart contracts offers an opportunity to rethink CCA thereby mitigating or even eliminating current CCA problems. Blockchain technology is one of the emerging digital innovations that offers the necessary features required for improved CCA. It offers traceability, immutability, security and decentralisation of digital records of transactions that can be automated by embedded smart contracts. According to Mason [2], a smart contract is an automated protocol of coded instructions that self-execute upon the fulfilment of pre-determined conditions. The resulting transactions are recorded digitally with benefits to contracting parties that include enhanced audibility, accountability, transparency, communication and clearly defined roles and responsibilities [3]. The unique characteristics of blockchain and smart contracts are together likely to form some of the key features required by a digital solution for CCA improvement.

This chapter focuses on a specific application of blockchain-based solutions, namely CCA. It first considers the contractual processes involved and the problems they typically entail, and then, by accessing the latest literature, reflecting on how CCA might be automated using blockchain to address the problems. To contextualise this, relevant contractual processes and provisions are extracted from *FIDIC*[4] *Red Book* (RB) 2017 [4]. This is a standard contract model produced by FIDIC and it is widely adopted worldwide for infrastructure projects procured through the design-bid-build procurement route. The applicability of selected blockchain-based solutions to these contractual processes and provisions is analysed. Finally, an agenda is established for researchers and contract drafting bodies, the underlying motivation being to transform CCA in line with the digitalisation of other construction project management processes.

11.2 Construction contract administration as a multi-functional system

In the delivery phase of construction projects, a contract refers to legally binding agreements that formalise inter-organisational relationships between the contracting parties; for example, between the employer and contractor [5]. These agreements specify a set of conditions of contract. For large and complex construction projects, independent associations have produced so-called 'standard forms of construction contract', such as the Joint Contracts Tribunal (JCT[5]) and the New Engineering Contract (NEC[6]) by the Institution of Civil Engineers (in the UK) and the American Institute of Architects[7] (USA). For a more international applicability, a standard form for international use was produced by FIDIC [6].

[4]FIDIC is a French acronym for 'Fédération Internationale des Ingénieurs-Conseils', which translates into English as 'the International Federation for Consulting Engineers'.
[5]https://www.jctltd.co.uk/category/contract-families
[6]https://www.neccontract.com/products/contracts
[7]https://learn.aiacontracts.com/articles/6150803-list-of-all-current-aia-contract-documents/

Ideally, these standard forms of contracts are often devised to specify the following six components:

- Stipulation of the rights, obligations and liabilities of the contracting parties;
- Allocation of risk between the contracting parties;
- Definition of remedies for materialised risks or breached obligations;
- Procedures to be followed by the contracting parties or their agents;
- Set timeframes to initiate and/or complete these procedures; and
- Dispute resolution mechanisms (e.g. adjudication).

Accommodating these components in a standard form of contract results in a complex variety of inter-related and cross-referenced contractual clauses, sub-clauses and provisions. These establish the various contractual procedures that must be adhered to for an effective and efficient CCA. Furthermore, these must be regularly revised to reflect changing business practices and evolving legal doctrine. CCA has been established as an operating system that applies the contractual provisions of the contract *ex post* at the project level [7–10]. This application ensures the fulfilment of defined contractual obligations and protection of defined contractual rights of the contracting parties. In so doing, the aim of CCA is to achieve intended outcomes for the parties; in particular their financial interests [7] in terms of employer's realisation of investments and contractor's profitability of its involvement in the project delivery [11]. To that effect, it can be established that the CCA system executes the day-to-day formal activities associated with various contractual mechanisms and processes. It is during this day-to-day administration that the agents acting on behalf of the involved organisations apply and refer to contractual provisions [12–14].

Given the complexity of construction contracts and their dispersed conditions, a rational approach to their functional classification is required. According to the functional approach to contract [15,16], CCA involves interactive application of three mechanisms application of three mechanisms – *control, coordination and adaptation* – that embody the aforementioned six key components of a construction contract [38]. A more detailed analysis of CCA functions is provided by Gunduz and Elsherbeny [9] who list 11 functions. These two classifications are combined as shown in Figure 11.1, with the slight modification of splitting the single function *claims and dispute resolution management* found in Gunduz and Elsherbeny [9] into two constituent parts. This modification results in the model, shown in Figure 11.1, of 12 CCA functions at the project level and three mechanisms at the interorganisational level. This will be used for subsequent analysis.

As Figure 11.1 shows, the 12 functions represent various contractual processes mapped to the three mechanisms of the functional approach to contract design all of which interact with one another. For example, the financial management process that deals with payment-related matters is classified under the *control* mechanism, but it also relies on other *coordination*-related processes including the documents and records process as well as the contract closeout process. Likewise, the claims management process is classified as an *adaptation* mechanism, but its execution hinges on other *control*- and *coordination*-related processes. Despite the presence

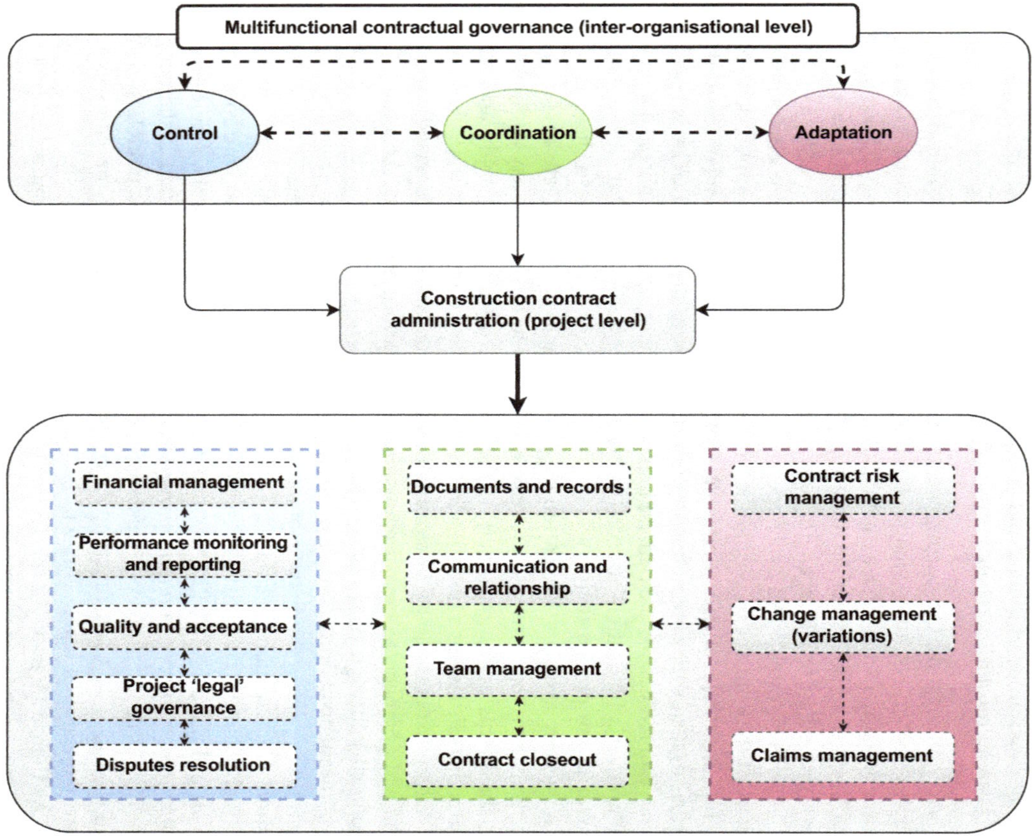

Figure 11.1 CCA functions (adapted from [9,16])

of processes dealing with the functions of contract design, CCA in real-world practice continues to face many inherent challenges, as next illuminated.

11.3 Challenges facing CCA (causes and effects of poor CCA)

At the project level, the resulting complexity can create confusion and poor performance among construction practitioners involved in carrying out the requisite contractual procedures. For the contractor, these duties (collectively known as contract administration) are carried out by a specialist employee. On the employer's side, however, these tasks are usually performed by an appointed agent, often the employer's lead consultant and labelled variously as *contract administrator*, *architect*, *project manager*, *employer's representative* or the *engineer*[8] depending on the particular standard form of contract. Interestingly, and somewhat controversially, this agent is frequently required to play a dual role: acting on behalf of the employer in dealing with the contractor but on occasion making 'independent' contractual decisions between the contractor and employer (see, e.g., Ndekugri *et al.* [17]). When these decisions are not in the contractor's favour or differ from its expectations, disputes can arise.

Examples of dispute cases from various jurisdictions around the world can be found in a list[9] produced by Howard Kennedy (a London-based law firm). Furthermore, Chappell [18], in his book, discusses 300 contractual issues while offering insights into their resolutions based on the body of case law. Both sources illustrate the practical challenges faced by CCA through numerous instances where disputes arose due to agents failing to: (i) apply contractual procedures, (ii) understand notice requirements, (iii) interpret contractual provisions, (iv) avoid valuation discrepancies and (v) implement agreed or adjudicated decisions. These failures compromise the efficiency and effectiveness of CCA (i.e. poor CCA). For example, in a highway construction project, poor CCA can lead to unclear project completion date and nonpayment-related disputes due to unresolved time extension claims. These represent the basis for the contractor's inability to execute remaining obligations (e.g. construction works) leading to poor project performance and a lack of trust between the organisations involved. The key message of this example is that poor CCA can lead to poor performance of construction projects and that this can have a 'ripple effect' on the industry itself. Hence, there is a clear need for improvement; and this demands revealing underlying factors that trigger such failures and their adverse effects. Based on a review of the relevant scholarship, Figure 11.2 shows the synthesised factors associated with poor CCA.

Figure 11.2 organises the synthesised factors into the following four groups:

- Root causes: behind each of the 'Causes' shown in Figure 11.2, there are undoubtedly 'Root Causes'. These factors have been abstracted to organisational,

[8]Henceforward the terminology adopted to describe the employer's agent will be that used in the FIDIC contract, namely, 'The Engineer'.

[9]https://insights.howardkennedy.com/post/102j8xv/table-of-fidic-cases-may-2024-update

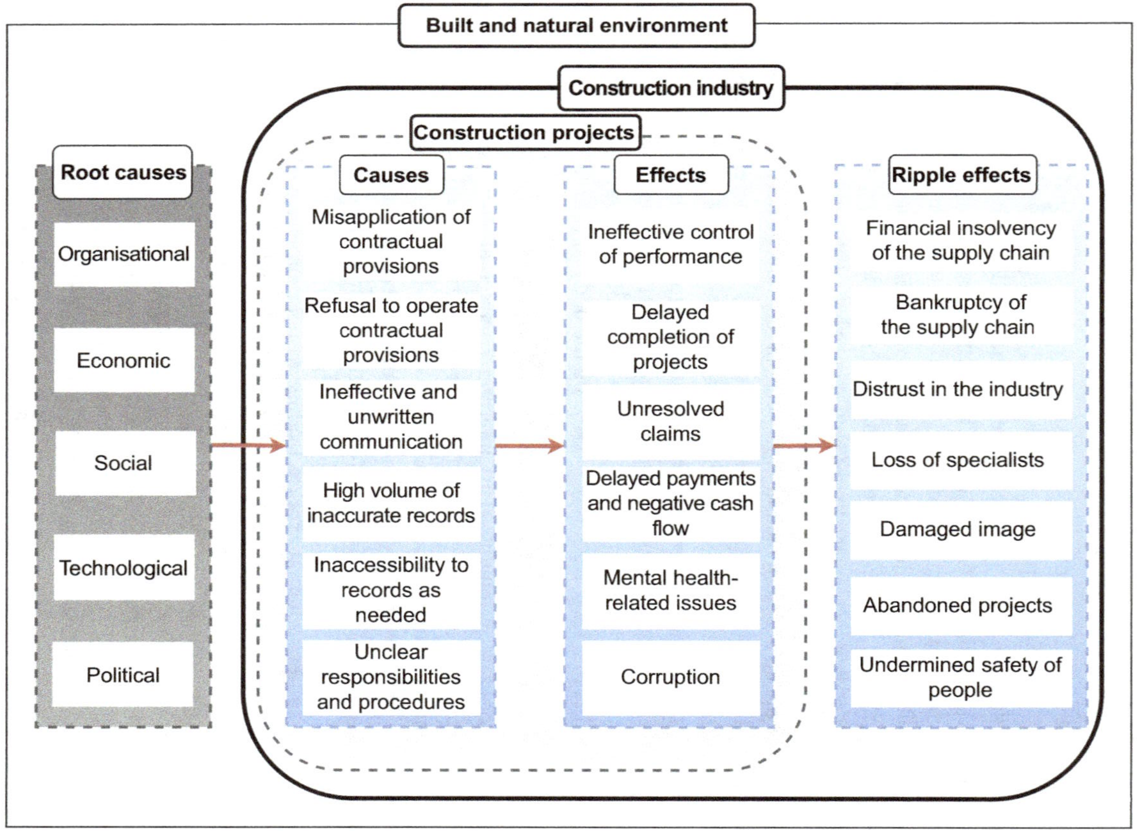

Figure 11.2 Causes and effects of poor CCA and their relation to the wider construction industry

economic, social, technological and political. An example of organisational factors is opportunistic behaviour [19]. According to Flyvbjerg [20], optimism bias and misinformation represent social factors, and technological factors include information asymmetry, while strategic misrepresentation may be an additional 'political' factor.

- Causes: this group demonstrates common challenges, encountered on construction projects, which tend to be often labelled as causes of poor CCA, many of them are behavioural. For example, misapplication of contractual provisions (by any party) might be the result of incompetence, ignorance or commercial opportunism. These factors can be encapsulated under the umbrella of unfulfillment and unclarity of the contractual components at the project level [21–23].
- Effects: this group indicates adverse factors that may directly or indirectly result from the identified causes. These factors impose adverse effects on the overall performance of construction projects, the organisations involved and their respective staff and agents [24–26]. A notable factor is the deterioration of mental health of staff involved in a poor CCA-related issue as pressure mounts on them from their respective organisations to resolve the issue [27]. Fraud or corruption are also notable factors [28].
- Ripple effects: the factors shown under this group can be considered as the externalisation of the effects from the project environment into the construction industry environment. The factors illustrated under this group represent indicators that propagate into the industry at large. Notably, poor CCA-related bankruptcy cases of construction organisations continue to rise worldwide and even the 'giant cohort' is not immune. For example, poor administration of variation works on multiple large projects led to non-payment to Carillion, which ultimately went into bankruptcy in 2018,[10] in turn precipitating the collapse of many within its supply chain. This gives plausible reasons for the overall industry's stakeholders to continue their distrust in the industry.

Indeed, the foregoing paragraphs imply that poor CCA represents a failure in the construction industry that imposes significant adverse consequences on its performance. Of course, contract associations, industry practice-orientated organisations and scholarly efforts have attempted to reduce or eliminate the occurrence of the highlighted causes and severity levels of their effects with the aim of improving CCA. Notable examples demonstrating such attempts include the following:

- Revising standard forms of contract to reflect 'best practices' as a result of practice-based cases and legal cases accumulated over the years from construction-related disputes. Examples include: the introduction of FIDIC RB 2017 based on the earlier 1999 version, the update of *FIDIC Green Book* in 2021[11] and the introduction of a new suite of contracts by NEC in 2017.[12]

[10]https://committees.parliament.uk/work/5425/carillion-joint-inquiry/news/98025/qatari-developer-disputes-swathes-of-carillion-evidence/

[11]https://infra.global/fidic-unveils-new-and-updated-green-book-short-form-of-contract/

[12]https://www.neccontract.com/products/contracts

- Visualising contractual procedures and text by means of diagrams to assist CCA practitioners in grasping contractual workflows. An example of such diagrams can be found in NEC publications of flow charts for engineering and construction contracts.[13] But this is infrequent with other standard forms of contract, including FIDIC.
- Publishing reports that statistically demonstrate causes behind poor CCA and occurrence of disputes with sets of recommendations. Notable examples being annual Arcadis reports on disputes.[14]
- Proposing the adoption of relational contracting as complementary or substitute to conventional 'transactional' contracting [29,30] which can be traced back to Macneil [31] and is arguably encapsulated under the concept of trust [32,33]. In construction this approach has taken several different forms (e.g. with early contractor involvement or risk/benefit sharing) but with limited implementation [34].
- Conducting academic research such as the analytical work presented in Chen *et al.* [16] and classification work proposed in Gunduz and Elsherbeny [9].
- Developing digital solutions at research level that support individual contractual processes coupled with proposing their adoption and implementation in the construction industry.
- Exploring the potential of emerging technologies by contract drafting associations. A recent example being the FIDIC Digital Transformation Committee[15] formed in April 2022.

Despite the preceding attempts, CCA in practice continues to face many inherent challenges implying limited effectiveness of such attempts. Therefore, it can be established that CCA requires novel solutions orientated towards preventing, reducing or mitigating the causes and effects highlighted above. These solutions should offer, *inter alia*, trusted information, transparency, immutability of records, enhanced accountability and auditability, and corruption-resistant mechanisms. These required aspects are embedded in the unique features offered by blockchain technology and its associated smart contracts.

11.4 Blockchain technology and smart contracts in the context of CCA

The previous chapters, and in particular Chapter 2, have shed light on this emerging technology and its distinctive features including the incorporation of 'smart contracts' – a technology in its own right.

[13]https://www.neccontract.com/products/contracts/nec4/user-guides-and-flow-charts/nec4-engineering-and-construction-contract-flow-charts

[14]https://www.arcadis.com/en-us/knowledge-hub/perspectives/north-america/united-states/2023/construction-disputes-report-2023

[15]https://fidic.org/about/committees/digital-transformation-dtc

Many industries, including construction, already use blockchain, as demonstrated in a range of use cases described in this book. A blockchain records transactions and validates digital events (e.g. information) conducted in the network in form of encrypted 'blocks' and 'chains' the entire recorded transactions chronology stored across multiple nodes [35]. Blockchain has been described as a novel type of database in the form of a ledger [36] that operates on the integration of multiple core technological components: (i) peer-to-peer networks of nodes (computers), (ii) a shared distributed ledger of digital information or business transactions, (iii) cryptography that generates time-stamped digital signatures and hash values of tamper-proof records, (iv) consensus mechanisms that govern transactions and validate them before they are appended to the ledger and (v) persistent coded rules called 'smart contracts' that digitally execute processes and/or business terms.

The integration of these components produces the innate characteristics of blockchain. Of relevance to this chapter, they include (i) traceability of data contained in transactions, (ii) immutability of recorded transactions, (iii) transparency of transaction context and its visibility, (iv) instant auditability of recorded transactions, (v) consistency and authentication of recorded transactions across nodes, (vi) standardisation of processes and terms and (vii) automation of business logic through the execution of 'smart contract' codes [37–39].

These characteristics could pave the way for the following improvements in CCA:

1. The representation of actual contractual processes in line with the conditions of contract (e.g. FIDIC RB 2017). Such a feature is achieved by smart contracts.
2. Streamlining of contractual obligations and their transparent communication to the organisations and actors involved. This improvement results from decentralised and distributed ledgers.
3. Irrevocable automated execution of contractual obligations while adhering to contractual logic. Both blockchain-based smart contracts and immutability enable the realisation of this feature.
4. The administration of individual contractual processes while allowing visibility and isolation of their corresponding tasks. This is accomplished by the unique identification of digital tokens that represents various blockchain-based processes.
5. The agents are enabled to fulfil their contractual obligations while simultaneously notifying their respective organisations and the employer of such completed obligations. Achieving this is the result of the embedded visibility of the expected and completed contractual tasks and the traceability of their corresponding transactions on blockchain.
6. A trusted single source accessible to all involved agents and organisations. The core characteristic that enables this is the inherent immutability.
7. Accessible records of executed tasks that are accurate, authentic, unalterable and complete. This is achieved through immutability of records in the form of transactions.
8. Moderated behaviour of involved agents and their respective organisations. The expected moderation of behaviour is a by-product of a blockchain-based

process of system. For example, if an expected task is not performed within a certain time limit, the programmed smart contract may be able to autonomously perform the task.

9. Enhanced visibility of executed obligations while carrying unalterable proof of their completion and enable extraction of relevant data associated with a given contractual process while maintaining ensured reliability.

These improvements would enable better execution of *control*, *coordination* and *adaptation* mechanisms and their associated 12 processes (see Figure 11.1) for the mutual benefit of all involved. By targeting the causes of poor CCA the ripple effects suffered by the industry at large would be reduced.

Researchers and developers have explored how blockchain and smart contracts might be applied to a wide range of processes in construction management. Section 11.5 provides examples of these applications and the benefits they profess. Following which, their fit to CCA problems is discussed.

11.5 Mapping retrieved studies to the identified CCA functions

This section seeks scholarly evidence of the workability and applicability of blockchain technology to transform various contractual processes by exploring connections between the multi-functional CCA classification and related blockchain-orientated solutions in the literature.

By searching the Scopus and ASCE libraries,[16] and applying the inclusion and exclusion criteria shown in Table 11.1, a systematic search yielded 24 studies that included a relevant blockchain proof of concept. These were then mapped against the 12 CCA processes recognised earlier to see which CCA processes had been addressed.

Since each proof-of-concept was considered to offer a high level of plausible evidence that a specific use case can be operated with blockchain technology [3] they were then mapped against the 12 previously identified CCA processes to assess which of the CCA processes were covered by the studies. Additionally, their applicability to real contractual conditions was examined, and for this, the FIDIC RB 2017 was used as an exemplar of real contractual conditions.

In what follows, each of the 12 CCA functions is introduced and briefly described with its typical challenges highlighted. Subsequently, under each heading, the outcomes of the blockchain proof-of-concept studies mapped to the respective CCA function are described. Then, the potential applicability of the mapped studies to relevant examples extracted from FIDIC RB 2017 is considered. In what follows, the chapter turns to the analysis. For clarity of presentation, the analysis of the 12 CCA functions is grouped into their corresponding groups: *control*, *coordination* and *adaptation*.

[16]Keywords used included *blockchain, smart contract, construction,* and *contract administration.*

Table 11.1 Inclusion and exclusion criteria for the retrieved studies

Inclusion criteria	Exclusion criteria	Reason behind exclusion
Studies that focus on prevalent practices in the construction phase.	Academic review and conference studies.	Review studies provide aggregation while rigorous conference studies are converted into journal articles.
Studies that have proposed a proof-of-concept simulation.	Studies that are irrelevant to the construction phase (e.g. procurement and tendering) or prevalent practices (e.g. enhancing BIM security, cryptocurrency-based payment).	This chapter focuses on contractual processes during the construction phase.
	Studies that adopted questionnaires, interviews or focus groups without demonstrating a specific blockchain-based solution.	Given that blockchain technology has not diffused in the construction industry, how accurate and well-informed the respondents are on a subject is questionable.
	Replicated studies the ideas of which are published by the same authors in different journals.	Considering identical ideas encapsulated in the same framework or simulation does not serve this study.
	Studies where blockchain is discussed in the conclusion.	These studies do not contribute to chapter aim.

11.5.1 Blockchain applications for control-orientated CCA functions

11.5.1.1 Financial management

The *financial management* function reflects the rights and obligations, procedures, specified time limits for contracting parties in respect of payment. This includes advanced, interim and final payments along with the release of retention money. However, parties tend to either misinterpret or misapply, deliberately neglect or refuse to operate the corresponding contractual procedures, usually with the aim of improving or protecting their own financial position [21]. For example, a client may pay late because of lack of funds or a contractor may overclaim to improve its cash flow. Despite the availability of legal frameworks to enforce payments in construction worldwide, the issue of late payment and non-payment has persisted [40]. In this chapter, eight studies had developed blockchain-based proof-of-concept simulations with the aim of improving *financial management* and securing timely payment to contractors and their supply chains.

Das *et al.* [35] developed a blockchain-based framework to execute contractual provisions for payment procedure using a mixture of manual and smart contract-based processes. Chong and Diamantopoulos [41] exploited building information modelling (BIM) to provide progress data of installed construction elements to a blockchain network that enabled certification and triggered secure smart contract-based payment to subcontractors. In [42] blockchain, BIM and reality capture technologies (i.e. robots, sensors) were integrated to provide data to smart contracts [42]. Unlike the semi-automated approach of Chong and Diamantopoulos [41], Hamledari and Fischer [42] eliminated the need for payment submission as it was triggered by reality capture technologies that connected on-site progress data with smart contracts. Sigalov *et al.* [43] demonstrated a semi-automated model that connected off-chain BIM with blockchain-based smart contracts for payment from employers to contractors. A BIM-Hyperledger Fabric blockchain integration with smart contract functions for payment was also used in Elghaish *et al.* [44] for Integrated Project Delivery projects, and in Elghaish *et al.* [45] for more traditional Design-Bid-Build projects. Wu *et al.* [46] used a permissioned blockchain with mobile-based application to develop a 'freezing and disbursement' payment system to tackle late payment. This had encouraging results when applied in a shadow system on an actual project. In one of the most recent studies, Cheng *et al.* [47] used a Hyperledger Fabric-based system to record and transfer cost-related data between contracting parties and their agents and used illustrative scenarios to demonstrate the potential benefits for improving trust and minimising cost-related disputes.

The solutions reported in the above research efforts have the potential to revolutionise current payment practices in the construction industry by allowing automatic execution of selected payment-related provisions and self-execution of payment via smart contracts. In the context of the FIDIC RB 2017, this would underpin the efficacy of sub-clauses *14.3* [Application for Interim Payment] and *14.9* [Release of Retention Money], the latter having long been observed to be a contentious issue.

An extract from *14.3* [Application for Interim Payment] stipulates:

> The Contractor shall submit a Statement to the Engineer after the end of the period of payment stated in the Contract Data (if not stated, after the end of each month). Each Statement shall: (a) be in a form acceptable to the Engineer; (b) be submitted in one paper-original, one electronic copy and additional paper copies (if any) as stated in the Contract Data; and (c) show in detail the amounts to which the Contractor considers that the Contractor is entitled, with supporting documents which shall include sufficient detail for the Engineer to investigate these amounts together with the relevant report on progress in accordance with Sub-Clause 4.20 [Progress Reports].

An extract from *14.9* [Release of Retention Money] states:

> After the issue of the Taking-Over Certificate for: (a) the Works, the Contractor shall include the first half of the Retention Money in a Statement; or (b) for a Section, the Contractor shall include the relevant percentage of the first half of the Retention Money in a Statement.
>
> After the latest of the expiry dates of the Defects Notification Periods (DNP), the Contractor shall include the second half of the Retention Money

> in a Statement promptly after such latest date. If a Taking-Over Certificate was (or was deemed to have been) issued for a Section, the Contractor shall include the relevant percentage of the second half of the Retention Money in a Statement promptly after the expiry date of the DNP for the Section.

Using blockchain would prevent or reduce the misinterpretation or misapplication of contractual-related provisions that have been encoded, verified and agreed upon during the initiation phase of a project before deployment to the blockchain system. The transparent and accessible information provided by a blockchain system would expose any failure to process or certify a contractor's payment and eliminate refusals to pay and delayed payments.

11.5.1.2 Performance monitoring and reporting management

Performance management is a core contractual process that contributes to the success of construction projects [48] and benefits all organisations involved [49]. In the traditional sense, this CCA function enables the employer or its agents to monitor the time and cost of the project (*monitoring quality*, will be considered in the next section). The contractor is typically required to prepare progress reports with supporting evidence to demonstrate ongoing compliance with planned time and project cost. In practice, the execution of this CCA function properly can be challenging when it is conducted manually, but digital technologies may assist [50]. BIM-based project performance monitoring tools and enterprise resources planning and reporting systems (e.g. Oracle Primavera) are commonly used but their benefits are hampered by reluctance or inability to share information, discrepancies in chronological records and lack of traceability, immutability, transparency and trust in the data input. Three studies show how blockchain may tackle these issues.

Wang *et al.* [51] used actual project data in a blockchain-based information model to enable reliable comparisons between planned and actual progress in precast operations. Another article proposed a 'two-layer adaptive blockchain-based supervision' model using on-chain and off-chain networks for off-site modular housing production, allowing stakeholders to access relevant records, while confidential data irrelevant to the project were kept unshared [36]. In a novel approach focused on thermal performance of buildings, Hunhevicz *et al.* [52] presented the use case of monitoring the operational energy effectiveness of a building procured on a performance-based contract. A smart contract directly checks the delivery of energy performance levels stipulated in the contract and executes payments from the contract bank account to the contractor and facility manager. This novel use case shows how blockchain can act as a core technology that enables daily monitoring of performance and can be adapted to other use cases for progress reporting on construction sites.

In the context of FIDIC RB 2017, sub-clauses *4.20* [Progress Reports] and *8.7* [Rate of Progress] can benefit from such solutions. For example, sub-clauses *4.20* [Progress Reports] stipulates the following:

> Monthly progress reports... shall be prepared by the Contractor and submitted to the Engineer... each progress report shall include: ...

> comparisons of actual and planned progress, with details of any events or circumstances which may adversely affect the completion of the Works in accordance with the Programme and the Time for Completion, and the measures being (or to be) adopted to overcome delays.

Of course, monthly progress reports are the aggregation of daily reports and records without which accurate monthly reports cannot be produced. Although this fact is stipulated in sub-clause *6.10* [contractor's records] in FIDIC RB 2017 as follows: '... for each work activity shown in the Programme, at each work location and for each day of work', it has been challenging in practice to comply with due to an array of reasons including inconsistent daily reporting practices adopted by construction teams and required efforts to record keeping while ensuring trusted information is maintained.

However, sub-clause *8.7* [Rate of Progress] stipulates:

> If, at any time: (a) actual progress is too slow to complete the Works or a Section (if any) within the relevant Time for Completion; and/or (b) progress has fallen (or will fall) behind the Programme (or the initial programme if it has not yet become the Programme) under Sub-Clause 8.3 [Programme],other than as a result of a cause listed in Sub-Clause 8.5 [Extension of Time for Completion], then the Engineer may instruct the Contractor to submit, under Sub-Clause 8.3 [Programme], a revised programme describing the revised methods which the Contractor proposes to adopt in order to expedite progress and complete the Works or a Section (if any) within the relevant Time for Completion.

The above studies demonstrate that blockchain-based systems could enhance the CCA performance monitoring and reporting function required by FIDIC sub-clauses *4.20* [Progress Reports] and *8.7* [Rate of Progress] in real time. This enhancement could eliminate the wait for monthly progress reports, and subsequent action by the engineer, then by the contractor, before a performance-related problem is addressed. The basis for this improvement stems from the fact that contractual provisions are converted into data parameters using smart contracts. The data parameters can be coded to reflect the required contractual progress reporting metrics. For example, the planned versus actual work percentages along with allocated resources and their productivity rates. In addition, such systems will act as a trusted source for information that simultaneously and contemporaneously connects the involved organisations providing timely updates. Furthermore, from a drafting efficiency perspective a prospective blockchain-based solution could integrate the two sub-clauses, whereas currently no cross-reference exists.

11.5.1.3 Quality and acceptance management

Managing *quality* in construction projects ensures that the built asset meets its structural and functional specifications over its design life [53]. It also reduces or can prevent the occurrence of defects and rework due to nonconformance with the specifications [54]. However, empirical observations reveal that quality management including recording and documentation of quality-related issues have been consistently reported as a CCA problem [55]. For example, throughout the construction phase of the Berlin Brandenburg Airport project, completed in 2020, some 120,000 construction-related defects were unearthed; in addition, over a half million installation-related problems were identified.[17] Another official audit report reveals that Delhi Mass Rapid Transit System suffered from poor quality of civil structure among other issues.[18]

Construction contracts require the contractor's work to conform with specifications, an example being FIDIC RB 2017, *4.9* [Quality Management and Compliance Verification] which states that:

> The QM System shall be in accordance with the details stated in the Specification (if any) and shall include the Contractor's procedures: (a) to ensure that all Notices and other communications under Sub-Clause 1.3 [Notices and Other Communications], Contractor's Documents, as-built records (if applicable), operation and maintenance manuals (if applicable), and contemporary records can be traced, with full certainty, to the Works, Goods, work, workmanship or test to which they relate; (b) to ensure proper coordination and management of interfaces between the stages of execution of the Works, and between Subcontractors; and (c) for the submission of Contractor's Documents to the Engineer for review.

Prescribed remedies for failure to do so include the right to reject completed works and issue a nonconformance report (NCR) as exemplified by FIDIC RB 2017, *7.5* [Defects and Rejection] which stipulates:

> If the Contractor fails to promptly submit a proposal (or revised proposal) for remedial work, or fails to carry out the proposed remedial work to which the Engineer has given (or is deemed to have given) a Notice of No-objection, the Engineer may: (a) instruct the Contractor under sub-paragraph (a) and/or (b) of Sub-Clause 7.6 [Remedial Work]; or (b) reject the Plant, Materials, Contractor's design (if any) or workmanship by giving a Notice to the Contractor, with reasons, in which case sub-paragraph (a) of Sub-Clause 11.4 [Failure to Remedy Defects] shall apply.

The preceding contractual provisions show the importance of effective *quality management* in construction projects. However, operating such contractual provisions in practice present significant issues that can affect quality inspection, including contractually loaded language for procedures, data entry redundancies, confusion and inefficient information management [56]. To reduce the adverse effects of such practical issues, three blockchain-based scholarly articles have proposed improvements.

[17]https://www.dw.com/en/berlins-new-airport-finally-opens-a-story-of-failure-and-embarrassment/a-55446329

[18]https://cag.gov.in/en/audit-report/details/114644

In the first, Wu *et al.* [57] presented a solution to improve on-site *quality management* using blockchain and smart contracts to automate quality inspection records securely. An actual project involving cast-in-situ bored piles was used by Sheng *et al.* [58] to demonstrate the effectiveness of blockchain immutability and smart contract automation in tackling the information flow of quality acceptance and associated NCRs. The third article [59] used a permissioned Hyperledger Fabric blockchain to support a remote real-time inspection and validation of work with mobile-based data entry. The practitioners involved welcomed the benefits of data traceability, record immutability, information transparency, that enabled remote inspection, but concerns were raised over cost and stability of the IT infrastructure and the absence of legal frameworks; concerns that are typical with the emergence of any new technology. The three scholarly articles demonstrate the potential of blockchain for improving *quality management* function of CCA and its applicability to the relevant FIDIC provisions noted as examples above.

11.5.1.4 Project 'legal' governance

Within the context of CCA functional classification, this CCA function aligns a project with its local legal ecosystem. It mainly focuses on the contractor's obligations to comply with legal requirements to apply for and obtain permissions from statutory authorities (e.g. local municipality and traffic police departments) and affected stakeholders (e.g. utility service providers and local communities) to commence construction work. Such requirements include providing temporary traffic diversions, accesses, health and safety plans, and environmental plans. In FIDIC RB 2017, for example, sub-clauses *1.13* [Compliance with Laws] and *2.2* [Assistance] both place obligations on the contract parties, and for jurisdictions where the protection of environment is of a primary importance, FIDIC RB contains sub-clause *4.18* [Protection of the Environment]. Typical obligations include obtaining building permits, adherence to regulations (including those relating to health and safety and the environment). For example, sub-clause *1.13* [Compliance with Laws] states:

> The Contractor shall give all notices, pay all taxes, duties and fees, and obtain all other permits, permissions, licences and/or approvals, as required by the Laws in relation to the execution of the Works.

And sub-clause *4.18* [Protection of the Environment] stipulates that:

> The Contractor shall ensure that emissions, surface discharges, effluent and any other pollutants from the Contractor's activities shall exceed neither the values indicated in the Specification, nor those prescribed by applicable Laws.

Two blockchain-based studies relate to this function. In Nawari and Ravindran [60], a solution for plan review and building permit approval was proposed. On receipt of an application by the local authority, the system performs an automated compliance check and using smart contract functions deployed on the blockchain. Based on entered data variables, if coded conditions inside the smart contracts are met, a corresponding transaction is generated to issue the building permit or otherwise notify the applicants of necessary revisions. The research team of Zhong *et al.* [61] proposed the automatic monitoring and evaluation of environmental performance based on using smart contracts on a Hyperledger Fabric blockchain with pollution data collected by sensors and compared with permitted levels. Warnings are triggered if these are exceeded. Prototypes such as these are considered applicable to the *project governance* CCA function exemplified by the FIDIC sub-clauses referred to above.

11.5.1.5 Disputes resolution management

Chen *et al.* [16] classify *disputes resolution management* under the *control* mechanism. They exemplify adjudication as one of the common *disputes resolution* processes that aims to curb opportunism. As noted earlier in Section 11.1, practice has revealed that claims[19] can evolve into disputes if they are not settled or if the decision outcome conflicts with the expectation of the party raising the claim. In FIDIC RB 2017 edition, this dispute process can be triggered by the contractor's *notice of dissatisfaction* under sub-clause *3.7.5* [Dissatisfaction with Engineer's determination]. This trigger initiates the process of dispute avoidance/adjudication board (DAAB) stipulated under sub-clauses *21.3* [Avoidance of Disputes] and *21.4* [Obtaining DAAB's Decision], an onerous process with multiple tasks allocated to the appointed board and the contracting parties. An example of such tasks can be seen in the following extract from sub-clause *3.7.5* [Dissatisfaction with Engineer's Determination]:

> If either Party is dissatisfied with a determination of the Engineer: (a) the dissatisfied Party may give a NOD[20] to the other Party, with a copy to the Engineer; (b) this NOD shall state that it is a 'Notice of Dissatisfaction with the Engineer's Determination' and shall set out the reason(s) for dissatisfaction; ... (d) thereafter, either Party may proceed under Sub-Clause 21.4 [Obtaining DAAB's Decision].

Only one recent study was found that applied blockchain to *dispute resolution* [62]. This offered a novel blockchain system as an alternative *dispute resolution* method which was reported to reduce the likelihood of payment-related disputes and manage their process transparently. Under the system, a contractor would invoke a smart contract function for the dispute (e.g. for rejected payment) to be referred to randomly selected, independent registered professionals in the blockchain network ('blockchain-based jurors'). A majority decision is then taken through a designed consensus mechanism on the Ethereum blockchain. Although this novel system appears to be ambitious with limited applicability in practice, the implementation of such blockchain systems could improve dispute outcomes and enable the contracting parties to obtain faster decisions. However, for such a novel blockchain-based *dispute management* system to be adopted in practice, it is likely that it would first require a blockchain-based *claim management* system to be in place.

[19]Claims management is an *adaptation* mechanism discussed in Section 11.5.3.3.
[20]NOC: Notice of Dissatisfaction.

11.5.2 Blockchain applications for coordination-orientated CCA functions

11.5.2.1 Document and record management

Construction projects have traditionally suffered from inadequate information sharing and poor management of documents and records [63–65]. Currently used decentralised document management systems such as cloud-based Aconex, Dropbox and Oracle Unifier cannot prevent alteration, deletion, accessibility denial and incorrect revision history. To overcome these shortcomings, blockchain technology was proposed as novel technology. To that end, five studies have been identified that demonstrate the potential of blockchain in resolving document and information-related issues in construction.

Ciotta *et al.* [66] developed a decentralised Ethereum-based application for documentation that could be used by multiple stakeholders, allowing them to have updated information on structural drawings and to prevent the exchange of incorrect versions of files. Through a proof-of-concept demonstration, the application was reported to have enabled the communication between different common data environments (CDEs) while ensuring transparency and reliability of the information. It is noted here that the traditional approach to CDEs operates in a centralised manner with shortcomings that hinder coordination of drawings across different organisations. In another research work [67], a prototype request for information (RFI) workflow was proposed where documents were immutably recorded on a blockchain platform. In the same research work, a scenario was illustrated to demonstrate how the proposed solution would address information-related challenges that might arise post construction during the defect liability period. Furthermore, Tao *et al.* [68] proposed a Hyperledger Fabric blockchain-based system integrated with the Interplanetary File System (IPFS). IPFS is an innovative protocol that builds an addressable and peer-to-peer file storage system without a centralised server. In IPFS, every uploaded file is encrypted into a cryptographic hash value called content identifier (CID) which represents the 'address' or 'hyperlink' of the file. Only agents who have the CID can access or download the file regardless of their location. As such, files on IPFS are distributed and cannot be manipulated by a central agent. Along the same line, if the contents of a file are changed, its cryptographic hash value changes accordingly which signals that modification has taken place. Similar integration was demonstrated by Das *et al.* [69] to support an RFI workflow process. The designed smart contracts focused on verifying data and users, enabling workflow progression and recording document change status through its lifecycle. Attempting to introduce a more automated solution to information management, Li *et al.* [70] integrated Hyperledger Fabric with Internet-of-Things (IoTs) for the secure management of information of modular construction elements. Their developed solution was demonstrated through a case study. The role of the IoTs, which were tagged on the elements and the transportation trucks, was to track the process status of the modular elements. All the IoTs events were captured, processed, validated and stored on the blockchain using smart contract-based computation. Their developed solution was evaluated by stakeholders of the actual project revealing effectiveness of such novel systems compared to current practices while considering blockchain as a promising digital infrastructure for such applications. Overall, this group of studies implies a value proposition in terms of connecting various organisations through standardised solutions for document management and data exchange while offering timely transparency of the process and its corresponding immutable transactions that prevalent centralised solutions have not afforded.

Based on the above, it can be inferred that blockchain can significantly improve the CCA function of *document and record management* both during and post construction. From the perspective of FIDIC 2017, a blockchain-enabled document and record management platform could support sub-clause *4.4* [Contractor's Documents] which demands the creation, retention and accessibility to a relatively high volume of documentary exchanges with signatures from multiple actors. An example of such demands is shown in the following provisions of this sub-clause:

> The Contractor shall prepare all Contractor's Documents and the Employer's Personnel shall have the right to inspect the preparation of all these documents, wherever they are being prepared... The Engineer shall, ..., give a Notice to the Contractor: (i) of No-objection (which may include comments concerning minor matters which will not substantially affect the Works); or (ii) that the Contractor's Document fails (to the extent stated) to comply with the Contract, with reasons.

11.5.2.2 Communication and relationship management

Poor communication and working relationships are widely recognised as problems in construction. Blockchain-enabled smart contracts can receive, verify, update and record information in form of immutable and transparent transactions, thus enhancing communication and working relationships. This is supported by two scholarly articles that focus on *communication* improvements achieved through blockchain-based solutions.

In the first, Yang *et al.* [71] used a permissioned blockchain for the material approval process and the ensuing availability of immutable and accessible records resulted in transparent communications among stakeholders. In fact, the use of permissioned blockchain enabled access to be restricted to those agents concerned with a specific process in order to maintain privacy of data. The authors demonstrated that improvement was achieved in respect of communicating the workflow of cladding materials from the concept design to the final approval while maintaining privacy through six separate channels on Hyperledger Fabric. All corresponding interaction events were recorded as transactions on the blockchain network. This recalls the Grenfell Tower Tragedy in 2017 one of its main causes was attributed to cladding materials approval process.[21] Hence, such blockchain-based solutions could be exploited in the future to enhance communication in

[21]https://www.grenfelltowerinquiry.org.uk/report/phase-2/overview.html

respect of critical material approvals. In the second, enhanced efficiency was reported in communicating decisions for essential actions among dispersed project participants [72]. In this second study, the authors asserted that blockchain-based traceable data communication could offer good potential benefits for supply chain management in a construction project. In particular, the tangible benefits would materialise in a project that involves team members with various backgrounds. As Schmidt and Wagnar [73] suggest, blockchain reduces behavioural uncertainty in transactional relationships. With a blockchain-based system, all communication threads associated with a given process are exchanged and recorded in the form of immutable transactions accessible to all team members involved in a given project.

The pertinent FIDIC 2017 provision is sub-clause *1.3* [Notices and Other Communications]. This sub-clause already recognises digital records and digital communication systems as contractually appropriate:

> where the electronic original is transmitted by the electronic address uniquely assigned to each of such authorised representatives... An electronically transmitted Notice or other communication is deemed to have been received on the day after transmission, provided no non-delivery notification was received by the sender. All Notices, and all other types of communication as referred to above, shall not be unreasonably withheld or delayed.

This establishes the prospect of blockchain-based communication systems for which the above studies can be a point of departure, and which would, as a result, enhance *communication and relationship management* among project organisations and their corresponding agents.

11.5.2.3 Team management

Team management is a CCA function for defining roles and responsibilities, assign accountability and resolving conflicts. No example was identified of a blockchain study that directly matched this function. There are three possible reasons behind this. First, the lack of diffusion of blockchain-based applications across industries hinders the investigation of the impacts of such applications on *team management*. Second, the current focus of scholarly and practical attempts tends towards improving processes and systems which involve *team management*. This, in turn, leads to the third possible reason: that improved *team management* is an embedded benefit achieved through the implementation of a blockchain-based solution. With the results demonstrated in the previous analysed studies, it is not difficult to envisage that this emerging technology could improve team management.

In the context of CCA, blockchain can enhance accountability and auditability [74] there are two critical characteristics that are believed to enhance *team management* from a contractual perspective. On accountability, the contractor's team is obliged to fulfil its contractual obligations towards the employer; hence, it is held accountable to explain and justify any conduct that contradicts those obligations. For example, a delay in completing a given critical construction activity may be caused by an unprocessed RFI on part of the employer's agents. In current practice, the accountability tends to be placed on the contractor's team despite the fact that it is on the employer's agents. With a blockchain-based system, this accountability is shifted from the contractor to the employer's agent (e.g. the engineer) who has not responded to the RFI. Another predicated scenario could be that smart contracts are programmed in such a way that if the responsible agent does not take action within a specified time limit, then the smart contracts would automatically take action on their behalf. In such a scenario, agents would be encouraged to act with due-diligence without procrastination to avoid accountability that could arise for delayed decisions.

This is where the auditability comes into play. Through blockchain, the employer can trace the corresponding transactions regarding that raised RFI to trace and audit the issue through the corresponding transactions while being ensured that what is recorded is trusted and authentic. The trust and authentication are reinforced through explicitly defined roles and responsibilities, the use of digital signatures and timestamps, and immutability of transactions. Of course, all these are in line with the construction contract that would be converted into a blockchain-based process coupled with smart contracts. To that end, such a system would remove uncertainty and conflict from the exercise by either party of its rights under FIDIC sub-clauses *2.3* [Employer's Personnel and Other Contractors] and *6.9* [Contractor's Personnel] to request removal of an individual for malpractice.

The following provisions extracted from sub-clause *2.3* are examples of such contractual rights:

> The Engineer may require the Contractor to remove (or cause to be removed) any person employed on the Site or Works, including the Contractor's Representative and Key Personnel (if any), who: (a) persists in any misconduct or lack of care; (b) carries out duties incompetently or negligently; (c) fails to comply with any provision of the Contract.

There are sometimes cases of unfair dismissal of the contractor's personnel from construction projects due to their competence in protecting their own organisations' contractual rights. The preceding example of the RFI process shows that with blockchain-based system in place, the engineer would be less likely to misuse the above provisions as a way of threatening the contractor's team. But it is acknowledged that there are limited case studies that focus on improved *team management* through the use of blockchain-based applications due to the lack of diffusion across industries. That said, it is not difficult to appreciate how such applications will change the traditional approaches to *team management* through decentralised applications with enhanced accountability and auditability.

11.5.2.4 Contract closeout management

Timely *closeout* a construction project is an important process that enables project owners to operate or occupy the built asset without delays that result in financial losses for the involved organisations [75]. At the *closeout* stage, a series of tasks must be performed to ensure that the project has met its intended purpose and that parties' rights and obligations have been satisfied. For example, according to

FIDIC 2017 clause *9* [Tests on Completion] the contractor carries out tests to ensure functional and structural stability before handover and the employer's personnel inspect the completed works before takeover pursuant to sub-clause *10.1* [Taking over the Works or Sections].

An example of clause *9* [Tests on Completion] being the following (*9.1* [Contractor's Obligations]):

> As soon as the Works or Section have, ..., passed the Tests on Completion, the Contractor shall submit a certified report of the results of these tests to the Engineer. The Engineer shall Review such a report and may give a Notice to the Contractor stating the extent to which the results of the tests do not comply with the Contract. If the Engineer does not give such a Notice within 14 days after receiving the results of the tests, the Engineer shall be deemed to have given a Notice of No-objection.

Likewise, sub-clause *10.1* [Taking over the Works or Sections] stipulates:

> the Works shall be taken over by the Employer when:(a) the Works have been completed in accordance with the Contract, including the passing of the Tests on Completion and except as allowed in....

This whole process needs a high volume of documentation together with its coordination by key staff. The elaborate contractual requirements for the *closeout* stage resonate with the capabilities of blockchain. As with the theme of *team management,* no evidence could be found in the literature for the use of blockchain during the closeout. Although its implementation would arguably contribute to more successful construction project *closeout.* For example, the contractual provisions of sub-clause *9.1* [Contractor's Obligations] could be converted into smart contracts that can be triggered by transactions of blockchain-based contractual processes. On that basis, it can be postulated that this contractual process, if it would operate on a blockchain system, would largely rely on other contractual processes using blockchain. For example, a blockchain-based *quality management* system would be used to report and close the list of outstanding works (i.e. repair works). Likewise, a blockchain-based *payment* system would include all records on payment. Along the same line, a blockchain-based application could generate taking over certificates after automatic verification of completed associated tasks. With this integrated approach to contractual processes, the *closeout management* process could be completed with reduced human efforts and errors. As a result, project owners could operate projects without the need for extending the closeout stage. Although specific examples of such a predicted application remain to be seen in the future in the discourse of blockchain in construction.

11.5.3 Blockchain applications for adaptation-orientated CCA functions

11.5.3.1 Contract risk management

It is an axiom that risks are allocated between contracting parties via the conditions of their contract, the provisions of law or both. This lays the foundation for the course of action when a given risk materialises. Where a risk has been allocated to the contractor, its control is deemed to be included in the contract sum and any subsequent delay would be subject to payment of delay damages by the contractor, as in FIDIC 2017 sub-clause *8.8* [Delay Damages]. Where a risk has not been allocated to the contractor, its consequences may entitle the contractor to compensation (with additional time and/or cost). The mechanisms for the latter are either a directed 'change order' (in the language of FIDIC called a 'variation') or through a claim with a variety of clauses relevant to these compensation mechanisms.[22] At time of writing, no research efforts have investigated how this CCA function could be improved by blockchain technology. However, it is believed that a specific blockchain-based risk management is not required. The main reason behind this belief is that various risk factors tend to be embedded in various contractual processes. For example, the provisions concerning delay damages stipulated under sub-clause *8.8* of FIDIC RB 2017 state:

> If the Contractor fails to comply with Sub-Clause 8.2 [Time for Completion], the Employer shall be entitled subject to Sub-Clause 20.2 [Claims For Payment and/or EOT] to payment of Delay Damages by the Contractor for this default. Delay Damages shall be the amount stated in the Contract Data, which shall be paid for every day which shall elapse between the relevant Time for Completion and the relevant Date of Completion of the Works or Section. The total amount due under this Sub-Clause shall not exceed the maximum amount of Delay Damages (if any) stated in the Contract Data.

Another related example is when the contractor encounters differing site conditions at site, the contractor may be entitled to both time extension and additional payment pursuant to FIDIC sub-clause *4.12* [Unforeseeable Physical Conditions]:

> If the Contractor encounters physical conditions which the Contractor considers to have been Unforeseeable and that will have an adverse effect on the progress and/or increase the Cost of the execution of the Works, the following procedure shall apply: 4.12.1 Contractor's Notice: After discovery of such physical conditions, the Contractor shall give a Notice to the Engineer,... 4.12.2 Engineer's inspection and investigation: The Engineer shall inspect and investigate the physical conditions within 7 days,... 4.12.4 Delay and/or Cost: If and to the extent that the Contractor suffers delay and/or incurs Cost due to these physical conditions, having complied with Sub-Clauses 4.12.1 to 4.12.3 above, the Contractor shall be entitled subject to Sub-Clause 20.2 [Claims For Payment and/or EOT] to EOT and/or payment of such Cost.

These provisions clearly reveal that *risk management* as an *adaptation* mechanism is separately exercised within various contractual processes; hence, it relies on other processes being blockchain-ed. For example, the application of

[22]Both *change management* and *claims management* are discussed in the next sections.

delay damages relies on two process: *performance monitoring and reporting* in relation to the time for completion and *claims management*. In the same way, the notification of the differing site conditions as a technical risk factor relies on the *claims management* process. Apart from that, the objective language of the provisions, denoted by 'if' and 'shall' lend themselves to be programmed using smart contracts. Thus, it is not unreasonable to predict that the implementation of blockchain-based contractual processes can, in fact, contribute to *risk management* from a contractual perspective.

11.5.3.2 Change management

Effective *change management* is a crucial but challenging function in CCA. However, no research effort was found that directly addresses this CCA function by exploiting blockchain. Nevertheless, as *change management* relies heavily on correct application of contractual provisions and adherence to performing obligations associated with other CCA functions (e.g. *performance monitoring and reporting management*, *document and record management*, and *communication and relationship management*), proposals identified previously under these functions can contribute. FIDIC 2017, sub-clause *13.3* [Variation Procedure] describes the mechanisms for managing changes on a construction project. It is argued that this procedure can be rationalised on a blockchain-based system to tackle the associated challenges. For example, the following provisions could theoretically be converted into smart contract functions:

> The Engineer may instruct a Variation by giving a Notice (describing the required change and stating any requirements for the recording of Costs) to the Contractor in accordance with Sub-Clause 3.5 [Engineer's Instructions]. The Contractor shall proceed with execution of the Variation and shall within 28 days (or other period proposed by the Contractor and agreed by the Engineer) of receiving the Engineer's instruction, submit to the Engineer detailed particulars including: (a) a description of the varied work performed or to be performed, including details of the resources and methods adopted or to be adopted by the Contractor.

With smart contracts reflecting the preceding contractual provisions, challenges facing *change management* in practice could be reduced. Such challenges include that a change may be the result of a combination of causes rather than a single clear factor, oral instructions to implement change and a direct instruction from the employer to the contractor to change the scope [76]. These instances often lead to claims and disputes due to disagreements after the completion of the work on the causes of the change and liability. A blockchain-based system could tackle such problems by connecting all involved agents in a single platform that contains the full history of the change including the instructions, causes, effects and responsibilities. For example, smart contracts ensure decisions are made by authorised agents, formalise contractual provisions and their corresponding transactions are transparent and accessible to them instantly [77]. The net foreseeable positive impact of such a system is the reduction of disagreements between the employer and contractor over variations, which frequently evolve into claims and disputes.

11.5.3.3 Claims management

The efficient and effective application of this CCA function first necessitates the identification and operation of the relevant sub-clauses. For claims application and settlement, a multistep approach is generally adopted with the involvement of various personnel from the contracting parties. In practice, *claims management* faces an array of process, content and behaviour-related challenges [78]. Process-related challenges may stem from: procedural invisibility and belated decisions/determination. Whereas content-related challenges include: misapplication of contractual clauses, incompleteness of relevant records and distortion of records/information. In terms of challenges triggered by behaviour, inadvertent and intentional inaction towards contractual administrative activities is frequently reported.

To enhance construction *claims management*, an optimal solution that addresses the relevant challenges needs to: (i) enhance visibility of the process, (ii) streamline the contents of construction claims, (iii) meet the contractual needs of contracting parties and (iv) moderate the behaviour of involved agents. Encapsulating these aspects in a single solution may now be possible with the advent of blockchain technology and its associated smart contracts.

A robust blockchain-based system that addresses such problems would be beneficial for FIDIC *claims management* process analysed in Section 2.4.2 of Chapter 2. However, none of the analysed studies has been found to address this. It can be inferred that the use of blockchain reported for other CCA functions (e.g. the *communication and relationship management* and *document and record management* processes) can indirectly support the *claims management* process. For example, in the case of a claim for an extension of time (EoT) in FIDIC 2017, both *Notice of Claim* (sub-clause *20.2.*1) and *Engineer's Initial Response* (sub-clause *20.2.*1) can be chronologically and immutably recorded on a blockchain platform. Other contractual obligations stipulated under sub-clause *20.2* [Claims for Payment and/or EoT] can be converted into blockchain-based systems.[23] Hence, administrative execution of FIDIC claims management procedure could be achieved more effectively and benefit the contracting parties and the employer's personnel.

Evidence of the benefits of blockchain-based applications to *claims management* is found in other industries. For example, Shetty *et al.* [79] collected systematic evidence of the effectiveness of blockchain-based solutions in the medical insurance industry. To that end, scholars in the field of blockchain in construction can learn from such evidence to develop similar solutions applicable to construction contracts. From a broader perspective of CCA and dispute resolution, the implementation of a blockchain-based *claims management* system would be likely to reduce the incidence of disputes. If disputes do proceed to arbitration or other

[23]These contractual provisions of FIDIC RB 2017 that deal with claims management are relatively long and contains multiple cross-references and contractually loaded language. Hence, they are not reported here.

external dispute forum, the records and evidence underpinned by blockchain could serve the proceedings without the need to re-verify, compile and cross-reference the information again.

11.6 Research agenda: challenges and opportunities to blockchain-enabled CCA

The previous section has systematically classified and evaluated the relevance of published blockchain studies to recognised CCA functions, using contextualised examples of sub-clauses and provisions from the *FIDIC Red Book* (2017 edition). Not all of the CCA challenges identified in Section 11.3 have been explicitly addressed in this literature, though some reasonable inferences can be made to support the value of adopting a blockchain-based solution for CCA. To examine the feasibility of doing so, the 'Adoption Model' proposed by Li *et al.* [80] is employed. This model consists of four dimensions – *technology*, *process*, *policy* and *society* – as lenses through which the viability of blockchain adoption can be viewed. As a result, nine adoption challenges are identified and related future research opportunities are proposed.

11.6.1 Technology-related challenges

11.6.1.1 Technical malfunction

The trusted acceptance of any automated system can be undermined by technical malfunction at either the blockchain level or smart contract level or both. At the blockchain level, a Hyperledger Fabric permissioned blockchain solution may encountered accessibility problems due to an error on the cloud where the solution is hosted. At the smart contract level, a problem may arise from inputting incorrect data type (e.g. date format) which results in preventing the generation of a corresponding transaction. If this, for example, prevented the certification of a contractor's application for payment the resulting delayed transfer of money may prompt a contractor's claim for 'financing charges',[24] representing negative impacts on both parties and on the project itself. Such technical risks could be addressed by specified 'back-up' such as reverting to the conventional paper-based payment processing mechanism on a temporary basis [81] or recording the errors using conventional approaches. Despite its inefficiency, this type of 'parallel working' as a potential mitigation strategy would probably be desirable for any early adopters of the system. How and when such temporary measures would become necessary remains uncertain, and an area to be explored. A user guideline and response plan should be developed to avoid the type of unintended consequences exposed by the Horizon IT Scandal in the UK.[25]

[24]See *FIDIC Red Book* 2017, sub-clause *14.8*: Delayed Payment.
[25]https://corporate.postoffice.co.uk/horizon-scandal-pages/context

11.6.1.2 Authenticity of oracles

The reliance on data from 'oracles' – off-chain human or artificial interventions required to trigger the self-execution of blockchain functions[26] – raises questions of verification and validation of these data, particularly as blockchain records are immutable and resulting smart contract transactions are irreversible. For example, in the novel solutions developed by Hamledari and Fisher [42] for *financial management* and Li *et al.* [70] for information management, there has been less focus on the aspects of data verification and validation handled by the employed IoTs as data feeders to the blockchain-based smart contracts. In other industries, such aspects have been technically investigated; for example, in the healthcare industries [82] and smart city engineering [83]. Future research must focus both on establishing protective protocols to ensure validity of transmitted data at the intersection between the off-chain and on-chain networks and their corresponding on-chain transactions. For example, a fog computing layer has been noted in the solution developed by Shukla *et al.* [82]. It should be noted that such potential mitigation strategies are required for full-automation solutions, and they appear to be required in the distant future after blockchain-based solutions have shown their future-proofness.

11.6.1.3 Blockchain platform lifespans

Certain contractual provisions and other legal provisions require the retention of records for many years after the completion of a project. This raises issues about the intended lifespan of associated blockchain systems that still require investigation, particularly for permissioned blockchain applications hosted on the cloud. Future research is needed to tackle this challenge, especially in construction projects procured through public–private partnerships arrangement where 25–30-year concession contracts are the norm.

11.6.2 Process-related challenges

11.6.2.1 Replicating the contractual logic

Converting the logic of a given contractual mechanism (such as interim payment of the contractor) into a self-executing smart contract is a considerable challenge. A typical example is provided by the payment provisions of the 2017 *FIDIC Red Book*. The payment process is stipulated in sub-clause *14.3* [Application for Interim Payment] and includes multiple stages. The first stage is set out in *12.1* [Works to be Measured], *12.2* [Method of Measurement] and *12.3* [Valuation of the Works]. This initial stage may require adjustment under *13.6* [Adjustments for Changes in Laws], *14.9* [Release of Retention Money], *14.2* [Advance Payment] and *14.5* [Plant and Materials intended for the Works]. The final stage may involve sub-clause *3.7* [Agreement or Determination] that deals with certifying provisional amounts for determined claims and variations which are still under negotiation. The complications are further compounded when contracts are terminated or when

[26]See Section 2.7 in Chapter 2 for more explanation on oracles working mechanisms.

previously paid amounts are to be deducted. Ensuring that such multiple, dispersed, yet inter-dependent contractual provisions are logically replicated in a blockchain-based payment system is a demanding task. Currently, there are limited initiatives that carry out such a task, albeit for relatively less complicated contractual provisions and processes. For example, the Weather Ledger project has attempted to do so for compensation events that are triggered by inclement weather events under the NEC model form of construction contract.[27] In addition, the Accord Project presents a number of templates to enable coding of contractual provisions in various industries into smart contracts.[28] Recently, a conceptual framework has been developed that demonstrates how the contractual time performance management process of FIDIC RB 2017 can be converted into blockchain-based smart contracts [84]. This might involve thousands of lines of code to reflect the contractual provisions and ensure full cross-references within a traditional construction contract: a task that would require effective communication between contract experts and blockchain developers.

11.6.2.2 Detailed unit rates

Many of the stages within the CCA payment process require applications to be accompanied by detailed itemisation of cost. Many unit rates for work items (in schedules or bills of quantity) are composite and may include elements for temporary works and testing. Furthermore, it is not uncommon for specific work items to be completed by multiple actors. For example, in a pipeline project, different work items may be allocated to different actors: the trench excavation and backfilling work may be allocated to a subcontractor, the pipe procurement and laying work may be carried out by the main contractor, and the dewatering work may be undertaken by a specialist subcontractor. At the same time, the bills of quantity stipulate that payment shall be made from the employer to the contractor based on a linear meter unit rate that include all these separate work items. For such scenarios, current practices involve the use of separate subcontracts with different payment mechanisms, and these have been problematic. Empirical observations report that such problems include underpayment, late payment and overlapping scope of subcontract works leading to payment application for the same work items by the multiple subcontractors involved [85]. Any blockchain-enabled payment system would require detailed payment rates of the work items to be in place. This is not an insurmountable problem *per se* as the information can be invoked from off-chain 'oracles' where it already exists. However, for a blockchain-based smart contract to achieve the required distribution of payment for such complex work items may be more challenging. A good point of departure for such future work can be the approach adopted by Cheng *et al.* [47] that used separate blockchain channels where multiple actors are involved in a project. However, the work focused on privacy of data and not disaggregation of payment rates.

[27] https://www.digicatapult.org.uk/wp-content/uploads/2021/11/Weather_Ledger_WP5_report-v2.pdf
[28] https://accordproject.org/

11.6.3 Policy-related challenges

11.6.3.1 Procurement route and contractual frameworks

A challenge to the wholesale adoption of blockchain-based CCA applications is the variety of procurement frameworks available. Few of the analysed studies had mentioned this and few had defined which procurement framework they were considering [44,45]. But none has developed a blockchain application for a contractual process in the context of a specific standard form of contract (e.g. FIDIC or JCT). Although it has frequently suggested that policymakers need to reform current procurement practices and standard contracts to align with technology, there is the counter-argument that procurement strategy should drive technology and not the reverse [86]. Research efforts are needed to explore how to align blockchain-based applications with the most commonly used procurement routes and standard forms of contract. These have evolved and developed over time and are unlikely to be readily discarded. Aligning blockchain applications as much as possible to current practice, rather than vice-versa would make its adoption more likely. To that end, blockchain applications can accommodate various procurement routes and offer governance underpinned by blockchain characteristics. This can be achieved by identifying actors and configuring as peers on the blockchain. For example, the *design-bid-build* route can be implemented in Hyperledger Fabric using three main actors (i.e. the employer, the engineer, the contractor) on a single channel. In addition, subcontractors can be included on the platform through secondary channels. This enables the separation of actors and ensures privacy of data that are irrelevant to the subcontractors, thereby enhancing the practical applicability of blockchain in line with relevant procurement practices.

11.6.4 Society-related challenges

11.6.4.1 Acceptance of the contracting parties

Under most contractual arrangements there are two parties to the main contract for execution of the works: the employer (or owner) and the contractor. Implementing a blockchain-based platform needs acceptance by both. This may be a challenge as at time of writing blockchain is still in its exploratory stage in the construction domain. To accelerate acceptance, both employers and contractors must be consulted, persuaded and involved in the development of such systems. Other industries have started to implement blockchain-based solutions with tangible benefits in overcoming persistent challenges. For example, MediLedger has been implemented in the healthcare industry in the USA.[29] Another example can be found in the automotive industry where Renault has adopted blockchain-based solutions for compliance certification in Europe.[30] Following such reported successful implementation in other industries, a future research agenda should involve

[29]https://www.chronicled.com/

[30]https://www.renaultgroup.com/en/news-on-air/news/xceed-a-new-blockchain-solution-for-renault-plants-in-europe/

'demonstrator projects' with evaluation of the attitudes of the contracting parties in the construction industry.

11.6.4.2 Implications for supply chain cash flow

It is not uncommon for construction projects to exceed the contractual time for completion without the contract works being fully or properly completed. The contractual responsibility may lie with the employer or the contractor, or it may be due to the occurrence of an 'external' event. A disagreement often follows, and it is common for the employer's financial department to stop processing contractor's payment applications pending settlement. A similar scenario often emerges with disagreements over defective work. In a fully automated blockchain-based payment system, this cannot be done without defining specific rules inside the smart contract functions. However, if such rules allow an upstream payment to be withheld, downstream subcontractor payments would not be processed even when a subcontractor's works had no relevance to the disagreement between the employer and contractor. The likelihood of such challenges can be reduced if other relevant contractual processes were implemented before a blockchain-based payment system. Thus, an issue for future research is the degree of flexibility or rigidity entailed in a blockchain-enabled payment system in relation to the overall CCA system.

11.6.4.3 Accessibility to dispute resolution boards

In construction, contractual disputes may arise due to many reasons, and it is important for contracts to include accessibility to external dispute resolution processes (e.g. mediation, adjudication, arbitration or ultimately the courts). Ignoring, for the moment, the possibility of the blockchain technology itself being a source of dispute, the ability of a blockchain platform to maintain the parties' accessibility to dispute resolution mechanisms has not been considered in the current literature. For example, those responsible for the various dispute resolution processes (e.g. adjudicators and arbitrators) may refuse or be incapable of engaging with a technology-driven system or depart from the conventional systems with which they are familiar. An optimistic solution would allow an intermediary expert to extract information relevant to the dispute. However, the mechanism allowing accessibility to this expert coupled with the admissibility of the blockchain records without further verification needs to be carefully considered. To this effect, a blockchain consensus mechanism protocol with a focus on accessibility and admissibility could be developed for construction projects that adopt blockchain technology. Despite the nonavailability of real-world examples of blockchain technology in *dispute resolution management*, the potential of the technology has been inescapable from relevant law bodies. For example, the Law Commission of England and Wales has produced it advice to Government on the use of smart contracts[31] which explains and demonstrates their working mechanisms and benefits to legal practice.

[31]https://lawcom.gov.uk/document/smart-contracts-2/

11.7 Insights and conclusions

The analysis of blockchain-based solutions in this chapter highlights the growing scholarly interest in its application to CCA, driven by the construction sector's broader digital transformation efforts. A further contribution has been to contextualise this evolving work within construction contract sub-clause and provisions using the most commonly used international standard form of contract, the *FIDIC Red Book* 2017 edition. This chapter has demonstrated the feasibility of codifying contractual provisions (exemplified by FIDIC RB 2017) into executable script using smart contracts. It has also taken account of the significant interconnection and cross-referencing that is common in standard form contracts and that individual sub-clauses cannot be codified in isolation. CCA practitioners can now grasp the benefits of the technology adopted here and its associated notion of smart contracts within a real contractual context. As demonstrated in Figure 11.2 in Section 11.3, poor CCA creates negative impacts on construction projects and on the industry at large. The argument has been presented for improving CCA through the adoption of blockchain technology. The potential for doing so is readily apparent in the analysed proof-of-concept studies.

However, challenges remain, as explained in Section 11.6., and a series of potential research questions have emerged for future scholars and contract drafting committees aiming to improve CCA through the adoption of blockchain technology and its associated smart contracts. For example, both replicating the contractual logic of processes stipulated in standard forms of contract and designing blockchain solutions for various procurement routes appear to be key research areas that could bring digital construction scholars and members of contract drafting committees together.

Blockchain smart contracts can mitigate misapplication or refusal to execute contract provisions by accurately encoding the corresponding clauses from paper-based contracts. As a result, the automated execution of the six associated contractual components (see Section 11.2 of this chapter) for CCA can be rationalised using on-chain smart contract codes and specific protocols for off-chain oracles that feed the smart contracts with data. This is coupled with instant visibility of executed tasks and rendered decisions for contractual procedures to all registered agents. In this way, a fully automated or semi-automated (i.e. with limited human involvement) blockchain system would prevents or mitigate the two problems in question (i.e. misapplication and/or refusal to execute contract provisions). This is achieved due to the blockchain traceability feature that makes visible the inactions of any actor responsible for the execution of a contractual task.

Based on the analysis of the proof-of-concept studies, it can be inferred that causes of poor CCA that stem from inaccurate records, inaccessibility to records and unclear responsibilities and procedures can all be tackled by using features offered by blockchain technology. Notably, traceability, immutability and decentralisation collectively pose a digital shield to any inadvertent or intentional deletion or modifications to relevant records for any given blockchain-based CCA procedure, while ensuring their unified distribution across the permissioned actors. As noted earlier, current computerised and cloud-based systems are unable to provide these features since deletion of any document or record is possible without notification. Hence, at both project and inter-organisational levels, a blockchain-based CCA establishes transparency of decisions, enhances trust, enables auditability and provides a means of accountability.

Our analysis reveals that blockchain research has primarily focused on financial management in CCA, with limited attention to other key processes. Of the 12 CCA functions identified, only seven have been explored in blockchain-related studies, with eight of the 24 studies concentrating on payment-related issues. This focus reflects blockchain's proven success in addressing financial challenges in other industries, encouraging scholars to adapt these solutions to tackle chronic non-payment and delayed payment problems in construction. However, there has been no study that models the whole payment cycle (i.e. from the advance payment to the final payment) of a given standard form of contract using a blockchain system. The absence of such a study is explained by the difficulty encountered in encoding the contractual logic of corresponding provisions and sub-clauses which tend to be dispersed across the conditions of contract (see Section 11.6.2.1). In future research directed towards this type of application, contract experts could be consulted to validate the corresponding contractual logic before converting it into smart contracts.

While claims management remains underexplored in blockchain research, its integration with existing solutions offers significant potential. For instance, a blockchain-based system could automatically flag delay claims by detecting schedule deviations in real time, eliminating the need to wait for monthly progress reports. This ensures the integrity of programme-related data and provides immediate visibility to stakeholders, reducing delays in addressing claims.

In turn, blockchain-based CCA governance could revolutionise *claims management* by offering a readily rationalised claims procedure showing a claim's status across its stages. This would include chronological versioning of records necessary for establishing evidence for a given claim enabling parties to reach an agreement on entitlement with records being held on unified permissioned ledgers. Thus, issues arising from records manipulation, discrepancy or confused or inadequate evidence would be eliminated. Blockchain-based smart contracts could be programmed to automatically or semi-automatically serve notices of claims to extensions of time and compensation for additional costs in line with the mechanisms stipulated in the paper-based conditions of contract. In addition, contractual obligations in respect of time limits could be programmed inside smart contracts in the form of code-based conditionals to execute actions if human agents neglect to do so. The 'wait-and see' approach that dominates claims management in practice would then dissipate under the rigour of blockchain-based CCA.

The realisation of blockchain-based *claims management* system would be confronted by challenges. As with a blockchain-based payment system, coding smart contracts to exactly simulate the contractual logic for the corresponding contract provisions and sub-clauses will be the most pronounced challenge. Such codes for contractual provisions stipulated in standard forms of contracts (e.g. FIDIC, JCT, AIA) have not yet been developed. Current research efforts are centred on converting parts of explicit contractual provisions into customised computable rules to reflect portions of the contractual reality and overcome the complexity embedded in contractual provisions. Current efforts are, however, being made to develop smart contract codes that reflect customised rules for contractual processes and for their interaction with oracles. This will pave the way towards achieving smart contract codes for the contractual conditions of commonly used standard forms of contract. As a result, the efficiency and effectiveness of the CCA processes identified earlier would be substantially improved in comparison with current practice. In a blockchain-based CCA system all these processes would interact in a dynamic manner. Together with other valuable digitalisation initiatives in the construction management discipline, the case of blockchain for CCA is likely to contribute to improving the construction industry. The emphasis placed here is that the technology is not proposed as a replacement for human agents or *tried and tested* contractual practices but as an enhancement.

This chapter highlights blockchain technology's potential to address chronic issues in CCA, providing preliminary evidence of its benefits and proposing a research agenda to tackle key challenges. Incremental adoption of blockchain could transform CCA by enhancing governance and addressing persistent failures, particularly in claims management. However, progress requires innovative thinking and recognition of blockchain's limitations as a complementary, rather than all-encompassing, solution.

It is recommended that scholars along with 'brave' early adopters can work together towards addressing the preceding proposed agenda given the potential of blockchain governance for CCA and the construction industry. In broader terms, the construction industry at large is moving towards the digitalisation of a wide array of processes and systems, while efforts devoted to the digitalisation of CCA, despite its crucial role in the successful delivery of construction projects, appear to be lagging behind. With a growing interest in improving contractual practices in the construction industry, scholars along with 'brave' early adopters can work together towards developing advanced solutions without the need for radical changes to prevailing practices. As a result, workability, acceptability and scalability of blockchain-based solutions for CCA in general and claims management in particular can be extended throughout the industry.

Construction contracts and their administration are a notoriously conservative domain that has progressed little alongside the digital transformation of the industry in general. Nevertheless, this chapter may serve to encourage policy makers and practitioners to further explore the benefits of blockchain to CCA. It is proposed that the perspectives of employers, contractors and policymakers should be included in future research efforts. This may help to drive adoption levels by achieving balanced benefits for all actors involved, while improving the operationalisation of procurement and contractual frameworks devised by policymakers. These actors have a significant role to perform, and this chapter is addressed as much to them as to academics, policymakers and theorists.

References

[1] Abrahamson M. The new Abrahamson principles: tools to assist in contract drafting – Part 2, contract management. *Construction Law International* 2007;2:6–9.

[2] Mason J. *Innovating Construction Law: Towards the Digital Age*. London: Routledge; 2021.

[3] Li J, and Kassem M. Applications of distributed ledger technology (DLT) and blockchain-enabled smart contracts in construction. *Automation in Construction* 2021;132:103955. https://doi.org/10.1016/j.autcon.2021.103955.

[4] FIDIC. *Conditions of Contract for Construction.* Lausanne: FIDIC; 2017.

[5] Hughes W, Champion R, and Murdoch J. *Construction Contracts: Law and Management.* 5th ed. London: Routledge; 2015.

[6] Bunni NG. *The FIDIC Forms of Contract*. New York: Wiley; 2005.

[7] Ning Y. Impact of quality performance ambiguity on contractor's opportunistic behaviors in person-to-organization projects: The mediating roles of contract design and application. *International Journal of Project Management* 2018;36:640–9. https://doi.org/10.1016/j.ijproman.2018.01.008.

[8] RICS. *Pathway Guide: Quantity Surveying and Construction*, London; 2018. Available at: https://www.rics.org/content/dam/ricsglobal/documents/join-rics/qs_and_construction.pdf.

[9] Gunduz M, and Elsherbeny HA. Operational framework for managing construction – Contract administration practitioners' perspective through modified Delphi method. *Journal of Construction Engineering and Management* 2020;146:04019110. https://doi.org/10.1061/(asce)co.1943-7862.0001768.

[10] Papajohn D, El Asmar M, Molenaar KR, and Alleman D. Comparing contract administration functions for alternative and traditional delivery of highway projects. *Journal of Management in Engineering* 2020; 36:04019038. https://doi.org/10.1061/(asce)me.1943-5479.0000727.

[11] Guccio C, Pignataro G, and Rizzo I. Do local governments do it better? Analysis of time performance in the execution of public works. *European Journal of Political Economy* 2014;34:237–52. https://doi.org/10.1016/j.ejpoleco.2014.01.010.

[12] Chong HY, Balamuralithara B, and Chong SC. Construction contract administration in Malaysia using DFD: A conceptual model. *Industrial Management and Data Systems* 2011;111:1449–64. https://doi.org/10.1108/02635571111182782.

[13] Fawzy SA, and El-adaway IH. Contract administration guidelines for managing conflicts, claims, and disputes under World Bank-funded projects. *Journal of Legal Affairs and Dispute Resolution in Engineering and Construction* 2012;4:101–10. https://doi.org/10.1061/(asce)la.1943-4170.0000091.

[14] Barakat M, Abdul-Malak M-A, and Khoury H. Pivotal new roles and changes introduced by the 2017 FIDIC's claim and dispute resolution mechanism. *Journal of Legal Affairs and Dispute Resolution in Engineering and*

Construction 2020;12(1):04519049. https://doi.org/10.1061/(asce)la.1943-4170.0000355.

[15] Schepker DJ, Oh WY, Martynov A, and Poppo L. The many futures of contracts: Moving beyond structure and safeguarding to coordination and adaptation. *Journal of Management* 2014;40:193–225. https://doi.org/10.1177/0149206313491289.

[16] Chen Y, Wang W, Zhang S, and You J. Understanding the multiple functions of construction contracts: The anatomy of FIDIC model contracts. *Construction Management and Economics* 2018;36:472–85. https://doi.org/10.1080/01446193.2018.1449955.

[17] Ndekugri I, Smith N, and Hughes W. The engineer under FIDIC's conditions of contract for construction. *Construction Management and Economics* 2007;25:791–9. https://doi.org/10.1080/01446190701411216.

[18] Chappell D. *Construction Contracts: Questions and Answers*. 4th ed. London: Routledge; 2020. https://doi.org/https://doi.org/10.1201/9781003080930.

[19] Williamson OE. *The Economic Institutions of Capitalism. Firms, Markets, Relational Contracting*. New York: Free Press; 1985.

[20] Flyvbjerg B. What you should know about megaprojects and why: An overview. *Project Management Journal* 2014;45:6–19. https://doi.org/10.1002/pmj.21409.

[21] Peters E, Subar K, and Martin H. Late payment and nonpayment within the construction industry: Causes, effects, and solutions. *Journal of Legal Affairs and Dispute Resolution in Engineering and Construction* 2019;11(3):04519013. https://doi.org/10.1061/(ASCE)LA.1943-4170.0000314.

[22] Jelodar MB, Yiu TW, and Wilkinson S. Dispute manifestation and relationship quality in practice. *Journal of Legal Affairs and Dispute Resolution in Engineering and Construction* 2016;8(1):C4515003. https://doi.org/10.1061/(asce)la.1943-4170.0000171.

[23] Khalef R, El-adaway IH, Assaad R, and Kieta N. Contract risk management: A comparative study of risk allocation in exculpatory clauses and their legal treatment. *Journal of Legal Affairs and Dispute Resolution in Engineering and Construction* 2021;13(1):04520036. https://doi.org/10.1061/(asce)la.1943-4170.0000430.

[24] Love P, Davis P, Ellis J, and Cheung SO. Dispute causation: Identification of pathogenic influences in construction. *Engineering, Construction and Architectural Management* 2010;17:404–23. https://doi.org/10.1108/09699981011056592.

[25] Odeyinka HA, Lowe J, and Kaka AP. Artificial neural network cost flow risk assessment model. *Construction Management and Economics* 2013;31:423–39. https://doi.org/10.1080/01446193.2013.802363.

[26] Gumusburun Ayalp G. Critical predictors of burnout among civil engineers at construction sites: A structural equation modelling. *Engineering, Construction and Architectural Management* 2022;29(9):3547–73. https://doi.org/10.1108/ECAM-12-2020-1066.

[27] Cattell K, Bowen P, and Edwards P. Stress among South African construction professionals: A job demand-control-support survey. *Construction*

Management and Economics 2016;34:700–23. https://doi.org/10.1080/01446193.2016.1203967.

[28] Chan APC, and Owusu EK. Corruption forms in the construction industry: Literature review. *Journal of Construction Engineering Management* 2017; 143:04017057. https://doi.org/10.1061/(asce)co.1943-7862.0001353.

[29] Poppo L, and Zenger T. Do formal contracts and relational governance function as substitutes or complements? *Strategic Management Journal* 2002;23:707–25. https://doi.org/10.1002/smj.249.

[30] Benítez-Ávila C, Hartmann A, Dewulf G, and Henseler J. Interplay of relational and contractual governance in public-private partnerships: The mediating role of relational norms, trust and partners' contribution. *International Journal of Project Management* 2018;36:429–43. https://doi.org/10.1016/j.ijproman.2017.12.005.

[31] Macneil IR. Economic analysis of contractual relations: Its shortfalls and the need for a rich classificatory apparatus. *Northwest University Law Review* 1981;75:1018–63.

[32] Rahman MM, and Kumaraswamy MM. Joint risk management through transactionally efficient relational contracting. *Construction Management and Economics* 2002;20:45–54. https://doi.org/10.1080/01446190110089682.

[33] Zhang SB, Chen J, and Fu Y. Contract complexity and trust in construction project subcontracting. *Engineering, Construction and Architectural Management* 2020;27:2477–500. https://doi.org/10.1108/ECAM-02-2019-0113.

[34] Lahdenperä P. Making sense of the multi-party contractual arrangements of project partnering, project alliancing and integrated project delivery. *Construction Management and Economics* 2012;30:57–79. https://doi.org/10.1080/01446193.2011.648947.

[35] Das M, Luo H, and Cheng JCP. Securing interim payments in construction projects through a blockchain-based framework. *Automation in Construction* 2020;118:103284. https://doi.org/10.1016/j.autcon.2020.103284.

[36] Li X, Wu L, Zhao R, Lu W, and Xue F. Two-layer adaptive blockchain-based supervision model for off-site modular housing production. *Computers in Industry* 2021;128:103437. https://doi.org/10.1016/j.compind.2021.103437.

[37] Anthony Jr B. Distributed ledger and decentralised technology adoption for smart digital transition in collaborative enterprise. *Enterprise Information Systems* 2021;17(4):465–98. https://doi.org/10.1080/17517575.2021.1989494.

[38] Msawil M, Greenwood D, and Kassem M. A Systematic evaluation of blockchain-enabled contract administration in construction projects. *Automation in Construction* 2022;143:104553. https://doi.org/10.1016/j.autcon.2022.104553.

[39] Elbashbishy TS, Ali GG, and El-adaway IH. Blockchain technology in the construction industry: Mapping current research trends using social network analysis and clustering. *Construction Management and Economics* 2022; 40:406–27. https://doi.org/10.1080/01446193.2022.2056216.

[40] Motawa I, and Kaka A. Modelling payment mechanisms for supply chain in construction. *Engineering, Construction and Architectural Management* 2009;16:325–36. https://doi.org/10.1108/09699980910970824.

[41] Chong HY, and Diamantopoulos A. Integrating advanced technologies to uphold security of payment: Data flow diagram. *Automation in Construction* 2020;114:103158. https://doi.org/10.1016/j.autcon.2020.103158.

[42] Hamledari H, and Fischer M. Construction payment automation using blockchain-enabled smart contracts and robotic reality capture technologies. *Automation in Construction* 2021;132:103926. https://doi.org/10.1016/j.autcon.2021.103926.

[43] Sigalov K, Ye X, König M, *et al.* Automated payment and contract management in the construction industry by integrating building information modeling and blockchain-based smart contracts. *Applied Sciences (Switzerland)* 2021;11(16):7653. https://doi.org/10.3390/app11167653.

[44] Elghaish F, Abrishami S, and Hosseini MR. Integrated project delivery with blockchain: An automated financial system. *Automation in Construction* 2020;114:103182. https://doi.org/10.1016/j.autcon.2020.103182.

[45] Elghaish F, Pour Rahimian F, Hosseini MR, Edwards D, and Shelbourn M. Financial management of construction projects: Hyperledger fabric and chaincode solutions. *Automation in Construction* 2022;137:104185. https://doi.org/10.1016/j.autcon.2022.104185.

[46] Wu L, Lu W, and Xu J. Blockchain-based smart contract for smart payment in construction: A focus on the payment freezing and disbursement cycle. *Frontiers of Engineering Management* 2022;9:177–95. https://doi.org/10.1007/s42524-021-0184-y.

[47] Cheng JCP, Liu H, Gan VJL, Das M, Tao X, and Zhou S. Construction cost management using blockchain and encryption. *Automation in Construction* 2023;152:104841. https://doi.org/10.1016/j.autcon.2023.104841.

[48] Hetemi E, Jerbrant A, and Mere JO. Exploring the emergence of lock-in in large-scale projects: A process view. *International Journal of Project Management* 2020;38:47–63. https://doi.org/10.1016/j.ijproman.2019.10.001.

[49] Maylor H, Geraldi J, Budzier A, Turner N, and Johnson M. Mind the gap: Towards performance measurement beyond a plan-execute logic. *International Journal of Project Management* 2023;41(4):102467. https://doi.org/10.1016/j.ijproman.2023.102467.

[50] Elghaish F, and Abrishami S. Developing a framework to revolutionise the 4D BIM process: IPD-based solution. *Construction Innovation* 2020;20:401–20. https://doi.org/10.1108/CI-11-2019-0127.

[51] Wang Z, Wang T, Hu H, Gong J, Ren X, and Xiao Q. Blockchain-based framework for improving supply chain traceability and information sharing in precast construction. *Automation in Construction* 2020;111:103063. https://doi.org/10.1016/j.autcon.2019.103063.

[52] Hunhevicz JJ, Motie M, and Hall DM. Digital building twins and blockchain for performance-based (smart) contracts. *Automation in Construction* 2022;133:103981. https://doi.org/10.1016/j.autcon.2021.103981.

[53] Love PED, Ika LA, Matthews J, and Fang W. Curbing poor-quality in large-scale transport infrastructure projects. *IEEE Transactions on Engineering Management* 2022;69:3171–83. https://doi.org/10.1109/TEM.2020.3031890.
[54] Koch C, and Schultz CS. The production of defects in construction – An agency dissonance. *Construction Management and Economics* 2019;37:499–512. https://doi.org/10.1080/01446193.2018.1519253.
[55] Love PED, Smith J, Ackermann F, Irani Z, and Teo P. The costs of rework: Insights from construction and opportunities for learning. *Production Planning and Control* 2018;29:1082–95. https://doi.org/10.1080/09537287.2018.1513177.
[56] Lin YC, Chang JX, and Su YC. Developing construction defect management system using BIM technology in quality inspection. *Journal of Civil Engineering and Management* 2016;22:903–14. https://doi.org/10.3846/13923730.2014.928362.
[57] Wu H, Zhong B, Li H, Guo J, and Wang Y. On-site construction quality inspection using blockchain and smart contracts. *Journal of Management in Engineering* 2021;37:04021065. https://doi.org/10.1061/(asce)me.1943-5479.0000967.
[58] Sheng D, Ding L, Zhong B, Love PED, Luo H, and Chen J. Construction quality information management with blockchains. *Automation in Construction* 2020;120:103373. https://doi.org/10.1016/j.autcon.2020.103373.
[59] Lu W, Wu L, Xu J, and Lou J. Construction E-Inspection 2.0 in the COVID-19 pandemic era: A blockchain-based technical solution. *Journal of Management in Engineering* 2022;38(4):04022032. https://doi.org/10.1061/(asce)me.1943-5479.0001063.
[60] Nawari NO, and Ravindran S. Blockchain and building information modeling (BIM): Review and applications in post-disaster recovery. *Buildings* 2019;9:149. https://doi.org/10.3390/BUILDINGS9060149.
[61] Zhong B, Guo J, Zhang L, Wu H, Li H, and Wang Y. A blockchain-based framework for on-site construction environmental monitoring: Proof of concept. *Building and Environment* 2022;217:109064. https://doi.org/10.1016/j.buildenv.2022.109064.
[62] Saygili M, Mert IE, and Tokdemir OB. A decentralized structure to reduce and resolve construction disputes in a hybrid blockchain network. *Automation in Construction* 2022;134:104056. https://doi.org/10.1016/j.autcon.2021.104056.
[63] Love PED, Zhou J, Matthews J, and Sing MCP. Retrospective future proofing of a copper mine: Quantification of errors and omissions in 'as-built' documentation. *Journal of Loss Prevention in the Process Industries* 2016;43:414–23. https://doi.org/10.1016/j.jlp.2016.06.011.
[64] Abdul-Malak M-A, and Bachnak T. Classification of contract-related documentation for evidentiary purposes in construction claims and disputes resolutions. *In Construction Research Congress 2020: Project Management and Controls, Materials, and Contracts*, 2020, pp. 1284–92.

[65] Al Qady M, and Kandil A. Automatic classification of project documents on the basis of text content. *Journal of Computing in Civil Engineering* 2015; 29:04014043. https://doi.org/10.1061/(asce)cp.1943-5487.0000338.
[66] Ciotta V, Mariniello G, Asprone D, Botta A, and Manfredi G. Integration of blockchains and smart contracts into construction information flows: Proof-of-concept. *Automation in Construction* 2021;132:103925. https://doi.org/10.1016/j.autcon.2021.103925.
[67] Erri P, Abhinaw S, Yiu TW, Zou Y, and Amor R. Blockchain-aided information exchange records for design liability control and improved security. *Automation in Construction* 2021;126:103667. https://doi.org/10.1016/j.autcon.2021.103667.
[68] Tao X, Das M, Liu Y, and Cheng JCP. Distributed common data environment using blockchain and interplanetary file system for secure BIM-based collaborative design. *Automation in Construction* 2021;130:103851. https://doi.org/10.1016/j.autcon.2021.103851.
[69] Das M, Tao X, Liu Y, and Cheng JCP. A blockchain-based integrated document management framework for construction applications. *Automation in Construction* 2022;133:104001. https://doi.org/10.1016/j.autcon.2021.104001.
[70] Li X, Lu W, Xue F, *et al.* Blockchain-enabled IoT-BIM platform for supply chain management in modular construction. *Journal of Construction Engineering and Management* 2022;148:04021195.
[71] Yang R, Wakefield R, Lyu S, *et al.* Public and private blockchain in construction business process and information integration. *Automation in Construction* 2020;118:103276. https://doi.org/10.1016/j.autcon.2020.103276.
[72] Lee D, Lee SH, Masoud N, Krishnan MS, and Li VC. Integrated digital twin and blockchain framework to support accountable information sharing in construction projects. *Automation in Construction* 2021;127:103688. https://doi.org/10.1016/j.autcon.2021.103688.
[73] Schmidt CG, and Wagner SM. Blockchain and supply chain relations: A transaction cost theory perspective. *Journal of Purchasing and Supply Management* 2019;25(4):100552. https://doi.org/10.1016/j.pursup.2019.100552.
[74] Hunhevicz JJ, and Hall DM. Do you need a blockchain in construction? Use case categories and decision framework for DLT design options. *Advanced Engineering Informatics* 2020;45:101094. https://doi.org/10.1016/j.aei.2020.101094.
[75] Tummalapudi M, Harper CM, Taylor TRB, Waddle S, and Catchings R. Causes, implications, and strategies for project closeout delays in highway construction. *Transportation Research Record* 2022;2676:479–90. https://doi.org/10.1177/03611981221087229.
[76] Sun M, and Meng X. Taxonomy for change causes and effects in construction projects. *International Journal of Project Management* 2009;27:560–72. https://doi.org/10.1016/j.ijproman.2008.10.005.
[77] Hunhevicz JJ, Hall DM, Brasey PA, Bonanomi MMM, and Fischer M. Decentralized project delivery on the crypto commons: Conceptualization,

governance mechanisms, and future research directions. *Project Leadership and Society* 2024;5:100132. https://doi.org/10.1016/j.plas.2024.100132.

[78] Aibinu AA. The relationship between distribution of control, fairness and potential for dispute in the claims handling process. *Construction Management and Economics* 2006; 24:45–54. https://doi.org/10.1080/14697010500226954.

[79] Shetty A, Shetty AD, Pai RY, *et al.* Block chain application in insurance services: A systematic review of the evidence. *Sage Open* 2022;12(1):1–15. https://doi.org/10.1177/21582440221079877.

[80] Li J, Greenwood D, and Kassem M. Blockchain in the built environment and construction industry: A systematic review, conceptual models and practical use cases. *Automation in Construction* 2019;102:288–307. https://doi.org/10.1016/j.autcon.2019.02.005.

[81] Mason J. Intelligent contracts and the construction industry. *Journal of Legal Affairs and Dispute Resolution in Engineering and Construction* 2017;9 (3):04517012. https://doi.org/10.1061/(asce)la.1943-4170.0000233.

[82] Shukla S, Thakur S, Hussain S, Breslin JG, and Jameel SM. Identification and authentication in healthcare internet-of-things using integrated fog computing based blockchain model. *Internet of Things (Netherlands)* 2021; 15:100422. https://doi.org/10.1016/j.iot.2021.100422.

[83] Rahman MS, Chamikara MAP, Khalil I, and Bouras A. Blockchain-of-blockchains: An interoperable blockchain platform for ensuring IoT data integrity in smart city. *Journal of Industrial Information Integration* 2022; 30:100408. https://doi.org/10.1016/j.jii.2022.100408.

[84] Msawil M, Li J, Elghaish F, Greenwood D, and Kassem M. Enhancing contractual time performance management: A Hyperledger fabric-based conceptual framework. *European Conference on Computing in Construction*, 2024.

[85] Martin L, and Benson L. Relationship quality in construction projects: A subcontractor perspective of principal contractor relationships. *International Journal of Project Management* 2021;39:633–45. https://doi.org/10.1016/j.ijproman.2021.05.002.

[86] Love PED, and Matthews J. The 'how' of benefits management for digital technology: From engineering to asset management. *Automation in Construction* 2019;107:102930. https://doi.org/10.1016/j.autcon.2019.102930.

Chapter 12

Legal risks for blockchain applications in the built environment: a legal perspective

Gavin P. Johnson[*]

12.1 Introduction

Blockchain, often described as a solution in search of a problem, is touted as a powerful tool to disintermediate needless complexity and reliance on 'trusted' third parties such as banks and other institutions of doubtful ethical and moral veracity. Some evangelise its future potential, some of whom may actually understand it too.

As with other technologies working at the edges of the public's perception, swathes of experts, opinions masquerading as facts, misunderstandings, confused and confusing explanations, and, frankly, fundamentally incorrect and misleading information run amok. As a friend memorably once said, 'it is time for the children to leave the playground and the adults to enter the room'. A mixed metaphor, but the sentiment is clear. Whilst there is much that could be done with the blockchain technological ecosystem, the question is: what should be done? It follows therefrom: how could it be done? It is the limits of these questions, from a legal perspective, that this chapter considers.

There are considerable legal risks for blockchain applications in the built environment. Interestingly, much of that risk does not stem from what I shall call the 'real', that is, real estate, property, construction and so on. Although these are each legally complex issues, there are mechanisms in place to control and mitigate risks, and specialists and experts with years of experience to manage them. Things do still go wrong (how else would lawyers make a living?), but in broad terms the risks are well known.

Furthermore, much of the risk does not stem from what I shall call the 'digital', that is, technology, software, data and so on. As with the 'real', these are similarly legally complex issues, but a wealth of legislation, strategies and redundancy and resilience mechanisms exist to govern and regulate the space, along with many domain specialists and experts from whom advice may be sought. Famously, or infamously, things do still go wrong, data is not secured properly, crypto exchanges

[*]Policy Advisory Committee, Construction Blockchain Consortium, University College London, UK

are hacked[1] [1] or mismanaged [2], smart contracts compromised and tokens transferred unlawfully.

The true challenges arise when the real and digital are connected. Whilst we may think of physical cash, handfuls of paper and metal, it has existed digitally for decades, in bank transactions, credit cards and now not even as a rectangle of plastic, but as an app on your smartphone. However, can the same be said of, say, a building? What is its digital equivalent? Whilst each of the real and digital operate in their own legal frameworks, interweaving those frameworks to create the types of solutions we imagine is not trivial. This is before the complications of implementing solutions across disparate legislations and jurisdictions.

Throughout this chapter, I footnote further reading that expands on a particular issue or topic, or that illuminates the wealth of opportunities around us if only we had the confidence to lean into our ideas and knew how to exploit them.

Before we begin our exploration, a note of caution. Laws and regulations in and around blockchain are broadly adapted from or interpreted in the context of current legal principles. As such, and as with any developing area, lawmakers could decide to change the laws and regulations or take a contrary view to one we might expect. This chapter should not be considered a definitive guide to the legal consequences of any given blockchain solution, but should cast light on the issues to which critical thought should be given.

Let us imagine a moment of insight has struck and you are about to build the next 'killer' blockchain application.[2] This chapter will explore the importance of ensuring a clear understanding of the legal frameworks that may govern the application, how the solution may be grounded in legal reality and what may be done for the solution's long-term resilience. We will consider and analyse the legal risks and offer practical insights and examples on how to ensure your revolutionary, disruptive blockchain application is built based on a strong, pragmatic, legal foundation.

A note of caution on the law. The law may or may not tell us what we think we already know. The law evolves, it both lags and leads change. Any new application must therefore be carefully considered for its legal context.

12.1.1 Background on blockchain technology

Whilst lawyers are not best placed to explain blockchain technology, there are several fundamental elements that must be appreciated to achieve a common understanding. The core challenge between client and lawyer is communication: 'we appear to be talking about the same thing, but are we in fact talking about the same thing?'

[1]A developer friend once said of cybersecurity, 'it's basically a racket'. Meaning that a certificate confirming software to be cyber secure was good only for the day on which it was given. The rate of development in the wider software space is furious and what was once considered secure has long since been hacked.

[2]For insight into the how of start-ups see Thiel P, Masters B. Zero to one: notes on startups or how to build the future. USA: Crown Business; 2014.

With this in mind, and forgive the simplified descriptions, it is important to know what it is we are talking about and what we mean when we say certain things. In part, this is why lawyers include 'definitions and interpretation' sections in contracts (another part is to reduce word count, we lawyers tend to the verbose[3]). Broadly, although with enhancement, I adopt the definitions of the Law Society of England and Wales as presented in recent publications [3] that, as a consolidation of inputs across legal and technical contributors, represents a baseline of common terms for understanding blockchain applications.

There are five key blockchain elements: (i) distributed ledger technology (DLT), (ii) blockchain, distinct from DLT, you can have one without the other, but they are intrinsically related, (iii) 'smart' and 'smart legal' contracts, (iv) tokens and tokenisation, which is where things get really interesting and (v) wallets and storage, you need to keep things somewhere and safe.

Other features, such as public and private blockchains, permissions, consensus protocols, nodes, cryptography, hash functions and other terms of which you may have heard, are beyond the scope of this chapter and more competently and eruditely addressed by other contributors.

12.1.1.1 DLT

An important distinction in ledger technology is between centralised, decentralised and distributed [4]. Historically, centralised ledgers were maintained, wherein there was only one copy of the ledger and all entries had to be made by a single, usually mandated, entity. There was one definitive record to which to refer: in modern parlance 'one version of the truth'. In the UK, an example relevant to the built environment is HM Land Registry (HMLR), which maintains a 'ledger' of land and property, and who owns it, for the whole jurisdiction.

More recently, decentralised ledgers have been utilised. In this case, trusted intermediaries are employed to keep and maintain records. A chain of trusted intermediaries may then connect two parties not previously known to each other. Such intermediators are brokers of trust and are collectively responsible for maintaining the ledger.

For our purposes, the drawbacks of centralised and decentralised ledgers gave rise to an important question: can the need for trust, and the risks inherent to that trust as a point of failure, be removed? This would simplify transactions and directly connect unknown parties without the need for an intermediary, thus disintermediating 'trust risks'[4] (from where the trope that blockchain is a tool for disintermediation has its origins). This is where DLT arises.

[3]This may be explained by lawyers historically being paid by the word, as well as the imprecision of language and its interpretation. Whilst lawyers cannot be blamed for the latter, the former is an example of incorrect incentives. We shall return to the concept of incentives later, but for an absurd example of how even the simplest of phrases are rife with ambiguity and imprecision, see 'I want to eat a burger', author unknown, date unknown, viewable at the fifth entry here: Translators Café. Re: Plain English or plain ignorance? https://www.translatorscafe.com/cafe/MegaBBS/forumthread8234msg141720.htm#141720 [Accessed 11 October 2024]

[4]Often expressed as, 'trust the message, not the messenger'.

DLT refers to a group of technologies that store, synchronise and maintain digital records across a network of computers. I imagine a spreadsheet and a list of entries, for example, time and date, value, transaction parties and so on. This record is maintained by all participants in the network, and all have an identical copy of the list of transactions.

12.1.1.2 Blockchain

It seems unavoidable to mention blockchain without also mentioning bitcoin and the 2008 Whitepaper published under the pseudonym Satoshi Nakamoto [5]. In simple terms, a blockchain collects together digital records in a 'block' of transactions (usually chronologically), encrypts them (known as hashing) and adds the new block of transactions to the previous block, thereby 'chaining' them together. Those readers who are aware of hashing will be aware of how such 'chains of blocks' are achieved, but this is a technical issue outside the scope of this chapter.

It is important to recognise that DLT does not need to be blockchain. For thousands of years, ledgers survived without encryption and chaining. Also, blockchain does not need to be DLT, transactions may be disparate, unconnected and disorganised and still be capable of being grouped and chained together (although one wonders why you would want encrypted unorganised data).

Onymous

Whilst a detailed treatment of the types of blockchain, such as open, closed, permissioned, permissionless, public, private and so on, is beyond the scope of this chapter, there is one point to note as it impacts legal considerations, which is 'onymity'.[5] You will have heard of anonymous and pseudonymous. These, literally, mean 'without name' and 'a false name', respectively.

In an anonymous system, the network participants may not be vetted or identifiable. Their identities may never be requested, and their locations obfuscated to further protect that anonymity. This has pros and cons, depending on the applicable case. The bottom line is that individual identity is hidden and although online activity is visible it cannot be traced back to that individual. This is an important consideration for privacy and data protection.

In a pseudonymous system, the network participants may be vetted and identifiable. This is important as, to comply with legal requirements, an individual may need to prove certain personal characteristics such as age to purchase age-restricted goods or access age-restricted services. The onus is generally on the supplier or provider to confirm these details and validate the individual as a lawful customer. However, the pseudonymous system may create a fictional persona (avatar) to represent the individual in the ecosystem, whilst still allowing them, as a 'confirmed lawful person' to conduct transactions and engage in that ecosystem's activities. Whilst such a created persona offers a degree of privacy on face value, transactions may be linked back to your real-world identity. There are many online resources that explore anonymity and pseudonymity in detail [6].

[5]There is no such word, but the root is 'onymous', from the Greek meaning 'having a name' or 'named'.

12.1.1.3 Smart and smart legal contracts

By way of anecdote, when I first started looking at blockchain and what it could do, a great deal of confusion was encountered around the concept of smart contracts. Many (non-lawyers) were insistent that this meant, to put it unkindly, the death of lawyers and, to put it kindly, these would eventually replace lawyers as a means of entering legally binding relationships between parties. As I explained much to their chagrin, the analysis disagrees.

Smart contracts

Smart contracts are defined as 'automated software code that executes on the occurrence of a specific or specified event or events' [7]. Whilst some suggest there remains debate about the nature of smart contracts being contracts, a first-year legal student applying contract formation analysis would rapidly conclude that they are not. Even on a trivial understanding, a smart contract is little more than a piece of code that executes an action when certain criteria are met [8], as the definition states. Smart contracts can and do execute and enforce the rules or conditions of an underlying, previously agreed, contract as an implementation of a term or terms of a contract. At their simplest, they effect transfers between ecosystem participants upon the occurrence of a trigger event or events. For example, a futures or forward options agreement, a predetermined price is met and the buy or sell order executes. That is it. As an implementation of a conditional requirement of a contract, they are not *the* contract. As the lawyers say, 'a smart contract is neither smart nor a contract'.

Smart legal contracts

If 'smart contracts' are not contracts, are 'smart legal contracts' contracts? Conceptually, yes. A smart legal contract is 'a legally binding contract in which some or all of the contractual terms are defined in and/or performed automatically by a computer program' [9]. Rather than being an automated execution of an element of a contract, it constitutes the entire agreement between the parties and meets the formation requirements discussed below.[6]

There are three categories of smart legal contracts: (i) natural language contract with automated performance (NLC), (ii) hybrid contract (HC) and (iii) solely code contract (SCC) [10].

Natural language contract. In a NLC, the contractual obligations are contained in a traditional contract, with the terms and conditions written in ordinary, albeit legal, language. Some or all of those obligations are performed automatically by computer code. Precisely, how those obligations are interpreted and translated into executable code, what the benchmark for such implementation is, or the reference standards for such, is not clear. Ultimately, those interpretations, translations, benchmarks and standards would need to be developed and determined in court. However, at least conceptually, the code operates solely to self-execute and/or automate part, parts, or all of that traditional contract [11].

[6]See Section 12.2.2.1.

You may see this type of smart legal contract referred to as an 'external' contract, as the code is written and executed outside of the parties' legally binding agreement. In many ways, this most closely reflects the simple smart contracts defined above, but with the operational elements of a contract automated.

Hybrid contract. In a HC, some contractual obligations are defined in a traditional NLC, and some are defined in computer code. Some or all the contractual obligations may be performed automatically by the code. There is a sliding scale as to the number of encoded and natural language terms.

You may think there is little difference between hybrid and natural language types. The distinction lies in the scale of automation and the pre-agreement between the parties of the incorporation of purely encoded terms, which may not be or have not been specifically negotiated. One may imagine a set or sets of encoded terms incorporated by reference, with which the functionality, inputs and outputs are well known, established and/or accepted as standard. This is similar to how lawyers draft standard form contracts that, in intention if not in practice, may simply be 'pulled-off-the-shelf' and applied to a given situation. An example of this in the UK are standard form building contracts such as the Joint Contracts Tribunal or New Engineering Contract suites. Internationally, the International Federation of Consulting Engineers standard forms may be recognisable.

Solely code contract. In a SCC, all the terms and conditions are defined in computer code and are performed automatically by that code. There is no natural language element. This type of smart legal contract, for the time being, would likely only be used to govern relatively simple agreements between parties, where the terms are standardised or regulated, deal with simple transactional matters and execute specific (well-known and accepted) restorative actions for a breach of the terms.

12.1.1.4 Tokens and tokenisation

There are many types, or 'species', of token. For simplicity, this chapter considers four fundamental (three general, one specific) types directly relevant to legal considerations. Tokens may incorporate multiple features and/or be complex hybrids with multiple functions. Tokens may evolve from one species to another. Each element, function and evolution of a token is important to note as the token that is delivered in a particular project may not be what was intended on a subsequent analysis.

Crypto-asset

In legal terminology tokens are to be formally defined in the Financial Services Markets Bill (FSMB).[7] FSMB, as well as many reports and consultations previously, adopts the term 'crypto-asset' as the preferred term for cryptocurrencies, security tokens and other tokenised representations of assets and financial instruments. A crypto-asset is formally defined as 'any cryptographically secured digital

[7]At the time of writing a proposed Act of Parliament to make provision about the regulation of financial services, markets and other connected purposes, including regulation for crypto-assets.

representation of value or contractual rights that – (a) can be transferred, stored or traded electronically, and that (b) uses technology supporting the recording or storage of data (which may include distributed ledger technology)'.

Utility tokens

A utility token may be defined as a token that grants the holder with access to specific goods or services [3]. A good example is, say, a ticket to an event. The token represents a right of access to the event. They may be traded, but do not typically fall within what is known as the 'regulatory perimeter' of a financial regulatory body.

Security tokens

A security token, as the name suggests, may be defined as a token that grants the holder specific rights or interests in a business [3]. A good example is, say, shares in a company or a financial instrument such as debt or profit share.

Exchange tokens

An exchange token may be described as a representation or method, of value transfer between two parties [3]. A good example is bitcoin, where the exchange of a bitcoin, bitcoins or parts thereof, are a means of exchanging, settling or recognising, a change in position between two or more parties as to value related to an exchange or investment. Exchange tokens are, typically, not controlled by a person or group, are not underpinned by assets and do not provide any rights or access to goods or services.

There exists a sub-set of exchange tokens known as 'stablecoins' [3]. A stablecoin may be defined as a token whose value is attached, linked or 'pegged' to a specific commodity (such as gold), a fiat currency (such as the US dollar) or group of assets (a basket of broadly price-stable assets), either directly or algorithmically, to produce a price-stable, minimally volatile unit.

Non-fungible tokens (NFTs)

Fungibility is the ability of a good or asset to be interchanged with others of the same type. A ten-pound note is functionally indistinguishable from any other ten-pound note. A bitcoin is similarly indistinguishable from any other bitcoin.

NFTs are a specific species of token that are not interchangeable. They are unique, and they are unique to the asset to which they relate. This results in two key determinations. First, this uniqueness may then be used to validate or guarantee, if the underlying contractual relationship obligates, the uniqueness of the asset to which the NFT relates. Second, holding the relevant NFT then, by definition, confirms ownership of the asset. If ownership is unambiguous, or legally certain, it should be enforceable. In very general and reductive terms, the law exists to address who rightfully owns what.

Fractionalisation

Lastly, on tokens, is 'fractionalisation'. As hinted at in the introduction to this sub-section, this can result in serious legal repercussions and any project must be alive to this risk. For present purposes, in simple terms this where the token is either 'split' into multiple parts after creation or is created as representative of a partial, or fraction, of the underlying asset. In either case, the act of fractionalising, to create a structure of shared ownership between multiple parties, may create either a share, as in a share or stock in the capital of a company, or a form of derivative. These are known as 'specified investments' [12] and are 'regulated activities' [13], meaning a project would need to be authorised by the relevant regulatory body in order to be able to deal in such investments, as they would then become.

12.1.1.5 Wallets and storage

If you own something, usually you need somewhere to keep it. In the blockchain ecosystem, tokens are ostensibly 'stored' in a 'wallet'. For the purposes of this chapter, we shall assume that this is the case as it is conceptually similar to how, say, your bank account holds your money.[8]

12.1.1.6 Summary of introduction

An encrypted blockchain DLT enables exclusive ownership of records and crypto-assets by ensuring that the right to send messages (or make transactions) on behalf of a public key relies on a private key, which is capable of being kept secret and known only to a single individual.

12.2 Legal risks associated with blockchain applications

This section considers the key legal risks to blockchain applications in the built environment [14]; we will consider their mitigation later. It should not be taken as exhaustive and is not a substitute for seeking legal advice on a specific project. Awareness of the legal space in which a project will operate provides two critical advantages.

First, internalities: preparedness, and planning; strategy. Any project should understand what the rules are, if only to work out how to distinguish from them without ending up in (serious) trouble.

Second, externalities: competitors, and regulators; market. Any project should understand the impact of the wider marketplace, competitor activities and responses, and what regulator activity may mean.

Ultimately, you know what you know, and you (probably) know what you do not know, the real obstacles will come from what you do not know that you do not know: the unknown unknowns.[9] Of course, the fourth permutation is what you do not know that you know; time to think.

[8]A detailed explanation is beyond the scope of this chapter. However, for those with a deeper understanding, the above analogy is not strictly correct. I understand that a wallet only contains private keys, which is a form of identification of the wallet itself. The wallet does not contain records of transactions or tokens. A wallet's function is to facilitate the sending of messages to the blockchain's network to update the ledger to record transaction events that, essentially, involve the wallet as sender or receiver of tokens. This association, albeit indirectly, gives rise to an entity's, or wallet controller's, right to claim ownership of crypto-assets.

[9]The original 'Johari Window' concept was developed by Luft J and Ingham H, two American psychologists, around 1955, but was infamously quoted by Donald Rumsfeld, former US Secretary of Defense, in a press briefing, 12 February 2002.

12.2.1 Overview of legal risks

In law, whilst the general principle always applies, there are always exceptions, exceptions to the exceptions and so on. Legal risks for blockchain applications fall into five broad categories: (i) contract, (ii) intellectual property rights (IPR), (iii) data protection and privacy, including cybersecurity, (iv) financial regulation [15],[10] and (v) commercial [16],[11] including consumer protection and competition.

This section is a perspective on the legal jurisdiction of England and Wales. Technology platforms operate globally. Any platform would inevitably seek to operate in markets worldwide. For completeness, but only at a high level and in specific instances, I shall compare the approaches taken by key 'blockchain jurisdictions', the US,[12] Switzerland[13] and Singapore.[14] The comparative analysis with the key blockchain jurisdictions is for interest only and appropriate advice should be sought from qualified lawyers in those jurisdictions before a blockchain application begins operations there.

12.2.2 Contracts, smart legal contracts and smart contracts

Contracts, smart legal contracts and smart contracts are all considered through the prism of traditional contract law principles. The enforceability and interpretation of smart 'legal' contracts are, and smart contracts may be, governed by contract law, including the formative elements of contracts such as (i) offer (one party must have made an unambiguous offer of or to do something for the other), (ii) acceptance (the other party must have accepted what was offered and the bargain struck), (iii) consideration (there must be an exchange of promises for value) and (iv) intention to create legal relations (the parties must have meant for a legally binding agreement to form between them).

As discussed above,[15] blockchain 'contracts' come in different, confusing and potentially ambiguous forms. There are (i) contracts, in the true legal sense of parties making binding promises to the benefit of the other, (ii) smart contracts, which may be described as executable provisions of an underlying contract and (iii) smart legal contracts, which operate in a hybrid permutation of encoded natural language and pure code, and any incremental form in between.

12.2.2.1 Contract formation

For a contract to be formed, the parties must meet the tests described above. A NLC *should* contain all the contractual terms in the relevant document. Similarly, a smart

[10]The CBC extensively considered financial regulations in the referred whitepaper, which represents an excellent primer for the subject.

[11]An interesting commercialisation example is Elon Musk talking about your vehicle earning money for you as a self-driving taxi, but the concept may be considered further to that of full vehicular autonomy. [16]

[12]Regulated by the Securities and Exchange Commission (SEC) and US Federal laws.

[13]Regulated by Financial Market Supervisory Authority (FINMA) and Swiss Federal laws.

[14]Regulated by the Monetary Authority of Singapore (MAS) and Singaporean law.

[15]See Section 12.1.1.3.

contract *should* execute as expected, assuming a standard of implementation. However, in the case of smart legal contracts, this may be complicated if the code and natural language elements are not set out in the same place,[16] non-standard code implementations have been used[17] and/or non-standard interpretations have been deployed.[18] As such, this form of contract is the most challenging from a legal perspective, both as to formation and interpretation.

Although expected to be rare in practice due to the complexity and nuance of language in contracts, and the difficulty of their transposition in to code, one can see the challenges this presents.[19]

For example, natural language terms may be set out in a separate natural language document or as comments in the code. In either case, there are legal risks associated with the status of such for the purposes of contract formation and the interpretation, particularly for dispute resolution. To avoid uncertainty, the status of any natural language comments in code should be made clear, including whether they form part of the contract, by inclusion of status obligations in the definitions and interpretation section. Comments in code are explored further below.[20]

Beyond the definition and description of what a smart (legal or not) contract is, the technicalities of the code and its execution are outside the scope of this chapter and fellow contributors are better placed to explain such. That said, even an executable operation by a smart contract must comply with the governing laws of the country within which it operates (or carries out an execution).

12.2.2.2 Asset entanglement

In a digital scenario, an NFT persists and confirms ownership regardless of any duplication of the underlying asset.[21] The holder of the NFT is the only one that may lawfully deal with the related asset.

In a physical scenario, the relationship between NFT and asset is less certain. How does a token in a wallet mean anything to a physical asset? Is the physical asset in question the *actual* asset purportedly represented by the token? How are they connected? How is that relationship authenticated? These are critical questions and present significant risks, particularly as we look to tokens to represent important and valuable (financial and emotional) assets such as our homes.

There is a distinction between digital assets, those native to a virtual existence, and real assets, those native to a physical existence. However, a true digital twin, once sufficiently well developed as a digital model, may carry enough data to be indisputably the single source of (data) truth for the specific real asset. With an appropriate legal framework, and regulatory approval, it may then be *de facto* the tokenised representation of the real asset.

[16]To mean, in a single document.

[17]To mean, generating 'known outcomes' from 'known inputs' commonly accepted by the relevant market.

[18]To mean, common, ordinary and/or expected meanings of legal terms accepted by the relevant market.

[19]See the 'I want to eat a burger' footnote above for an extreme example of language and interpretation.

[20]See Section 12.2.2.3.

[21]The 'Copy-Paste' and 'Right-click, Save As. . . ' hackers amongst us take note.

In essence, blockchain technology provides the medium for the digital-physical link, facilitates the collection and transfer of data, immutably recorded and persists for the life of the asset, becoming indistinguishable from the asset. The result is not only a life cycle management tool for real assets, but the one source of truth for the asset. The model 'knows' everything, from concept, to design, to construction (and all the materials, equipment, plant, goods and services employed), to completion and throughout its operative life. Conceptually this is 'whole (of) life governance', but it is also, potentially, capable of becoming the record of ownership for the asset.

12.2.2.3 Comments

Software programmers, as good practice, place comments (non-executing text) in their code. This helps them recall and others understand what the code is doing, or is supposed to do, in various parts. The difficulty in law arises from the rules of contract interpretation. In general terms everything within the contract document must be considered when interpreting the meaning and purpose of that contract, both in terms of the effects of specific clauses and the overall effect of the contract.

The question is: what is the status of 'coder' comments to the contract? Even if intended to be reminders or simply explanatory, the unintended consequence may be a conflict between the interpretation of the comment and the effect of execution. Even more concerning: what is the position if the comments indicate a level of incompleteness or appear to be posing a question? Legal principles of contract formation, such as 'consensus ad idem' (a meeting of minds), require that the contract be clear, certain, contain all the terms, or at least all the terms may be interpreted, to ensure that the contract is not void for ambiguity. If either case arises, what then? At best, a dispute as to the nature and basis of the contract. At worst, no contract. How this is then unwound by the courts depends on an analysis of the position between the parties.

12.2.2.4 Privity

Privity concerns the right to enforce the contract. Save for specifically conferred rights, wherein a third party may benefit under a contract and so may claim a right to enforce at least the specific parts to which it is a beneficiary, a contract may not be enforced by a party that is not a party to that contract.[22]

The question is, if a contract is deployed on a public blockchain, is privity retained if the contract were exercisable or inspectable by any number of parties, or required intermediary executors to complete transfers, such as consensus mechanism resource providers.[23] To facilitate the execution a term of a contract between two parties, do you need to know what that term is and the reason for execution?

[22]There are such things as 'unilateral' contracts. Lawyers will recall Carlill v Carbolic Smoke Ball [1892] EWCA Civ 1 as one of the first contract law principles they studied. However, this is outside the scope of this chapter.

[23]An analysis of consensus mechanisms and their resource providers is a complex technical issue beyond the scope of this analysis, but it is an issue to consider.

Perhaps not, but the parties to that transaction are relying on an automated intermediating system to complete the performance of their agreement. This is potentially a complex analysis beyond the scope of this chapter, but as more contractual elements are translated onto digital platforms, and reliance on the performance of that system grows, there appears to be a critical issue between deployment and consensus.

12.2.2.5 Confidentiality

Most, if not all, contracts contain confidentiality provisions. It is usually a breach of contract, enforceable by judicial injunction and a concomitant indemnity for damages,[24] if a party fails to observe confidentiality, subject to specific exceptions around the nature of the breach, information released and status of the information.

If terms of a contract are discoverable and these can be traced back to the relevant parties, is confidentiality lost? Frustratingly, it depends. This is the lawyer's stock answer (albeit an honest one) when asked for advice on the interpretation and application of the law in a general case, when that interpretation and application requires specific details. A paper-based contract would be filed away, inaccessible to anybody without access to the place of storage. Depending on the type of blockchain adopted, a digital-based contract may not enjoy the same level of protection.

It may be the case that, by necessity, a form of smart or smart legal contract is placed on a blockchain application to manage and administer the contract between parties. Others may deliberately attempt to discover that contract. Neither party to the contract has intended that contract to become public, so are unlikely to be in breach. However, if they used a public non-permissioned blockchain then, however inadvertent, there was a foreseeable risk of discovery. It may be that any contract, smart or otherwise, on a blockchain is discoverable by any participant to that ecosystem. There are tools that allow transactions to be seen, and tracked, across a blockchain.[25]

If transactions can be traced, what does this mean for individual privacy, for a business' research and development, and competition (competitors would be interested in understanding what their peers are doing if they could gain a competitive edge). Taken to the extreme, what might it mean for democracy? If blockchain transactions could be discovered and, as some suggest, blockchain were used to 'secure' voting rights, might your right to anonymously cast your ballot be lost? Is this an erosion of democracy?

Some may argue it is important that some transactions could be traced, such as in situations of fraud. The difficulty is implementation. The blockchain application would need to be designed, fundamentally, to permit tracking. Who decides the context within which that tracking applies? How are transactions identified as, in fact, fraudulent? An accusation is not enough, the legal system runs on the principle

[24] In law 'damages' broadly means recovery of financial losses, that is, money.

[25] See, for example, Etherscan, which lists block transactions allowing tracking of wallet-to-wallet transactions. Etherscan. https://etherscan.io [Accessed 11 October 2024].

of 'innocent until proven guilty' and the law generally does not permit 'fishing expeditions'[26] (a demand for disclosure of information in the hope that something useful is found). It is for the claimant to make out their claim, not for a defendant to make the claimants claim for them. There is no requirement to deliberately shoot oneself in the foot, so-to-speak, although doing so, inadvertently or not, may be a different matter.

12.2.2.6 Summary of contracts, smart legal contracts and smart contracts

The complexity of the law when applied to inadequately defined contracts is not trivial to navigate. Even with the best of intentions, without careful foresight, planning and clear legal advice, contracts of the complexity as those operating in the built environment will be challenging to robustly implement as HCs. Not everything can, or perhaps should, be reduced to code. Human intervention may be needed, particularly in cases of unintended or unwanted outcomes. A curious statement considering the need to avoid ambiguity, but the law recognises equitable principles. These are principles of 'fair treatment' before the courts that temper the rigidity and, in certain circumstances, harshness of the common law. These principles rely more on conscience, or a sense of what is right, and are difficult if not impossible to encapsulate definitively.[27] The law will not 'save you from a bad bargain', but it may temper an unfortunate one.

12.2.3 Intellectual property rights (IPR)

Blockchain applications may involve the creation, use or transfer of IPR such as patents, copyrights or trademarks. IPR laws govern the protection and enforcement of these rights.

IPR are only transferrable by deed. Without this there may only be some form of licence, limited in rights and, therefore, how the licence holder may deal with the underlying asset. Ownership is retained, by default, by the creator or current rights owner. This is an important consideration as, for example, with the interest in NFTs, it is not always clear what rights are intended to be transferred with, or 'attached' to, the NFT when purchased.

12.2.3.1 Secrecy

IPR, like confidentiality, in certain instances only 'works' if nobody else knows about it.[28] In the UK, trade secrets are protected in law [17]. Again, as with confidentiality, but potentially ruinous for a business, what happens if such secret materials are stored on a blockchain? Due to the nature of tokens and cryptographic

[26]Cf. Norwich Pharmacal Co v Commissioners of Customers and Excise [1974] AC 133 and the stringent tests required to be met for an innocent third party with knowledge of unlawful conduct.

[27]Contrast this with the absolutes of '0' and '1'. How does a binary system deal with non-0 and non-1 values?

[28]Think of the recipe for Coca-Cola. Broadly, the IPR in Coca-Cola is a 'trade secret', that is, nobody can replicate Coca-Cola because the 'recipe' is simply not known other than to a select few Coca-Cola employees.

encryption, the data may not be stored directly on the blockchain, nor may it be discoverable. However, all projects should be mindful of what information is to be stored and the accessibility of that information.

12.2.3.2 Data hosting and storage

By way of example, imagine you are an NFT owner. As we discovered above, your wallet does not hold the NFT itself, it merely points to where it may be, which in turn points to where the asset may be. Overall, a lot of pointing. What happens if the hosting platform or storage server is disconnected, the web service is terminated or perhaps the creator 'forgets' to pay their hosting fees?

If the asset is unavailable, is everything lost? If the image file and/or NFT disappear, what are you left? A transaction record of something you own, but cannot access, nor recover.[29] If the digital asset becomes inaccessible, the link will break and the NFT that remains may be worthless as it would no longer be associated with the digital asset. Since the NFT is unique and may not be replaced, the NFT purchaser might be left without recourse. Based on the use case of the NFT, this could result in business interruption, breaches of regulation and loss of data.

Ordinarily, we would turn to the governing contract to see what course of action should be taken. Without a governing contract, or terms that may be implied to form a contract, the recourse or route for recovery, or damages, is difficult if not impossible to implement.

12.2.3.3 Royalties

Smart contracts governing NFTs could allow for the distribution of funds for the payment of royalties to the creator each time the work is resold. However, these automated resale royalty payments might not occur unless the NFT is resold through the same platform, as how else could resale be 'known' by the underlying network that is managing the assets.

12.2.3.4 Underlying IPR

People all over the world are participating in the NFT market. However, many are not familiar with the legal restrictions relating to the underlying asset, which leads to potential IPR infringements. For example, an NFT purchaser might assume that they purchased exclusivity to the associated underlying artwork. In reality, the original creator is the copyright owner and retains the exclusive right to copy, distribute, modify and publicly perform or display the art (unless specifically and expressly granted to someone else). The purchaser of the NFT typically only receives the token and the right to the underlying asset for personal use. A

[29]An NFT and its digital asset should either be maintained separately or always move together. If separate, the NFT should be 'stored' (appreciating the technical inaccuracy of that statement) on the blockchain and, amongst other legal features, contain metadata information on where the associated digital asset is located, and be backed up with authorised and authenticated 'copies'. If together, the NFT and asset should relocate as one, wallet to wallet. How this may apply to a physical asset, is simpler and more complex. Simpler because an asset, such as a building, is difficult to 'lose'. More complex as the physical-digital association is not trivially established.

purchaser may have limited avenues for recourse without a clear and certain contract in place.

12.2.3.5 Summary

Specifically, NFTs have the potential to influence a transition to a digital world. NFTs are making it possible for creators to imbue properties like scarcity, uniqueness and proof of ownership to digital assets. They have inspired new methods and means of monetisation of items that were previously lost to the depths of the internet. The NFT market is still nascent, however, and significant infrastructure will likely continue to be built in the form of tokenisation platforms, distribution channels, custodial solutions, e-commerce integrations and more user-friendly interfaces. As the market continues to evolve, so do the legal and regulatory issues.

12.2.4 Data protection, privacy and cybersecurity

Software platforms often involve the processing and storage of personal data. Depending on the jurisdiction, data protection and privacy laws, such as the General Data Protection Regulation [18] (GDPR) in the European Union (EU), impose obligations on entities handling personal data, including consent requirements, data security measures and individuals' rights. An extraordinary defence of the rights of the individual? An extraordinary barrier to growth and an administrative nightmare for commerce? In part both and neither. Data protection and privacy may be simplified to 'data responsibility'. A failure to act responsibly with data is an accusation levelled at individuals, businesses and governments alike.

Individually, why save years of old messages, emails and images that are never looked at? Why do we need to share so much online? But then who wants to do a 'data cleanse' and make conscious decisions about years' worth of data? We all *should* take responsibility and go through this process, as difficult as it may be.

Businesses, why collect so much user personal data, with which you will do nothing, and to whom you will never sell anything? It is a liability. Identify, expunge.

Governments, forever the nosey-parkers masquerading as 'for-your-own-good'-ers, consider the liability. Are you overreaching on unsubstantiated grounds? Identify the person, natural and legal, and regulate the system. Hold back from trying to administer or manage the system. Let free enterprise move in to fulfil demand and comply with your regulations.

All this highlights a more fundamental problem with law and the 'push' economics[30] of modern commerce. Whilst I risk digressing into a treatise on the relative merits of a 'push' vs. 'pull' economics, the need for data protection is

[30]A 'push' market is characterised as a system that initiates production in anticipation of future demand, but also may create an apparent and false demand to create the market it needs to sell its foresighted products into. Nobody knew they wanted it, but people now feel that everybody else has it and they should get it too.

significantly reduced if 'pull' economics[31,32] were adopted,[33] which would solve most, if not all, of the discussion that follows.

12.2.4.1 Data protection

By design blockchain is data secure. All transaction information is hashed, and so unintelligible to a human reader, but also 'un-de-hashable'.[34] At least until quantum computing arrives, even then it may not be trivial to extract meaningful data.

12.2.4.2 GDPR

The EU GDPR, although modified in its adoption into UK law, created several data identity rights for individuals. Whilst a more complete explanation and analysis is readily available from online sources, the key rights for the purposes of this chapter are (i) the right to access and (ii) the right to erasure. The other rights essentially address dealing with personal data and, whilst important, are less relevant to our considerations.

Access concerns the rights of an individual to receive a copy of their personal data. For blockchain, assuming it is not truly anonymous, this should be relatively straightforward. Identify each instance of involvement in a transaction across the ledger and report. Why is this important? You need to know what you know, and you need to be compliant with granting access rights to data subjects.

The difficult part is erasure. Blockchain is immutable, it does not forget. If it forgot it would not be the ledger of record. And herein lies the contradiction, and conflict, between law and technology in their respective current forms.

Data protection laws give individuals the right to erasure of their personal data, but the immutable nature of the blockchain poses an obstacle to enforcement of this right. Data protection laws also provide individuals with the right to rectify inaccuracies in their personal data, but blockchain technology makes this retrospectively impractical. As such, tokens that contain personal information may breach data protection laws. Would this mean blockchain systems are unlawful? Possibly.

At first glance it appears paradoxical, how is the contradiction of a right to be forgotten enforced if you are permanently recorded, or at least discoverable, on a

[31]A 'pull' market is characterised as a system that initiates production as a reaction to present demand, thereby responding to an actual need in a market for a certain product. Everybody knows they need it and collectively create a demand market for a supply to feed. A pull system is characterised by self-sovereign identity with the data subject at the heart of the system.

[32]For insight into pull economics see Siegel D. *Pull: the power of the semantic web to transform your business.* New York: Portfolio; 2010.

[33]Although people don't always know what it is they want and need someone to show them that they need it. Steve Jobs, paraphrasing Henry Ford, famously said, 'a lot of times, people don't know what they want until you show it to them'.

[34]A hash cannot be reversed due to the nature of the one-way mathematical operation that created it. To 'break' or 'hack' a hash you would need to perfectly guess the data that went into it and reperform the hash process to confirm that that was the input data. A computationally impractical task to be achieved in a reasonable time frame based on current computing power.

blockchain? A possible solution may be in 'segregated identity'. In much the same way as multi-factor authentication (MFA) works today, so could a virtual persona, for the purposes of blockchain transaction records, be cryptographically dissociated for the purposes of the record, yet retained for user authentication at the point of access.

12.2.4.3 Identity

How is identity confirmed, if needed, on an encrypted ledger? It may be that the only solution to the acquisition of data is to reverse how, when and why data is not acquired, but released.

In a push economy, a website for the goods or services you wish to purchase requires you to sign up, create an account, fill in a form, create a password and so on. It then monitors your purchases and habits to suggest additional goods and services it thinks it can push to you. Aside from the security nightmare of endless accounts for things you neither need nor want beyond that single purchase, there are endless emails resetting forgotten passwords.

In a pull economy, the position is reversed. There is no account to be set up. You grant the website restricted (and time limited) access to your own self-sovereign identity wallet, confirm you are you and the address for delivery. Once confirmed access is terminated. No copies are taken; all the website needs to know is that whatever needed to be confirmed has been confirmed.

Identity data is not 'leaked' through websites or resold to third parties, GDPR no longer applies as you are your own personal data manager. Data is secure.

12.2.4.4 Confidentiality (in a data context)

Confidentiality is a critical issue. Taken to a 1984 dystopian extreme, if all your purchases and online habits are traceable, then would governments or corporations not be tempted to track your purchases, habits and locations under the guise of security, taxation or improving user experience?

12.2.5 Financial regulation

Blockchain applications involving cryptocurrencies, initial coin offerings (ICOs), security tokens or other financial solutions may be subject to financial regulations.

This is potentially the most significant part of any legal analysis performed on a blockchain applications. Why? Because this is easy to get wrong and if you do get this wrong you could find yourself in considerable difficultly with a regulator for providing services you had not realised you were providing, are not regulated or authorised to provide, and criminal prosecution may follow.

12.2.5.1 Payments

The core nature of money to our everyday lives, and the abuse of our dependence on it, has resulted in ever increasing attempts to regulate the ever-increasing complexity of the financial markets. This section could be expanded to fill many volumes of detailed requirements, but the basic message is that professional advice should always be sought. In the UK, modern financial regulation stems from the Financial Services and Markets Act 2000 (FSMA).[35] Thereafter a great deal of harmonisation at European level occurred.

12.2.5.2 Token analysis

Before we contemplate what may apply, we must first determine if they could apply. Tokens define the purpose of a blockchain ecosystem, and it is there we begin.

As defined above[36] there are three general species of token. In summary, exchange tokens are largely unregulated, utility tokens are potentially regulated depending on their characteristics and security tokens are regulated as their underlying assets are financial instruments that fall within the regulatory perimeter.[37] The specific details of this are not addressed here, but keep in mind that if you have a token that is related to an asset or financial instrument, it is likely a security. Ultimately it depends on what the token represents.

By way of example of the inadvertent creation of a security, most NFTs today are likely not securities for the purposes of financial regulation, but if it is a security then you need to think about the regulatory landscape and whether authorisation is required. Non-fractionalised art is generally excepted as a security and IPR is the key legal risk. But what if the ownership of an artwork is fractionalised between several people? They all cannot keep the asset at home and so must ‘share’ it. What happens if the artwork is lent out to a gallery or exhibition? How is any profit shared between the owners? The act of ‘fractionalising’[38] creates additional legal considerations. How are rights enforced for dealing in the asset? How are individual owner’s rights dealing with their ‘share’ managed? What rules apply on the division of income generated by the asset? What was an ownership NFT has transformed into a security. Having done so, you may have needed to be regulated to carry out the fractionalisation in the first place as this almost certainly falls within the regulatory perimeter.

Fractionalisation of an ordinarily unregulated token or tokenised asset may create a security. Beware the law of unintended consequences (as much as the actual law).

12.2.5.3 Know your client (KYC) and anti-money laundering (AML)

Ordinary payments for everyday goods and services vastly outnumber those used for illicit purposes. Such payments are easily checked, in fact your bank will check

[35] Pronounced ‘fizz-ma’, the Act created the Financial Services Authority (now Financial Conduct Authority) as a regulator for insurance, investment and banking, and the Financial Ombudsman Service to resolve disputes.

[36] See Section 12.1.1.4.

[37] The ‘regulatory perimeter’ defines the limits of what the regulator does and does not regulate. In the UK, the FCA’s perimeter report is regularly updated to define what is in and out of their scope https://www.fca.org.uk/publications/annual-reports/perimeter-report [Accessed 11 October 2024].

[38] See Section 12.1.1.4.6.

to whom you are making and from whom you are receiving payments as a matter of course. Suspicious activity may be reported.

KYC is the process by which you prove that you are you. For example, when you want to open a bank account you provide photographic identification and proofs of address to enable the bank to check who you are, that you are a real person (according to a government), that you live somewhere (so they can find you) and perform credit checks (to make sure you are apparently trustworthy).

AML is the process by which an intermediary must be certain that they are not handling funds obtained from or to be used for illicit purposes. For the nefarious actor to succeed, they and the movement of their funds must appear inconsequential, or they must find a method of payment that does not have much oversight. Money laundering employs methods of obfuscation to disguise the entry and exit points from a system of funds obtained from or used for illicit purposes. Operating a money laundering scheme through the banking system can be difficult and fraught with risk of discovery. Would it not be better to use an anonymous system of cryptographically secured communication? This is why cryptocurrencies have attracted considerable negative press over the years.

The Anti-Money Laundering Directive 5 (AML5)[39] specifically brings in scope crypto-assets. In summary, a blockchain platform must have oversight of the inbound funds from users, known as a source of wealth declaration, and monitor transaction volumes and values.

12.2.5.4 Markets in Financial Instruments Directive II (MiFID II)

MiFID II[40] is an EU regulatory framework implemented to govern financial markets and protect investors. It came into force whilst the UK was part of the EU and has been broadly transposed into UK law. It is a comprehensive regulatory framework that seeks to improve investor protection, enhance market transparency and promote the integrity and efficiency of financial markets in the EU. It sets out various obligations for investment firms, trading venues and other market participants.

MiFID II does not specifically mention or address crypto-assets as its primary focus is traditional financial instruments and activities. However, it is important to note that the regulatory landscape for crypto-assets and related activities is constantly evolving. As such, certain aspects of MiFID II may indirectly apply to certain crypto-asset-related activities and services, depending on their specific characteristics. For example, if a crypto-asset is classified as a financial instrument

[39]In the UK, AML5 was implemented as part of the Money Laundering and Terrorist Financing (Amendment) Regulations 2019, amending the Money Laundering, Terrorist Financing and Transfer of Funds (Information on the Payer) Regulations 2017.

[40]In the UK, MiFID II was implemented as part of the Financial Services and Markets Act 2000 (Regulated Activities) (Amendment) Orders (Nos. 1 and 2) 2017, Financial Services and Markets Act 2000 (Markets in Financial Instruments) (Nos. 1 and 2) Regulations 2017, the Data Reporting Services Regulations 2017 and Financial Services and Markets Act 2000 (Prospectus and Markets in Financial Instruments) Regulations 2018.

under MiFID II, the relevant provisions of the directive would be applicable. This determination depends on the specific features and legal classification of the crypto-asset in question.

12.2.5.5 Payment Services Directive 2 (PSD2)

PSD2[41] is an EU directive to regulate payment services, competition, innovation and security in the payment industry. It promotes open banking, strengthens security through strong customer authentication, and allows licensed third-party providers (with consent) to access customer account information and initiate payments. Furthermore, it enhances consumer protection and data privacy, prohibits surcharge fees and fosters innovation and competition in the payment industry.

12.2.5.6 e-Money

E-money,[42] or electronic money, refers to digital or electronic forms of currency that are stored and transferred electronically. It is a digital representation of value issued by an entity, typically a financial institution, which can be used for various electronic transactions. E-money provides a convenient and secure alternative to traditional physical cash and enables digital payments, online purchases and money transfers. It is governed by specific laws and regulations to ensure consumer protection and financial stability.

Together with PSD2, e-money has opened the potential for monetary tokens and, specifically, the potential creation of Central Bank Digital Currencies (CBDCs).

12.2.5.7 Statutory mechanisms

Whilst not strictly financial regulations, certain payments are required to follow statutory prescribed mechanisms. The classic example in the UK is the changes introduced by the Housing Grants, Construction and Regeneration Act 1996 [19]. This law radically changed how parties to construction contracts were required to pay each other. Outlawed were 'pay-when-paid' arrangements, and 'pay-on-invoice' arrangements become mandatory. Under the mechanism, on submission of a valid and payable invoice by the sub-contracted party, the contracted party could not withhold payment pending payment of their invoice by the upstream supply chain. That is not to say payment could never be withheld as a process was also formalised by which payment could be withheld provided that appropriate notice and reasoning was given. The example illustrates that in certain circumstances the law mandates specific processes.

12.2.5.8 Markets in crypto-assets (MiCA)

MiCA [20] is an EU Regulation to regulate crypto-assets within the EU. It introduces a comprehensive framework for issuers, service providers and digital asset marketplaces. MiCA intends to create legal clarity, enhance investor protection and

[41]In the UK, PSD2 was implemented as part of the Payment Services Regulations 2017.

[42]In the UK, E-money was implemented as part of the Electronic Money Regulations 2011.

promote market integrity in the rapidly evolving crypto-asset market. It proposes requirements for authorisation, capital, investor protection measures, governance and supervision. Post the UK's departure from the EU, MiCA is not directly applicable. However, the FSMB is presently going through review in the House of Lords and, if adopted, will broadly align with MiCA to ensure the UK may still sell financial services into the bloc.

12.2.5.9 Summary of legal risks associated with blockchain applications

For blockchain applications with financial elements there are a wealth of regulations that govern their activities. Navigating them can be challenging, but failure to do so is a greater risk with significant fines for non-compliance and, potentially, termination of the project as unlawful.

12.2.6 Commercial

Blockchain applications targeting consumers may be subject to consumer protection laws. These laws safeguard consumers from unfair practices, deceptive advertising, fraud and other similar adverse behaviours, to ensure that consumers are treated fairly and have access to accurate information.

Commercial law considerations will depend on the specific market within which the blockchain application operates. However, there are several generally applicable areas to consider.

12.2.6.1 Consumer protection

Consumers are afforded enhanced rights by virtue of their relative vulnerability when engaging with significantly larger, more powerful and better resourced corporate entities. Consider, when you sign up online to a service, do you read the terms and conditions? Yes? Good for you, they are a riveting read and you should know what you are getting into. In the more likely case that the answer is no, I do not blame you. They are catastrophically long, borderline impenetrable with legal terminology, and, realistically, are you going to try and negotiate those terms for something more favourable to you? Yes? Again, good for you, defend your rights. In the more likely case that the answer is again no, after all who has the time, requisite legal knowledge and sufficient personal economic power, then the law is there to save you from such a horribly one-sided deal.

If the terms are fixed and non-negotiable, then the law on unfair contract terms [21], consumer rights [22] and consumer credit [23] may apply. These are designed to ensure contract terms that are unfair cannot be enforced against a consumer where they result in an imbalance of rights and obligations. In reality, if you have a problem with a large corporate entity, it is in their interests to resolve it as quickly and fairly as possible, with as little fuss as possible, lest you take to social media and spread the bad news as far and wide as possible; simple reputation management.

The point is balance and fairness. Tie users up in legal knots, terrify them into submission, if necessary. However, it may be better to take a more balanced view and engage meaningfully.

12.2.6.2 Commercial property

As a matter of practicality, what happens if the private keys to a wallet are lost, stolen or a confidence trickster swindles them? How can ownership be determined and enforced?

Ownership in the 'it exists in a blockchain connected wallet' sense is a contractual agreement. Goods or services (digital or physical, assume parity) have been purchased. Value has been exchanged, such as payment made, and representative tokens are recorded as associated with a specific wallet. If you no longer hold the private keys to access the wallet in which the pointers to your assets reside, and you can no longer access those pointers, then can you be said to own the contents of that wallet? Ownership denotes an ability to deal with the asset in question, such as sell, mortgage, loan or rent out, leverage as collateral and so on. If you cannot deal with the asset, do you still own it?

If ownership is defined by access to the wallet, and tokens do not actually reside in the wallet but are pointed to by the wallet, what happens if those pointers change where they point to? If a blockchain is attacked and the information manipulated or rewritten the 'single point of truth' is no longer reliable as to what is that truth. Imagine HMLR was placed on a blockchain, was attacked and you discover you no longer own your home. The effects can be far reaching and catastrophic.

Playing devil's advocate for a moment, if tokenising is such a great a risk should we tokenise anything? Tokens are likely no riskier than other forms of asset holding mediums and blockchains are harder to attack and change. Personal responsibility plays a part (you lock your front door when you leave home, so why should your computer security be any less?), but mechanisms of governance, recovery and complaints and dispute resolution should be developed and implemented to, so far as possible, protect users and protect them, if mostly from themselves.

12.2.6.3 Competition

Over time governments have moved to legislate against anti-competitive or abusive behaviours by dominant or monopolistic market entities [24]. The premise being that monopolies or dominant players will adversely affect market dynamics, offer lower quality goods or poorer services, with higher prices. The overall result being detrimental outcomes for consumers.

Such anti-competitive analyses are complex to make out and most large technology companies have faced them at least once in their history [25]. However, this is not to say a new blockchain application should not aim to become the monopoly in its sector. Is it not easier to grow faster if there is nobody else to lure away your users with competing products?

Perhaps an application is not unique and there are close competitors, or an application involves collaboration among a consortium of competitors. Either case may result in competition law risks, known as 'antitrust' in other jurisdictions. These laws prohibit anti-competitive practices such as price-fixing, market allocation or abuse of dominant positions. Assuming the intent is not to collude or form a cartel to control the market, then why not 'team up' to tackle the problem?

12.2.6.4 Summary of commercial issues

For all blockchain applications, there are a myriad of legal requirements that govern their basic interactions with users. Whilst competition considerations may be more remote, at least initially, consumer protection should be at the forefront of every founder's mind. After all, what is a project without any users or user trust?

12.2.7 *Jurisdictional and cross-border legal issues*

As mentioned in the introduction to this section, but reiterated for completeness, it is important to note that the applicability of laws may vary depending on the jurisdiction and specific circumstances of the blockchain application. This section provides a summary insight into the legal areas considered above. The applicable laws in other jurisdictions must be approached as for the legal frameworks of the UK, with preparedness and focus on what needs to be achieved. Always seek advice from legal professionals familiar with both blockchain technology and relevant laws to ensure compliance and mitigate legal risks. Be mindful that pivoting to the art of the possible may be required.

12.2.7.1 Contract

Whilst the UK does not have specific legislation for smart contracts, existing frameworks recognise their potential enforceability. The principles of formation, interpretation and enforcement apply 'mutatis mutandis'.[43] Electronic and digital signatures are recognised for execution.

In the US, whilst specific requirements may vary from state to state, smart contracts are generally recognised as enforceable contracts, provided they meet the requirements of contract law.

In Singapore, smart contracts are recognised as enforceable within the existing legal framework. Specifically, the common law principles of contract law combined with the Electronic Transactions Act regulate their enforceability.

In Switzerland, a civil code-based legal system in contrast to the other jurisdictions above, a favourable regulatory environment for blockchain technology has been established. Smart contracts are recognised as valid, enforceable and are governed by contract law principles and the Code of Obligations as to their formation, interpretation and performance.

12.2.7.2 Multi-jurisdictional IPR

In the UK, well-established IPR laws for copyright, trademarks and patents apply as derived from both legislation and international treaties. Patents and trademarks require registration to ensure protection, but copyright arises automatically upon creation. Extra-jurisdictional protections are enabled through the UK's membership of the European Patent Convention (EPC) and EU Intellectual Property Office (IPO).

[43]Mutatis mutandis meaning 'all necessary changes having been made'. In essence, the main point remains the same although some changes may be necessary to take account of different situations.

The US has a comprehensive IPR system that includes patents, trademarks and copyright. Similarities to UK law include the formal recognition of trade secrets. Differences to UK law include the patentability of certain software features and follow-on royalty rights[44] are not recognised. As in the UK, patents and trademarks require registration to ensure protection, but copyright is automatic upon creation. Rights are registered with the US Patent and Trademark Office to secure legal rights.

Singapore has robust legal frameworks for patents, trademarks and copyright, which broadly follow principles similar to the UK. Patents and trademarks require registration, whilst copyright arises automatically on creation. Patents and trademarks are registered with the IPO of Singapore.

Switzerland has well-established IPR frameworks for patents, trademarks and copyright that require registration or arises on creation as with the other jurisdictions. Patents and trademarks are registered with the Swiss Federal Institute of Intellectual Property. Switzerland is also a member of the EPC.

12.2.7.3 GDPR

The UK adopted GDPR into law through the Data Protection Act 2018. After leaving the EU, the UK maintains its own data protection regime called the UK GDPR, which is largely aligns with the EU GDPR.

The US does not have a federal data protection law like GDPR and relies on sector-specific regulations to address data protection and privacy, such as the Health Insurance Portability and Accountability Act for healthcare data and the Gramm-Leach-Bliley Act for financial information. These are, perhaps, not as well coordinated in general principles as GDPR. Some states, such as California with the California Consumer Privacy Act and Virginia with the Virginia Consumer Data Protection Act, have enacted their own comprehensive data protection laws at the state level.

Singapore has the Personal Data Protection Act (PDPA), which governs the collection, use and disclosure of personal data. It many ways, PDPA closely parallels the principles of GDPR, in that the PDPA focuses on consent, purpose limitation and reasonable security practices for personal data protection. The PDPA applies to organisations operating in Singapore and the collection of personal data in Singapore, regardless of whether the data processing occurs locally or abroad.

Switzerland has the Federal Act on Data Protection (FADP), which regulates the processing of personal data by both private and public entities. The FADP emphasises the protection of individual privacy and provides provisions on data processing, data transfers and data security. The FADP does not directly align with GDPR but provides sufficient protection to enable data transfers with the EU.

12.2.7.4 Securities

As discussed above, the UK's securities laws are primarily governed by FSMA and the FCA is the regulatory body responsible for overseeing financial markets.

[44]Follow-on royalties rights concern royalties due on resale of the underlying creative works. There is no recourse for unpaid resale royalties.

Additional regulations have been implemented, such as MiFID II, to harmonise across the EU.

US securities laws are extensive and primarily derive from the Securities Act 1933, the Securities Exchange Act 1934 and the Investment Company Act 1940. The SEC is the regulatory body responsible for enforcement. The US has a comprehensive disclosure-based regulatory regime that requires issuers to register securities offerings with the SEC or qualify for exemptions such as private placements, crowdfunding and offerings to accredited investors.

Lamentably, you will note that US securities laws stem from the Great Depression almost one hundred years ago. Little has been done to update the approach of the SEC in contrast to the approach taken in Europe. Take, for example, the legal test to determine whether or not a security is a security. In the US, the test was established in the case of the SEC v. W. J. Howey Co. in 1946.[45] Yes, you read that correctly. The test that could determine whether a token is or is not a security is almost 80 years old and predates the creation of the first decentralised blockchain (2008), the world wide web (1989), the internet TCP/IP (1983), the single-chip processor (1971), public-key cryptography (1970s), the personal computer (1957), the operating system (1956) and even the transistor (1947). Definitely relevant, although for the law a 1946 case is rather recent.

A more recent case is CFTC v Ooki DAO [26]. The CFTC won in default judgment (essentially meaning the defendant did not show up to defend the action) against the decentralised autonomous organisation (DAO). This has profound consequences for DAOs generally. Whilst the legal details are complex, in summary, Ooki DAO was determined to be a person in law and so could be held liable for unlawful acts. Ooki DAO was found to have been operating an illegal trading platform engaged in retail commodity transactions and unlawfully acting as a futures commission merchant. As an unincorporated association (a partnership, with legal personality), all members, governance token holders are now jointly and severally liable for $643,542. The court ordered permanent trading and registration bans, removal from webhosting or domain name registration services and total removal of content from the Internet. With few exceptions, for example, Wyoming, DAOs are not recognised legal entities and face the risk of being treated as de facto unincorporated partnership with unlimited liability for individual DAO members and token holders. DAOs must re-assess their platforms, processes and legal status, whether in the US or the EU, and particularly as MiCA[46] [20] will soon come into

[45]Although not directly relevant to our analysis, the case details are of interest. Howey sold tracts of citrus groves to buyers in Florida who then leased the land back to Howey. Company staff tended the groves and sold the fruit on behalf of the owners. The parties shared the revenue. The buyers did not tend the land themselves. The Howey Test determines what qualifies as an 'investment contract' and would therefore be subject to US securities laws. An investment contract exists if there is an 'investment of money in a common enterprise with a reasonable expectation of profits to be derived from the efforts of others'. The Howey Test is important for categorising blockchain and digital currency projects with investors, users, regulators and other project stakeholders as certain cryptocurrencies and ICOs may be found to meet the definition under the test.

[46]See Section 12.2.5.3.

force. Ultimately, DAOs will face serious legal and technical challenges to meet the 'fully decentralised' requirements and will need to rethink any centralised ownership or governance protocols that may drag them within the MiCA regulatory perimeter.

Securities laws in Singapore are primarily governed by the Securities and Futures Act and related regulations. MAS is the regulatory authority responsible for overseeing securities markets. Akin to the US, Singapore has a disclosure-based regulatory framework that requires the registration of securities offerings, unless an exemption applies. MAS has introduced initiatives like the Exempt Capital Markets Services regime and the Financial Advisers Act to provide flexibility for certain types of securities activities.

Switzerland has regulatory frameworks for securities, with different laws and regulations governing securities issuance and trading. The main legislation includes the Swiss Code of Obligations, the Swiss Financial Market Infrastructure Act and the Swiss Stock Exchange Act. FINMA is the regulatory authority responsible for overseeing securities markets in Switzerland.

12.2.7.5 Commercial

The UK has a well-established legal system for commercial transactions. The Sale of Goods Act 1979 and the Supply of Goods and Services Act 1982 provide legal frameworks for the sale of goods and provision of services, to protect consumers. The Competition Act 1998 prohibits anti-competitive agreements, abuse of dominant market positions, and regulates mergers and acquisitions.

The US has a diverse legal framework for commercial transactions, including the Uniform Commercial Code (UCC), which has been adopted (with variations) by all 50 states. The UCC governs various aspects of commercial transactions, including sales of goods, leases, negotiable instruments and secured transactions. Additionally, there are federal laws addressing specific areas of commercial law, such as the Uniform Electronic Transactions Act and the Electronic Signatures in Global and National Commerce Act. The US has comprehensive competition laws, including the Sherman Act, the Clayton Act and the Federal Trade Commission Act, which prohibit anti-competitive agreements, monopolistic practices and mergers that substantially lessen competition or create a monopoly.

Singapore has a modern and comprehensive legal framework for commercial transactions that broadly parallels the UK, with comparable legislation to regulate the sale of goods and services. Singapore has adopted the United Nations Convention on Contracts for the International Sale of Goods to govern international commercial contracts. Additionally, Singapore has competition laws regulated by a competition act, which prohibits anti-competitive agreements, abuse of dominant market positions, regulates mergers that may substantially lessen competition, promote competitive markets, protect consumer welfare and foster economic efficiency.

Switzerland has a civil law system that governs commercial transactions. The Swiss Code of Obligations is the main legislation addressing commercial matters, including contracts, sales and leases. Swiss competition laws are governed by the

Federal Act on Cartels and Other Restraints of Competition, which prohibits anti-competitive agreements, abuse of dominant market positions, regulates mergers that lead to a significant restriction of competition, aims to ensure fair competition, protect consumer interests and maintain a level playing field for businesses.

12.2.7.6 Summary of jurisdictional and cross-border legal issues

Although there are differences between jurisdictions, there are also considerable similarities and harmonisation, particularly around contracts, consumer protection and competition. In any event, careful analysis is needed for all activities to be or expected to be carried out in a given jurisdiction.

12.2.8 Summary of legal risks associated with blockchain applications

Reducing the legal considerations to a handful of key takeaways is not trivial, but there are key practical steps any new blockchain application can take. At risk of over-simplifying:

1. Law and questions in law are about precision. Whilst we can talk in generalities, specifics require an actual question to arise in the context of the blockchain application. When taking legal advice, be as specific as possible in your inquiry.
2. The law can help frame a project and structure its activities lawfully. Weak, unlawful ideas may be struck down early. Do not waste time.
3. The more you know, the more knowledge you have of the law, the better you can see the opportunities to reinvent products and services. You will know which boundaries to push, and how to push them, to achieve change for the better.
4. The law is a tool. Applied correctly to the right situations it can help you to build a better application, with greater or stronger resilience, faster.
5. Philosophically, any new idea, new implementation is a journey. Ask who you want to be on that journey with. Engage a great lawyer, with an entrepreneurial outlook. Build the future together.

12.3 Cases, case studies and examples of legal risks

This section considers real-world examples and projects of legal risks in built environment blockchain projects, the problems they seek to address and the legal risks they and their users face. We shall then consider real projects building blockchain applications for the built environment.

12.3.1 Real-world examples of legal risks and their consequences

Following the five areas of legal risks discussed above,[47] real-world 'hypotheticals' in the context of the built environment may be imagined as arising in a

[47]See Section 12.2.1.

series of disputes. Lawyers use hypotheticals in conjunction with statutes, regulations and case law[48] to help interpret and understand the law and make educated predictions on how similar issues in new circumstances may be addressed.

12.3.1.1 Contracts, smart legal contracts and smart contracts

Hypothetical: a construction project utilises smart legal contracts to automate periodic payments upon completion of project milestones. For the purposes of this hypothetical, we will assume the payment mechanism is legally compliant with UK law. However, a difference arises regarding the interpretation of contract terms, resulting in a delay to payment and a contract dispute.

Risk: the law regarding payments in construction contracts is mandated in legislation. A failure to comply with the prescribed requirements is unlawful. Although the factual matrix needs to be established, immediate questions are, what does the smart legal contract say and where does it say it? Who was responsible for interpreting and implementing the terms, and were those interpretations and implementations agreed? Who encoded the terms and was an error made when they did so? The questions are manifold, and it may not be trivial to resolve this.

Conclusion: regardless of a failure to implement the smart legal contract correctly, which may or may not be the case, to mitigate legal risks and minimise contractual disputes in smart contract-based construction projects, stakeholders should (i) use clear and precise contractual terms, (ii) thoroughly review the contract and negotiation and (iii) include appropriate dispute resolution mechanisms. By taking these precautions, stakeholders can reduce the likelihood of contract disputes, streamline payment processes and mitigate the legal risks associated with such disputes.

12.3.1.2 Practical IPR

Hypothetical: a construction company develops a proprietary blockchain platform to streamline project management processes. Another company copies the platform's code and functionality without permission.

Risk: a direct copy-paste is copyright infringement and would lead a cause of action. By engaging in such activities, the copying company exposes itself to potential liability for copyright infringement, which can lead to legal consequences and damages. It is crucial for companies to protect their IPR and take appropriate legal action when unauthorised use occurs.

Conclusion: unfortunately, in the UK, IPR protection for software is limited to copyright. In programming there are multiple ways to achieve the same output from the same inputs, and sometimes there is only one way to achieve the desired output from the given input. For want of a blatant copy-paste, a claim may be difficult to make out.

[48]Decisions and judgments of the courts.

12.3.1.3 Data protection, privacy and cybersecurity

Hypothetical: a blockchain-based smart home system collects and stores personal data, including the inhabitants' daily routines and energy usage. A data breach occurs, exposing sensitive information.

Risk: collecting and storing personal data requires compliance with data protection laws, such as GDPR. The company is obligated to comply with such laws and protect the data they acquire. A breach may lead to loss of privacy, identity theft or worse resulting in legal consequences. The legal consequences may be severe, with heavy regulator fines, loss of reputation and personal claims made by the data subjects.

Conclusion: the system could be designed to mitigate legal risks and ensure compliance-by-design. Companies handling personal data in blockchain-based systems should (i) consider the extent to which personal data may need to be collected, (ii) implement robust security measures, such as encryption and MFA, (iii) conduct regular security audits to identify security flaws and system vulnerabilities, (iv) obtain proper consent from individuals and (v) adhere to applicable data protection laws. In the event of a data breach, prompt notification to affected individuals and cooperation with regulatory authorities can help minimise legal consequences and maintain user relationships.

Hypothetical: a blockchain-based property title registry is compromised by in a cyber-attack, leading to unauthorised modifications of property ownership records.

Risk: the affected property owners take action against the registry for negligence in implementing robust security measures and inadequate protection of their property rights. For the registry, negligence and inadequate security is inexcusable when required to protect the integrity and confidentiality of critical records. Unauthorised modifications will lead to confusion, disputes and loss of property rights. Legal actions will become complex, time consuming and expensive, if many people are affected and damages are awarded in any claims.

Conclusion: to mitigate legal risks, property title registries must prioritise robust security measures to protect the integrity of their systems. This includes implementing strong access controls, encryption mechanisms, regular security audits and continuous monitoring to detect and respond to potential cyber threats. By taking these precautions, the registry could reduce the likelihood of a successful cyber-attack and mitigate potential legal consequences.

12.3.1.4 Financial regulation

Hypothetical: a company launches an ICO to raise funds for a real estate project by issuing tokens to multinational investors.

Risk: the tokens offered by the company following token analysis may be considered securities under financial regulations. A failure to comply with securities laws may result in penalties, investor protection concerns (wherein the company fails to adequately inform investors what it is they are investing in and trading) and regulatory challenges if compliance requirements are not met.

Conclusion: to avoid adverse consequences and mitigate legal risks, the company should (i) take legal advice from experienced professionals, (ii) ensure compliance with regulatory requirements and (iii) carry out appropriate due diligence to determine the token classification. By taking proactive steps to address financial regulation and comply with applicable laws, the company can mitigate legal risks and improve trust with both investors and regulators.

12.3.1.5 Commercial

Hypothetical: an End User Licence Agreement (EULA) for a new DAO metaverse platform uses blockchain to secure ownership rights and facilitate transfers in its virtual world, but incorporates unfair obligations within its terms. The hype-marketing campaign is inspiring a fear of missing out in users who are desperate to be part of this new community. Users must accept the terms that run to several hundred clauses.

Risk: unlawful terms wrongfully infringe the rights of users, and the aggressive marketing campaign may be manipulative and misleading. Unfair contract provisions may give rise to challenges and be rendered unenforceable. The platform's reputation may be damaged beyond repair if trust is lost.

Conclusion: to mitigate legal risks the platform should review its terms and conditions and rectify any provisions that may be unlawful. By being transparent and fair in dealing with its users, engaging with regulators and industry bodies to adopt best practices and spell out the rights that the platform grants, the DAO may mitigate and even avoid any adverse legal consequences.

12.3.1.6 Jurisdictional and cross-border legal issues

Hypothetical: a multinational real estate development project uses blockchain to secure investment transactions. However, conflicts arise due to differing legal frameworks and jurisdictional challenges, leading to uncertainties regarding contract enforcement and dispute resolution across borders.

Risk: legal uncertainty in critical assets, such as real estate, could result in loss of investment, allegations of fraud and misuse of funds. On one side, the parties receiving the investment will seek correction and damages for any losses. On the other, a failure to deal appropriately with investors, mishandling funds, may result in regulator action, potentially incurring fines and other penalties.

Conclusion: to mitigate legal risks arising from conflicting legal frameworks and jurisdictional challenges stakeholders should (i) seek legal advice, (ii) carry out appropriate due diligence in the relevant jurisdiction, (iii) select carefully the choice of law and jurisdiction clauses (although real estate tends to be a reserved legal matter to the specific jurisdiction in which the real estate resides) and (iv) comply with any regulatory requirements of the jurisdiction. By proactively addressing conflicting legal frameworks and jurisdictional challenges, stakeholders can mitigate legal risks, enhance contract enforceability and facilitate effective dispute resolution.

12.3.2 Case studies of blockchain applications that have faced legal challenges

The hypotheticals are entertaining as esoteric thought experiments. They are designed as extreme examples of the legal risks and challenges a project may face. Building on the overview above,[49] and at risk of dating the analysis, we consider several real blockchain projects that are tackling some of those described use cases as well as situations where projects have faced legal difficulties. Whilst many are casualties of common start-up symptoms, such as insufficient funding, failure to address a real need, or lack of user buy-in, some came to a (seemingly insurmountable) legal roadblock. However, for every challenged project, there are others in the same space tackling different problems with greater success. In the subsections below, I example at least one project this was challenged and one that is challenging.

12.3.2.1 Payments

This area is popular with both new entrants and casualties. Several promising projects, some involving large and well-resourced enterprises, with clear vision and technical ability, have stopped reporting progress or ceased trading, without managing to deliver an operational solution. The reasons why are not always clear, but gleaning from various sources, it appears key issues revolved around financial regulation of a crypto-asset-based solution.

One casualty may have been Builderium [27], a marketplace platform to connect clients, contractors and trade workers, facilitate coordination and manage payment. Using reverse auction concepts [28], projects would be incentivised to be completed on time and on budget, without compromising quality.

Whilst there are many projects that offer payment solutions generally, albeit cryptocurrency focused, there are enterprise grade solutions specifically designed for construction and engineering. One solution is provided by Oracle [29,30],[50] which offers a blockchain-based platform focussed on streamlining payment processes, contract management and supply chain operations. The platform uses blockchain to increase efficiency, improve transparency, automate payment workflows for timely and accurate payment, secure transactions and enhance trust among project stakeholders.

12.3.2.2 Project management

Project management interacts with almost every aspect of a built environment project. This presents multiple opportunities for solutions but risks a nascent

[49]See Sections 12.1.1 and 12.2.

[50]A solution for individuals and contractors is Latium. Although not exclusively focused on construction, Latium is a blockchain-based platform that allows employers to make payments, both in fiat and cryptocurrencies, to contractors and freelancers for agreed tasks. It provides a decentralised and transparent system for tracking task completion and issuing payments, which can be beneficial for contractors and subcontractors involved in construction projects.

solution attempting to tackle too much and losing focus on solving a critical problem.

An example of a blockchain project management application that has closed is Structor.io [31]. Structor was a collaborative project management platform for the construction industry. It enabled users to manage their project and commercial workflows with tools for projects, time, workflow, risk and asset performance. Unfortunately, the product was launched just before the COVID pandemic, which likely forestalled adoption. Further legal process issues may have also arisen around the complexity of transposing construction contracts into the necessary event-decision-based workflows.

By contrast, the BuildCoin Foundation [32] focuses on leveraging blockchain technology to address challenges faced by the construction industry, particularly in developing countries. The platform provides funding [33],[51] visibility and accountability for construction projects, with subject matter expertise and transparent construction project management.

12.3.2.3 Procurement and supply chain management, and provenance

In our increasingly sustainability conscious time, consumers demand to know more about the origins of the goods and services they purchase. The same is true for materials and equipment used in construction projects.

An example of a procurement blockchain project that faced challenges is BlockArray [34]. BlockArray aimed to utilise blockchain technology to improve transparency and efficiency in supply chain processes. The platform aimed to provide real-time tracking and verification of goods, enabling visibility of movement and authenticity of products throughout the supply chain. Unfortunately, BlockArray was neither widely adopted nor attracted a critical mass of users. Achieving critical mass and building a thriving ecosystem is crucial for any project, and the limited adoption hindered BlockArray's growth. The legal complexity of integrating multiple data sources, inter-system and cross-jurisdictional data may also have hindered data integrity and scalability.

Although the complexity of supply chains remains, several projects are targeting either specific goods or sector specific supply chains.

Circulor [35] and Provenance [36][52] are blockchain-based platform that specialise in materials traceability, particularly in industries such as automotive,

[51]Specifically on funding arrangements, LendLedger is a blockchain-based platform that focuses on providing financial services to small and medium-sized enterprises in the construction and building materials sector. By utilising blockchain technology, they aim to improve access to financing for construction projects.

[52]Provenance is a blockchain platform that aims to enhance supply chain transparency and traceability, but also materials provenance. Whilst the team has pivoted away from construction materials, the platform continues to enable companies to trace and verify the origin, environmental impact and social responsibility of materials used in products, tracks materials, verify certification and improve overall accountability.

electronics and mining. The platform enables companies to track and verify the origin and sustainability of materials used in their supply chains [37][53][38][54]. Open Mineral [39] leverages blockchain technology for tracing and ensuring the ethical sourcing of minerals [40][55] as used in various industries. The platform provides transparency and trust in the supply chain, helping companies make responsible sourcing decisions. In each case blockchain technology is being applied to enhance transparency, traceability and efficiency, to track and verify material provenance and promote sustainability and ethical practices in supply chains.

12.3.2.4 Digital twins, BIM and real estate

BIM is an entire book to itself and there are projects underway to integrate BIM with blockchain. For simplicity, this section considers the interconnection of digital twins, real estate and NFTs. Why NFTs? NFT's experienced exponential growth[56] in use since popularised by various projects between 2017 [41] and 2021 [42]. Given their rapid rise, and the 'goldrush' that follows such hype, it is hardly surprising that legal problems followed.

An example of a real estate blockchain-based platform that faced challenges is ATLANT [43]. ATLANT attempted to create a decentralised real estate platform that tokenised ownership and trading of real estate assets using blockchain. Despite its initial vision, ATLANT faced significant difficulties, struggling to gain widespread adoption or attract a critical mass of users. Market inertia from traditional practices and convincing owners and investors to adopt the technology was a significant hurdle. Additionally, real estate is subject to complex legal frameworks that vary between jurisdictions. Challenges around compliance, in particular property ownership verification, hindered the platform's ability to operate effectively in different regions. Further, whilst the base technical challenges of scalability, security and user-friendly interface were present, ATLANT also faced operational issues, such as liquidity concerns and business model sustainability. Lastly, although not from an anti-competitive perspective, competition from similar platforms, and traditional established competitors, prevented ATLANT taking a significant market share.

[53]IBM Food Trust, whilst strictly not a built environment solution, is a blockchain-based platform that enhances transparency and traceability in the food supply chain. It enables participants to track the journey of food products from farm to table, ensuring food safety and reducing inefficiencies.

[54]Specifically on supply chains, TradeLens is a blockchain-based digital platform developed by IBM and Maersk for global supply chain management in the shipping and logistics industry. It improves transparency, efficiency and collaboration among stakeholders, enabling real-time tracking of shipments and documentation.

[55]A parallel application of ethical sourcing for fine goods is Everledger, which utilises blockchain technology to track and verify the provenance of various assets, including diamonds, gemstones and fine wines. The platform ensures transparency and authenticity throughout the supply chain, combating counterfeit goods and promoting ethical sourcing.

[56]Although daily trading volumes of NFTs collapsed during 2022 following various adverse marketplace conditions, suspension of cryptocurrency trading, fakes, plagiarism and accusations of fraud.

Despite the example above, digital twins [44][57,58] and real estate [45][59][46][60] [47][61] continue to see interest in revolutionising the property sector with blockchain enabled PropTech, to improve asset management, transactions and record keeping.

Propy [48] is a global property marketplace built on blockchain technology. It facilitates the purchase and sale of real estate properties, offering a secure and transparent platform for transactions. How it achieves this, however, is not by tokenising the property itself, but rather the shares of a holding company that owns the asset. In creating an NFT of those shares, the platform may also facilitate asset fractionalisation for multi-party real estate investment opportunities.

It is interesting to note that other platforms [49] similarly purporting to have created tradable tokenised real estate, have taken a similar approach and tokenised the shares in a company that owns the underlying asset. Sell the company and the asset follows, but strictly the underlying asset is not for sale in and of itself. There are reasons a deal would be structured this way, not least of which is tax efficiency. Whilst perhaps not strictly misrepresentations, as the websites make clear if you dig into them far enough, they could be clearer. The transaction is not 'true' in the sense of a direct exchange of asset ownership. Query also whether the additional intermediation is in fact beneficial to an already complex, multi-party process and overcrowded space.

Propy achieved its first NFT 'property' sale in 2017, but in 2021 an entrepreneurially minded real estate agent, Shane Dulgeroff, created an NFT representing a property for sale in California [50]. However, the token was attached to a crypto

[57]Mattereum is a blockchain-based platform that attempts to bridge digital and physical. The platform allows users to create digital representations, or 'smart property', of physical assets on a blockchain. These representations are backed by smart (legal) contracts, providing a mechanism for tracking ownership, transferring assets and resolving disputes in a decentralised manner. Combining the blockchain and smart (legal) contracts with IoT, creates a secure and transparent system for verifying and transferring ownership. The aim is to provide increased efficiency, transparency and trust in various industries, and address challenges associated with cross-border transactions such as asset provenance and financing.

[58]As of 2023, the case of Nike v StockX, aside from potential IPR infringements, concerned the trading of NFTs used as a method of authentication (twinning) and exchange for the sale of trainers stored in a vault. The question arose as to what the NFTs are: are they a 'ticket to depict ownership' or a 'digital asset in their own right'? It seems that the NFT simply points to the associated asset, which is held custodially on behalf of the token holders. The case continues. Nike, Inc. v. StockX LLC, 22-CV-00983 (VEC)(SN) (S.D.N.Y. Jan. 9, 2023).

[59]Coadjute is a platform for property transactions in real estate. The platform enhances collaboration, efficiency and transparency among stakeholders in property transactions such as buyers, sellers, agents, lenders and solicitors. The platform leverages blockchain to create a secure and immutable record of property-related data, such as listings, transactions and legal documents, reducing physical paperwork, improving data accuracy and expediting the overall process. Real-time communication and data sharing among participants, enables seamless coordination, reduces delays, eliminates duplicate data entry, automates manual tasks and presents a unified view of a transaction for all parties.

[60]ChromaWay develops blockchain solutions for real estate. One platform, Postchain, enables the creation of decentralised applications for property management and transactions.

[61]Ubitquity provides blockchain solutions to manage real estate title and land records. The platform aims to increase efficiency, reduce costs and enhance the security of property transactions.

artwork, and it was the artwork that was auctioned. Whoever won the auction would receive the NFT, the art and ownership of the house. The auction closed without a bid. There are multiple legal implications that would preclude a sale in this manner, not least of which are title transfer, financing, purchase of a potential security and taxation. Nonetheless, it was a creative application of the technology and demonstrates a desire to reshape and reimagine a traditional sector, highly resistant to change.

After years of development in the NFT space, in 2022, following the case of Osbourne v OpenSea [51], NFTs have been recognised as legal property in the UK. Property is capable of being owned and dealt with in legal proceedings, such as confiscation as the proceeds of crime, or frozen in a financial restrictions order. This is a key development as it means legal recognition of rights of ownership, albeit of the token and not necessarily the underlying asset.

12.3.2.5 Integrated smart cities

Although not strictly limited to blockchain, as smart cities integrate multiple technologies and digital solutions, there are many elements to which blockchain solutions are applied. It is too early to tell if such city-scale projects may be classified as successes, but we may compare current projects.

Seoul, South Korea, has launched several blockchain pilot projects, now culminating in its Metaverse Seoul Master Plan [52], to improve citizen services and enhance administrative processes. Blockchain is a cornerstone technology to secure data, create citizen digital IDs for self-authentication and enable financial transactions. In contrast, Songdo, South Korea, a city built from scratch with smart city technologies at its core, incorporating elements such as advanced communication networks, energy-efficient buildings, automated waste management systems and smart transportation infrastructure, has seen slow uptake, with only around 50% of the planned population taking up residence.

Singapore integrates smart technologies for urban planning, transportation, energy management and public services, using IoT, data analytics and real-time monitoring to optimise resources, improve sustainability and enhance the urban experience. More recently Singapore has been actively exploring blockchain technology for smart city applications through initiatives such as Project Ubin [53] and Ubin+ [54], which are exploring the use of blockchain for digital identity, trade finance and cross-border payments, to enhance efficiency, transparency and security in city processes.

Barcelona, Spain has implemented many initiatives through its Smart City [55] and Digital City [56] programmes, including intelligent lighting systems, smart parking, waste management and environmental monitoring, by leveraging IoT devices, data analytics and citizen engagement platforms to optimise resource usage, enhance mobility and improve urban services. Barcelona has explored the use of blockchain for data sharing and authentication for transportation, energy and tourism and has partnered with blockchain startups to support innovative projects.

12.3.2.6 Energy generation and use

The energy sector has seen a surge in blockchain-based solutions to replace outdated systems, from the management of renewable energy installations to energy trading. Whilst such solutions are dealing with macro-energy (generation and distribution), this is filtering down to micro-energy (device charging and inter-device sharing).

An example of an energy management blockchain-based platform that faced challenges is LO3 Energy [57] and the Brooklyn Microgrid. The Brooklyn Microgrid aimed to create a peer-to-peer energy trading platform using blockchain technology, allowing residents to buy and sell renewable energy produced from solar panels. Despite its innovative concept, the Brooklyn Microgrid faced various difficulties. The scalability of the blockchain technology used proved unsustainable. As the number of participants and transactions increased, the infrastructure faced limitations in processing speed and capacity, which hindered the platform's ability to accommodate a larger user base and scale up operations effectively. Additionally, the energy sector is subject to complex regulatory frameworks that vary across jurisdictions. Adapting the blockchain platform to comply with existing regulations, obtaining necessary approvals and ensuring legal compliance proved insurmountable at the time and the uncertainty impeded project growth and viability. In parallel, energy markets are highly regulated and dominated by established, monopolistic, suppliers. Overcoming the inertia of resistance to change in traditional energy markets, with producers and consumers hesitant to participate in a peer-to-peer trading platform, added to the slow transformational change.

Perhaps inspired by the Brooklyn Microgrid, even if indirectly, several projects are addressing specific aspects of the energy market.

Grid Singularity [58], aside from having a great name for marketing, focuses on developing blockchain solutions for energy markets [59][62] and grid management [60].[63] Electron [61] is a blockchain-based platform focused on the energy and EV sectors. They aim to create a decentralised ecosystem that enables transparent and secure data sharing and transactions between EVs, charging infrastructure and energy grids. These platforms aim to enable decentralised energy trading, enhance transparency, efficiency and interoperability, and promote renewable energy initiatives in the energy sector. Further, specifically for the EV sector, they aim to enhance charging infrastructure, enable secure transactions and explore innovative concepts like vehicle-to-grid integration.

12.3.2.7 Transport and utilities

Transport and transportation are the arteries carrying the lifeblood of cities, intercity economies and inter-continental tourism. Simplifying these systems and improving the experience would be welcomed by commuters and tourists alike. Water, less poetically, is the lifeblood of, well, life.

[62]PowerLedger is a blockchain-based platform for peer-to-peer energy trading. Individuals and businesses can buy and sell renewable energy directly, promoting decentralisation and sustainability.

[63]Enel Green Power is an energy generation company that provides energy solutions throughout the energy vertical. One solution is the integration of blockchain technology into electric vehicle charging. By utilising blockchain, they aim to enhance the management and optimisation of charging infrastructure, enabling smart and efficient charging networks.

An example of a blockchain-based platform in the transportation sector that faced challenges, and in a happy non-failure twist pivoted to carbon trading, is DOVU [62]. As to transportation, DOVU aimed to create a blockchain-powered marketplace for mobility-related data, incentivising users to share their transportation data in exchange for tokens. Whilst the data protection and privacy concerns alone are sufficient to derail this concept, DOVU did make some progress and partnerships with the transportation industry. A key challenge, as already mooted, was obtaining access to a critical mass of transportation data. Despite efforts to collaborate with transportation companies and incentivise data sharing, acquiring the necessary data sets proved challenging. Limited data for a platform reliant on data hindered the functionality and sharing-value proposition. Additionally, encouraging users to actively share transportation data and engage with the platform proved similarly challenging due to the inherent privacy concerns and misaligned incentives for contributors. Without a critical mass of users, a vibrant and valuable ecosystem was not possible. Specifically on the transportation data, as sensitive information, regulatory requirements as to privacy, consent and compliance are stringent. Further difficulties encountered were the practicalities of integrating diverse and legacy system data, and fragmented third-party applications to produce a seamless user experience and system interoperability, critical for an effective solution. In true entrepreneurial style, the DOVU team pivoted to carbon. The basic principles of data tokenisation, traded for value, in a marketplace are broadly similar, whether transport or carbon.

Taking a different approach, Mobi [63] is a blockchain consortium focused on the mobility and transportation industry. It brings together various stakeholders, including manufacturers, technology companies and mobility service providers, to explore and develop blockchain-based solutions to improve transportation services and efficiency, interoperability and data sharing (and incentivisation for data sharing) through collaboration.

As to utilities, in particular water, OriginClear [64] and WaterRecharge [65] are blockchain platforms that aim to improve water management, conservation, water rights and water trading. Each facilitates the tracking and monitoring of water usage, enables efficient and secure water trading, and tracks water usage in real time, promoting transparency in water transactions. Ultimately encouraging sustainable practices, the platforms can be used by various stakeholders, including water utilities, farmers and industrial users.

12.3.2.8 Consumer protection

Whilst not strictly a real-world example, but a notable experiment, was web researchers testing users' interactions with a new fictional metaverse platform [66].

Whilst signing up to an exciting new metaverse platform, users were asked to accept the platform's terms. This, typically, was a long, complicated and densely worded legalese document. Did people read this? No, they wanted to get on with building their corner of the world. Should they have read this? Yes.

It is important to note that metaverse platforms are contract governed worlds as between platform and user. They are distinct from real-world law, meaning they are effectively an authoritarian regime and are not new realms of liberal democracy and freedom. User terms can be changed arbitrarily at the platform's discretion and may disproportionately punish ordinarily lawful behaviour, for trivial or emotional reasons.

The terms in question contained what we shall call a 'first-born assumption' clause, which required users to handover their first-born child to the platform. Think Truman Show, but worse. Although the reality is that this would not be enforced or enforceable, the extreme nature of such clauses and greater authoritarian terms in EULA's are commonplace. As a user you are beholden to the platform's policies, which may be more restrictive than the comparable position in law as to, for example, freedoms of speech, expression or behaviour. Even trivial transgressions, determined by unqualified collective judgement,[64] may result in exile from the platform and loss of all acquired items.

Whilst reading this section you may have noticed a dearth of legal cases providing legal principles and opinions on the issues. This is not unintentional. Although blockchain is well over a decade in use, the law is only just beginning to catch up. To date, the UK has seen only around three dozen cases reported, but is at the jurisdictional forefront in this regard.

The cases that have been reported fall broadly into three categories: (i) theft or loss of crypto-assets or tokens including fraud, generally around bitcoin (due to its apparent value), NFTs (for the same) and ICOs, (ii) commercial terms, broadly contractual and IPR disputes, and (iii) various claims by the self-proclaimed creator of bitcoin. Whilst there are technical elements to each claim, where losses are concerned and tracing analyses are carried out for service of freezing orders and attempts at recovery, the cases do not concern the technology itself, but the legal frameworks within which they operate.

The leading case for loss of crypto-assets, tracing and attempts to recover assets is noted above.[65]

On contract terms, the case of Soleymani v Nifty Gateway [67] considered the nature of NFTs and a platform's terms of use. It is notable as a warning on consumer protection. On NFTs, the court was asked whether NFTs were the art or simply a unique code stored on a blockchain that permits access to and authenticates art. The answer lies in the statement. The NFT is not the asset, it merely points to an asset. On the terms of use, the website did not make clear to users the basis of the auction resulting in Soleymani believing he had overpaid for a particular artist's NFT that he won.

On IPR, the cases of Wirex Ltd v Cryptocarbon Global Ltd [68] concerned a trademark infringement dispute on a credit card reward scheme, and Crypton Digital Assets Ltd v Blockchain Luxembourg SA [69] concerned the wrongful appropriation of an in-development blockchain platform. Noted above[66] was

[64]To such collective judges, remember that just because you do not like something, does not mean it's unlawful. The law does not pass judgment on hurt feelings.

[65]See Section 12.3.2.4.

[66]See Sections 12.3.1.2 and 12.4.3.5.

copyright infringement for software code, which was directly tested in the latter case. However, these cases centred on well-defined IPR and not blockchain technology *per se*.

Whilst on consumer protection disputes, an alternative blockchain-based dispute resolution protocol such as Kleros [70], independent from traditional legal systems, may be able to help. Kleros aims to provide fast, transparent and decentralised arbitration for disputes. The platform uses blockchain technology and crowdsourced jurors to deliver a fair and unbiased resolution process. Users submit disputes, such as contract, e-commerce or IPR disputes, to be resolved on the platform. Jurors are randomly selected from a pool of network token holders based on their domain specific expertise to ensure disputes are addressed by those with relevant knowledge and who may provide informed judgments. Parties submit evidence and present arguments. The jurors review the evidence and make decisions based on their understanding of the case and applicable rules, reaching a final decision through a voting mechanism weighted by tokens held. The token-based system aims to incentivise jurors to act honestly and impartially. The protocol ensures transparency, as all rulings and decisions are recorded on the blockchain, making them publicly accessible and auditable.

12.3.3 Summary of legal risk cases and examples

Engaging in hypotheticals helps founders and innovators consider the legal risks that a project may encounter. Compared and contrasted with the lessons learned from past projects and tested against the latest legal judgments will prepare a project and develop resilience to future hurdles. Being prepared for those risks, taking legal advice and having an action plan ready to mitigate those risks may mean the difference between success and failure.

12.4 Best practices and solutions to mitigate legal risks

Over the course of this chapter, we have formalised our understanding of the underlying technology, considered key areas of law that govern activities in the various spaces and hypothesised scenarios whilst also looking at the challenges faced by real blockchain projects in the built environment.

The core of the challenge is that the built environment is not 'digitally native'. On one side, we have the physical realm: physical assets or processes that we want to digitise and digitalise. On the other, we have the digital realm: efficient streamlined standardised[67] processes, single points of truth, proofs of ownership and embedded legal attributes. We must build the physical-digital bridge.

By implementing appropriate strategies, organisations can proactively identify and address legal risks associated with blockchain applications, enhance compliance, security and increase the likelihood of project success. Knowing what we

[67]Ontologically defined and distinguished concepts and relationships, and taxonomically rationalised.

now know, what are the best practices that should be considered and applied to achieve this? In this section we will consider best (legal) practices.

12.4.1 Overview of best practices

The best practices and solutions outlined here are intended to provide a foundation for mitigating legal risks in blockchain applications within the built environment. It is essential to adapt and tailor these practices to the specific context, legal frameworks and requirements of a project to ensure a comprehensive risk mitigation strategy.

Best practices, at their simplest, come down to preparedness in three key areas: (i) risk mitigation, (ii) legal compliance and (iii) smart (legal) contract design.

Risk mitigation is an exercise in (i) legal due diligence: assess legal risks and comply with laws, (ii) cybersecurity: implement robust protocols, (iii) regulatory compliance: know the laws applicable to the application, (iv) clear contracts: define rights, obligations and resolution mechanisms to prevent conflicts and ensure enforceability and (v) audits and assessments: periodic reviews and audits to identify vulnerabilities, compliance gaps and ongoing 'live' risk mitigation.

Legal compliance: (i) understanding the relevant laws: stay up to date with changes, (ii) implementing KYC/AML procedures: prevent fraudulent activities, (iii) compliance with data protection: safeguard personal data, (iv) adherence to sector specific regulations: market principles and codes of best practice and (v) engagement with regulators: build constructive partnerships, seeking guidance and approvals to meet compliance requirements and mitigate legal risks.

Smart (legal) contract design: (i) enforceability: contracts should be drafted as 'compliant-by-design', (ii) reviewed and tested: identify and rectify any vulnerabilities or errors, (iii) dispute resolution mechanisms: built in mechanisms for dispute resolution, to help resolve conflicts efficiently and reduce legal risks, (iv) governance: transparent and well-defined frameworks for decision-making processes, consensus mechanisms and user participation, to enhance accountability and reduce legal uncertainties and (v) future flexibility: designed to accommodate an evolving legal and regulatory landscape, with updates and adjustments to comply with future changes (as well as practically for the inevitably pivoting).

12.4.1.1 Incentives, a side note

A further area to consider, although beyond the scope of this chapter, is incentives or behavioural economics. How do you achieve best practices when those who use your solution are not under your control? Much is made of large technology companies, but in particular social media platforms, for acquiring excessive amounts of data that is used to predict, and encourage, certain future behaviours. A notorious example is the 'like' button, which releases dopamine when hit resulting in you either wanting more 'likes' or to give more 'likes' [71].[68] This is monitored by the system and your data flow (what you see in your feeds) is reconstructed to further

[68]See the commentary of Palihapitiya C.

embed and reinforce the addictive behaviour. You see what you want to see, expend dwell-time on that and/or 'like' it, and that is negatively reinforced with you being shown more of the same. Unfair and manipulative? Perhaps. Adverse to mental health? Likely, based on current studies. Powerful? Most definitely. Now, imagine that could be repurposed for something that is socially, environmentally and constructively useful.

12.4.2 Strategies for risk mitigation

These strategies are for consideration only and may not all be relevant or suitable to a particular project. Apply judiciously (pun intended).

12.4.2.1 Risk assessment and management

Foresight, preparedness or simply a best guess.

First, assess. Carry out a careful, thoughtful and comprehensive risk assessment to identify potential legal risks and vulnerabilities associated with the application. There are many strategies and approaches to a risk assessment, but all that is really needed is thorough, diligent and questioning perspective. What could go wrong? Who could get it wrong? Where could it go wrong? When could it go wrong? Why would it go wrong? How could it go wrong?[69]

Second, prioritise. Prioritise risks based on their likelihood and potential impact on the project. Some techniques add or multiply an assigned 'likelihood' and (adverse) 'impact' value, or apply weighting to values, to facilitate ranking and prioritisation. Two additional factors to consider in the analysis matrix are time and resolution: how long may it take to address the issue, and what may the process be?

Third, manage. Develop a risk management plan that outlines the strategies to identify the risk and mitigate its impact, along with protocols or 'standard operating procedures' that may implemented in response to a risk. Knowing what to do, having a plan and executing that alleviates response anxiety and focusses minds on dealing with the situation.

12.4.2.2 Legal and regulatory compliance

Although touched upon above, as part of an overall strategy for risk mitigation, understanding the relevant sector, actors, stakeholders, movements in commercial approach and changes in law are all important elements in maintaining and deploying a cohesive overall strategy.

First, currency. Stay updated not only on relevant laws, regulations and industry standards applicable to blockchain and crypto-asset applications, but also, at a high level, on the wider built environment.

Second, compliance. Ensure compliance with data protection, privacy, property, construction and other applicable regulatory requirements. It may seem obvious but ask the question. It is better to ask the 'dumb' question and get an

[69]There are many risk management approaches, from the 'five bums on a bench' to the 'five why's'. All are fundamentally concerned with questioning and reasoning.

answer, then never know. Besides, there are no 'dumb' questions when it comes to understanding your project.

Third, advice. Engage legal professionals with expertise in blockchain and the built environment to provide guidance and ensure compliance. Engage with strategy advisors, mentors, knowledgeable colleagues, friends and acquaintances. Your network is much larger than you think – six degrees of separation and you know the world.

12.4.2.3 Secure development practices

Agile, lean, scrum and all those other working methodologies play an important role in the development cycles of a tech startup. However, as with any business, there are several simple principles that should always be adopted.

First, security. Just as you would secure a building with ID card barriers, security guards and even locks on doors, so you should cyber-secure your application. Implement robust cybersecurity measures to protect the blockchain application from unauthorised access, data breaches and cyber-attacks.

Second, discipline in approach. Adopt secure coding practices and conduct regular security audits and assessments. Part of a defence to legal risks may be to show not just what was done, but how it was done. Keep diligent records, however difficult the frenzied energy of a new exciting project.

Third, digital locks. An issue associated with security, but legally distinct is data protection. Adopt and implement encryption and other security technologies to protect sensitive data and transactions.

12.4.2.4 Smart contract auditing and code review

Check, review, test, repeat. Noted above was the concept of contract certainty. Eliminate ambiguity.

First, auditing. Conduct thorough audits and code reviews of smart contracts and smart legal contracts to identify and rectify any vulnerabilities or flaws. The position is less certain for NLCs as the process for distillation into code is not clear.

Second, engage experts. Engage with third-party auditors or security experts to ensure the integrity and security of smart contracts. Security should 'wrap around' the smart (legal) contract to protect inputs, outputs and its operation.

Third, test. Test the smart contracts in different scenarios and environments to verify their functionality and mitigate potential risks. Attempt false inputs, bad data and anything else an inexperienced user or nefarious actor may try, deliberate or inadvertent.

12.4.2.5 Transparency and user education

There is an African proverb, 'if you want to go fast, go alone, if you want to go far, go together'. Take users with you on the journey, explain the evolution and invite participation.

First, transparency. Foster transparency by providing clear and easily understandable information about the blockchain application, its functionalities and associated risks.

Second, educate. Educate users, stakeholders and participants about the risks and benefits of the blockchain application.

Third, communicate. Clearly communicate terms, conditions and potential risks to users through user agreements, terms of service or other relevant documents.

All of these may be summarised as 'trust' and how you create, nurture and maintain that with users.

12.4.2.6 Governance and compliance mechanisms

Whatever the application, being compliant, having strong oversight and transparent reporting underpin confidence and trust in any platform.

First, governance. Establish clear governance frameworks to ensure accountability, decision-making processes and compliance with legal and regulatory requirements.

Second, controls. Implement internal controls and procedures to monitor and enforce compliance within the blockchain application. Internal codes of practice, behaviours and leaders who exemplify that ethos will be reflected externally.

Third, adapt. Regularly review and update governance mechanisms to adapt to changing legal and regulatory landscapes. To paraphrase the words of Charles Darwin, 'it is not the strongest of the species that survives, nor the most intelligent; it is the one most adaptable to change' [72].[70] Although one may argue adaptability is a function of strength and intelligence.

12.4.2.7 Continuous monitoring and auditing

Legal risks are never 'settled' or 'solved'. They are ever present and ever changing.

First, monitor. Implement ongoing monitoring and audit processes to detect and mitigate risks.

Second, review. Regularly review the effectiveness of risk mitigation strategies and update them.

Third, inform. Stay informed about industry developments, best practices and emerging legal trends related to blockchain applications in the built environment.

12.4.3 Regulatory compliance, legal considerations and practical tips

Regulation. Considered in the light of practicalities, no one person can do and be everything. What immediate steps should a nascent blockchain application take?

12.4.3.1 Understanding applicable laws and regulations

Elaborating on best practices, where and on what should any nascent blockchain application spend its precious time?

First, and it is no small undertaking, become familiar with the relevant laws and regulations that govern, or may impact, the specific blockchain application. Areas of law, as discussed in more detail above, may include contracts, intellectual property, data protection, financial regulation and consumer protection.

[70]This is not a direct quote by Darwin, but an interpretation of an inference, as explained.

Second, news. Subscribing to regular newsletters and information feeds from relevant industry bodies should provide a wealth of news, articles, opinion pieces, events and upcoming changes. Most law firms publish newsletters on general legal developments, but also specialist teams will produce area specific articles. Typically, significant events are widely reported and opined.

12.4.3.2 KYC and AML

Many projects, but particularly those operating in the financial space or that have a financial element of functionality to them, require that the platform know who their users are, whether supply-side, demand-side or both. The reasons for this are perhaps obvious, but money laundering concerns are generally foremost and the reputational damage, let alone penalties and fines for not conforming to regulations or preventing laundering, can be project-ending.

First, onboarding. Implementing KYC and AML procedures are essential to verify the identity of users.

Second, follow them. Obvious, but as noted above, legal risks are a permanent travelling companion. KYC/AML requirements are designed to prevent fraud and fraudulent activities, ensure compliance and mitigate the risk of money laundering and terrorist financing.

12.4.3.3 Data protection and privacy

Above we considered data protection and privacy issues, their inherent risks and where mistakes may have been made historically.

First, compliance. Review, understand and implement compliance procedures with data protection and privacy laws, such as GDPR. Design them into the system to ensure 'compliance-by-design'.

Second, dealing. Ensure that any personal data collected and processed through the blockchain application is handled securely, with appropriate consent and transparency. Give users notice when personal data is acquired, the reasons for its retention and then delete it when it is no longer required.

Third, protection. If personal data must be retained for compliance, reporting or future reprocessing, take steps to protect that data. Consider encryption protocols, pseudonymity or anonymity protocols or techniques to protect, in the first instance, and obfuscate, in the second if a breach were to occur.

Minimise the risk by acquiring the minimum amount necessary to function. Contrary to social media, true value is not in user data, but what users can achieve with the platform.

12.4.3.4 Property and construction regulations

In the UK, modern property law arises from the Norman Conquest over 1,000 years ago (this counts as 'fairly recent' in property law terms). This has had the less fortunate side-effect of creating a complex system of layered, interconnected and conflicting laws. Property holds a special place in society, which is reflected in the amount of and complexity in the law. There is a reason for the proverb 'an Englishman's home is his castle', property is that important.

Construction also holds a special place in law. Surprisingly, construction (and shipping) case law accounts for a significant proportion of contract law principles that lawyers apply today. As such, construction has its own legislation to regulate the sector.

First, the web of law. Beyond direct property and construction legislation, there applies local planning permissions and building regulations, working licences, materials supply and standards, employment practices, project finance and that is before taxation is considered. These rules vary country to country, with requirements specific to the jurisdiction in which the blockchain application is deployed.

Second, ownership. Even before the project has been built, there are complex considerations for the property transaction, transfer of ownership or rights of development, purchase finance and post-project refinance. Any blockchain application, albeit with the laudable intent to disintermediate the complexity of such a market, cannot avoid compliance with the market's requirements.

12.4.3.5 IPR in application

IPR are a series of fundamental rights that protect the product of human endeavour. Protecting these rights, for example for an artist, musician or professional is crucial to their profession and existence.

First, respect. Respect for IPR in the development and deployment of the application. New ideas and their implementation is challenging. Look at the work of others, be inspired by it and acknowledge it.

Second, permission. Seek permissions or licenses to use the intellectual property of the rights owners if you wish to use it within the blockchain application.

Third, protect. Software is dealt with differently in different jurisdictions but tends to enjoy comparatively weak protection compared to other IPR. Protect what you can, where you can, with as many rights as you can. Refrain from oversharing if you wish to commercially exploit software.

12.4.3.6 Jurisdictional considerations

Whilst limited inter-jurisdictional comparisons were drawn above, there are commonalities as well as differences. Types of law, common or civil, tend to approach similar legal problems in similar ways. Whilst the detail may vary, the overall effects tend to be broadly consistent.

First, understand. Understand jurisdictional similarities and differences that may arise when implementing blockchain applications across borders.

Second, conflicts. Whilst understanding the similarities and differences is a first step, analysing where laws conflict, or where regulatory implementation requires a difference process, is critical when operating in multiple jurisdictions.

Third, expertise. I have written it before and will write it again, there are legal experts available out there with deep knowledge of international laws and cross-border regulations that can help navigate the complexities. Many lawyers enjoy applying the law to innovative ideas and see the relentless advancement of technology as an opportunity, not a threat to be extinguished.

12.4.3.7 Regulatory engagement and compliance

Finally, open the conversation.

First, engagement. Many regulatory authorities routinely offer guidance, have innovation teams, sandboxes and other services designed to engage with new technologies in their market (perhaps a 'if you cannot beat them join them' reaction). In theory, at least, early engagement can ensure compliance with applicable regulations. Forewarned is forearmed.

Second, participation. Participate in industry discussions and collaborations to contribute to the development of regulatory frameworks for blockchain applications in the built environment. The UK is working on crypto-asset regulations, the Bank of England has a working group for CBDCs and the EU is introducing MiCA. The world is alive to the potential of blockchain applications.

Third, communication. Maintaining an open dialogue with regulators to get ahead of and address any concerns or inquiries related to the blockchain application will prevent delays in deployment.

12.4.4 Smart contract design and dispute resolution mechanisms

The success or failure of a blockchain application comes down to what it does and how it does it. Does what it does solve a need or real-world problem? Is the best way to solve that problem with a blockchain application and smart contracts? If the answer is yes to both of those questions, then how do we ensure that what is built works, is lawful and legally enforceable?

12.4.4.1 Legal enforceability

Smart contract design should be such that it is legally enforceable. What does this mean?

First, ensure we are talking about the same thing and the right thing: the contract, the smart legal contract or the smart contract. Parties must be 'of one mind' to ensure that what is built is what was intended to be built, putting aside the inevitable future pivots.

Second, (i) a smart contract must, as a matter of function, fact and the implementation of an underlying 'real' contract (smart legal or not), execute the exchange of value as expected and as intended, (ii) a smart legal contract must, regardless of the scale of hybridisation, incorporate all key contract principles, regulatory and jurisdictional requirements, as well as ensure any automated execution of terms and conditions produces the expected and intended effect and (iii) a contract must, frankly, be a contract.

Third, it sounds simple to say but, clearly defining the rights, obligations and conditions of the parties that are party to the arrangement is critical. Often starting with a blank sheet of paper and writing or drawing what 'should' happen will illuminate pitfalls along the way and help precisely define expectation and intention.

Fourth, seek legal advice. Again, it sounds simple to say, but ensuring that the input and output of a smart contract and that the language and provisions expressed in a smart legal contract in fact align with legal requirements and, as such, would be enforceable in court.

Fifth, this may seem as though I am advocating lawyers should start reviewing code, or at least reviewing the natural language provisions and implementation of code, to assess outcomes based on specified inputs. The answer is, yes, that is exactly what I am proposing. How else can a blockchain application, or any other similar non-blockchain implementation, de-risk? Lawyers should embrace technology and see how it can improve their own work and the wider profession.

12.4.4.2 Code review and testing

At the risk of continuing to state the obvious, review what has been written and test it. Find willing volunteers to mercilessly attempt to derail the entire application. Often, more so than not, it is when we break things that we understand how they actually work and how to improve them. Learn by doing, learn by failing, but always persevere.

One expression of the Silicon Valley startup mantra is 'fail fast, fail often'. I agree with the intent, if disagree with the interpretation. Be agile, be flexible, learn fast, so as not to fail often.

First, conduct code reviews and testing of the implemented elements to identify vulnerabilities, bugs or errors that produce unexpected or unintended outcomes. These could, as an inaccurate expression of the intention of the contract, create legal risks and lead to disputes.

Second, on an informal level ask your community for help, ask peers to review code, and, subject to IPR protection, an AI review tool could also be employed.[71] All of this is to narrow the risks. Alternatively, and as an application matures, on a formal level, engage third-parties, auditors or security experts to review, test and validate the contract implementation for security, accuracy, reliability, repeatability and any other 'abilities' needed.

Third, beware the edge cases. Colloquially when it comes to IT, there exists edge user problems.[72] Fair or unfair, there is an unpredictability as to how users will interact with a new application and will try and do surprising things, such as entering incorrect data, mistyping inputs, clicking the wrong button at the wrong time, the list is endless. Evaluating these scenarios will help ensure robust functionality and mitigate potential risks, preventing the 'garbage-in-garbage-out', 'sea-of-nonsensical-data'.

12.4.4.3 Security and privacy

When people think about contracts, it evokes notions of security, certainty, that the power of the law is looking after them and their interests. The formality of 'executing' a contract, signing your name, demonstrating a commitment to the

[71] Although any project should beware of the risks inherent with this.

[72] Colloquially known in IT circles as: PEBCAK – 'problem exists between chair and keyboard', PICNIC – 'problem in chair not in computer' or 1D-10-T: 'idiot' failures.

principles agreed are all, and are all designed, to impart a sense of elevated importance.

Is the same feeling evoked when people think about the applications on their computers or smartphones? I am not sure (especially if you just button bash your way through the legals to get into the app as quickly as possible). Make access easy, definitely, but some interaction to confirm the gravity and seriousness may incentivise better behaviours.

First, any application, not just a blockchain one, should implement robust security measures to protect its operations from unauthorised access, tampering and nefarious actors.

Second, beyond the general system security, consider using independent encryption techniques or other security protocols to secure sensitive (both personal and commercial) data, within the contract and as may be collected, processed or produced by it.

Third, consider privacy and data privacy (each as distinct from security), to ensure that any personal, or other personally sensitive data, is appropriately protected, with the barest minimum collected, used, processed and stored, in full compliance with relevant data protection laws.

12.4.4.4 Flexibility and upgradability

The law changes, evolves, step-changes sporadically and shifts in unpredictable directions. Planning for the future and designing in legal flexibility will save hours of wasted development.

First, flexibility. Contracts should be designed for flexibility, upgradability or whole replacement to accommodate changes in legal requirements, or more immediately likely, pivots in business needs.

Second, modification. Mechanisms that enable modification or amendment of contract terms without compromising their integrity or disrupting ongoing transactions should be implemented. I understand that this is technically difficult, may require chain-forking and so on, but could system design allow for networks of interacting smart contracts to accommodate this?

Third, communication. Changing an ongoing contract after execution can require significant legal time to be expended. Modifications to contracts must be communicated transparently to all parties involved, who may or may not accept them even if they are necessary to the intermediating function that enables the transactions.

12.4.4.5 Dispute resolution

It is likely that the most significant use of any contract is for enforcement of promises made under it. Parties want to ensure that in the event one party does not live up to their end of the bargain, the other can force them to do so or recover their losses (damages) for that failure.

Dispute resolution comes in many forms, each with their own application suitable to certain circumstances. Particular to the built environment are additional unique alternative dispute resolution requirements. For construction contracts, adjudication is a statutory mandated form of alternative dispute resolution. Construction adjudication is permitted at any time on the occurrence of a dispute or 'difference' between parties to a contract. For real estate, the land and property tribunal is the first referral forum. The purpose of this section is not to explore the specifics of each form of alternative dispute resolution, but to provide a general case or framework of understanding.

Most contracts provide contractual rights, or obligations, to attempt some form of alternative dispute resolution before the courts become involved. To manage this, specific clauses are included to manage the escalation of a dispute through various forums. The intent is to resolve a dispute early, with minimal cost and time consumed, to avoid catastrophically expensive litigation.

First, forum. Provisions must be included in contracts that obligate parties to engage in a form or forms of escalating alternative dispute resolution, usually negotiation, followed by mediation,[73] followed by adjudication or arbitration,[74] followed lastly by referral to the courts for a final decision.

Second, an alternative to the alternatives. A novel approach may be to explore the use of online dispute resolution platforms that themselves leverage blockchain technology to provide efficient and transparent dispute resolution [70].[75] These platforms may facilitate the resolution process by automating certain steps, provide a secure and immutable record of the dispute and its resolution. Such platforms are in their infancy, but represent an interesting possibility for faster, and cheaper, dispute resolution.

Third, hybrid. We have HCs, why not hybrid dispute resolution. A hybrid dispute resolution model would combine traditional methods with blockchain-based mechanisms. For example, a traditional arbitration could be supported by blockchain technology to enhance transparency of information between the parties and enforceability of the decision.

Fourth, enforcement. Dispute success is pyrrhic without the award that comes with the decision. Any contract should include provisions for the enforcement of an outcome. Automated and automatically executing mechanisms for the decision or that facilitate the transfer of awarded assets could be incorporated into the system's contracts design to enable this.

Fifth, phoenix. As a phoenix from the ashes so can the project be reborn. Rarely is careful thought given to what follows, what lessons may be learned, what changes may be made. Can the system evolve to avoid further or future disputes? Disputes are inherently difficult and negatively experienced, even for the victors.

[73]Mediation is a forum in which a neutral third party facilitates negotiations to identify mutually acceptable solutions. These can be much more time and cost efficient, and so appealing from a business expediency point of view. However, a settlement agreement may not necessarily be binding, the forum may not be suitable for complex questions of law and it requires the parties to be proactive, pragmatic and compromise.

[74]Arbitration is a formal forum in which a dispute is submitted to an arbitrator or panel of arbitrators for a binding decision. The process parallels a formal court proceeding and can be more expensive, but the decisions remain confidential, whereas court decisions are public (justice must be seen to be done).

[75]See, for example, as referred in Section 12.2.6.1.

This may be assuaged if useful, and valuable learning outcomes are identified and positive change is implemented.

12.5 Conclusion

Legal journeys, like rollercoasters, are full of twists, turns and the occasional loop-de-loop. Properly navigated, and advised, legal frameworks act to support and enhance a blockchain application, it would be 'compliant-by-design'. Connect this to the inherent transparency, immutability, robustness, democratising and disintermediating nature of blockchain and we have kismet!

Tables 12.1 and 12.2 summarise the previously described legal risks and best practices.

12.5.1 Summary of legal risks

By implementing best practices, organisations can mitigate legal risks associated with blockchain applications. It is essential to adapt and tailor these practices to the specific context, legal frameworks and requirements of each project to ensure comprehensive risk mitigation and compliance. Consulting with legal professionals,

Table 12.1 Summary of legal risks

Legal risks	Description
Intellectual property	Infringement, copying or unauthorised use of proprietary blockchain platforms, code or functionalities can lead to copyright and potentially other intellectual rights infringements.
Data breaches and privacy	Inadequate security measures or vulnerabilities in blockchain applications can result in data breaches, leading to legal consequences related to data protection and privacy laws. A breach of data protection can now run to fines of €20 million or 4% of global turnover.
Unauthorised modifications and ownership disputes	Cyber-attacks targeting blockchain-based databases, such as property title registries, asset ownership records or containing other sensitive and valuable data, can result in unauthorised modifications of ownership records, leading to claims for negligence and inadequate protection of property rights.
Contract disputes	Disputes arising from the interpretation of smart contracts, and smart legal contract terms can lead to delays, financial disputes and legal actions seeking remedies and resolutions.
Jurisdictional challenges	Differences in legal frameworks and jurisdictional challenges can create uncertainties regarding contract enforcement and dispute resolution when operating blockchain applications across borders.

Table 12.2 Summary of legal risks

Best practices	The five principles for this practice
Risk mitigation strategies	• conduct thorough due diligence and risk assessments, • implement robust cybersecurity, • ensure regulatory compliance with law and regulation, • establish clear contract terms and dispute resolution mechanisms and • conduct regular audits and assessments.
Regulatory compliance and legal considerations	• stay up to date on law and regulation, • implement KYC/AML procedures to prevent fraudulent activities, • comply with data protection and privacy laws, • adhere to property and construction regulations and • engage with regulatory authorities for guidance and approvals.
Smart contract design	• ensure smart contracts are legally enforceable, • review and test smart contract code for vulnerabilities, • include dispute resolution mechanisms in smart contracts, • establish clear governance frameworks, and • accommodate future legal changes.
Dispute resolution mechanisms	• avoid disputes, • if unavoidable, incorporate escalation mechanisms to include mediation, arbitration, and/or online dispute resolution, • explore hybrid models that combine traditional and blockchain-based dispute resolution methods, • in any event streamline the process as far as possible and • failing avoidance, ensure enforceability of dispute resolution outcomes.

specialising in blockchain, technology and the built environment, can provide invaluable guidance throughout development and deployment.

12.5.2 Outlook on legal risks

The law is as ever changing as the sands and dunes of the desert. It may seem immobile, permanent, but it shifts grain by grain in small and subtle movements blown by the winds of change.

12.5.2.1 Evolving regulatory landscape

As blockchain technology continues to advance and gain traction in the built environment, regulatory frameworks and legal considerations will also evolve.

New technology is a harbinger of the unknown, of unintended consequences, of unexpected uses. Governments and regulatory authorities may introduce new laws, guidelines and standards specific to blockchain applications, addressing issues such as data protection, privacy, property rights and contract enforcement. The Law Society, the Law Commission, the FCA and industry bodies have produced many reports and commentaries over the last decade or so opining on issues, analysing the space and indicating their preferred 'directions of travel'. The government has 'crypto' working groups and lobbyist bodies have sprung up to 'encourage' favourable (to them at least) regulations. It will be crucial for organisations to stay informed about these developments and adapt their blockchain applications to comply with the changing regulatory landscape.

12.5.2.2 Data protection and privacy challenges

With the increasing use of blockchain applications in the built environment, the collection and processing of sensitive data will become increasingly prevalent. This raises concerns about data protection and privacy. Organisations will need to carefully navigate legal requirements, such as the GDPR, equivalent local implementations of data laws and other data protection regulations, to ensure that personal and sensitive data is handled securely and in compliance with applicable laws. Privacy-enhancing technologies and techniques, such as zero-knowledge proofs and decentralised identity solutions, may play a significant role in addressing these challenges.

12.5.2.3 Interoperability and standardisation

The built environment involves multiple stakeholders, including contractors, architects, manufacturers and suppliers, financers, purchasers, tenants, owners, operators, governing bodies, regulators, the list goes on, and each of whom may need, and ultimately use, different blockchain platforms or applications to suit their particular business requirements. Achieving interoperability and standardisation of communication among these systems will be essential to ensure seamless data exchange and contract execution. Legal risks may arise if the interoperability and integration of different blockchain applications are not properly addressed, potentially leading to inconsistencies, corrupted data, regulatory non-compliance or, in a worst case, disputes.

12.5.2.4 Smart contract governance and liability

As smart contracts become more prevalent in the built environment, challenges related to their governance and liability may arise. Determining the legal status and enforceability of smart contracts and smart legal contracts, as well as liability for errors or failures in their execution, will be an important area of legal consideration. As legal challenges and disputes rise, it will be incumbent on law makers to legislate and regulate to protect users and participants. In the meantime, organisations will need to establish clear governance frameworks, contract terms and dispute resolution mechanisms to mitigate risks and provide legal certainty in the event of disputes or contractual issues.

12.5.2.5 Tokenisation and asset ownership

Tokenisation enables the representation of real-world assets, such as properties or infrastructure projects, on blockchain networks. However, legal complexities may arise concerning the ownership, transfer and legal status of tokenised assets. Questions regarding property rights, regulatory compliance and jurisdictional challenges may require further legal clarification and frameworks to ensure the smooth adoption and operation of tokenised assets in the built environment. As for the section above, law makers may be required to intervene to regulate the activities of blockchain applications and the wider ecosystem and/or recognise standards for physical-digital interconnection.

12.5.2.6 Liability and accountability

In blockchain applications involving multiple participants and stakeholders, questions of liability and accountability may arise. Determining responsibility for errors, omissions or breaches in blockchain-based systems, particularly in complex projects, may require legal frameworks and contractual provisions to allocate responsibility and address potential disputes.

12.5.2.7 Summary of outlook on legal risks

Overall, the future is bright. Legal risks in blockchain applications in the built environment will be a dynamic and evolving landscape. Organisations will need to actively monitor legal developments, engage with regulators and collaborate with legal professionals specialising in blockchain and the built environment to navigate these risks effectively. By staying informed, engaged and proactive, organisations can position themselves to leverage the benefits of blockchain technology whilst managing legal risks and ensuring compliance.

12.5.3 Final thoughts and recommendations

Blockchain technology and its legal implications are continuously evolving. It is crucial for stakeholders to stay informed about the latest developments, regulatory changes and best practices in blockchain and the built environment. Engage with legal professionals who specialise in blockchain and the relevant industry to gain insights and guidance on legal risks, compliance requirements and contractual obligations.

12.5.3.1 Conduct comprehensive risk assessments

Before implementing a blockchain application, conduct thorough risk assessments to identify potential legal risks and develop mitigation strategies. Consider all aspects, including contract disputes, IPR, data protection, regulatory compliance and jurisdictional challenges. This will help stakeholders proactively address and minimise legal risks throughout the project lifecycle.

12.5.3.2 Engage and collaborate with regulators

Engaging with regulators and industry-specific regulatory bodies is crucial to gain clarity on legal requirements and compliance expectations. Actively participate in industry discussions and collaborate with regulators to contribute to the development of regulatory frameworks and standards for blockchain applications in the built environment. Open lines of communication with regulatory authorities can help address any concerns, seek guidance and ensure compliance.

12.5.3.3 Implement robust security measures

Given the potential for cyber-attacks and data breaches, stakeholders must prioritise cybersecurity and data protection. Implement robust security measures, such as encryption, access controls and regular security audits, to safeguard sensitive data and protect against unauthorised access or modification. Consider industry best practices and engage with cybersecurity experts to enhance the security posture of blockchain applications.

12.5.3.4 Establish clear governance frameworks

Clearly define governance mechanisms, roles and responsibilities within the blockchain application. Establish transparent rules for decision-making, dispute resolution and contract enforcement. Smart contract and smart legal contract design should incorporate appropriate mechanisms for consensus, upgrades and dispute resolution. Consult legal professionals to ensure the governance frameworks align with legal requirements and provide clarity for all participants.

12.5.3.5 Continuously monitor legal developments

The legal landscape surrounding blockchain and the built environment is evolving rapidly. Stay updated on new laws, regulations and case precedents that may impact blockchain applications. Regularly review and update contractual agreements, terms of service and privacy policies to ensure ongoing compliance with legal requirements and to address any emerging risks or obligations.

12.5.3.6 Foster collaboration and industry standards

Collaboration among stakeholders is essential to address common legal challenges and promote industry-wide best practices. Engage in industry associations, consortiums or working groups to share knowledge, experiences and lessons learned. Working together, stakeholders can establish industry standards, guidelines and frameworks that enhance legal certainty, interoperability and the overall success of blockchain applications in the built environment.

By adopting best practices and these recommendations, stakeholders can navigate the legal landscape, mitigate risks and foster a secure and compliant ecosystem for blockchain applications in the built environment. By proactively addressing legal risks and collaborating with legal experts and regulators, stakeholders can unlock the full potential of blockchain technology whilst ensuring legal compliance and protecting the interests of all parties involved.

References

[1] Wikipedia. Mt. Gox. https://en.wikipedia.org/wiki/Mt._Gox [Accessed 11 October 2024], Investopedia. What was Mt. Gox? Definition, history,

collapse and future. https://www.investopedia.com/terms/m/mt-gox.asp [Accessed 11 October 2024].
[2] Reiff, N, Rasure, E, and Velasquez, V. The collapse of FTX: What went wrong with the crypto exchange? https://www.investopedia.com/what-went-wrong-with-ftx-6828447 [Accessed 11 October 2024].
[3] *Blockchain: Legal & Regulatory Guidance*. 2nd ed. London: The Law Society and Tech London Advocates; 2022.
[4] Diagrams and explanations may appear elsewhere in this book, but for an example see *Centralized vs Decentralized vs Distributed Network: An Overview*. Blockchain Engineer; 2023. https://blockchainengineer.com/centralized-vs-decentralized-vs-distributed-network/ [Accessed 11 October 2024].
[5] Nakamoto, S. *Bitcoin: A Peer-to-Peer Electronic Cash System*. Bitcoin. https://www.bitcoin.org; 2008.
[6] *Pseudonymity and Anonymity: Be Untraceable in the Blockchain World*. ImmuneBytes; 2023. https://www.immunebytes.com/blog/pseudonymity-and-anonymity-be-untraceable-in-the-blockchain-world/ [Accessed 11 October 2024].
[7] *Blockchain for Lawyers* eBook. 1st ed. Australia: Legaler; 2018.
[8] Buterin V, co-creator of the Ethereum blockchain, once tweeted that he'd have preferred 'smart contracts' to be called something more 'boring and technical' like 'persistent scripts'. 2018. https://bitcoinexchangeguide.com/ethers-vitalik-buterin-regrets-using-nick-szabos-smart-contract-term-likes-persistent-scripts/ [Accessed 30 June 2023].
[9] *Smart Legal Contracts: Advice to Government*. CP 563, Law Com No. 401. London: Law Commission; 2021.
[10] Mendelson, B. *Smart Legal Contracts: Explanatory Graphic*. London: LexisNexis and Founders Law; 2022 Maintained. [Accessed 30 June 2023].
[11] Mendelson, B. *Smart Legal Contracts*. London: LexisNexis and Founders Law; 2022 Maintained. [Accessed 30 June 2023].
[12] Financial Services and Markets Act ('FSMA') 2000 (Regulated Activities) Order 2001 (SI 2001/544), Part III.
[13] FSMA 2000, s.22 et seq.
[14] Penzes, B. *Blockchain Technology in the Construction Industry, Digital Transformation for High Productivity*. ICE, Institution of Civil Engineers; 2018. http://dx.doi.org/10.13140/RG.2.2.14164.45443.
[15] Maciel, A, Garbutt, L, Saxon, R, Johnson, G, Papadonikolaki, E, and Campbell-Turner, B. *Blockchain and Construction Cash Flow White Paper*. London: Construction Blockchain Consortium; 2020.
[16] Income Craze. *Make passive income with Tesla's Robotaxi network*. Published by Medium; 2021. https://incomecraze.medium.com/make-passive-income-with-teslas-robotaxi-network-80bfca249c56 [Accessed 11 October 2024].
[17] Trade Secrets (Enforcement, etc.) Regulations 2018 (SI 2018/597).
[18] The General Data Protection Regulation 2016/679 and, as enacted into UK law, Data Protection Act 2018.

[19] HGCRA 1996, as modified by the Local Democracy, Economic Development and Construction Act 2009.
[20] Regulation 2020/0265 (COD) for Markets in Crypto-assets, amending Directive (EU) 2019/1937.
[21] Unfair Contract Terms Act 1977.
[22] Consumer Rights Act 2015.
[23] Consumer Credit Act 1974, as amended by Enterprise Act 2002 and Consumer Credit Act 2006.
[24] Competition Act 1998 and Enterprise Act 2002.
[25] See, for example, Microsoft Corp. v Commission (2007; T-201/04), Google LLC v Commission (2021; T-616/17), Intel v Commission (EU:T:2014:547, T-286/09), and Apple Inc. v Commission (2020; AT.40437).
[26] Commodity Futures Trading Commission v Ooki DAO, case no. 3:22-cv-05416-WHO.
[27] Builderium. https://builderium.io or https://builderium.com [both no longer accessible].
[28] Investopedia. What is a reverse auction? How it works, example, and risks. https://www.investopedia.com/terms/r/reverse-auction.asp [Accessed 11 October 2024].
[29] Oracle. Construction and engineering products and Oracle blockchain. https://www.oracle.com/uk/construction-engineering/products/ and https://www.oracle.com/blockchain/ [Accessed 11 October 2024].
[30] Latium. https://latium.org/ [Accessed 11 October 2024].
[31] Structor. https://www.structor.io [no longer accessible].
[32] Build Coin Foundation. https://www.buildcoinfoundation.org [Accessed 11 October 2024].
[33] Lendledger. https://www.lendledger.io [no longer accessible].
[34] BlockArray and FreightTrust. https://www.blockarray.com and https://freighttrust.net [both no longer accessible].
[35] Circulor. https://www.circulor.com [Accessed 11 October 2024].
[36] Provenance. https://www.provenance.org [Accessed 11 October 2024].
[37] IBM. Food Trust. https://www.ibm.com/products/supply-chain-intelligence-suite/food-trust [Accessed 11 October 2024].
[38] TradeLens. https://www.tradelens.com [Accessed 11 October 2024].
[39] Open Mineral. https://openmineral.com [Accessed 11 October 2024].
[40] Everledger. https://everledger.io [Accessed 11 October 2024].
[41] See, for example, CryptoKitties. 2017. https://www.cryptokitties.co [Accessed 11 October 2024].
[42] See, for example, Bored Ape Yacht Club. 2021. http://www.boredapeyachtclub.com [Accessed 11 October 2024].
[43] ATLANT. https://atlant.io (the site no longer supports the project) [Accessed 11 October 2024].
[44] Mattereum. https://mattereum.com [Accessed 11 October 2024].
[45] Coadjute. https://coadjute.com [Accessed 11 October 2024].
[46] ChromaWay. https://chromaway.com [Accessed 11 October 2024].
[47] Ubitquity. https://www.ubitquity.io [Accessed 11 October 2024].

[48] Propy. https://propy.com [Accessed 11 October 2024].

[49] See, for example, DigiShares or RedSwan CRE. https://digishares.io and https://redswan.io. [Both accessed 11 October 2024].

[50] CNN. *He tried to cash in on the NFT craze by auctioning a house. It didn't work.* https://edition.cnn.com/2021/04/27/success/nft-real-estate-auction-california-home/index.html [Accessed 11 October 2024].

[51] Osbourne v (1) Persons Unknown and (2) Ozone Networks Inc t/a OpenSea [2022] EWHC 1021 (Comm).

[52] Seoul Metropolitan Government. *Metaverse Seoul, the new continent of Seoul.* https://english.seoul.go.kr/policy/smart-city/metaverse-blockchain/ [Accessed 11 October 2024].

[53] Monetary Authority of Singapore. *Project Ubin: central bank digital money using distributed ledger technology*. https://www.mas.gov.sg/schemes-and-initiatives/Project-Ubin [Accessed 11 October 2024].

[54] Monetary Authority of Singapore. *Ubin+: advancing cross-border connectivity with wholesale digital currencies*. 2022. https://www.mas.gov.sg/schemes-and-initiatives/ubin-plus [Accessed 11 October 2024].

[55] Field Actions Science Reports. *Barcelona's smart city vision: an opportunity for transformation*. Special Issue 16, 2017, pp. 70–75. https://journals.openedition.org/factsreports/4367 [Accessed 11 October 2024].

[56] Ajuntament de Barcelona. *Barcelona digital city: map of resources to boost digitalisation in the city*. https://ajuntament.barcelona.cat/digital/en/about-us [Accessed 11 October 2024].

[57] LO3 Energy. https://lo3energy.com [no longer accessible].

[58] Grid Singularity. https://gridsingularity.com [Accessed 11 October 2024].

[59] PowerLedger. https://www.powerledger.io [Accessed 11 October 2024].

[60] Enel Green Power. https://www.enelgreenpower.com [Accessed 11 October 2024].

[61] Electron. https://electron.net [Accessed 11 October 2024].

[62] Dovu. formerly https://dovu.io, now https://dovu.earth [Accessed 11 October 2024].

[63] Mobi. https://dlt.mobi [Accessed 11 October 2024].

[64] OriginClear. https://www.originclear.com [Accessed 11 October 2024].

[65] WaterRecharge. https://waterrecharge.org [no longer accessible].

[66] Obar JA, and Oeldorf-Hirsch A. The biggest lie on the Internet: ignoring the privacy policies and terms of service policies of social networking services. *Information, Communication & Society*. 2020;23(1):128–147.

[67] Soleymani v Nifty Gateway LLC [2022] EWCA Civ 1297.

[68] Wirex Ltd v Cryptocarbon Global Ltd [2021] EWHC 617 (IPEC).

[69] Crypton Digital Assets Ltd v Blockchain Luxembourg SA [2021] EWHC 1172 (Ch).

[70] Kleros. https://kleros.io [Accessed 11 October 2024].

[71] Orlowski J, The Social Dilemma. 2020.

[72] University of Cambridge. *Darwin Correspondence Project: the evolution of a misquotation*. https://www.darwinproject.ac.uk/people/about-darwin/six-things-darwin-never-said/evolution-misquotation [Accessed 11 October 2024].

Chapter 13

Conclusions: cross-cutting themes, challenges and recommendations

Mohamad Kassem[*], *Abel Maciel*[†] *and Daniel Hall*[‡]

This concluding chapter summarises the key insights presented throughout this book, connecting individual contributions to a broader discourse on blockchain's role in addressing inefficiencies in the architecture, engineering, construction and operations (AECO) industry. It highlights blockchain's foundational principles, socio-technical dimensions and applications such as supply chain management, construction contract administration (CCA), property tokenisation and integration with digital twins and IoT systems.

Cross-cutting themes are identified, including blockchain's potential to enhance transparency, decentralise governance, support lifecycle management and enable tokenisation for inclusive and sustainable practices. This chapter also examines key challenges to adoption, such as legal and regulatory hurdles, scalability and performance limitations, integration with legacy systems and stakeholder resistance. These challenges are contextualised within the fragmented and complex nature of the built environment. Additionally, the chapter explores the dynamics of innovation ecosystems, analysing how open, closed and hybrid models influence the scalability and diffusion of blockchain in the AECO sector.

This chapter adapts recommendations from the World Economic Forum's (WEF) blockchain framework, tailoring them to support blockchain adoption within the AECO sector. These include strategies for stakeholder engagement, regulatory alignment, pilot projects and the development of interoperable and scalable systems.

Finally, the chapter calls for a collective effort among academia, industry and policymakers to position blockchain not as a standalone solution but as a catalyst for innovation, inclusivity and resilience, extending its impact beyond the built environment.

[*]School of Engineering, Newcastle University, UK
[†]Bartlett School of Architecture, University College London, UK
[‡]Department of Management in the Built Environment, Faculty of Architecture and the Built Environment, Delft University of Technology (TU Delft), Netherlands

13.1 Synthesis of chapters

This book explored the transformative potential of blockchain technology in the built environment through 11 core chapters. Each chapter addressed a distinct aspect of blockchain's application, from foundational concepts to practical implementations and sector-specific challenges. A summary of the scope and contributions of each chapter is included in the following paragraphs.

Chapter 2 provides a foundational understanding of blockchain technology and explores its application in construction projects. It provides the technical foundations of blockchain technology, defining distributed ledger technologies (DLT) and highlighting blockchain's key attributes such as immutability, transparency and security. It categorises blockchains into permissionless and permissioned types, comparing platforms like Ethereum and Hyperledger Fabric. The chapter introduces smart contracts as programmable tools for automating processes and discusses oracles as connectors between blockchain and real-world data. The chapter also discusses oracles and their significance in blockchain-based applications. It categorises oracles as human, software and hardware, providing examples of how different types of oracles are utilised in construction projects. By showcasing examples in tendering, material management and energy monitoring, it illustrates blockchain's potential to enhance transparency and efficiency in the built environment.

Chapter 3 introduces the socio-technical systems (STS) theory as a lens to explore blockchain adoption in the construction sector. It outlines how blockchain technology, characterised by its transparency, immutability and decentralised nature, addresses long-standing inefficiencies in construction. However, the chapter emphasises that blockchain adoption requires more than technical implementation – it must consider the interplay between technological and social factors. The chapter reviews STS theory's origins, its relevance to technological innovation and its application to the built environment, particularly digital technologies like building information modelling (BIM). It highlights conceptual models like the multi-level perspective (MLP) and presents a socio-technical framework tailored for blockchain adoption, combining dimensions of technology, process, policy and society. By providing a roadmap and specific models for blockchain integration, Chapter 3 establishes a comprehensive approach to navigating adoption barriers and leveraging blockchain's potential to transform construction processes and relationships among stakeholders.

Chapter 4 addresses the challenges of ensuring data security in AECO projects, particularly in the context of increasingly digital workflows and collaborative environments. It emphasises the importance of securing BIM models, given their rich visual and semantic data content. The chapter explores the principles of data security, including confidentiality, integrity and availability, outlining their relevance to AECO projects. It discusses various approaches to data storage, advocating for the use of distributed technologies like blockchain and the interplanetary file system (IPFS) to enhance data security. A universal distributed framework is proposed, integrating these technologies to facilitate secure data sharing, lifecycle transparency and selective data visibility while maintaining functionality. By addressing the interlinked facets of data confidentiality, digital trust and data utility, the chapter sets a foundation for more secure and efficient data management practices in AECO projects. It also outlines potential future research directions, including the integration of deep learning models for sensitive data classification and improved inclusivity for smaller project stakeholders.

Chapter 5 focuses on the tokenisation of property rights and its implications for the real estate sector, emphasising its ability to enhance liquidity, accessibility and efficiency. It highlights the potential benefits of tokenisation such as reduced transaction costs, increased market accessibility and fractional ownership. It outlines the steps involved in tokenising property, emphasising the need for secure data management and identity verification. The chapter proposes a property tokenisation management system, detailing its key components and functionalities. It emphasises the importance of data organisation within this system, including user accreditation, property documentation management and token and smart contract association. It also describes the system's reporting capabilities, its data warehousing functionalities and the integration of GIS tools for property visualisation and urban analytics.

Chapter 6 introduces the concept of digital building logbooks (DBLs) and their role in facilitating a circular economy within the built environment. It underscores the importance of DBLs in capturing and managing building information throughout its lifecycle, including design, construction, operation, maintenance and demolition. It discusses the challenges of DBL implementation, particularly the need for secure and reliable data storage, integration of legacy information and user-friendly interfaces. The chapter proposes a blockchain-based DBL architecture, outlining a three-layered system consisting of a data layer, a logic and services layer and a user interface layer. It advocates for a hybrid data storage system, combining decentralised storage using the IPFS with centralised databases for legacy information. The chapter concludes by emphasising the potential impacts of blockchain-enabled DBLs, including fostering circular economies, advancing energy-efficient renovations and creating decentralised data marketplaces. It also identifies future research directions such as developing standardised frameworks, integrating legacy data and aligning blockchain-based solutions with legal and regulatory frameworks. This vision positions DBLs as critical tools for driving transparency, efficiency and sustainability in the built environment.

Chapter 7 focuses on the application of blockchain technology to facilitate renewable energy integration within seaport microgrids. It provides a practical example of how blockchain can be used to manage energy trading among buildings within a microgrid, using the Port of Milford Haven in South Wales as a case study. The chapter investigates the scalability and potential viability of using Ethereum smart contracts to prioritise renewable energy trading within the microgrid, examining the impact of Ethereum gas prices on transaction costs. It describes the energy consumption and production patterns of the buildings included in the case study, detailing their specific energy needs and the presence of renewable energy sources. Results indicate that while blockchain can enhance energy efficiency and transparency, challenges such as Ethereum's price volatility, high gas fees and public adoption barriers must be addressed. The findings underscore blockchain's potential to decentralise energy systems, improve renewable energy adoption and foster community engagement. However, the chapter also calls for further research into private blockchain solutions, legal frameworks and the scalability of blockchain-enabled microgrids in energy-intensive industries.

Chapter 8 explores the application of blockchain and distributed technologies to support co-authorship and collaborative design within architectural practice. It acknowledges the shift from sole authorship to shared effort across disciplines in contemporary architectural projects, emphasising the need for effective communication protocols and knowledge-sharing mechanisms. It reviews existing research on collective authorship in architectural design, highlighting key themes such as remote collaboration, communication protocols and knowledge sharing. The chapter draws connections between stigmergic principles and collaborative design, evaluating research articles based on their incorporation of communication, feedback loops, decentralisation and the use of a shared substrate. Using practical prototypes such as blockchain-integrated BIM and DAO frameworks, the chapter illustrates how decentralised systems can facilitate design optimisation, shape grammar applications and project governance. It also addresses challenges such as trust, scalability and the need for robust data infrastructure, proposing innovative solutions like 'soulbound' tokens for skill recognition and incentivisation. By combining technological insights with socio-technical principles like stigmergy and the viable system model, the chapter proposes a vision for transforming architectural practice into a non-extractive, community-driven discipline. Finally, the chapter underscores the importance of further research and experimentation to unlock the full potential of blockchain in design governance.

Chapter 9 focuses on applying blockchain technology to enhance construction supply chain management (CSCM). The chapter first identifies challenges within current CSCM approaches, including fragmented information sharing, lack of trust among stakeholders and the absence of a single source of truth. To address these issues, the chapter presents a blockchain-based CSCM system implemented in a real-world case study involving modular construction between Hong Kong and Mainland China. This pilot project demonstrates how blockchain improves supply chain transparency, ensures data immutability and facilitates efficient quality control processes. Additionally, the chapter provides practical guidelines for implementing blockchain systems, emphasising key considerations like stakeholder management, scalability and legal compliance. While the chapter underscores blockchain's transformative potential in CSCM, it also highlights unresolved challenges such as scalability limitations, regulatory uncertainty and resistance to change. It concludes by discussing future opportunities, including integrating blockchain with IoT, AI and DAOs, to further enhance supply chain resilience and efficiency.

Chapter 10 examines the integration of BIM, digital twins and enterprise common data environments (CDEs) to address data observability challenges in large-scale infrastructure projects. It emphasises the transition from traditional, fragmented data management approaches to comprehensive enterprise-grade solutions capable of managing diverse, interconnected data sources. It highlights the need for a transition from traditional fragmented data management approaches to comprehensive enterprise-grade solutions capable of managing diverse interconnected data sources. Blockchain and DLTs are presented as tools to enhance data integrity, traceability and security and as enablers of seamless integration within enterprise CDEs. By examining real-world case studies and theoretical frameworks, the chapter underscores the need for innovative data governance strategies and advocates for the adoption of decentralised technologies to optimise project outcomes and align with enterprise objectives.

Chapter 11 explores the potential of blockchain technology to transform CCA. The chapter maps blockchain applications to 12 core CCA functions, categorised into control, coordination and adaptation mechanisms. Examples include financial management, quality assurance, document management and dispute resolution. It demonstrates how blockchain can enhance these processes by automating contractual obligations, ensuring data integrity and fostering accountability among stakeholders. Case studies and proof-of-concept implementations are used to validate blockchain's applicability, particularly in areas like payment automation, performance monitoring and document traceability. While blockchain presents opportunities for improving CCA, the chapter also identifies challenges such as replicating complex contractual logic and ensuring the authenticity of off-chain data. It concludes by advocating for further research and collaboration to address these challenges and unlock blockchain's full potential in streamlining and enhancing CCA practices.

Chapter 12 provides a detailed analysis of the legal and regulatory considerations surrounding blockchain applications, particularly within the built environment. It highlights the importance of understanding the legal frameworks governing these applications and the need to ensure their grounding in legal reality for long-term resilience. The chapter examines the legal implications of tokenisation, considering various types of tokens and their potential to represent property rights or financial instruments. It also investigates the legal aspects of smart contracts, highlighting the distinction between 'smart contracts' and 'smart legal contracts' and the potential legal risks associated with different types of smart contracts. The chapter exposes the complexities of tokenisation and fractionalisation, and the regulatory challenges of applying existing laws like the General Data Protection Regulation (GDPR) and financial regulations to blockchain-based systems. Practical examples, such as the risks of pseudonymity, asset entanglement and data immutability highlight the need for careful planning and strategic design to navigate these challenges. Based on an analysis of international regulatory approaches, the chapter provides a roadmap for stakeholders to adopt blockchain technology responsibly. The chapter stresses the importance of collaboration between legal experts, technologists and policymakers to address gaps and foster innovation while maintaining trust and accountability.

Building on these summaries and their respective full chapter, the following section analyses cross-cutting themes that emerge across the book, connecting individual chapters to broader discussions on blockchain's role in transforming the built environment.

13.2 Cross-cutting themes

This book examines the transformative potential of blockchain and DLT in addressing systemic challenges within the construction industry and the built environment. This section highlights the cross-cutting themes that emerge across the chapters. A summary can be seen in Table 13.1.

13.2.1 Blockchain as a systematic challenge

Blockchain technology, with its inherent attributes of transparency, immutability and decentralisation, offers **transformative solutions at the system level.** The built environment is challenged by long-standing inefficiencies, fragmentation into silos and trust deficits. Blockchain can offer a novel approach to these challenges, but requires a full system's deployment, which can be quite challenging. Chapter 3 argues that blockchain can automate processes via smart contracts and help to reduce or eliminate these bottlenecks such as delays in payment processing, documentation

Table 13.1 Summary of themes found in this book

Theme	**Mechanism**	**Contributing chapters**
Blockchain as a systematic solution	Addressing systematic inefficiencies, silos and trust deficits through transparency, immutability and decentralisation	2, 3, 9, 11
Blockchain and socio-technical systems	Balancing technology with social, organisational and governance factors; reconfiguring roles in construction contract administration.	3, 8, 11, 12
Integration with digital technologies	Enhancing GIS, IoT, digital twins and common data environments for traceability, transparency and project management.	5, 9, 10
Data security and digital trust	Ensuring confidentiality, integrity and availability through decentralised systems; enhancing trust via tamper-proof ledgers and smart contracts.	4, 9, 11
Tokenisation and Democratisation	Enabling fractional ownership, broader market participation and sustainability investments through tokenised assets and systems.	5, 6, 7, 12
Collaborative and Decentralised Governance	Decentralising decision-making and enhancing stakeholder coordination, shared authorship and contract administration.	3, 8, 11
Sustainability and circular economy	Promoting green technologies, lifecycle tracking, renewable energy trading and resource efficiency through blockchain applications.	5, 6, 7

errors and misaligned projects schedules. Chapter 9 illustrates how blockchain can bridge these silos by providing a unified platform for information sharing and collaboration in supply chain management. Chapter 11 emphasises trust deficits as a critical challenge in construction, where disputes and opportunistic behaviours are common. This book shows how blockchain's transparency and immutability offer potential system's level approaches to ensure that transactions and changes are recorded and verifiable, improving integration and trust among all stakeholders.

13.2.2 Blockchain and socio-technical environments

Blockchain implementation requires a holistic approach that **balances technological innovation with social, organisational and governance factors**. Chapter 3 explicitly situates blockchain adoption within STS theory, emphasising the interplay between technology, processes and societal factors. It advocates for a MLP that integrates dimensions such as policy, organisational culture and stakeholder dynamics alongside technical capabilities.

Chapter 11 examines blockchain's impact on CCA through a socio-technical lens, highlighting how the technology reconfigures traditional roles and interactions among stakeholders. Blockchain's transparency and immutability enhance trust and accountability, reducing opportunistic behaviours and facilitating smoother coordination. The chapter also stresses the importance of aligning blockchain solutions with existing organisational processes to ensure successful integration.

Chapter 12 contributes to the STS theme by examining how blockchain adoption in construction must align with legal and regulatory frameworks to address jurisdictional inconsistencies, operational risks and data protection laws. It highlights the integration of technological innovations such as smart legal contracts and tokenisation with solutions like cryptographic protocols and identity segregation to balance automation with compliance. The chapter explores cross-border regulatory challenges and financial frameworks, emphasising the importance of adapting blockchain applications to diverse legal and cultural contexts for operational resilience.

By framing blockchain as a STS, these chapters collectively highlight the importance of integrating human, organisational and technological dimensions to unlock the full potential of blockchain in the built environment.

13.2.3 Blockchain integration with digital technologies

The integration of blockchain with key digital technologies in construction and the built environment elevates it from a standalone solution to a *foundational element of a connected and interoperable digital ecosystem*.

Blockchain's integration with GIS is highlighted in Chapter 5, particularly in the context of property tokenisation and urban analytics. Blockchain enhances GIS by providing a secure platform for managing geospatial data linked to property assets. This integration supports applications such as urban planning, infrastructure visualisation and environmental impact assessment. Tokenised property systems benefit from this synergy as blockchain ensures the integrity of ownership data while GIS provides spatial insights for decision-making. Chapter 9 demonstrates how IoT-enabled sensors combined with blockchain enhance traceability and transparency in modular construction supply chains. Chapter 10 explores the integration of blockchain with digital twins and CDEs, highlighting the importance of real-time data observability and enhanced project management in large-scale infrastructure projects. By providing immutable data trails and ensuring secure access control, blockchain enhances the functionality of CDEs in managing complex, multi-stakeholder infrastructure projects. This integration addresses challenges such as data fragmentation, version control and unauthorised access, creating a more robust data management ecosystem.

These synergies highlight blockchain's potential as a key component of integrated digital ecosystems. This theme further underscores two key aspects: the necessity of interdisciplinary approaches to technology adoption and the imperative to address the challenges of technological complexity and scalability inherent in the integration of diverse technologies.

13.2.4 Data security and digital trust

Data security and trust enhancement are key advantages of a blockchain-based system. Chapter 4 critiques centralised data security models and advocates for decentralised systems using blockchain and encryption technologies to ensure data confidentiality, integrity and availability in AECO projects. By removing reliance on single points of failure, blockchain enhances the resilience of data systems, particularly in projects involving sensitive BIM models or other high-value data assets. Chapter 9 applies blockchain's tamper-proof ledger to enhance trust among supply chain stakeholders. Blockchain's immutability ensures that all transactions and records are verifiable, creating a single source of truth for stakeholders. This is particularly valuable in cross-border modular construction projects, where trust is often compromised due to inconsistent or inaccessible data. By providing a transparent ledger of material provenance and movement, blockchain enables seamless collaboration and reduces the risk of fraud or miscommunication. Chapter 11 highlights how blockchain-based smart contracts improve trust in CCA by automating compliance and reducing disputes. Self-executing agreements can enforce terms and conditions and enhance compliance. Securing contractual obligations on an immutable ledger can also help in reducing opportunistic behaviours and fostering reliable stakeholder relationships.

This theme suggests that blockchain can bring some key contributions to the built environment, including decentralised resilience (i.e. addressing vulnerabilities in centralised systems by decentralising control and ensuring data confidentiality, integrity and availability, Chapter 4), trust in supply chains (fostering collaboration and reducing risks through tamper-proof auditable records, Chapter 9), automated compliance (streamlining contract administration by automating obligations and reducing disputes, Chapter 11), accountability and transparency (recording actions and transactions on an immutable ledger to promote stakeholder accountability, Chapters 9 and 11) and balancing transparency and privacy (using encryption and selective visibility to ensure data privacy while maintaining secure collaboration, Chapter 4).

13.2.5 Tokenisation and democratisation of access

Blockchain has the potential to *make ownership and investment more accessible and inclusive* across the built environment, through tokenisation and democratisation of access. Blockchain can be used to reduce financial barriers and foster broader participation in markets traditionally dominated by larger investors.

Chapter 5 discusses fractional ownership of real estate assets through blockchain-based tokens, making investment accessible to a broader demographic. By converting property ownership into blockchain-based tokens, investments can be divided into smaller tradable units, making them accessible to a broader demographic. This approach reduces entry barriers for small-scale investors and promotes more inclusive participation in real estate markets. Tokenisation also enhances liquidity by creating secondary markets for trading these digital assets, aligning with the broader goals of democratising property ownership. Chapter 7 extends this concept to renewable energy microgrids, enabling community participation in energy trading through tokenised systems. Tokenised systems allow individuals and organisations to trade renewable energy units, prioritising sustainability and local energy use. Chapter 6 incorporates tokenisation into the framework of DBLs, proposing it as an approach to incentivise investment in energy retrofitting and other sustainability initiatives. Blockchain tokens linked to DBLs can represent credits for energy savings or funding opportunities for green upgrades, aligning financial incentives with environmental goals. Chapter 12 examines legal and regulatory challenges for fractional ownership can be addressed. It highlights conflicts between blockchain's immutability and data protection laws, proposing solutions like cryptographic protocols to ensure compliance. The chapter explores tokenised asset classification under financial regulations, focusing on frameworks like the EU's MiCA and addressing risks such as anti-money laundering and know your customer compliance.

These contributions suggest the potential role of blockchain in transforming traditional economic models in the built environment and contributing towards more inclusive and sustainable systems.

13.2.6 Collaborative and decentralised governance

Blockchain can enhance collaboration, accountability and shared decision-making by *decentralising traditional hierarchical processes and transforming governance structures* in the built environment.

Chapter 8 illustrates how blockchain enables decentralised governance in architectural design workflows. By decentralising authority, blockchain empowers diverse stakeholders to contribute equitably to design processes, enabling shared authorship and reducing reliance on traditional, top-down governance structures.

Chapter 3 situates decentralised governance within the broader STS framework for blockchain adoption. It highlights the importance of governance as a socio-technical dimension, particularly in BIM-integrated systems where multiple stakeholders interact. Blockchain's ability to establish decentralised data ownership and transparent decision-making mechanisms is presented as key to overcoming challenges in stakeholder coordination and trust.

Chapter 11 extends the concept of decentralised governance to CCA. Blockchain's transparency and immutability enable the decentralisation of contractual functions, such as financial management, progress tracking and quality control. This decentralised approach is expected to reduce disputes, streamline coordination and enhance compliance, transforming traditional contract administration practices.

13.2.7 Sustainability and circular economy

Blockchain can play a key future role in a more sustainable and circular built environment, by aligning financial and operational practices with environmental goals.

Chapter 5 discusses how tokenisation of real estate assets can promote energy-efficient retrofitting by facilitating investment in green technologies. Chapter 6 introduces blockchain-enabled DBLs to track building lifecycle data and support circular economy principles. Chapter 7 explores blockchain's role in renewable energy trading within seaport microgrids, emphasising its potential to decarbonise energy systems and reduce operational emissions.

These cross-cutting themes demonstrate blockchain's multifaceted impact across the built environment. They connect blockchain's applications and technical features, such as data security and interoperability, with broader societal goals, including sustainability, inclusivity and collaborative governance. By addressing these recurring aspects across the chapters, this book establishes a transformative and innovative technology within the built environment. However, the adoption and implementation of blockchain are accompanied by significant challenges that are examined in detail in the following section.

13.3 Key challenges

Although blockchain demonstrates significant potential for the built environment, as previously discussed, its widespread adoption faces some critical challenges. These challenges encompass legal and regulatory constraints, scalability and performance, integration with legacy systems, data quality and reliability, fragmentation of standards and protocols, ethical and social concerns, and stakeholder resistance and adoption barriers. These challenges underline the intricate nature of implementing blockchain in a fragmented and highly regulated sector. The following subsections provide a detailed examination of these challenges.

13.3.1 Legal and regulatory challenges

Legal and regulatory considerations constitute a significant barrier to the adoption of blockchain technology, both broadly and in relation to the specific applications discussed in this book. The implementation of tokenisation and fractional ownership, as discussed in Chapter 5, faces jurisdictional inconsistencies and the lack of standardised property and financial laws across regions. Similarly, Chapter 11 highlights the uncertainties surrounding the enforceability of smart contracts in CCA, including issues related to digital signatures, liability and dispute resolution. Chapter 12 further expands on broader regulatory concerns, including compliance with data protection laws such as GDPR, token classification and financial reporting obligations. These legal complexities underscore the importance of harmonising regulatory frameworks to support technology adoption while preserving accountability and compliance.

13.3.2 Scalability and performance limitations

Blockchain systems are often challenged by efficient scalability, particularly in applications requiring high transaction throughput or real-time data updates. Chapter 2 outlines the technical limitations of existing blockchain platforms, such as slow transaction speeds and high operational costs, which are also evident in supply chain applications (Chapter 9) and energy microgrids (Chapter 7). Improving blockchain scalability and reducing associated costs are critical steps for enabling its adoption in large-scale time-sensitive projects.

13.3.3 Integration with legacy systems

The integration of blockchain with existing legacy systems presents significant technical and operational challenges. Chapter 6 highlights difficulties in incorporating blockchain into traditional data management systems, such as DBLs, while Chapter 10 emphasises the need for interoperability between blockchain, BIM and digital twin systems. Additionally, Chapter 5 points to the challenges of aligning blockchain-based tokenisation platforms with legacy property registries and financial systems. Ensuring seamless integration is essential to leveraging blockchain's full potential while maintaining its interface with existing processes and practices.

13.3.4 Data quality and reliability

The effectiveness of blockchain systems depends heavily on the accuracy and reliability of data inputs that are often provided through external oracles (software, hardware, humans). Chapter 2 highlights the challenges of ensuring trustworthy oracle connections between off-chain sources and blockchain systems, emphasising that the accuracy of data feeds directly impacts the integrity of blockchain processes. Similarly, Chapters 9 and 11 identify data quality issues as potential risks in supply chain management and CCA, respectively, where inaccurate inputs can undermine the integrity of smart contract automation. Developing reliable oracle solutions is critical to addressing these vulnerabilities. Chapters 2 and 9 raise the need for AEC-specific oracles capable of validating and verifying data from diverse sources such as IoT devices, project schedules and material inventories.

13.3.5 Fragmentation of standards and protocols

The absence of standardised blockchain protocols creates fragmentation, impeding interoperability and scalability across platforms. Chapter 5 discusses the challenges posed by inconsistent standards in tokenised property systems. Token classification, data structures and smart contract functionalities often vary across platforms, making it difficult to establish a unified marketplace for tokenised real estate. Chapter 10 highlights similar issues facing the integration of blockchain with BIM and digital systems. Chapter 12 discusses the broader implications of fragmented standards on regulatory compliance. Blockchain systems operating under varying protocols may struggle to meet legal requirements for data protection, financial reporting and auditability (e.g. classification of tokens and smart contracts differing across jurisdictions create regulatory uncertainty and complicate cross-border implementations).

13.3.6 Ethical and social implications

Blockchain adoption raises important ethical and social considerations including issues of privacy, equity and unintended consequences. Chapter 8 highlights potential inequities in decentralised governance models, where power dynamics may still favour certain stakeholders. The chapter calls for equitable governance in decentralised governance models to ensure fair participation and prevent the consolidation of power among a few dominant stakeholders. Chapter 3 identifies the broader ethical responsibility of balancing technological innovation with social equity. Blockchain, as a STS, can inadvertently exclude stakeholders who lack access to necessary resources such as technological infrastructure, funding or expertise. The chapter calls for a socio-technical approach to blockchain adoption that prioritises education, capacity-building and stakeholder engagement to bridge these gaps and foster inclusive participation. Chapter 12 discusses the risks of privacy breaches and data misuse in blockchain systems, identifying the need for privacy-preserving technologies such as zero-knowledge proofs and selective encryption to address these risks and ensure ethical data management.

13.3.7 Stakeholder resistance and adoption barriers

Resistance from stakeholders is another critical barrier to blockchain adoption. Chapter 3 identifies the lack of pervasive awareness and the limited technical understanding of blockchain systems as a barrier causing hesitation and resistance to adoption among project participants. Chapter 11 highlights the trust deficits and competitive concerns that often accompany the introduction of blockchain. Traditional stakeholders may view blockchain systems, especially smart contracts, as disruptive to their roles or authority (e.g. loss of control and fear of transparency). Addressing these adoption barriers requires targeted education, demonstration projects and clear communication of blockchain's value propositions.

These challenges collectively highlight the multifaceted barriers to blockchain adoption in the built environment. The complexities of legal and regulatory frameworks, limitations related to scalability, challenges associated with integration and resistance from various stakeholders necessitate a unified endeavour among technologists, policymakers and industry leaders. Tackling these challenges through targeted research, collaborative efforts and the development of flexible and responsive frameworks – that respond dynamically to the unique and changing needs of the built environment – will be crucial for unlocking blockchain's potential.

13.4 Ecosystem barrier to blockchain diffusion in the built environment

Blockchain is a systemic innovation that affects entire ecosystems rather than isolated entities [1]. As a result, its success will heavily depend on the ecosystem dynamics and the effectiveness of orchestrators within the prevalent ecosystem. For instance, Papadonikolaki *et al.* [1] found that the innovation ecosystem for construction blockchain is moving towards the open innovation ecosystem. In an open innovation ecosystem, open-source consortia are identified as the preferred orchestrators due to their ability to facilitate collaboration and innovation by providing a platform for various stakeholders to contribute and share resources [1]. However, the study also identified elements of closed and hybrid innovation ecosystems for construction blockchain. In the closed ecosystem, contractors were identified as key orchestrators with responsibility for managing and controlling the innovation processes within a more restricted and controlled environment. However, contractors' effectiveness as orchestrators is challenged by: (1) narrow profit margin which makes it challenging for them to invest in and lead the blockchain innovation ecosystem. They need to ensure that any orchestration efforts lead to efficiency gains and quicker processes to justify the investment [1]; (2) inability to exert influence over the entire ecosystem [1]; and (3) they are often seen as traditional stakeholders who may view blockchain systems and smart contracts as disruptive to their roles or authority (e.g. loss of control and fear of transparency), as identified in Chapter 11. In between, the two ecosystems (i.e. closed and open), there is the hybrid ecosystem that requires orchestrators to switch between different modes of orchestration – ranging from more controlled, closed approaches to more open, collaborative ones – depending on the challenges and needs of the network [2]. While hybrid ecosystems for blockchain have challenges such as managing the balance between open and closed efforts to ensure that the ecosystem remains dynamic and responsive to changes, they offer the opportunity for orchestrators to leverage diverse capabilities and resources from both internal and external sources, fostering innovation and growth [2]. Within the hybrid construction blockchain ecosystem, Papadonikolaki *et al.* [1] identify contractors, software vendors, consortia, SMEs and insurance companies as key orchestrators, with each playing a role based on their strengths and capacities to manage and facilitate innovation.

Another significant hurdle for blockchain adoption in the AECO sector is the weak ecosystem of start-ups. Start-ups are uniquely positioned to address the complexities of technology adoption in the AECO industry [1] due to their agility in rapid prototyping and iteration of solutions. Start-ups are also best positioned to address the specialised and niche blockchain applications tailored to the intricate workflows of AECO projects. To date, research on which blockchain innovation ecosystem is best for fostering the growth of construction blockchain start-ups is insufficient. However, emerging evidence suggests that closed systems – found to be prevalent in the current construction blockchain innovation ecosystem [1] – deter the development of a vibrant start-up ecosystem. For instance, Papadonikolaki *et al.* [1] observed that although hybrid elements exist within construction blockchain innovation ecosystems, the predominant inclination towards closed systems restricts the potential for start-ups to thrive. Blockchain technologies rely on interconnected networks of developers, infrastructure providers and end-users, where start-ups play a critical role in amplifying network effects. Sadeghi *et al.* [3] suggest that establishing a supportive ecosystem, particularly one that includes active participation from start-ups, could significantly accelerate the adoption and diffusion of blockchain in the built environment. Start-ups accelerate the development of interoperable tools, standardisation efforts and shared knowledge. However, the current predominantly closed innovation ecosystem for construction blockchain appears to undermine its potential within the built environment.

Finally, blockchain adoption in the construction sector is significantly hindered by a lack of readiness for business model transformation. Challenges identified earlier in this chapter, such as legal and regulatory uncertainties and the absence of standardised frameworks, further exacerbate this issue [4,5]. Additionally, the construction sector's traditional competitive culture and inadequate levels of collaboration contribute to an environment that perpetuates the currently prevalent closed-ecosystem model, as previously discussed. Consequently, many blockchain applications remain in the early stages of prototype development. However, with the ongoing shift towards open and hybrid innovation models [1], greater scaling and commercialisation are anticipated in the near future [6].

13.5 Recommendations

Building upon the challenges and barriers identified throughout the book and synthesised in this chapter, this section presents a structured set of recommendations. These recommendations are adapted from the WEF blockchain framework [7]. While the original framework is not explicitly tailored to construction, its recommendations have been refined to address the specific context of the AECO sector (Table 13.2).

13.6 Final reflection

As blockchain technology matures, its integration into the built environment has the potential to reshape how we design, construct and manage our buildings and infrastructure. Building a functional and effective blockchain ecosystem for the AECO industry necessitates a collective effort to move beyond proof-of-concept stages.

Table 13.2 Recommendations for blockchain adoption in the AECO sector

Theme	Action items
Strategic alignment	• Identify clear objectives and use cases for blockchain implementation, ensuring the technology addresses specific challenges or pain points such as inefficiencies, lack of trust or manual errors in AECO processes. • Explore opportunities where blockchain can drive systemic and disruptive transformations such as redefining procurement models, enabling decentralised decision-making or fostering new collaborative business models that challenge traditional practices. • Conduct thorough feasibility assessments to ensure blockchain is the appropriate solution for the challenges identified.
Stakeholder engagement	• Involve relevant stakeholders including clients/employers/owners, designers, contractors, regulators and end-users early in the adoption process to align and understand expectations and foster trust. • Facilitate open discussions to ensure alignment on the value blockchain brings to each stakeholder's operations.
Regulatory and legal readiness	• Engage with regulators to clarify legal uncertainties and encourage supportive policies for blockchain adoption (e.g. evaluate existing regulatory frameworks that govern data privacy, smart contracts and blockchain usage in the AECO sector). • Collaborate with policymakers to support the development of regulations that support blockchain while ensuring accountability and data security.
Technology and interoperability	• Prioritise interoperability to ensure the blockchain system can interact seamlessly with other digital technologies (BIM and IoT) and legacy systems. • Develop standardised protocols and interoperability frameworks for oracles that can feed verified AECO data (e.g. from supply chains, construction schedules and performance monitoring systems) into blockchain platforms. • Select a blockchain platform based on specific use-case requirements such as scalability, security, data throughput and ease of integration with existing technologies like BIM and IoT.
Governance framework	• Establish governance structures tailored to blockchain networks in the AECO sector, incorporating decision-making protocols, conflict resolution mechanisms and defined participant roles. • Ensure transparency, accountability and equitable participation while enabling the alignment of network rules with existing regulatory and contractual requirements in AECO projects.

(Continues)

Table 13.2 (Continued)

Theme	Action items
Pilot projects and prototyping	• Implement pilot projects to test blockchain applications in controlled low-risk environments but real-world scenarios (project bidding, procurement or compliance tracking). • Use these pilots to identify potential barriers and refine the system.
Risk and impact assessment	• Conduct comprehensive risk assessments to identify technical, operational and reputational risks associated with blockchain deployment. • Use impact metrics (e.g. improved supply chain transparency, reduced payment delays or enhanced compliance tracking) to evaluate blockchain's contributions to operational efficiency, cost reduction and trust-building in AECO processes.
Collaboration and ecosystem building	• Foster collaboration among AECO industry stakeholders to share resources, knowledge and best practices for blockchain adoption accelerating the identification of use cases, streamlining implementation efforts and reducing duplication of resources. • Promote the development of industry consortia or working groups to drive standardisation and collective innovation in blockchain for AECO.
Performance monitoring	• Establish clear key performance indicators (e.g. measuring improvements in supply chain visibility, reduction in payment processing times, enhanced data security and compliance with project requirements) to track the blockchain system's efficiency, transparency and cost-effectiveness over time. • Collect feedback from stakeholders in AECO projects to improve and adapt the blockchain system to establish whether the blockchain solution aligns with evolving industry needs and expectations.
Scaling and long-term strategy	• Develop a long-term roadmap (with milestones, timelines and resource allocation) for scaling blockchain applications in the AECO sector, transitioning from pilot projects to broader implementation across the value chain (e.g. a staged approach implementing the system gradually to selected applications such as supply chain management, project collaboration and compliance tracking). • Use insights from pilot projects to identify best practices, address common barriers and refine strategies for deployment. Focus on building scalable, interoperable and user-friendly systems that deliver measurable benefits across all stages of the AECO value chain.

While the applications discussed in this book were primarily at the proof-of-concept stage, many were tested with real-world industry data and case studies, demonstrating significant positive impacts across several key areas: enhanced transparency and traceability (Chapters 5 and 9); improved efficiency and automation (Chapters 7 and 11); data security and reliability (Chapters 9 and 10); inclusivity and democratisation (Chapters 5, 6 and 8); lifecycle and sustainability management (Chapter 6); and trust and accountability (Chapters 9 and 11).

To translate these promising initial results into widespread industry transformation, a coordinated and collaborative approach is essential. Collaboration among industry professionals, policymakers and academic researchers – through initiatives such as fostering partnerships, developing standardised interoperability protocols, piloting projects and addressing regulatory challenges – will be crucial for successful blockchain adoption and long-term integration of blockchain technology into the built environment.

This book encourages stakeholders to regard blockchain not as a standalone solution but as an integral component of a broader collaborative effort to modernise the design, construction and operation of the physical built environment. While challenges persist, the opportunities for improvement through blockchain technology are considerable, as explored throughout this book.

This book also makes important contributions to the field of digitalisation in the AECO industry. Its originality lies in effectively bridging theoretical frameworks with practical applications of blockchain in the AECO. Key contributions include:

- **Novel frameworks** such as the socio-technical blockchain adoption frameworks (Chapter 3), decentralised governance models (Chapter 8), legal and regulatory frameworks (Chapter 12) and tokenisation frameworks (Chapters 5 and 6).
- **Emerging applications** of blockchain-enabled supply chain systems (Chapter 9), energy microgrids and renewable energy trading (Chapter 7), CCA (Chapter 11), data security in BIM and digital twins (Chapter 10), collaborative design in architecture (Chapter 8) and property tokenisation (Chapter 5),
- Detailed exploration of **legal risks and compliance challenges** (Chapter 12).
- **Novel technical approaches** involving blockchain integration with IoT, BIM and microgrid systems demonstrating the potential of the resultant solution in transforming existing practices and enabling new business models and organisational paradigms.

Finally, the findings of this book extend *beyond the scope of blockchain and the built environment*, providing valuable insights into broader discussions on digital transformation and sustainable development. These implications include fostering circular economies, enabling inclusive investment through tokenisation, enhancing trust and collaboration and promoting broader participation.

While many still argue that *blockchain is a solution in search of a problem*, this book seeks to provide a balanced perspective by integrating evidence-based insights from practical applications with acknowledgement of its limitations and challenges. In doing so, it avoids speculative claims and focuses on the real-world potential of blockchain, which, although slowly emerging, shows promise.

We hope this book motivates stakeholders across academia, industry and policy to engage actively with blockchain's potential, not just as a technology but as a future catalyst for innovation, inclusivity and resilience.

References

[1] Papadonikolaki E, Tezel A, Yitmen I and Hilletofth P. Blockchain innovation ecosystems orchestration in construction. *Industrial Management & Data Systems*. 2022;123(2):672–694. https://doi.org/10.1108/IMDS-03-2022-0134.

[2] Reypens C, Lievens A and Blazevic V. Hybrid orchestration in multi-stakeholder innovation networks: practices of mobilizing multiple, diverse stakeholders across organizational boundaries, *Organization Studies*. 2019; 42(1):61–83. https://doi.org/10.1177/0170840619868268.

[3] Sadeghi A, Vafadarnikjoo A, Borrmann A and Kerschbaum F. A framework for trustworthy data management using blockchain in construction ecosystems. *IEEE Access*. 2023;11:61034–61048. Available from: https://ieeexplore.ieee.org/document/10125997.

[4] Xu Y, Chong HY and Chi M. Modelling the blockchain adoption barriers in the AEC industry. *Engineering, Construction and Architectural Management*. 2023;30(1):125–153. https://doi.org/10.1108/ECAM-04-2021-0335.

[5] Wu H, Zhong W, Zhong B, Li H, Guo J and Mehmood I. Barrier identification, analysis and solutions of blockchain adoption in construction: a fuzzy DEMATEL and TOE integrated method. *Engineering, Construction and Architectural Management*. 2025;32(1):409–426. https://doi.org/10.1108/ECAM-02-2023-0168.

[6] Nguyen B, Buscher V, Cavendish W, Gerber D, Leung S, Krzyzaniak A and Flapper T. *Blockchain and the Built Environment*. London: Arup; 2019.

[7] World Economic Forum. *Redesigning Trust: Blockchain Deployment Toolkit*. Geneva: World Economic Forum; 2020 [cited 2024 Dec 4]. Available from: https://widgets.weforum.org/blockchain-toolkit/pdf/WEF_Redesigning_Trust_Blockchain_Deployment%20Toolkit.pdf [accessed 1 December 2024].

Index

accountability 85, 314
Aconex 85
actors 47
adaptation mechanism 335
Add Currency function 188, 191
Add Energy function 188
adoption barriers 422–3
adoption model 338
Advanced Encryption Standard (AES) 88
Aerospace Technology Initiative (ATI) 71
Alibaba Cloud 133
Amazon Web Services (AWS) 133–4, 137
anti-money laundering (AML) 370–1
Anti-Money Laundering Directive 5 (AML5) 371
antitrust 374
application programming interfaces (APIs) 17–18, 128, 162
architecture decentralised autonomous organisation (ArchiDAO) 6, 227
  non-fungible token 230–1
  proposed structures 231
  soulbound token 231–2
  token prototype 211
architecture, engineering and construction (AEC) 166, 210
  industry and current political economy 208
architecture, engineering, construction and operations (AECO) projects 3, 8, 83, 411
  context and challenges 84–7
  data security requirements for 87
    data confidentiality 87
    digital trust 87–90
  information security requirements for 87
  universal distributed framework for secure collaboration in 91
    blockchain network 92–4
    encryption modules 96–8
    selectively transparent data models 98–100
    storage network 94–6
Argent 135
artificial intelligence (AI) 16, 261, 280
Asia-Pacific region 112
asset entanglement 362–3
asset information management systems (AIMS) 278
asset management 94
asset performance analytics 286
AstraKode 14
asymmetric cryptography 97
asymmetric encryption 84, 91, 103
  public key of 97
atomicity, consistency, isolation, durability (ACID) 280
auditing 395
authorised personnel (AP) 249

bare owner 132
benefits management (BM) 67, 71
Big Data 212
BIM 360 85
Bitcoin 2, 11, 38, 152, 264
blockchain 152–3, 208–10, 221, 353, 356, 417
 barriers and facilitators to blockchain adoption 56
  business and governance models 60
  data, privacy and security 57–8
  investment and costs 59–60
  policy 58–9
  process 59
  R&D and use cases 58
  society 60–2
 challenges and barriers 201–2
 comparison to related studies 200–1
 databases 153
 for energy transactions 185
  framework specification 191–2
  scenarios of analysis 188–91
 evaluation and results 192–200
 governance technology 210–11
 infrastructure 221–4
 integration with digital technologies 417–18
 for microgrids 183–5
 platform lifespans 339
 renewable energy in industry 180
  microgrids for renewable energy distribution 182–3
  for port industry 180–2
 and socio-technical environments 417
 as systematic challenge 416–17
blockchain-enabled CCA, challenges and opportunities to 338
 policy-related challenges 341
  procurement route and contractual frameworks 341
 process-related challenges 339
  detailed unit rates 340
  replicating contractual logic 339–40
 society-related challenges 341
  acceptance of contracting parties 341–2
  implications for supply chain cash flow 342
 technology-related challenges 338
  authenticity of oracles 339
  blockchain platform lifespans 339
  technical malfunction 338
blockchain ledger 94
blockchain network 92–4
blockchain smart contracts 93–4
blockchain technology 3, 5, 9, 10–11, 115, 133, 170, 177, 179, 241, 302, 313–14
 integration of 139
BlockDAGs 274, 303
Blowfish algorithm 97
building information management (BIM) 16
building information modelling (BIM) 3, 33, 36, 83, 111, 118, 148–9, 216, 241, 273, 294, 323, 385–7
 conceptual modelling in 292
 data lineage and observability 290–5
 geometric information in 99
 semantic information in 99
building management systems (BMS) 83
building service inspector (BSI) 249
business models 4, 60, 113

California Consumer Privacy Act of 2018 (CCPA) 282
cellular automata (CA) algorithm 213
Central Bank Digital Currencies (CBDCs) 372
Chainlink 247
change management 336–7
ChatGPT 296
circular economy 171, 420
claims management 337–8
client representative (CL) 249
closeout management process 334
cloud-based CDE systems 275
cloud computing 280
  role of 135
  use of 133–4
Command Line Interface (CLI) 135
commercial property 374
common data environments (CDEs) 6, 84, 273, 330
  enterprise total lifecycle data governance 278
    enterprise digital twins 279
  evolution of data architectures 280–2
  methodology 282
    levels and annotation 285
  trajectory of 275
    particulars of BIM CDEs 276–7
    providing for digital twins in 277–8
communication and relationship management 331–2
competition 374
computer-aided design (CAD) 281
confidentiality 364–5, 369
consortium blockchain platforms 11
construction contract administration (CCA) 7, 313, 411
  blockchain applications for adaptation-orientated CCA functions 334
    change management 336–7
    claims management 337–8
    contract risk management 334–6
  blockchain applications for control-orientated CCA functions 323
    disputes resolution management 329
    financial management 323–5
    performance monitoring and reporting management 325–6
    project 'legal' governance 328–9
    quality and acceptance management 326–8
  blockchain applications for coordination-orientated CCA functions 330
    communication and relationship management 331–2
    contract closeout management 333–4
    document and record management 330–1
    team management 332–3
  blockchain technology and smart contracts in 320–2
  challenges and opportunities to blockchain-enabled CCA 338
    policy-related challenges 341
    process-related challenges 339–40
    society-related challenges 341–2
    technology-related challenges 338–9
  challenges facing CCA 317–20
  as multi-functional system 314–17
construction industry 33
construction sector 1

construction supply chain (CSC) 240
construction supply chain management (CSCM) 6, 240
  benefits of using blockchain in 241
    enhanced information sharing and stakeholder collaboration 241–2
    enhanced security and risk management 243–4
    improved sustainability and operational performance 244–5
    improved transparency and traceability 242–3
  challenges 263
    feasibility and scalability issues 264
    regulatory uncertainty 265–6
    resistance to change 264–5
    significant investment 265
    system security and data privacy 263–4
  implementing blockchain in 245
    cost–benefit balance 246
    human resource readiness 246
    interoperability 247
    legal compliance 247
    scalability 247
    stakeholder management 245–6
    steps for successful implementation with real-life case study 247–60
  prospects 261
    decentralised autonomous organisations 262–3
    digital identity management 262
    integration with IoT and AI 261–2
consumers 196
  daily costs for producers and 196
  protection 373, 389–91
  yearly costs for producers and 196
content identifiers (CID) 156, 330
continuous monitoring 395
contract administration 317
contract formation 361–2
contract management 2
contract risk management 334–6
cost analyses 183
cost-benefit analysis 6
cost–benefit balance 246
counterargument 18
COVID-19 pandemic 240
critical success factors (CSFs) 62
  for adoption of blockchain technologies 62–3
cross-cutting themes 411
  blockchain
    integration with digital technologies 417–18
    and socio-technical environments 417
    as systematic challenge 416–17
  collaborative and decentralised governance 419–20
  data security and digital trust 418–19
  ecosystem barrier to blockchain diffusion in built environment 423–4
  final reflection 424–8
  key challenges 420
    data quality and reliability 421
    ethical and social implications 422
    fragmentation of standards and protocols 422
    integration with legacy systems 421
    legal and regulatory challenges 420–1
    scalability and performance limitations 421

stakeholder resistance and adoption barriers 422–3
recommendations 424
sustainability and circular economy 420
synthesis of 412–16
tokenisation and democratisation of access 419
crypto-asset 358–9
cryptocurrency 12
cryptoeconomics 154
cryptographic algorithm 249
cryptographically secure digital building twins 154
cryptography 10, 253
customer relationship management (CRM) 276
cyberattacks 58, 83

DAOstack 221–4
data availability 89
database management 216
data confidentiality 87, 96
data fabrics 281
data hosting and storage 366
data hub architecture 279
data integrity 89, 96
data interoperability 148
data lakes 280
data layer 156–9
data lineage 274
data mesh architecture 305
data meshes 281
data orchestration 277
data privacy 148, 263–4
data processing 17
data protection 368
data quality and reliability 421
data security 148
data storage 208–10
data traceability 89–90
data utility 90–1
data warehouse 120
decentralised applications (DApps) 12, 135, 218
decentralised autonomous organisations (DAOs) 5, 54, 208, 210, 262–3, 304
decentralised data marketplace 161
decentralised finance (DeFi) 12, 113, 209
decentralised ledger technology 208–10
decentralised marketplaces 154
decentralised storage 153
decentralising architectural design
building decentralised autonomous organisation as living lab 227
ArchiDAO 227–9
ArchiDAO non-fungible token 230–1
ArchiDAO proposed structures 231
ArchiDAO soulbound token 231–2
stigmergy 229–30
crises through development 207–8
nature of design in AEC industry and current political economy 208
governance technology 210–11
interlude 211–17
optimisation 218
DAOs and shape grammars further development 226–7
DAOstack 221–4
prototype 1 218–20
prototype 2 220–1

prototype 3 224–5
shape grammars and blockchain 221
primitives 208
blockchain, decentralised ledger technology and data storage 208–10
decentralised autonomous organisations 210
decentralized model 184
decision-making process 85, 161, 224
Decision-Making Trial and Evaluation Laboratory (DEMATEL) method 55
decision trees 11
decryption 96
Department for Transport (DfT) 285
Diffie–Hellman key exchange protocol 97
diffusion of innovations (DOIs) 36
digital accessibility 13
digital age 38
digital asset 16, 21
digital building logbooks (DBLs) 2, 4, 148
administration costs 148
conceptualisation 155
conceptual architecture 156–64
motivation and objective 155–6
data interoperability, privacy and security 148
distributed ledger technologies, blockchains and smart contracts 152–3
identified gaps to implementing 150
implementation challenges for 149
incentives 148
other key enabling technologies 153
cryptoeconomics and decentralised marketplaces 154
cryptographically secure digital building twins 154
decentralised storage 153
possible technical approaches to 149–52
potential impacts and future research directions 164
accelerating business case for renovation and reuse 166–7
enabling better security and privacy of data 170
enabling new peer-to-peer business models 167
fostering new governmental policies, incentives and regulations 168–9
improving existing legal structures 167–8
moving towards more circular Europe 169–70
self-infrastructuring backbone 166
digital enabler module 120
digital identity management 6, 262
digital innovation 1, 113–14
digitalisation 57, 112
DigitalOcean 133
digital signatures 11, 97
digital technologies 33
digital tokens 111
digital transformation 118, 133
digital transition 127
digital trust 85, 88
data availability 89
data integrity 89
data traceability 89–90
data utility 90–1
digital twins (DTs) 6, 44, 154, 166, 273, 296–7, 385–7
directed acyclic graphs (DAGs) 281, 282

dispute avoidance/adjudication board (DAAB) 329
dispute resolution management 342
dispute resolution method 329
disputes resolution management 329
distributed ledgers 92
distributed ledger technologies (DLTs) 2, 9, 32, 115, 152–3, 274, 355–6
  actors model 47
  adoption studies at different levels 52
    consortia level 54
    market level 53–4
    organisation and project 55–6
    supply chain level 54–5
  blockchain oracles 16–18
  blockchain technology 10
    benefits and capabilities of blockchain-based applications 18–20
    blockchain-based application 20–2
    generic view of 10
    prevailing blockchain platforms 12–14
    types 11
  definition 9–10
  evidence of blockchain applicability in built environment 22–4
  four-dimensional model 44
    policy 46
    process 45–6
    socicty 46
    technology 45
  smart contracts 14–15
distributed network 92
distributed storage technologies 89
diverse tests 213
DLT Actors Model 3
domain-oriented decentralisation 305
DOVU 389

e-commerce 154
economic impact 129–33
economic sustainability 140
edge computing 280
e-InStar 254, 258
Einstein's framework 288
electronic IDentification, Authentication and trust Services (eIDAS) 117
e-money 372
encryption modules 96–8
encryption protocols 3, 96, 98
End User Licence Agreement (EULA) 382
energy consumption 110, 203
energy demand 177
Energy Efficiency Directive (EED) 147
energy generation and use 387–8
Energy Performance of Buildings Directive (EPBD) 166
energy retrofit module 120
energy transaction function 188
energy usage 57
enterprise CDEs (E-CDEs) 273
  observing provenance in 285
    BIM data lineage and observability 290–5
    digital twins data lineage and observability 295–9
    enterprise data lineage and observability 286–90
    summary of enterprise 299–301
  prospecting for enterprise digital twin data architectures 302
    attributions 303–5
    functions 302
    sequences 302–3

enterprise content management (ECM) 278
enterprise data lineage and observability 286–90
enterprise digital twins 279
enterprise resource planning (ERP) 241, 276
environmental impact 178
ERC-20 protocol 114
ERC-721 protocol 114, 117
ERC-1238 protocol 114
Eternalism 298
Ethereum-based smart contracts 5
Ethereum blockchain 329
Ethereum ecosystem 232
Ethereum network 185
Ethereum platforms 11
Ethereum smart contracts 183–4
Ethereum virtual machine (EVM) 12
EU INTERREG piSCES project 185
European Commission 148
European Green Deal 169
European Union (EU) 367
exchange tokens 359
extension of time (EoT) 337

Federal Act on Data Protection (FADP) 376
federated computational governance 305
financial management 323–5
financial regulation 381–2
Financial Services and Markets Act 2000 (FSMA) 370
Financial Services Markets Bill (FSMB) 358
financial value analysis 129
Fourth Industrial Revolution (4IR) 5, 208, 240
fractionalisation 359–60
fractional ownership 113–14, 122–4
Fungible Tokens Generator 161

gas price 186
General Data Protection Regulation (GDPR) 7, 59, 148, 282, 367–9, 375–6
geographical information systems (GIS) 111, 118, 241, 278
global warming 178
Google Cloud Platform 133
Google Gemini 296
governance and compliance mechanisms 395
governance models 60
governmental supervision unit (GSU) 245
Grasshopper environment 226
greenhouse gas (GHG) emissions 177, 182
Gwei 186

Halting rules 223
hash algorithm 253
hash values 241
HM Land Registry (HMLR) 355
human resource readiness 246
Human Sciences and Advanced Technology (HUSAT) 35
hybrid contract (HC) 358
Hyperledger Fabric 2, 11, 13, 14, 92, 251–2, 330, 341

IBM Cloud 133
identity 369
Industry 4.0 240
Industry Foundation Classes (IFC) 149
Information Delivery Systems 148
information management systems (IMS) 278

information technology (IT) 35
infrastructure 65
infrastructure as a service (IaaS) 134, 251–2
initial coin offerings (ICOs) 369
integrated project delivery (IPD) 86
integrated smart cities 387
intellectual property rights 365
 data hosting and storage 366
 royalties 366
 secrecy 365–6
 underlying IPR 366–7
intellectual property rights (IPR) 361
interlude 211–17
Internet-of-Things (IoTs) 134, 330
interoperability 127, 247
inter-organisational system (IOS) 40
interplanetary file system
 configuration of IPFS clusters for distributed data management in AECO projects 96
 functions and characteristic features of 95
InterPlanetary File System (IPFS) 3, 56, 84, 92, 153, 156, 209, 330

J Shed 189

key performance indexes 151
know your client (KYC) 370–1
K Shed 189

land access and requirements management systems (LARMS) 302
Leavitt's diamond 35
legal and regulatory compliance 393–4
legal compliance 247
legal risks 353
 background on blockchain technology 354
 blockchain 356
 DLT 355–6
 smart contracts 357
 smart legal contracts 357
 tokens and tokenisation 358–60
 wallets and storage 360
 best practices and solutions to mitigate legal risks 391
 incentives, side note 392–3
 regulatory compliance, legal considerations and practical tips 395–8
 smart contract design and dispute resolution mechanisms 398–402
 strategies for risk mitigation 393–5
 cases and examples 391
 case studies of blockchain applications that have faced legal challenges 383–91
 commercial 373
 commercial property 374
 competition 374
 consumer protection 373
 summary of commercial issues 375
 confidentiality 369
 data protection 368
 final thoughts and recommendations 405
 conduct comprehensive risk assessments 405
 continuously monitor legal developments 406
 engage and collaborate with regulators 405–6
 establish clear governance frameworks 406

foster collaboration and industry standards 406
implement robust security measures 406
financial regulation 369
e-money 372
know your client and anti-money laundering 370–1
legal risks associated with blockchain applications 373
markets in crypto-assets 372–3
Markets in Financial Instruments Directive II 371–2
payments 369–70
Payment Services Directive 2, 372
statutory mechanisms 372
token analysis 370
GDPR 368–9
identity 369
intellectual property rights 365
data hosting and storage 366
royalties 366
secrecy 365–6
underlying IPR 366–7
jurisdictional and cross-border legal issues 375
commercial 378–9
contract 375
GDPR 375–6
multi-jurisdictional IPR 375–6
securities 376–8
summary of jurisdictional and cross-border legal issues 379
legal risks associated with blockchain applications 360
contracts, smart legal contracts and smart contracts 361–5
overview of legal risks 361
outlook on 403
data protection and privacy challenges 404
evolving regulatory landscape 403–4
interoperability and standardisation 404
liability and accountability 405
smart contract governance and liability 404
summary of outlook on legal risks 405
tokenisation and asset ownership 405
real-world examples of legal risks and their consequences 379
commercial 382
contracts, smart legal contracts and smart contracts 380
data protection, privacy and cybersecurity 381
financial regulation 381–2
jurisdictional and cross-border legal issues 382
practical IPR 380
summary of 402–3
summary of legal risks associated with blockchain applications 379
linked data approach 151
Linux Foundation 12
logbook management 161
logic layer 160–3

machine learning (ML) 134, 280
macro roadmap 63
benefits management 67
infrastructure 65
regulatory frameworks 65–6
research and development 66–7
scope of requirements 64

- sector engagement 65
- taskforce 64
- upskill workforce 66

macrosocial systems 33
man-in-the-middle attacks 97–8
markets in crypto-assets (MiCA) 372–3
Markets in Financial Instruments Directive II (MiFID II) 371–2
Merkle trees 152, 209
meso roadmap 67
- adopt 69
- assess 68
- develop 69
- pilot 69
- refine 69
- scope 68

MetaMask 135
microgrids 178–9, 202
- blockchain for 183–5
- smart contract framework 187

Microsoft Azure 133
Microsoft Copilot 296
Milford Haven Port microgrid 180, 182, 190
minimal path approach 213
modular integrated construction (MIC) 240
multi-factor authentication (MFA) 369
multi-level perspective (MLP) model 33, 35

natural language contract (NLC) 357–8
negative utilitarianism 289
nodes 93, 133
non-cooperative Nash equilibrium 221
non-Euclidean geometry 287
non-fungible tokens (NFTs) 110, 113, 136, 161, 230–1, 359
non-repudiation 85
Notary Council 121
novel frameworks 427
nuclear power plants 182

off-chain component 17
on-chain component 17
onymous 356
open Building Information Modelling (openBIM) approach 148–9
Oracle Cloud 133
oracles 15, 340
- authenticity of 339
- hardware oracles 16
- human oracles 16
- software oracles 16

Packaway building 188
Pangea scheme 116
payments 369–70
Payment Services Directive 2 (PSD2) 372
Pearl River Delta (PRD) 248
peer-to-peer (P2P) networks 10, 149, 388
performance management 325
permissioned platforms 11
permissionless platforms 11
Personal Data Protection Act (PDPA) 376
platforms as a service 134
policy 58–9
Polybase 232
- for soulbound NFT process 233

power grids 179
Practical Byzantine Fault Tolerance 264
practical IPR 380
predictive analytics 289
private blockchains 184

private key 97
privity 363–4
procurement 384–5
project 'legal' governance 328–9
project management 383–4
proof-of-concept (PoC) 65
proof of stake (PoS) consensus mechanism 136, 185
proof of stakeholders 253
proof-of-work (PoW) algorithm 39, 169
property master data 120
property tokenization 110
  digital architecture implementation for 133
    Ethereum as potential blockchain platform 135–6
    proposed technical architecture 136–8
    role of cloud computing 135
    use of cloud computing 133–4
  digital workflow for 125
  fractional ownership and digital innovation 113–14
  framework for 114
    barriers to adoption 121–2
    economic impact 129–33
    as financing mechanism to upkeep built assets 128–9
    organization of information in property tokenization management system 118–20
    possible business models with property tokenization platform 122–4
    product and process innovation 127–8
    technological chain for implementation of property tokenization 124–7
  as paradigm shift 111–13
  transformative potential of 140
provenance 384–5
pseudonymous system 356
public key 97
public-key cryptography 96
public-private key cryptography 152

quality and acceptance management 326–8
quantum field theory (QFT) 288

Rainbow 135
Raspberry Pi 243
real estate 385–7
registered structural engineer (RSE) 249
regulatory frameworks 65–6
regulatory perimeter 359
Remix IDE 191–2
renewable energy
  in industry 180
    microgrids for renewable energy distribution 182–3
    for port industry 180–2
  potential of 181
request for information (RFI) 330
research and development (R&D) 58, 66–7
return on investment (ROI) 249
risk assessment 393
risk management 243–4, 335, 393
Rivest, Shamir and Adleman (RSA) algorithm 97
roadmaps
  to adoption 63
    linking meso and macro roadmaps 70–1
    macro roadmap 63–7
    meso roadmap 67–9
royalties 366

scalability 247
scope of requirements (SoR) 63–4, 70
seaport industry 178
seaports 202
secrecy 365–6
secret-key cryptography 96
sector engagement (SE) 65, 70
secure development practices 394
security protocols 88
security tokens 359
self-executing smart contract 339
self-infrastructuring 166
self-serve data infrastructure 305
semantics analysis 284
services layer 160–3
SHA-256 98
shape grammars 221
small and medium-sized enterprises (SMEs) 59
smart contract auditing 394
smart contract functions (SCFs) 13, 18
smart contracts (SCs) 3, 9, 14–15, 32, 92, 93–4, 112, 124, 133, 152–3, 179, 183, 186, 243, 304, 354, 357
  for access and ownership management 161
  costs of smart contract functions 194
  for data-based performance assessment 161
  development lifecycle of 15
  gas prices of smart contract functions 192
  for platform governance and legal compliance 161
smart legal contracts 357
social sustainability 117
society 60–2
socio-technical environments 417
socio-technical landscape 33
socio-technical regimes 34
socio-technical systems (STS) theory 3, 32
  and adoption of building information modelling 36–8
  and adoption of technological innovations 33–5
  approaches in built environment literature 35–6
  blockchain 38
    and DLT adoption 52–63
    socio-technical considerations for DLT and SCs in built environment 40–1
    trust perspective of blockchain-based systems 41–3
  conceptual foundations 32–3
  within organisation 35
  roadmaps to adoption 63
    linking meso and macro roadmaps 70–1
    macro roadmap 63–7
    meso roadmap 67–9
  socio-technical model of DLT and SCs in built environment 43
    application of DLT four-dimensional model and DLT actors model 47–52
    DLT actors model 47
    DLT four-dimensional model 44–7
software as a service (SaaS) 134, 151
software programmers 363
solely code contract (SCC) 358
solidity energy trading smart contract 183
solidity programming language 186
soulbound token 231–2

source of wealth declaration 371
sovereign identity 117
specified investments 360
stablecoins 359
stakeholders 245, 266
  management 245–6
  resistance 422–3
statutory mechanisms 372
stigmergy 229–30
storage network 94–6
Superbonus 128
supply chain management (SCM) 2, 94, 239, 384–5
supply chain process 93, 239
sustainability 420
sustainable construction materials 245
symmetric encryption 84, 91, 97
system security 263–4

taskforce (TF) 63–4, 70
task-technology fit (TTF) model 40
team management 332–3
technology acceptance model (TAM) 63
technology-organisation-environment (TOE) model 56
3D assembly checkpoint 254
3D modelling 287
token analysis 370
tokenisation 110, 358–60
tokens 358–60
  asset-backed tokens 115
  currency tokens 114
  security tokens 114
  utility tokens 115
traceability 242–3
transparency 242–3, 314, 394–5
transport 388–9
Transport Layer Security (TLS) 88
Twitter 14

Uniform Commercial Code (UCC) 378
unit/product management 120
unmanned aerial vehicles (UAVs) 3, 83
upskill (US) workforce 66, 71
user authenticity 96
user-centric interface 164
user education 394–5
user layer 163–4
user master data 120
user workflow diagram 252
usufructuary 132
utilities 388–9
utility tokens 359

wallets 12, 360
Weather Ledger project 16, 43
Web 2.0 61
web-layer 163
workflow mapping 251
World Economic Forum's (WEF) 411
World Wide Web 209

zero-knowledge proofs 5, 156, 170, 232